THE CHLOROPHYLLS

List of Contributors

M. B. ALLEN

S. ARONOFF

N. K. BOARDMAN

LAWRENCE BOGORAD

L. J. BOUCHER

WARREN L. BUTLER

RODERICK K. CLAYTON

G. COHEN-BAZIRE

R. C. DOUGHERTY

J. C. GOEDHEER

A. S. HOLT

J. J. KATZ

BACON KE

WALTER LWOWSKI

JOHN M. OLSON

RODERIC B. PARK

G. R. SEELY

W. R. SISTROM

ELIZABETH K. STANTON

HAROLD H. STRAIN

WALTER A. SVEC

LEO P. VERNON

The Chlorophylls

Edited by

LEO P. VERNON and GILBERT R. SEELY

Charles F. Kettering Research Laboratory
Yellow Springs, Ohio

1966

ACADEMIC PRESS New York and London

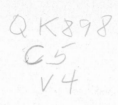
ACADEMIC PRESS INC.
111 Fifth Avenue, New York, New York 10003

United Kingdom Edition published by
ACADEMIC PRESS INC. (LONDON) LTD.
Berkeley Square House, London W.1

LIBRARY OF CONGRESS CATALOG CARD NUMBER: 66-24176

PRINTED IN THE UNITED STATES OF AMERICA

Contributors

Numbers in parentheses indicate the pages on which the authors' contributions begin.

M. B. ALLEN, *Laboratory of Physical Biology, National Institute of Arthritis and Metabolic Diseases, Bethesda, Maryland* (511)

S. ARONOFF, *Department of Biochemistry and Biophysics, Iowa State University, Ames, Iowa* (3)

N. K. BOARDMAN, *Commonwealth Scientific and Industrial Research Organization, Division of Plant Industry, Canberra, Australia* (437)

LAWRENCE BOGORAD, *Department of Botany, The University of Chicago, Chicago, Illinois* (481)

L. J. BOUCHER, *Argonne National Laboratory, Argonne, Illinois* (185)

WARREN L. BUTLER, *Department of Biology, University of California, San Diego-La Jolla, California* (343)

RODERICK K. CLAYTON, *Charles F. Kettering Research Laboratory, Yellow Springs, Ohio* (609)

G. COHEN-BAZIRE, *Department of Bacteriology and Immunology, University of California, Berkeley, California* (313)

R. C. DOUGHERTY, *Argonne National Laboratory, Argonne, Illinois* (185)

J. C. GOEDHEER, *Biophysical Research Group, Physics Institute, University of Utrecht, the Netherlands* (147, 399)

A. S. HOLT, *Division of Biosciences, National Research Council, Ottawa, Canada* (111)

J. J. KATZ, *Argonne National Laboratory, Argonne, Illinois* (185)

BACON KE, *Charles F. Kettering Research Laboratory, Yellow Springs, Ohio* (253, 427, 569)

WALTER LWOWSKI, *Department of Chemistry, Yale University, New Haven, Connecticut* (119)

v

JAN 4 - 1967

JOHN M. OLSON, *Biology Department, Brookhaven National Laboratory, Upton, New York* (381, 413)

RODERIC B. PARK, *Department of Biology and Lawrence Radiation Laboratory, University of California, Berkeley, California* (283)

G. R. SEELY, *Charles F. Kettering Research Laboratory, Yellow Springs, Ohio* (67, 523)

W. R. SISTROM, *Department of Biology, University of Oregon, Eugene, Oregon* (313)

ELIZABETH K. STANTON, *Biology Department, Brookhaven National Laboratory, Upton, New York* (381)

HAROLD H. STRAIN, *Argonne National Laboratory, Argonne, Illinois* (21)

WALTER A. SVEC, *Argonne National Laboratory, Argonne, Illinois* (21)

LEO P. VERNON, *Charles F. Kettering Research Laboratory, Yellow Springs, Ohio* (569)

Preface

Since publication of the more recent comprehensive reviews on chlorophyll and its role in photosynthesis (i.e., "Photosynthesis," by E. Rabinowitch, and Volume V of the "Handbuch der Pflanzenphysiologie") there have been many new developments, including new techniques, in the expanding study of chlorophylls in living and nonliving environments. Our purpose in organizing this treatise has been to review these developments, and at the same time to summarize the state of knowledge in the more established areas of the physics, chemistry, and biology of chlorophylls. We have not attempted to cover photosynthesis per se, except for those aspects in which the chlorophylls are directly involved. We hope this work will help those investigating selected aspects of chlorophyll to keep abreast of other methods and approaches and will provide the interested scientist with a modern, conceptually organized treatment of the subject.

The treatise is divided into four sections. The first deals with the chlorophylls as chemical entities, and treats their isolation, analysis, chemistry, and synthesis. The second concerns chlorophylls in real and colloidal solution and in the solid state *in vitro*, and includes the effects of aggregation on visible, infrared, and NMR spectral properties. The third section treats the biosynthesis, organization, and properties of chlorophylls in the plant and bacterial cell, and the fourth is concerned with the photochemical and photophysical behavior of chlorophylls *in vitro* and *in vivo*.

One question of nomenclature deserves comment. Correspondence with our contributors revealed considerable dissatisfaction with the usual name "chlorobium chlorophyll 660 (or 650)" now applied to the principal chlorophylls of the green bacteria; these names are both cumbersome and inaccurate. Several alternative designations were suggested, but none proved acceptable to all concerned. Not wishing to compound the confusion, we requested of the authors that the usual name be employed throughout the book for the sake of uniformity, leaving the responsibility of nomenclature to some future committee which will hopefully put more order into this area.

Our sincere thanks are extended to the authors for their splendid contributions and for their enthusiastic cooperation in every phase of

vii

preparation of this treatise. We wish to express our appreciation to Mrs. Bessie Knedler and Mrs. Helga Smith for typing our own manuscripts and performing other secretarial duties. Finally, we wish to acknowledge our debt of gratitude to the late Charles F. Kettering whose curiosity about photosynthesis and many other aspects of the world around him and whose willingness to support their investigation have made possible many discoveries in the field of science.

LEO P. VERNON
GILBERT R. SEELY

Charles F. Kettering Research Laboratory
Yellow Springs, Ohio
May, 1966

Contents

Section I
Isolation and Chemistry

1. The Chlorophylls—An Introductory Survey

S. Aronoff

2. Extraction, Separation, Estimation, and Isolation of the Chlorophylls

Harold H. Strain and Walter A. Svec

3. The Structure and Chemistry of Functional Groups

G. R. Seely

4. Recently Characterized Chlorophylls

A. S. Holt

5. The Synthesis of Chlorophyll *a*

Walter Lwowski

Section II

Physical Properties in Solution and in Aggregates

6. Visible Absorption and Fluorescence of Chlorophyll and Its Aggregates in Solution

J. C. Goedheer

7. Infrared and Nuclear Magnetic Resonance Spectroscopy of Chlorophyll

J. J. Katz, R. C. Dougherty, and L. J. Boucher

8. Some Properties of Chlorophyll Monolayers and Crystalline Chlorophyll

Bacon Ke

SECTION III
State of the Chlorophylls in the Cell

9. Chloroplast Structure

Roderic B. Park

10. The Procaryotic Photosynthetic Apparatus

G. Cohen-Bazire and W. R. Sistrom

11. Spectral Characteristics of Chlorophyll in Green Plants

Warren L. Butler

12. Absorption and Fluorescence Spectra of Bacterial Chlorophylls *in Situ*

John M. Olson and Elizabeth K. Stanton

13. Chlorophyll-Protein Complexes

Part I. *Complexes derived from Green Plants* 399

J. C. Goedheer

Part II. *Complexes derived from Green Photosynthetic Bacteria* ... 413

John M. Olson

Part III. *Optical Rotatory Dispersion of Chlorophyll-Containing Particles from Green Plants and Photosynthetic Bacteria* · 427

Bacon Ke

19. Physical Processes Involving Chlorophyll *in Vivo*

Roderick K. Clayton

List of Abbreviations

A(od)	absorbance (optical density)
ADP	adenosine diphosphate
ALA	δ-aminolevulinic acid
ATP	adenosine triphosphate
BChl	bacteriochlorophyll
Chl	chlorophyll
Chlide	chlorophyllide
Copro and Coprogen	coproporphyrin and coproporphyrinogen
Cyt	cytochrome
DCMU	3-(3,4-dichlorophenyl)-1,1-dimethylurea
DPIP, $DPIPH_2$	2,6-dichlorophenol indophenol and its reduced form
ev	electron-volt
ESR (EPR)	electron spin resonance
FMN, $FMNH_2$	flavin mononucleotide and its reduced form
Hemato	hematoporphyrin
MB	Methylene blue
mv	millivolt
NAD, NADH (DPN, DPNH)	nicotinamide adenine dinucleotide and its reduced form
NADP, NADPH (TPN, TPNH)	nicotinamide adenine dinucleotide phosphate and its reduced form
NMR	nuclear magnetic resonance
ORD	optical rotatory dispersion
P700	Reaction center chlorophyll of chloroplasts
P870 and P890	Reaction center chlorophylls of bacteria
PBG	porphobilinogen
PChl	protochlorophyll
Pchlide	protochlorophyllide
Pi	orthophosphate
PMS	phenazine methosulfate (methyl phenazonium methosulfate)
PPNR	photosynthetic pyridine nucleotide reductase, Ferredoxin
PQ	plastoquinone
Proto and Protogen	protoporphyrin and protoporphyrinogen
PS1 and PS2	pigment systems 1 and 2 of chloroplasts
TBP	tetrabenzporphin
TMPD	N,N,N',N'-tetramethyl-p-phenylenediamine
Tris	tris(hydroxymethyl)aminomethane
UQ (CoQ)	ubiquinone
Uro and Urogen	uroporphyrin and uroporphyrinogen.

Section I
Isolation and Chemistry

The Chlorophylls—An Introductory Survey

S. ARONOFF

Department of Biochemistry and Biophysics, Iowa State University, Ames, Iowa

I. Introduction

The name "chlorophyll" was given initially (*1*) to those green pigments involved in the photosynthesis of higher plants. Subsequently it has been extended to all classes of photosynthetic porphyrin pigments. Functionally, this appears to be proper, since the chlorophylls have not been shown to have any other function except—possibly indirectly—in some types of bacterial phototaxis (*2*). On the other hand, it excludes phycocyanin and phycoerythrin, the closely related bilin-proteinoids which, under certain conditions of light and culture of organism, may become the major pigment in the action spectrum of photosynthesis (*3*).

The early history of the chemistry of chlorophyll has been written by Willstätter and Stoll (*4*). It begins with Berzelius (*5*), who observed the retention of color in a leaf extract despite the action of strong alkali and acid. The conversion of chlorophyll to red pigments prompted Verdeil (*6*) to suggest a relationship between chlorophyll and heme, especially since his analysis of chlorophyll erroneously showed it to contain iron (as well as potassium and phosphorus), a conclusion which was maintained until the researches of Willstätter.

The partitioning of the plastid pigments between an ethereal solution containing the yellow carotenoids and an acidic aqueous solution of the blue-green pheophytins and pheophorbides (which he called "phyllocyanin") was first accomplished by Fremy (*7*), although he apparently thought that "green" chlorophyll itself was a mixture of the two types of pigments, i.e., the blue-green and the yellow.

At about the same time Stokes (8) suggested, as a result of his spectroscopic observations, that chlorophyll consisted of two components: "It may be mentioned in passing, that the green fluorescent residue [i.e., after extraction of the carotenoids] is still a mixture, consisting of two different substances, both green, and both exhibiting a red fluorescence." At the same time, he initiated the approach of nonhydrolytic partitioning of chlorophyll between immiscible solvents. This procedure was extended appreciably by Sorby (9); as a result he first noted the blue color of chlorophyll a. However, physical proof of the existence of two chlorophylls in green leaves did not come until much later, i.e., until the advent of (adsorption) chromatography, in the work by Tswett (10).

In the meantime, the major chemical advances were being performed by Hoppe-Seyler (11); his acidic degradation of (allomerized) chlorophyll to a red pigment (which he called phylloporphyrin) that bore a strong spectroscopic resemblance to hematoporphyrin strengthened the earlier hypothesis of a homology between chlorophyll and heme. Completely identical degradation products were not obtained until Nencki and co-workers (12, 13) degraded the porphyrins reductively to the pyrrole level and isolated "hemopyrrole" [shown by Willstätter and Asahina (14) to be a mixture of homologous pyrroles] from both chlorophyll (i.e., "phyllocyanin") and hemin.

Two additional bits of history are of contemporary interest. One was the formation of "crystalline chlorophyll," i.e., ethyl chlorophyllide, by the action of ethanol on leaves, first observed by Borodin (15) and extended by Monteverde (16), the latter having actually isolated these crystals and determined their spectroscopic properties. Of even more current interest was the prediction by Nencki (17) that the similar chemical properties of chlorophyll and hemin denote a common origin of plant and animal life and that comparison of similar compounds of flora and fauna provides an insight into chemical and organismal evolution.

The modern era of the study of chlorophyll chemistry was initiated by Willstätter and his school, most of his researches being summarized in "Untersuchungen über Chlorophyll" (4). We are indebted to Willstätter for our understanding of the gross properties of chlorophyll, its preparation and its degradation, as well as of phytol. Willstätter first obtained the correct empirical formulas for the chlorophylls, showing them to be magnesium complexes devoid of iron and phosphorus. Willstätter was responsible for the discovery of chlorophyllase and utilized it in the esterification of the chlorophyllides. However, our detailed understanding of the structure of chlorophyll resulted from the studies of H. Fischer and co-workers, who were the first to delineate the structure of the porphyrin ring, both deductively and by synthesis (*18*), as well as the fine structure of the degradation products which Willstätter had prepared.

A variety of chlorophylls have been described: chlorophylls *a*, *b*, *c*, and *d;* bacteriochlorophylls *a*, and *b;* chlorobium chlorophylls 660 and 650; their immediate precursors and degradation products, e.g., the pheophytins and pheophorbides. Of these chlorophylls, only three, chlorophylls *a* and *b* and bacteriochlorophyll *a*, are known definitively; that is to say, the proposed structures of the others are assumed to varying degrees (see Section II). The lack of knowledge of the precise structures, plus the acknowledged influence of the medium upon the absorption spectrum of the chlorophylls, has led to a rash of equivocation in nomenclature by wavelength absorption of the far-red maximum. This is the case with the chlorobium chlorophylls indicated above, and with the sequence of maximum changes observed in the absorption spectra of leaves during chlorophyll *a* formation. In some cases, e.g., bacteriochlorophyll *a* in the Thiorhodaceae, the multiplicity of absorption bands in the far red *in vivo* is known to involve the same pigment in different environments, whereas with chlorobium chlorophylls a multiplicity of pigments is found by chromatography (6 in one case and 7 in the other) (*18a*). The chlorobium chlorophylls, though known to be diverse chemically, are nevertheless difficult to distinguish spectroscopically, a difficulty which does not permit their individual recognition *in vivo*.

It is therefore not surprising that the nomenclature of the chlorophylls is somewhat untidy. There are few areas of human endeavor in which bias is exerted more strongly than in nomenclature, since it arises in most instances from a Messianic conviction of its validity or desirability. The nomenclature of the chlorophylls is no exception and has, in addition, its own brand of academic humor. Despite this warning, there is an occasional necessity to suggest the possibility of a generalization of nomenclature. Thus Jensen *et al.* (*18a*) propose that all photosynthetic porphyrins be designated either as (a) chlorophylls or (b) bacteriochlorophylls. However, it is merely a matter of choice whether one prefers to designate the class according to its phylogeny (since the latter pigments appear to be confined to the bacteria) or whether one prefers the chemical basis for taxonomy. If the chemical basis is selected, there would seem to be no serious objection to designation of three major classes as (a) chlorophylls, (b) bacteriochlorophylls, and (c) chlorobium chlorophylls with the corresponding abbreviations: Chl, Bchl, and Cchl. The chemical basis for this nomenclature will be discussed in Section II. Briefly stated, however, Chl and Bchl are differentiated by their state of reduction, Chl being a dihydroporphine and Bchl a tetrahydroporphine. Bchl and Cchl are distinguished by the state of oxidation of carbon 6d (= carbon 10 in Fischer's numbering), being bonded in the former to a carbomethoxy group and to a hydrogen in the latter. Possible additional criteria (though these are not yet universal) are the presence of substituents on the δ-methine carbon of the Cchls 660, and the long-chain ester (phytol in the Bchls and farnesol in the Cchls).

II. Chemical Structures

Every branch of knowledge has its specialized language, and within the separate areas are dialects which may be so unique as to be understandable virtually only by initiates. The chemistry of the porphyrins is among these, and consequently it is necessary to provide a crude linguistic map of the territory.

First, we assume that the term porphyrins includes the entire class of closed, completely conjugated tetrapyrroles. The parent compound of this class is porphine (I), and all other subclasses of porphyrins are

Ia   Ib   Ic

Porphine (1 tautomeric and 2 canonical forms)

referred to the state of oxidation of this compound. Thus, we may speak of di-, tetra-, or hexahydroporphines, where reduction occurs only on the periphery of the pyrrolic rings. (Reduction at the methine carbons results in a class of compounds known as *porphyrinogens*.) The dihydroporphines are also known as *chlorins* (II) and the tetrahydroporphines sometimes as *bacteriochlorins* (III). Finally, a common feature of the chlorophylls, distinguishing them from nonphotosynthetic porphyrins, is the cyclopentanone ring, conjoint with ring III. The porphine form (IV) of the structure is designated as *pyroporphine* (IVa), and the chlorin form correspondingly as pyrochlorin (IVb). The prefix *pyro* arises from the chemical origin of the compound; resulting from pyrolysis, e.g., of pheophorbide *a* (V), from which pyropheophorbide *a* (VI) is obtained.

All naturally occuring porphyrins have a propionic acid residue at

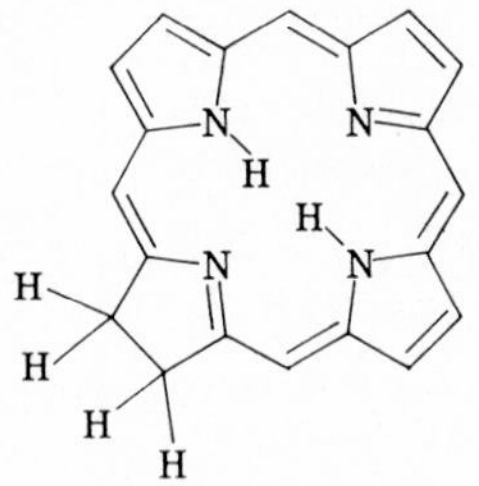

Dihydroporphine (chlorin)

II

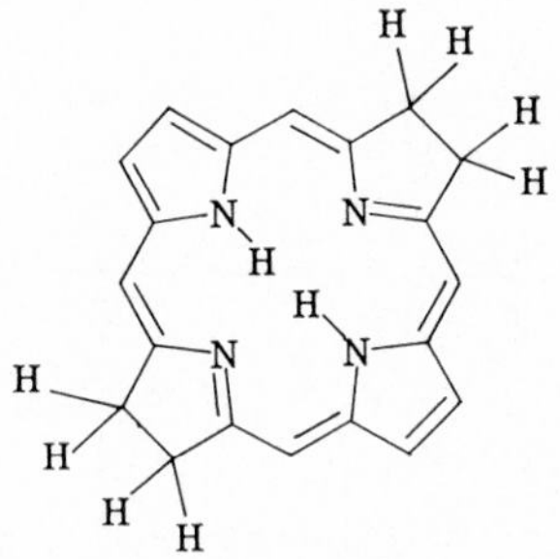

Tetrahydroporphine
(dihydrochlorin,
or bacteriochlorin)

III

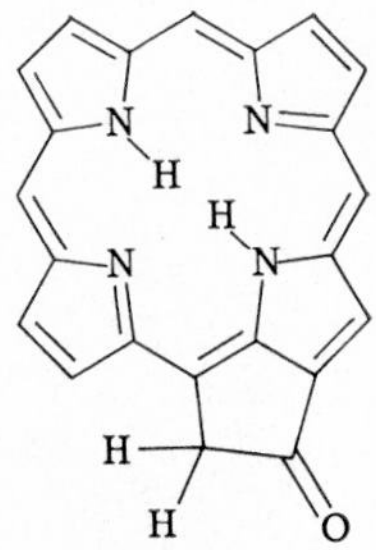

Pyroporphine

IVa

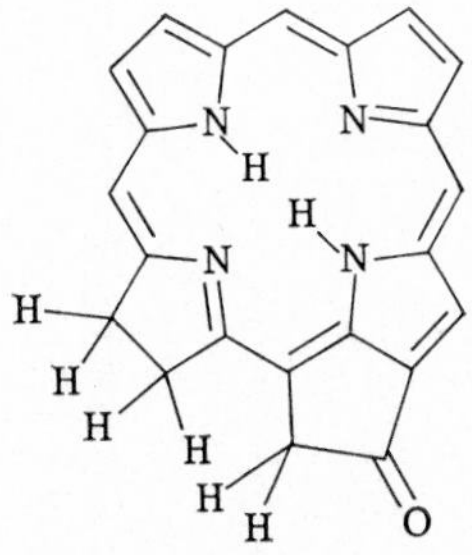

Pyrochlorin

IVb

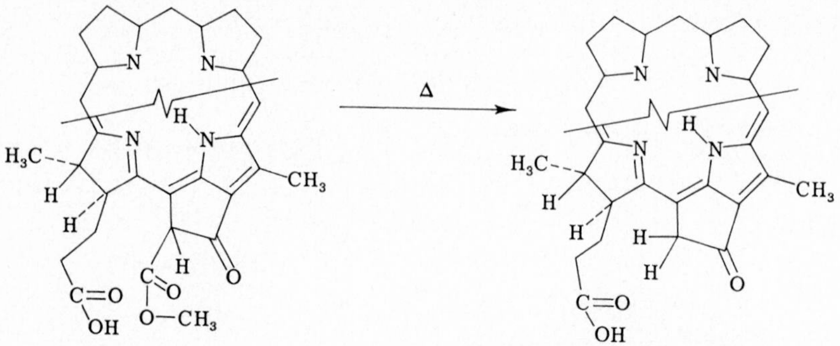

Pheophorbide
V

Pyropheophorbide
VI

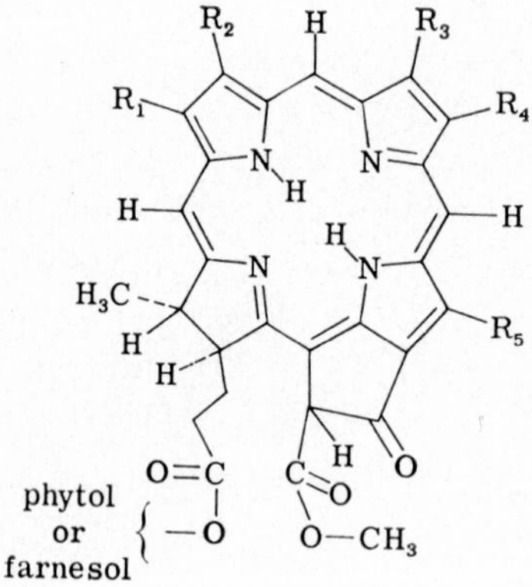

Phytin
VII

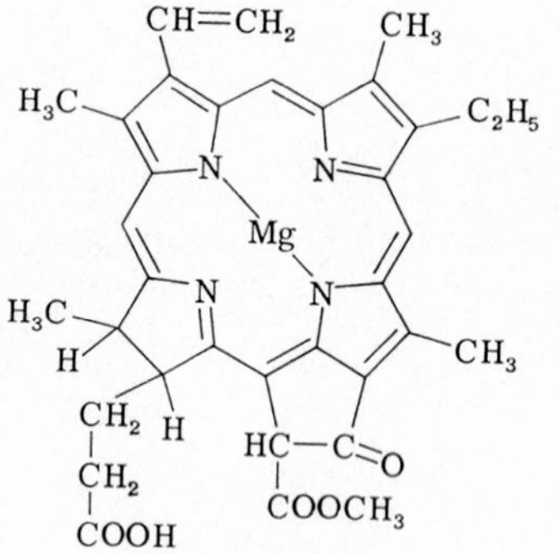

Chlorophyllide *a*
VIII

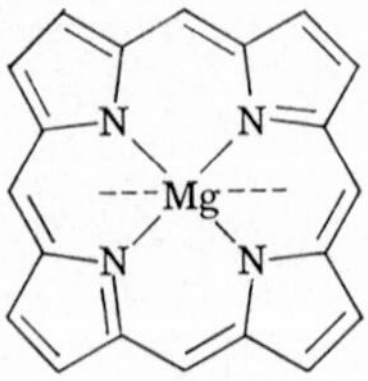

a Phyllin
IX

position 7. In the chlorophylls, this position is esterified with a long-chain alcohol (either phytol or farnesol). As illustrated above, the corresponding free acid is known as a *phorbide* (V) if it does not contain magnesium. When in the form of the naturally occurring ester, the compound is called a *pheophytin* (VII). Synthetic esters, e.g., of methanol or ethanol are known, respectively, as methyl and ethyl pheophorbide. Finally, if a phorbide is liganded with magnesium (i.e., magnesium having replaced the central hydrogens) it is known as a *chlorophyllide* (VIII). In the absence of other identifying characteristics, magnesium porphyrin chelates are known as *phyllins* (IX). The nomenclature of the

TABLE I

NOMENCLATURE OF THE CHLOROPHYLLS[a]

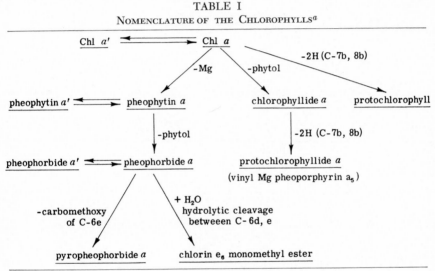

[a] See structure (XII).

porphyrins related to Chl is summarized in Table I. Chl a' represents a structural isomer of Chl a (Chapter 2).

There is no universally accepted numbering system for the identification of the individual carbon and nitrogen atoms of the porphyrins. If a biogenetic enumeration were utilized, in which advantage is taken of biologically equivalent atoms, then Wittenberg and Shemin's (19) system has obvious meaning (see X). It has the disadvantage of complete divorce from the extensive literature utilizing the Fischer 1-to-8 system (18) (see also Chapter 3). In any event, there is as yet no certainty of the biogenetic equivalence of the four porphobilinogens in chlorophyll formation, as in heme (see Section III). For all these reasons, but primarily to bridge the gap with the classical literature, the enumeration shown in (XI) has been suggested (20). It will be observed that in both

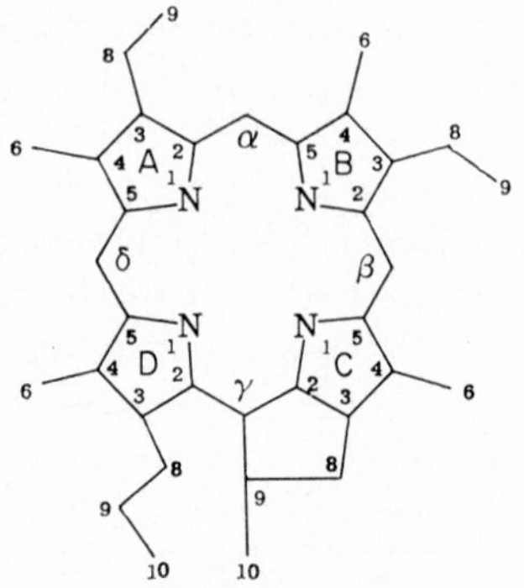

Wittenberg and Shemin's
numbering system

X

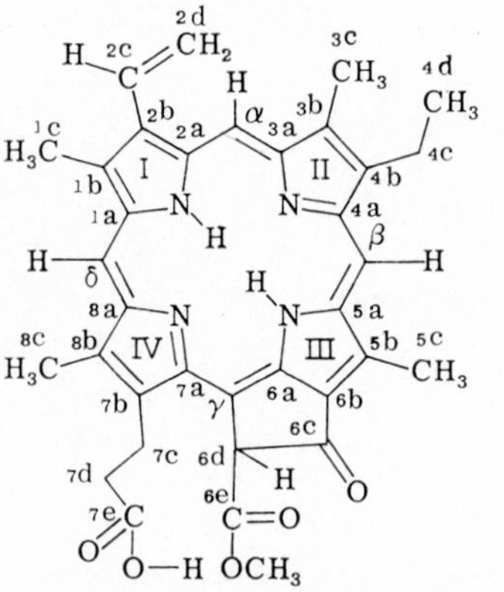

Numbering system for
naturally occurring porphyrins

XI

(X) and (XI) the numbering of the carbons of the cyclopentanone ring differs from Fischer's. Fischer's classical carbons 9 and 10 correspond to 6d and 6e in the system of (XI). Carbon 6e, it will be noted, was not numbered in the classical system.

A. Major Classes of Chlorophylls

1. THE CHLOROPHYLLS

The chlorophylls included formally within this group are Chls *a*, *b*, *c*, and *d*. Chl *c* probably will be excluded eventually, because of the virtual certainty (see below) that it is a porphine and not a chlorin. The formal structures for Chls *a* and *b* are depicted in (XII). Extensive reviews are available, see e.g. Aronoff (*20*), delineating the logic involved in the deduction of the structure from primarily degradative experiments. More recently, the validity of the formulation has been demonstrated by its total synthesis (*21*), starting from substituted, free pyrroles and involving (XIII) as the initial porphyrin (see Chapter 5). From (XIII) it was possible to obtain (in a number of steps) isopurpurin 5-methyl ester (XIV), which was of import in the sequence of syntheses as the first synthetic porphyrin identical with a chlorophyll degradation compound. From (XIV) it was then possible to form chlorin e$_6$ (XV), which, in turn,

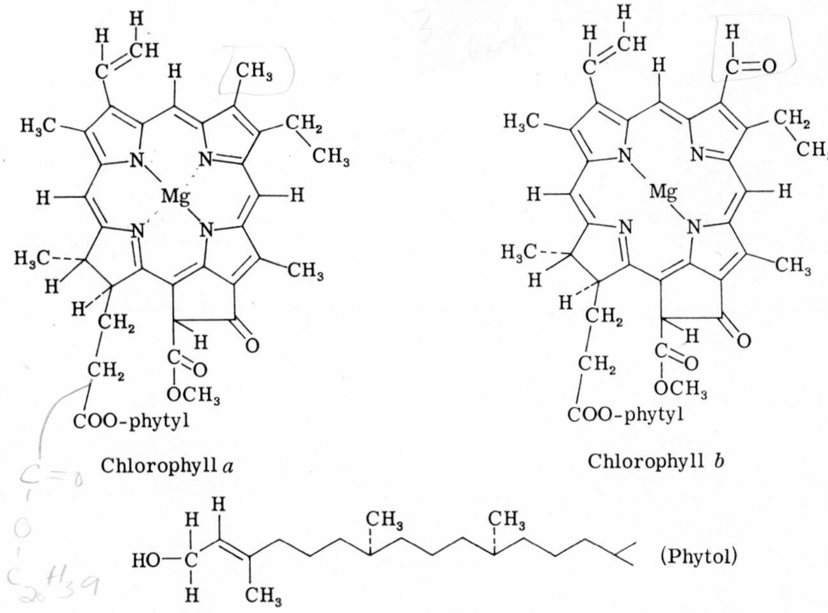

Chlorophyll *a*

Chlorophyll *b*

(Phytol)

XII

is readily converted to pheophorbide. Phytol, whose absolute configuration has also been determined recently (*22, 23*) as 3,D-7,D-11,15-tetramethyl hexadec-*trans*-2-en-1-ol could now be inserted chemically or enzymatically, and the magnesium be added by a Grignard reaction. There is therefore no longer any doubt as to the validity of the structures given for chlorophylls *a* and *b*.

The chemistry of Chl *c* is known to only a very limited extent (*24*). The absence of a prominent peak in the red suggests that it is more probably a porphine than a chlorin. Were Chl *c* a chlorin, it should presumably be oxidizable to a porphine, e.g., by quinone; unfortunately, there is no published evidence of this. Quantitative analysis of the magnesium content (atomic absorption spectroscopy) has resulted in a value of 2.31 ± 0.045%. Assuming one atom per molecule, this is equivalent to a molecular weight of 1013–1152, with a mean of 1052, or about twice that expected of the phytol-free compound. While earlier suggestions were made (25) that Chl *c* might be phytol free because of its relatively low acid number of 12, there is unfortunately no direct evidence on this point from the purified preparation of Jeffrey. It would be logical to re-determine molecular weights following enzymatic or acid hydrolysis. The presence of long-chain fatty alcohols is also difficult to reconcile with

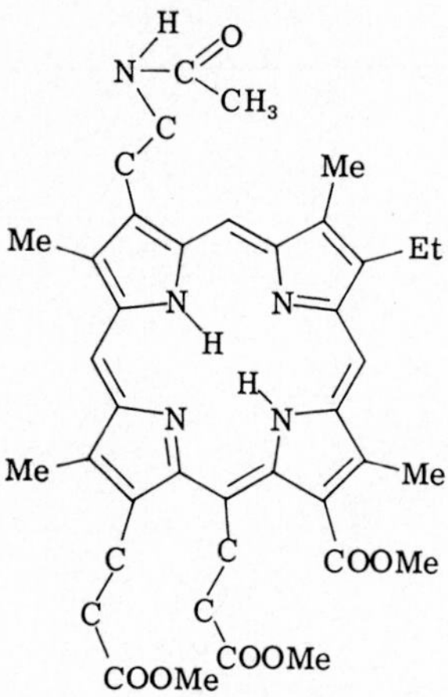

Woodward's porphyrin

XIII

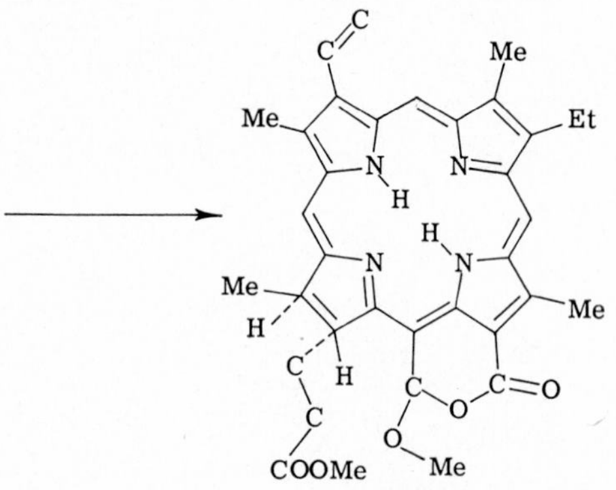

Isopurpurin 5-methyl ester

XIV

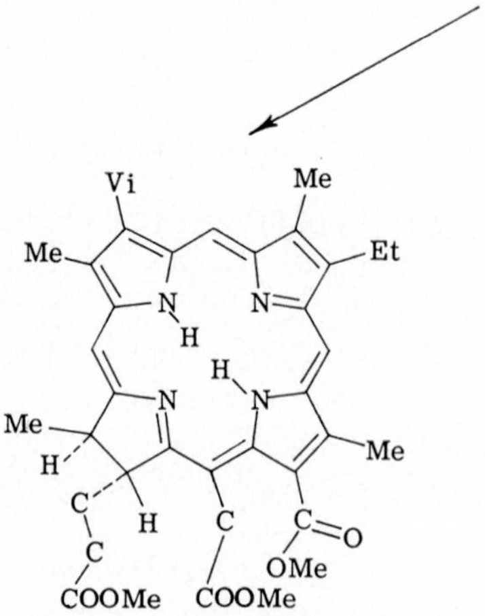

Chlorin e$_6$ trimethyl ester

XV

its relative ease of crystallization. This chlorophyll is found in numerous marine algae and diatoms, apparently serving as the accessory pigment, instead of Chl *b*. However, if it is truly functional and a porphine, it contrasts with all the other chlorophylls, which are either chlorins or dihydrochlorins.

Chl *d* is a minor chlorophyll component of some red algae, where it may occur in the presence of Chl *b*, which is usually absent when accessory pigments occur. It is thought to be 2-devinyl-2-formylchlorophyll *a* (*26*), since its absorption spectrum is identical with that of the compound derived from permanganate oxidation of Chl *a*. Furthermore the absorption spectrum of the reduction product resulting from the action of sodium borohydride resembles that of 2-devinyl-2-hydroxy-methylchlorophyll *a*. There is no reason to believe that the variability of distribution of the pigment or indeed, its very presence, is artifactual.

2. THE BACTERIOCHLOROPHYLLS

It has long been known that the Bchl occurring in the Thiorhodaceae is a tetrahydroporphine, or dihydrochlorin. It can be converted into a chlorin by oxidation with appropriate quinones (e.g., 2,3-dichloro-5,6-dicyanobenzoquinone). This Bchl, now called Bchl *a*, has the structure 2-devinyl-2-acetyl-3,4-dihydrochlorophyll *a* which agrees with the original proposal of the Hans Fischer school. However, the precise position of the "extra" H's as being on ring II, in positions 3,4, has been verified only recently (*27*). Bchl *a* from *Chromatium* was converted to the bacteriochlorin e_6 trimethyl ester and this, in turn, oxidized with chromium trioxide to the corresponding maleimides and succinimides. It had been

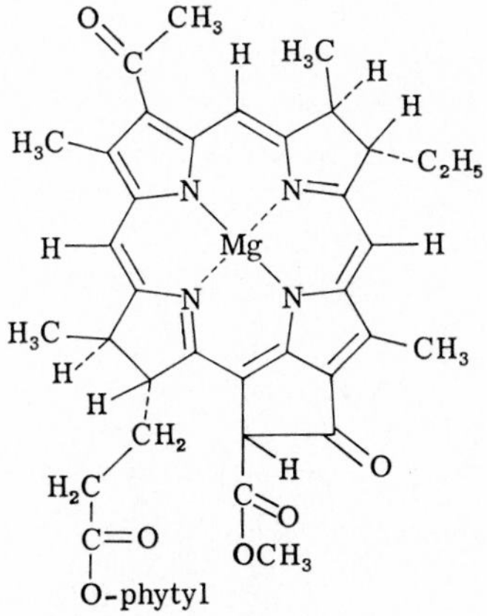

Bacteriochlorophyll *a*

XVI

found earlier that Chl *a* oxidation resulted in the formation of a methyl, propionic acid succinimide (i.e., *trans*-hemotricarboxylic imide or *trans*-dihydrohematinimide) whereas the corresponding porphines yielded hematinic acid. The finding of major amounts of methyl, ethyl succinimide (as well as methyl, ethyl maleimide) in addition to dihydrohematinimide, showed that the additional hydrogens arose from ring II. Furthermore, infrared spectroscopy of its crystalline *p*-bromophenacyl ester showed it to be *transoid*, rather than *cisoid*. Bchl *a* thus has the structure of (XVI). At the moment no facts concerning Bchl *b* are available beyond mention of its existence in isolated species of Athiorhodaceae, e.g., a *Rhodopseudomonas* species (*3*).

3. The Chlorobium Chlorophylls

The green sulfur bacteria, such as *Chlorobium* and *Chloropseudomonas,* contain Cchl along with a trace of Bchl *a*. Some strains of this group contain a pigment complex with major absorption of the extracted pigments in ether at 660 mμ, while others have an equally complex group with a maximum at 650 mμ. The latter may be separated chemically into at least 6 different pigments (*28, 29*), all of which appear to have the general structure of the magnesium chelate of 2-devinyl-2-α-hydroxyethylpyropheophorbide *a* farnesyl ester (XVII). The 660 group, separable chromatographically into at least 7 components, appear to have the same

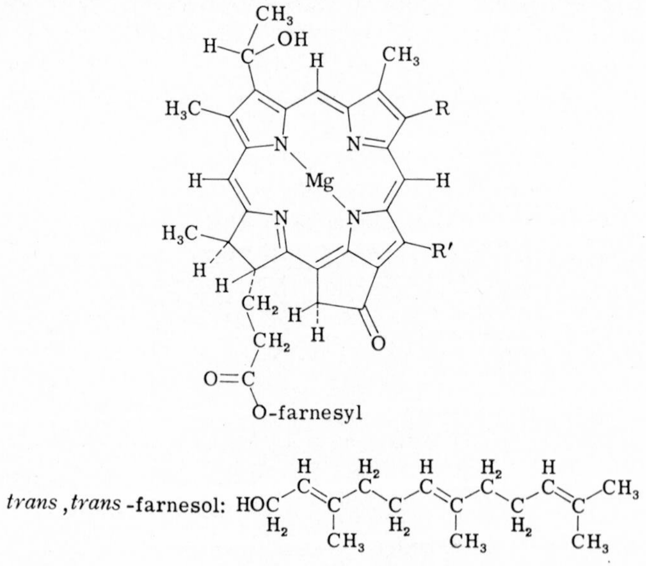

Chlorobium chlorophyll

XVII

general structure, with the additional feature of a δ-methyl group. Details will be found in Chapter 4.

B. Spectroscopy

Interpretations of the spectroscopy of the porphyrins in the visible region are based almost entirely on molecular symmetry. In general, it is believed that there is a short and a long axis (XVIII) even in otherwise symmetrical porphines, dictated by the opposite arrangement of the central hydrogens. The symmetry is further lost in the di- and tetrahydro-

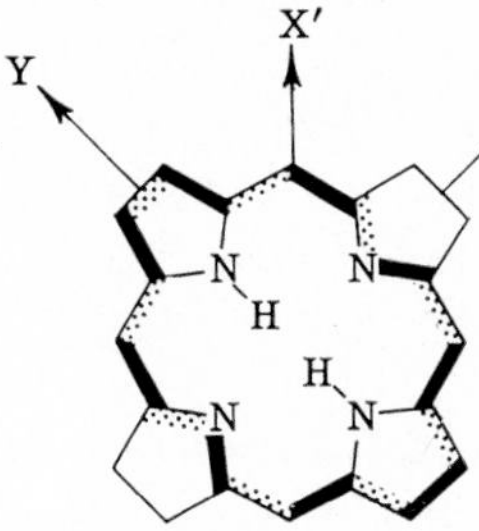

XVIII

porphines, resulting in absorption spectra by which the initiated can recognize them at a glance. Details of spectroscopy are given in Chapters 6 and 7.

C. Biogenesis

The biogenesis of the heme porphyrins is now a classic aspect of biochemistry, proceeding from the condensation of succinyl coenzyme A (CoA) with glycine to form δ-aminolevulinic acid (δ-ALA) which, in turn, is dimerized to porphobilinogen. Four molecules of the latter are condensed to the various porphyrinogens, from which the various porphines are derived by oxidation: uroporphine III, coproporphine III, and eventually protoporphine. Protoporphine is then chelated with iron to provide heme (30). It has been inferred that a similar pathway exists for chlorophyll formation in the plant and that protoporphine is the bifurcation point for the chlorophylls and the hemes (31), and a general pathway for subsequent chlorophyll biogenesis has been proposed (32).

Three general approaches to the study of the biogenesis of the chlorophylls have been utilized: (a) the use of tracers, (b) the use of mutants, and (c) the use of spectroscopy (for the later stages). By the use of tracers (33) it has been shown that (at least in the soybean) there is no turnover (i.e., neither synthesis nor degradation) of chlorophyll in a mature leaf, but the kinetics of carbon incorporation is rather complex (34).

As a result of the chlorophyll mutation studies, algae were found which, following the blockage of chlorophyll formation, produced copious amounts of protoporphine IX and its monoester and, in another case, magnesium protoporphine (35) and its monoester. It is this mutant, along with the *de facto* production of protoporphine, which provides the major evidence that the latter compound is common to the biosynthesis of chlorophyll and hemes. Inasmuch as trace amounts of protochlorophyllide, chlorophyllide, and protochlorophyll may be found in many higher plants, if we may assume Granick's proposed pathway, there remains but the elucidation of the steps between magnesium protoporphine and magnesium vinylpheoporphine a_5 (protochlorophyllide). This would include the stages of reduction of carbons 4c,d to ethyl and the conversion of carbons 6c,d,e to the cyclopentanone ring, followed by methyl esterification of 6e.

There is now reasonable agreement that the final stages of chlorophyll synthesis proceed for the most part through the steps: protochlorophyllide → chlorophyllide → chlorophyll *a*. The evidence for this sequence (in contrast to that postulated earlier of protochlorophyllide → protochlorophyll → chlorophyll), arises both from spectroscopic and from radiotracer studies.

III. Function

The major problem concerning the function of chlorophyll is whether it serves merely as an energy collector and transmitter, or whether it is also involved in energy transduction, e.g., electron donation and acceptance, or protonation and deprotonation.

Elementary calculation of the photon flux in photosynthesis at moderate light intensities (e.g., 1000–2000 ft.-candles), coupled with the energetic demand of at least 4 photons to reduce one CO_2 to the CH_2O level, shows that the photons absorbed by diverse chlorophyll molecules in a brief time (e.g., 1 μsec), must be utilized cooperatively. How this energy is transferred *in vivo* is still a matter of conjecture (see Chapter 19). In part, this is a matter of the lack of knowledge of the organization of chlorphyll within the lamellae of the chloroplasts (Chapters 9 and 10), i.e., the degree of ordering, the extent of aggregation, the geometry of protective carotenoid groups, its possible association with proteins, etc. The reactions which may be shown for chlorophyll *in vitro* do not necessarily hold for chlorophyll *in vivo*. For example, the spectroscopic identification of the triplet state appears to be impossible unless chloroplasts are nonfunctional (36). The oxidation of chlorins to porphines by quinones, virtually a quantitative reaction *in vitro,* does not appear to occur *in vivo.* Many of the direct photochemical redox reactions of isolated chlorophyll, now known collectively as Krasnovskii reactions, are generally not direct reactions when they also occur *in vivo.*

Three general methods by which energy transfer between accessory pigments and the chlorophylls, or between the chlorophylls themselves, may occur are resonance energy transfer (37), semiconduction (38, 39), and exciton transfer (40). Only the first two have been shown to exist. There is no experimental evidence for the last, though if the chlorophylls around the collection point were highly organized, there would be no basis for excluding it.

The collecting point for the energy (or the "reaction center") appears, at the present, to be a pigment with a red absorption maximum at about 700 mμ and is therefore designated as P700 (41). The exact nature of this pigment is still unknown. Its longer wavelength absorption has suggested that it may be a Chl *a* polymer, more specifically a dimer. Such dimers were originally postulated by Brody (42) from studies of concentrated alcoholic solutions of chlorophyll. However, the existence of dimers in ethanolic solutions has been questioned severely by Closs *et al.* (43), whose infrared and osmometric studies suggested depolymerization of porphyrins by alcohol (see Chapter 7). That chlorophyll dimerizes in apolar solutions such as benzene was shown osmometrically (44), but the spectroscopy of such concentrated solutions is not consistent with a dimeric nature for P700 (45).

On the other hand, although Bannister's "colloidal chlorophyll" is a possible candidate by virtue of the position of its red maximum, there is no knowledge of the structure of that micelle. Recently, chlorophyll and bacteriochlorophyll proteins have been isolated (46, 47). It is possible that the former is an artifact resulting from the use of Triton X-100 in its preparation.

However, porphyrins in general are notorious for their ability to adsorb on proteins, and consequently the demonstration of the *in vivo* existence of such a complex would be a most difficult technical problem.

An intriguing mystery is that of the role of carotenoids, which are ubiquitous to all natural photosynthetic organisms. When they are absent, either as the result of a mutation or by growth of the organism in diphenylamine (48), the chlorophyll is extremely susceptible to photooxidation. (Many "chlorophyll" mutants are actually carotenoid mutants, the chlorophyll being destroyed photooxidatively in the field.) Not all the carotenoid is effective in this manner; presumably that which is effective has a special geometric relationship to chlorophyll. Which of the various carotenoids plays this role *in vivo* is not known, but experiments with model systems have shown that the chain length of the carotenoid must be adequate, as must the degree of unsaturation (49). Its protective effect is thought to be exerted by quenching of the triplet state of chlorophyll, though this phenomenon is usually associated with free radicals, such as O_2 and semiquinones.

The present scheme for the function of chlorophyll in photosynthesis envisages it as photoionizing, i.e., ejecting an electron when in the photoexcited state. This is thought to occur in two different environments (e.g., with diverse accessory pigments and different electron-transfer components). These two systems of components have different loci for their absorption maximum in the red (system I, 683 mμ and system II, 673 mμ in green plants). Thus, diverse partial photochemistry may be made to occur according to the wavelength used. The presumed product in each of the photochemical steps is an oxidized chlorophyll and an electron; the chlorophyllonium ions of the two systems are of appreciably different redox potential. In neither case is the immediate electron acceptor known, although in one case the ultimate acceptor appears to be ferredoxin (and then NADP), while in the other it may be a quinone (and then the cytochrome electron transport system, i.e., cytochrome f, and possibly cytochrome b_6; see Chapter 18). Oxidized chlorophyll would be reduced to its neutral form by oxidizing an aqua-dismutase on the one hand (with the evolution of oxygen), while in the other system it oxidizes the cytochrome(s). Consequently, contemporary theory suggests that chlorophyll acts both as an energy transmitter and a transducer, though the latter takes the form of a special molecule—special by virtue of its unique locus and association with some other moiety.

It may be of historic value to point out that other hypotheses have been entertained in the past. Chlorophyll has been suggested as a direct hydrogen donor, either via its enolizable H on carbon 6d (nee 10), or the two on carbons 7b and 8b, or that it represents the oxidized form of a bacteriochlorophyll (i.e., with additional H's on 3b and 4b). In other theories, the ability of chlorophyll to react with CO_2 *in vitro* (i.e., a weak acid-weak base reaction) was extended to the concept of a decarboxychlorophyll being the initial CO_2 acceptor for photosynthesis. For a review of the above, see Aronoff (50). Finally, an intriguing concept, not yet shown invalid, is that chlorophyll itself may be involved in photophosphorylation, via a phosphoenolic chlorophyll (at carbon 6d, in its enolic form) with photosynthesis being thought of as a dismutation of phosphoric acid, rather than water (51).

REFERENCES

(1) F. Pelletier and J. B. Caventou, *Ann. Chim. Phys.* [2] **9**, 194-196 (1818).

(2) A. Manten, Thesis, University of Utrecht (1948).

(3) F. T. Haxo and L. R. Blinks, *J. Gen. Physiol.* **33**, 389 (1950).

(4) R. Willstätter and A. Stoll, "Untersuchungen über Chlorophyll." Springer, Berlin, 1913.

(5) J. Berzelius, *Ann. Chem.* **27**, 296 (1838).

(6) F. Verdeil, *J. Prakt. Chem.* **33**, 478 (1844).

(7) E. Fremy, *Compt. Rend.* **50**, 405 (1860); **61**, 188 (1865); **84**, 983 (1877); *Ann. Chim. Phys.* [4] **7**, 78 (1866).

(8) G. G. Stokes, *Ann. Physik* [2] **4**, 220 (1854); *Proc. Roy Soc.* **13**, 144 (1864); *J. Chem. Soc.* **17**, 304 (1864).

(9) H. C. Sorby, *Proc. Roy Soc.* **21**, 442 (1873).

(10) M. Tswett, *Ber. Deut. Botan. Ges.* **24**, 316 and 384 (1906); **25**, 140 (1907); *Biochem. Z.* **10**, 414 (1908).

(11) F. Hoppe-Seyler, *Z. Physiol. Chem.* **3**, 339 (1879); **4**, 193 (1880); **5**, 75 (1881).

(12) M. Nencki and J. Zaleski, *Ber. Deut. Chem. Ges.* **34**, 997 (1901).

(13) M. Nencki and L. Marchlewski, *Ber. Deut. Chem. Ges.* **34**, 1687 (1901).

(14) R. Willstätter and Y. Asahina, *Ann. Chem.* **385**, 188 (1911).

(15) A. Borodin, *Botan. Ztg.* **40**, 608 (1882).

(16) N. A. Monteverde, *Acta Horti Petropolitani* **13**, 148 (1893).

(17) M. Nencki, *Ber. Deut. Chem. Ges.* **29**, 2877 (1896).

(18) H. Fischer and H. Orth, "Die Chemie des Pyrrols," Vol. 2, Part I. Akad. Verlagsges., Leipzig, 1937; H. Fischer and A. Stern, *ibid*, Vol. 2, Part II (1940).

(18a) A. Jensen, O. Aasmundrud, and K. E. Eimhjellen, *Biochim. Biophys. Acta* **88**, 466 (1964).

(19) T. Wittenberg and D. Shemin, *J. Biol. Chem.* **185**, 103 (1950).

(20) S. Aronoff, *in* "Handbuch der Pflanzenphysiologie" (W. Ruhland, ed.), Vol. I, p. 234. Springer, Berlin, 1960.

(21) R. B. Woodward, *J. Am. Chem. Soc.* **82**, 3800 (1960); *Pure Appl. Chem.* **2**, 383 (1961); *Angew. Chem.* **2**, 651 (1960).

(22) J. W. K. Burrell, L. M. Jackman, and B. C. L. Weedon, *Proc. Chem. Soc.* p. 263 (1959).

(23) P. Crabbe, C. Djerassi, E. J. Eisenbraun, and S. Liu, *Proc. Chem. Soc.* p. 264 (1959).

(24) S. W. Jeffrey, *Biochem. J.* **86**, 313 (1963).

(25) S. Granick, *J. Biol. Chem.* **179**, 505 (1949).

(26) A. S. Holt, *Can. J. Botany* **39**, 327 (1961).

(27) J. H. Golden, R. P. Linstead, and G. H. Whitham, *J. Chem. Soc.* p. 1725 (1958).

(28) A. S. Holt, D. W. Hughes, H. J. Kende, and J. W. Purdie, *Plant Cell Physiol.* (*Tokyo*) **4**, 49 (1963).

(29) D. W. Hughes and A. S. Holt, *Can. J. Chem.* **40**, 171 (1962).

(30) S. Granick, *Harvey Lectures* **44**, 220 (1959).

(31) S. Granick, *J. Biol. Chem.* **172**, 717 (1948).

(32) S. Granick, *J. Biol. Chem.* **183**, 173 (1950).

(33) S. Aronoff, *in* "Radiation Biology and Medicine" (W. D. Claus, ed.), p. 633. Addison-Wesley, Reading, Massachusetts, 1958.

(34) A. A. Shlyk and E. Fradkin, *Biofizika* **7**, 281–290 (1962).

(35) S. Granick, *J. Biol. Chem.* **230**, 1168 (1961).

(36) A. Müller, D. C. Fork, and H. T. Witt, *Z. Naturforsch.* **13b**, 142 (1963).

(37) L. N. M. Duysens, Ph.D. Thesis, University of Utrecht (1957).

(38) W. Arnold and H. K. Maclay, *Brookhaven Symp. Biol.* **11** (**BNL 512** (**C28**)), 1 (1958).

(39) R. C. Nelson, *J. Chem. Phys.* **27**, 864 (1957).

(40) A. S. Davydov, "Theory of Molecular Excitons" (translated by M. Kasha and M. Oppenheimer, Jr.). McGraw-Hill, New York, 1962.

(41) B. Kok, *Biochim. Biophys. Acta* **22**, 399 (1956).

(42) S. S. Brody and M. Brody, *Nature* **189**, 547 (1961); *Biochim. Biophys. Acta* **54**, 495 (1961).

(43) G. L. Closs, J. J. Katz, F. C. Pennington, M. R. Thomas, and H. H. Strain, *J. Am. Chem. Soc.* **85**, 3809 (1963).

(44) S. Aronoff, *Plant Physiol.* **38**, 628 (1963).

(45) M. Nissen, M.S. Thesis, Iowa State University (1965).

(46) J. S. Kahn, *Biochim. Biophys. Acta* **79**, 234 (1964).

(47) T. T. Bannister, *Physiol. Rev.* **1**, 115 (1963); T. T. Bannister and J. E. Bernardini, *Photochem. Photobiol.* **1**, 535 (1963).

(48) R. C. Fuller and I. C. Anderson, *Nature* **181**, 250 (1958).

(49) H. Claes, *Z. Naturforsch.* **16b**, 445 (1961).

(50) S. Aronoff, *Botan. Rev.* **23**, 65 (1957).

(51) M. Calvin, *in* "Horizons in Biochemistry" (M. Kasha and B. Pullman, eds.), pp. 23–57. Academic Press, New York, 1962.

—2—

Extraction, Separation, Estimation, and Isolation of the Chlorophylls*

HAROLD H. STRAIN AND WALTER A. SVEC

Argonne National Laboratory, Argonne, Illinois

* Based on work carried out under the auspices of the U. S. Atomic Energy Commission.

21

I. Basis of Interest in Analytical Methods

A. Significance of Chlorophylls

Interest in analytical methods for the detection, estimation, and identification of the chlorophylls now arises in many diverse areas of investigation. Greenness or chlorophyll content is an indication of the quality of foods and fodders (*1*). It reveals the ripeness and quality of many fruits and vegetables (*2*). It reflects the keeping qualities of some foods (*3*), and it indicates various changes during the preparation, preservation, and storage of green vegetables (*4–9*).

In photosynthesis, the chlorophylls absorb the sunlight that is indispensable to the unique autotrophic activities of plants (*10–13*). They link the inorganic and the organic worlds. In the sea and on the land, the chlorophyll content of the native plants is a key to the production of oxygen (*14*) and of organic matter (*15–22*). Among agricultural plants, there is a relationship between the chlorophyll content (chlorophyll synthesis) and crop production, especially as influenced by the inorganic nutrition (*23–35*) and by pesticides (*36*).

The capacity of plants to maintain a functional pigment system with little variation throughout the long course of organic evolution has become of great interest to geneticists, taxonomists, and evolutionists (*37*). The occurrence of the same principal chlorophyll in all autotrophic (oxygen-producing) plants that have been examined points to a common origin for all these organisms (*37, 38*). Variations of the chlorophylls and the associated carotenoid pigments are related to the taxonomic classification of the organisms (*37–45*) (see Chapter 16). In the proliferation of plant material, the maintenance of these pigment systems depends upon the direct transmission of the chloroplast material to each new cell. Although remarkably constant in normal plants, the pigment system may be altered by irradiation, various chemical reagents (*28, 31, 46*), and genetic effects (*47–52*).

B. Limitations of Analytical Methods

Progress in many studies of the chlorophylls depends upon the use of analytical methods for the identification and estimation of these pigments. As a consequence, there has been an increasing need for reliable analytical procedures (*19, 53, 54*).

In spite of a great deal of experimentation concerning the extraction and estimation of the chlorophylls, no one method combines simplicity, wide applicability, ready reproducibility, and high sensitivity. In fact, so many variable conditions affect the pigments, their extractability and their reactions, that the prospect for the selection of a single analytical procedure that will estimate all the green components of various plants and plant products with high precision is not very promising, as may be inferred from diverse investigations (*38–66*).

Several extensive reviews with numerous citations have already been devoted to the analytical chemistry of the chlorophylls (*38, 62–66*), to their structural chemistry and synthesis (*55–61*), and to their occurrence in various plants (*37–43, 63, 65, 67, 68*). To conserve space and avoid undue repetition, only the more recent and especially pertinent references are cited in this chapter. The objective of this summary is not to reproduce all the material that is readily available in the reviews (*38, 63, 65, 67*) and bibliographies (*19, 69*), but to evaluate the analytical procedures and consider the validity of the conclusions that result from the application of these methods. To this end, the properties of the natural chlorophylls most useful for analytical determinations will be considered. These properties include separability by extraction and chromatography, and the spectral absorption characteristics of the highly purified pigments and their frequently encountered alteration products.

II. Nature of Chlorophylls

A. Properties of Chlorophylls

From the analytical standpoint, the principal photosynthetically functional green pigments extractable from autotrophic (oxygen-producing) and photoheterotrophic organisms (*68*) with organic solvents are regarded as chlorophylls. These pigments are products of the autotrophic and photoheterotrophic growth of various organisms. By these criteria, the fully deuterated green pigments produced by algae grown autotrophically in heavy water (*70*) are truly chlorophylls (*71–74*).

The various green and gray-brown products formed from the natural green pigments when cells are injured or killed (75–77), subjected to various reagents (46, 78–80), or exposed to various unfavorable conditions (77, 81) are regarded as chlorophyll alteration products. Because the chlorophylls are exceptionally labile, it is virtually impossible to extract and isolate them without the formation of some of their alteration products (19, 38, 57, 73, 75, 76, 82–85). At the present stage of progress, it is frequently difficult to determine whether these additional pigments are constituents of the growing plant or artifacts.

Most analytical methods are not suited to the identification and estimation of chlorophylls within the chloroplasts. Chemical investigations are restricted largely to pigments isolated in a state of high purity. Physical investigations must be standardized primarily in relation to the pure pigments. Physical measurements such as the spectral absorption characteristics are often difficult to interpret because chlorophylls in the chloroplasts and in solid and colloidal form exhibit absorption maxima displaced toward the red region of the spectrum relative to the maxima of solutions in organic solvents (11, 86–98).

All the chlorophylls are tetrapyrrolic pigments. All contain magnesium. All exhibit pronounced absorption bands in the blue-green and red regions of the visible spectrum. All exhibit similar, though not identical, solubility in various solvents. Many are isomerized spontaneously in solution, especially at elevated temperature (38, 99, 100). They are decarbomethoxylated upon prolonged heating of the solutions (73). They are decomposed by acids, alkalies, oxidizing reagents, hydrolytic enzymes, oxidative enzymes, strong adsorbents, and intense light (38, 57, 73, 76, 77, 101, 102). They are oxidized (allomerized) rapidly when dissolved in relatively inert solvents, such as alcohols in the presence of air (38, 57, 76). But when isolated in the solid state and stored in evacuated and sealed ampules, the chlorophylls may be preserved for long periods without change or alteration.

B. Properties Useful for Analysis

Many of the properties of the chlorophylls may be utilized in methods for their examination, detection, and estimation. With many of these properties, such as the magnesium or nitrogen content, nuclear magnetic resonance properties, isomerization effects, and most chemical reactions, the pigments must be separated from one another and from other plant constituents that may interfere with the determinations (65, 103). With other properties, the pigments need not be separated from the mixtures found in plant extracts. Of these properties, the fluorescence (63, 65, 104, 105) and the spectral absorption by the chlorophylls (63, 65, 67) and the pheophytins (*65, 106, 107*) in the visible region of the spectrum are most commonly employed. A few properties, such as absorption spectra and fluorescence, may be utilized for the estimation of total chlorophyll in plant cells (see Section IV).

The adsorbability of the chlorophylls, as employed in chromatographic systems, serves more analytical purposes than any other property (*38, 64–66, 108, 109*). It provides the most effective methods for the detection and isolation of the natural green pigments and for the separation of these pigments from their various alteration products (*38, 65, 73, 76, 77, 110, 111*). It also facilitates the purification and isolation of these pigments (see Section VIII). It provides a basis for the description of the pigments and for their identification by comparison with authentic substances (*38, 65, 109*). In conjunction with spectral absorption and nuclear magnetic resonance (see Chapter 7), chromatographic methods provide the key to many analytical, preparatory, structural, chemical, and physiological investigations (*38, 65, 71–74, 84, 112–115*).

In spite of their great contributions to investigations of the chlorophylls, chromatographic methods have several significant limitations. They do not separate some of the colorless substances encountered in plant material from the pigments, hence pigments eluted after chromatographic separation are usually contaminated with these colorless substances (*71, 72, 84, 115, 116*). These and other colorless substances from the adsorbent and solvent may affect the spectral curves of the pigments in the nuclear magnetic resonance (*113*), infrared, and ultraviolet regions and may give erroneously low values for spectral absorption coefficients based upon the weight of pigment obtained by evaporation of the solution (*117*).

In many chromatographic separations, green zones of the common chlorophylls *a* and *b* serve as reference standards. Additional green zones may be misinterpreted as indicative of "new" chlorophylls, although in fact they may be due to chlorophyll alteration products (*76, 77, 83–85, 99, 100, 118, 119*) (Section V) or to anomalous zone formation (*109*). Such misinterpretations may be avoided only by special experiments designed to determine the origin of the pigments forming the additional zones.

C. Examination by Partition and Chromatography

Chlorophylls may be separated from one another and from many other plant constituents by partition between immiscible solvents. Separations may involve only one or a few partitions using selected solvent pairs, or they may be based upon multiple partitions (Craig procedure)

(*120*). Most of the chloroplast pigments may also be separated by chromatography and by various combinations of these methods. The most effective procedure for the separation of chlorophylls from one another, their various alteration products, the carotenoid pigments, and most colorless cellular constituents is chromatography with very mild adsorbents, such as cellulose, starch, or powdered sugar. A typical separation of the chloroplast pigments of spore-producing and seed-producing plants by column chromatography is shown in Fig. 1.

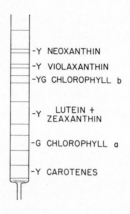

-Y NEOXANTHIN

-Y VIOLAXANTHIN

-YG CHLOROPHYLL b

-Y LUTEIN +
    ZEAXANTHIN

-G CHLOROPHYLL a

-Y CAROTENES

FIG. 1.   Leaf pigments separated in a column of powdered sugar washed with petroleum ether plus 0.5% *n*-propanol (*38*).

Chromatographic separations may be qualitative or quantitative, on an ultramicroscale or on a preparative scale. Qualitative methods reveal the number, the sequence and the identity of the pigments. Quantitative methods provide individual pigments that may be estimated fluorimetrically (*63, 121*), colorimetrically, or spectrophotometrically (*63, 65, 122–127*). As preparatory methods, they provide individual pigments in sufficient quantity for isolation in the solid state and in sufficient purity for determination of various physical and chemical properties (see Section VIII).

Chromatographic methods (*128*) may be employed in many different modifications, namely, columnar chromatography (*38, 65, 129*), one-way paper chromatography (*109–111, 116, 121, 127, 130*), two-way paper chromatography (*109, 115, 131*) as in Fig. 2, radial paper chromatography as in Fig. 3 (*109, 132*), also with acceleration by centrifugal force (*133*), and thin-layer chromatography (*109, 134–138*). Partition chromatography with polyethylene has also been used (*112, 139*). Contrary to some of the statements in the literature, columnar separations are as fast

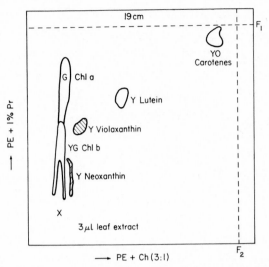

FIG. 2.   Leaf pigments separated by two-way chromatography on paper washed first with petroleum ether plus 1% *n*-propanol, then transversely with petroleum ether plus chloroform (3:1) (*109*).

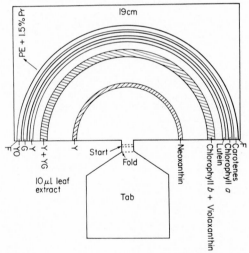

FIG. 3.   Leaf pigments separated by radial chromatography on paper (*109*).

or faster than separations in paper and in thin layers. In columns, the pigments are less exposed to air and oxygen than in paper and in thin layers of the adsorbent. In one-way and in two-way separations, colorless substances affect the separations, distorting the zones of the separated pigments. The migration of the zones relative to the migration of the

TABLE I

CHROMATOGRAPHIC SEQUENCE OF THE CHLOROPHYLLS AND SOME OF THEIR ISOMERS IN COLUMNS OF POWDERED SUGAR WITH PETROLEUM ETHER PLUS 0.5–2.0% n-PROPANOL AS THE WASH LIQUID

Chlorophylls	Color of zones	λ_{max} (mμ)				Solvent	Reference
Chlorophyll c (most adsorbed)[a]	Light yellow green	447.3		578.5	626.5	Ether	(139a)
Chlorobium chlorophyll 650	Green		425.0		650.0	Ether	(68)
Chlorobium chlorophyll 660	Green		430.0		660.0	Ether	(139a)
Deuteriochlorophyll b	Yellow green		451.0		640.5	Ether	(139a)
Chlorophyll b	Yellow green		452.5		642.0	Ether	(72)
	Yellow green		470.0		650.0	Methanol	(41)
Chlorophyll d	Yellow green		445.0		686.0	Ether	(100)
	Yellow green		456.0		696.0	Methanol	(100)
Bacteriochlorophyll	Purplish gray	357.5			770.0	Ether	(139a)
Deuteriochlorophyll b'	Yellow green		—		—	—	(139a)
Chlorophyll b'	Yellow green		—		650.5	Methanol	(41)
Chlorophyll d'	Yellow green		—		697.0	Methanol	(100)
Isochlorophyll d	Green		—		661.0	Methanol	(100)
Deuteriochlorophyll a	Green		428.0		659.0	Ether	(72)
Chlorophyll a	Green		428.5		660.5	Ether	(72)
	Green		432.5		665.0	Methanol	(41)
Protochlorophyll	Green	535.0	432.0	571.0	623.0	Ether	(63, 65)
Isochlorophyll d'	Green		—		661.5	Methanol	(100)
Deuteriochlorophyll a'	Green		—		—	—	(139a)
Chlorophyll a' (least adsorbed)	Green		428.5		661.0	Ether	(139a)

[a] Brackets indicate pigments incompletely separated under the adsorption conditions.

solvent, the R value, is frequently reported, yet current experience shows that the R value increases with the concentration of the leaf extract and with the presence of colorless impurities (*109, 116, 124*).

The selectivity or resolving power of these chromatographic methods is a function of the adsorbent and the solvent. The sequence of the separated pigments, often used as a basis for their description and identification, is shown in Figs. 1–3 and Tables I and II. In these tables, pigments incompletely separated under the adsorption conditions are indi-

TABLE II

CHROMATOGRAPHIC SEQUENCE OF CHLOROPHYLLS *a* AND *b* AND SOME OF THEIR
ALTERATION PRODUCTS IN COLUMNS OF POWDERED SUGAR WITH PETROLEUM
ETHER PLUS 0.5%–2.0% *n*-PROPANOL AS THE WASH LIQUID

Pigments	λ_{max} (mμ)(ether)		Reference
Pyrochlorophyll *b* (most adsorbed)	453.5	640.5	(*73, 139a*)
Chlorophyll *b*	452.5	642.0	(*72*)
⌈ Chlorophyll *b'*	—	—	(*139a*)
⌊ Methyl pyrochlorophyllide *a*	428.0	659.0	(*73*)
Methyl chlorophyllide *a*	427.5	660.5	(*73*)
Methyl chlorophyllide *a'*	—	—	(*139a*)
Methyl pheophorbide *a*	408.5	667.0	(*73*)
Methyl pyropheophorbide *a*	408.5	667.0	(*73*)
Pyrochlorophyll *a*	429.0	659.5	(*73*)
Chlorophyll *a*	428.5	660.5	(*72*)
Chlorophyll *a'*	428.5	661.0	(*139a*)
Pheophytin *a*	409.0	667.0	(*73*)
Pyropheophytin *a* (least adsorbed)	409.0	667.0	(*73*)

cated by brackets. Not only the separability, but also the chromatographic sequence, varies with the adsorbent and the solvent (*38, 66*). When petroleum ether plus 0.5% *n*-propanol is employed as the wash liquid for paper or columns of powdered sugar or cellulose, adequate separation of the pigments is usually obtained, as shown in Figs. 1–3. The *n*-propanol produces a better separation of the chlorophylls from the xanthophylls than do the other aliphatic alcohols, including isopropanol. Methanol may separate from the petroleum ether as a distinct phase, and butanols and higher alcohols are difficult to remove from the elutriates by washing with water, as is required for readsorption or crystallization of the pigments (*38*).

Many other conditions have been described for the separation of chlorophylls by chromatography in paper and in thin layers of cellulose (*134*), sugar, and diatomaceous earth (kieselguhr G) (*116, 135–138*). The greatest disadvantages of all these chromatographic methods are the loss of some 10% of the pigments during their separation, elution, and recovery (*122, 134, 140*) and the alteration of the pigments by re-active adsorbents, such as diatomaceous earth (*109*).

Figures 1–3 illustrate the chromatographic sequence of the green and yellow pigments obtained from seed plants, spore-producing plants, and many green algae. Analogous chromatographic patterns were observed with the extracts of algae and bacteria belonging to diverse taxonomic groups (*37–40, 131–133, 141, 142*).

In spite of their high resolving power, chromatographic methods do not separate the deuterated chlorophylls from the ordinary chlorophylls (*71*). They do not separate some of the chlorobium chlorophylls (*40, 59*) (see Section III).

The selection of chromatographic methods is largely empirical. For most qualitative and preparative work, chromatographic methods pro-vide adequate resolving power, but for precise quantitative work, the elution and recovery of the separated pigments (about 90%) may not be sufficiently complete (*122, 142*). For preparative work, the limited capacity of the mild adsorbents poses problems in the use of large quan-tities of the adsorbents and the easily inflammable, nonpolar solvents (see Section VIII).

The formation of separate zones in chromatographic systems has al-most universally been accepted as proof of the presence of a corre-sponding number of constituents in the mixtures being separated. There are, however, many examples of the double zoning or even multiple zoning of the individual chloroplast pigments (*109*). Conversely, the formation of single zones when mixtures of deuterated and ordinary chlo-rophylls, or chlorobium chlorophylls, are adsorbed is no indication that the various molecular species are identical (see Section III). These anomalous effects illustrate some of the precautions that must be exer-cised in the interpretation of the chromatographic observations.

III. Individual Chlorophylls

A. Properties Required for Analytical Determinations

Estimation of the chlorophylls in plant material requires establish-ment of the relationship between the property employed for measure-ment and a unit weight of the pigment. Specifically, this requires knowl-edge of the percentage composition (empirical analysis) when the content of nitrogen or magnesium is employed. It requires knowledge of the specific absorption coefficients when the spectral absorption properties are utilized. The most direct way to determine the empirical composition and the specific absorption coefficients is to isolate the pigments and to make the appropriate measurements. Indirectly, the pigments may be estimated by one method, as by the magnesium content of their solution, *provided the magnesium content of the pure pigment has been determined*. Under these circumstances, additional properties such as the absorption coefficients may be determined relative to the magnesium content.

TABLE III

SPECTRAL PROPERTIES OF ORDINARY CHLOROPHYLLS (CHL *a*, CHL *b*) AND OF
DEUTERIOCHLOROPHYLLS (D-CHL *a*, D-CHL *b*) WITH VALUES
REPORTED BY VARIOUS AUTHORITIES[a]

Properties	Chl *a*	D-Chl *a*	Chl *b*	D-Chl *b*	Reference
λ_{max}, red	660.5	659.0	642.0	640.5	(72)
λ_{max}, red	660.0	—	642.5	—	(143)
λ_{max}, red	662.0	—	644.0	—	(65)
Abs. coef., red	96.6	88.6	61.8	57.9	(72)
Abs. coef., red	102.1	—	56.8	—	(143)
Abs. coef., red	100.9	—	62.0	—	(65)
Mol. abs. coef., red	86,300	85,600	56,100	56,700	(72)
λ_{max}, blue	428.5	428.0	452.5	451.0	(72)
λ_{max}, blue	429.0	—	453.0	—	(143)
λ_{max}, blue	430.0	—	455.0	—	(65)
Abs. coef., blue	125.1	116.1	175.3	165.7	(72)
Abs. coef., blue	135.0	—	171.0	—	(143)
Abs. coef., blue	131.5	—	174.8	—	(65)
Mol. abs., blue	111,700	112,200	159,100	162,000	(72)
Ratio abs., blue/red	1.30	1.31	2.84	2.86	(72)
Ratio abs., blue/red	1.32	—	2.82	—	(143)
Ratio abs., blue/red	1.30	—	2.82	—	(65)
Ratio abs., blue/red	*ca.* 1.31 (cf. Table VI)	—	—	—	(139)
Ratio abs., blue/red	1.19	—	—	—	(112)

[a] Solvent, diethyl ether; temperature, 25° C.

Qualitatively, some evidence may be gained about the pigments from the absorption at one wavelength relative to that at another (*139*), usually the ratio of absorption at the "blue" maximum to that at the "red" maximum, as in Table III. A more extensive spectrophotometric comparison, and one which is independent of pigment concentration, is obtained by superposition of plots of log (log I_0-log I) against the wavelength (*41*). I_0 and I are the incident and transmitted light intensities. These methods provide an indication of the presence (or absence) of pigments that differ spectroscopically from the one under consideration. They provide no indication whatever concerning the presence of colorless substances or of pigments with almost identical spectra, such as

chlorophylls a and a', chlorophyllide a, methyl chlorophyllide a, deuteriochlorophyll a, deuteriochlorophyll a' (see Table II) (38, 112, 115,
139).

Although quantitative determinations of the chlorophylls may be
undertaken without isolation of the pigments, the results must be based
upon properties previously reported for the purified, individual pigments.
Moreover, it is extremely helpful to know which chlorophylls may be
expected in various plant materials. For these reasons, the occurrence
and some properties of the chlorophylls most significant for analytical
work are summarized here.

B. Occurrence and Properties

The number of natural chlorophylls is not large. Chlorophyll a (Chl
a) is first in order of abundance, followed by Chl b, Chl c, Chl d, protochlorophyll, bacteriochlorophyll, and chlorobium chlorophylls. Fully
deuterated chlorophylls are formed only when algae are grown autotrophically in heavy water, D_2O (71). The abundance of Chl c must be
placed below that of Chl b because it occurs in small proportions (19,
53, 142, 144–146) and because the total mass of the marine plants, which
includes the diatoms and dinoflagellates (the principal sources of Chl c),
is much smaller than that of the land plants (41).

The isomers of Chl a, Chl b and Chl d are frequently observed in
the extracts of plants. With careful, rapid extraction and adsorption, not
more than traces of these isomers are detectable (38, 83, 109). Consequently the chlorophyll isomers are considered as chlorophyll alteration
products (Section V, B).

1. Chlorophyll a

Chl a, $C_{55}H_{72}O_5N_4Mg$, occurs in all plants that produce oxygen by
photosynthesis (38, 44, 45, 65). In one investigation, based upon chromatographic studies of about 500 plant species, it was found as the principal
chlorophyll in all plants above the chemo-autotrophic bacteria (37, 38,
41, 42). In another investigation of about 200 species of Australian plants,
Chl a was also found as the principal green pigment in the algae, the
spore-producing, and the seed-producing plants (44, 45, 83). Thus far
about 1000 species have been examined (19, 53, 83, 105, 108–110, 116,
121–133, 135, 137, 138, 140–142, 144–148). Chl a occurs unaccompanied
by other chlorophylls in blue-green algae (Cyanophyceae), yellow-green
algae (Xanthophyceae or Heterokontae), and some red algae (Rhodophyceae) (38, 39, 44, 65). It is accompanied by small quantities of Chl d
in a few red algae (38, 100). It is always found with a little Chl c in

diatoms (Bacillariophyceae), a symbiotic alga of a sea anemone, dino-flagellates (Dinophyceae), and all brown algae (Phaeophyceae) (*19, 38, 131, 114*). It is accompanied by Chl *b* in *Euglena,* green algae (Chlo-rophyceae), spore-producing and seed-producing plants with the ex-ception of a few mutants (*28, 38, 47, 49–52*).

The chromatographic sequence of Chl *a* relative to other chlorophylls and their alteration products is summarized in Tables I and II. Addi-tional sequences with the pigments in plant extracts are indicated by Figs. 1–3. Similar chromatograms with the extracts of various kinds of plants have also been described (*38, 117*).

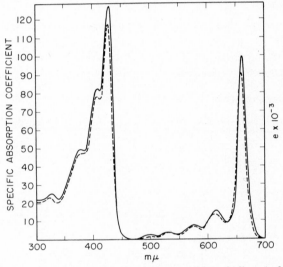

FIG. 4.   Spectral absorption curves for ordinary chlorophyll *a* (solid line) and for fully deuterated chlorophyll *a* (broken line) in ether.

Values for the specific spectral absorption coefficients, which form the basis of spectrophotometric methods for estimation of Chl *a,* are summarized in Table III (*72*) and plotted in Fig. 4. The results by various workers (*65, 72, 112, 139, 143, 149–151*) are in moderately good agreement except for the ratio of the absorption in the red spectral region to that in the blue region reported by one group (*112*) (see Section VII, C, p. 51 and Table III, p. 31).

2. DEUTERIOCHLOROPHYLL *a*

Deuteriochlorophyll *a*, $C_{55}D_{72}O_5N_4Mg$, occurs, unaccompanied by other chlorophylls, in blue-green algae (*Synechococcus lividus* and *Phor-midium luridium*) and, accompanied by deuteriochlorophyll *b*, in green algae (*Chlorella vulgaris* and *Scenedesmus obliquus*) grown in pure heavy water (see Section VIII, C) (*71*). As indicated in Table I, the chromatographic behavior is virtually identical with that of Chl *a*. As shown by Fig. 4, the spectral absorption properties are similar to those of Chl *a*. Because the deuterium is twice as heavy as hydrogen, the percentage of deuterium is nearly twice that of the hydrogen in the ordinary Chl *a*. The specific absorption coefficients are lower than those of ordinary Chl *a*, but the molecular absorption coefficients are equal (Table III) (*72*). Relative to the ordinary Chl *a*, the absorption maxima are shifted to slightly shorter wavelengths, an effect attributable to the deuterium atoms linked to the carbon atoms in the conjugated system of double bonds (*72*).

Deuteriochlorophyll *a* yields a fleeting yellow color changing to green when a solution in ether is mixed with an equal volume of methanol plus potassium hydroxide (10%). In this respect, chlorophylls *a*, *a'*, *b*, and *b'* and deuteriochlorophylls *a*, *a'*, *b*, and *b'* are alike.

3. CHLOROPHYLL *b*

Chl *b*, $C_{55}H_{70}O_6N_4Mg$, occurs as the minor chlorophyll in *Euglena*, the green algae, and the higher plants, as noted above. Its chromato-

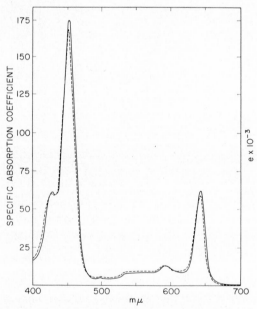

FIG. 5. Spectral absorption curves for ordinary chlorophyll *b* (solid line) and for fully deuterated chlorophyll *b* (broken line) in ether.

graphic behavior is indicated in Tables I and II and Figs. 1–3. The spectral absorption coefficients are summarized in Fig. 5 and in Table III (72).

4. DEUTERIOCHLOROPHYLL *b*

Deuteriochlorophyll *b*, $C_{55}D_{70}O_6N_4Mg$, accompanies the deuteriochlorophyll *a* in the green algae grown in heavy water (71). Chromatographically (Table I) and spectroscopically (Fig. 5 and Table III), it resembles Chl *b* (72).

5. CHLOROPHYLL *c*

Chl *c*, formula unknown, is the minor chlorophyll accompanying Chl *a* in diatoms, dinoflagellates, brown algae, and a certain symbiotic alga of sea anemones (19, 38, 44, 65, 117, 131, 144, 148, 152–156).

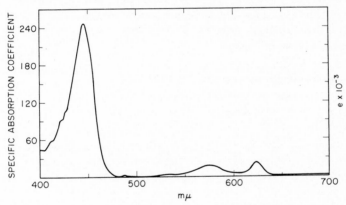

FIG. 6. Spectral absorption curves for chlorophyll *c* in ether.

With respect to the spectrum (Fig. 6), adsorbability (Table I), solubility, and reaction with alkali, Chl *c* differs a great deal from all the other chlorophylls. Spectral absorption coefficients for recrystallized

TABLE IV

CHLOROPHYLL *c*: WAVELENGTHS OF THE ABSORPTION MAXIMA AND SPECIFIC ABSORPTION COEFFICIENTS (*e*) IN ETHER

λ_{max} (mμ)	e ($\times 10^{-3}$)	λ_{max} (mμ)	e ($\times 10^{-3}$)	λ_{max} (mμ)	e ($\times 10^{-3}$)	Reference
446.0	—	579.5	—	627.0	—	(152)
448.0	—	578.0	—	628.0	—	(155)
447.0	227.0	579.5	20.6	628.0	22.0	(65)
449.0	169.7	580.0	15.7	628.0	15.8	(117)
447.3	262.1	578.5	27.4	626.5	29.6	(139a)

Chl c (see Section VIII, D), dissolved in ether, are much higher than those reported before (Table IV), owing, no doubt, to difficulties in the purification of this pigment and to arbitrary assumptions about the molecular weight (63, 65).

Dissolved in ether, Chl c is rendered insoluble by shaking with dilute aqueous sodium bicarbonate and sodium hydroxide solutions, and it collects at the interface between the two phases. On neutralization with dilute acetic acid, Chl c redissolves in the ether. When dissolved in ether and treated with potassium hydroxide in methanol (10%), Chl c turns brown instantly, and in a matter of seconds, purplish pink. This pink color persists for days without return of the green color.

6. CHLOROPHYLL d

Chl d, $C_{54}H_{70}O_6N_4Mg$ (proposed) (157, 158), occurs as the minor chlorophyll accompanying Chl a in some red algae (38, 39, 44, 65, 100, 158, 159). It has been isolated by chromatography and has also been prepared by oxidation of Chl a with permanganate (157, 158). The spectral absorption maxima in ether occur at 445 and 686 mμ (100). Specific absorption coefficients have been reported, based upon magnesium determinations and an assumed molecular weight equal to that of Chl a (65) and only two units less than that for the proposed structure (157). The proposed structural formula fails to explain some of the properties of this chlorophyll (56), for example, the formation of isochlorophyll d with spectral absorption properties almost identical with those of chlorophyll a (100). The suggestion that Chl d might be formed as a postmortem product in red algae (159) has never been substantiated (44).

7. PROTOCHLOROPHYLL

Protochlorophyll, $C_{55}H_{70}O_5N_4Mg$, occurs in very small quantities, along with protochlorophyllide and vinyl pheoporphyrin a_5 in the yellow leaves of seedlings grown in the dark. It also occurs in the inner seed coat of pumpkin seeds (24, 38, 57, 63, 65, 67, 160–168). The location of protochlorophyll [obtained from the inner seed coats of manroot, *Echinocystis fabaceae*, (Cucurbitaceae)] relative to other chlorophylls in the chromatographic column is shown in Table I (38). Spectral absorption properties have been reported in detail (63, 65).

8. BACTERIOCHLOROPHYLL

Bacteriochlorophyll, $C_{55}H_{74}O_6N_4Mg$, is the major chlorophyll of various photosynthetic bacteria (Thiorhodaceae, Athiorhodaceae, and Hyphomicrobiaceae) (see Section VIII, G) (40, 57, 63, 65, 67, 98, 169–

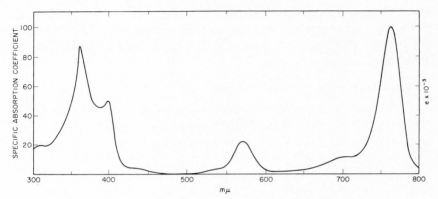

FIG. 7. Spectral absorption curve for bacteriochlorophyll in ether.

175). It accompanies chlorobium chlorophylls in some of the green sulfur bacteria (Chlorobacteriaceae) (40, 173). Bchl has also been reported to occur in a second form, Bchl b, in but one species, Rhodopseudomonas sp. (169).

As an aid in the detection and isolation of Bchl, its adsorbability relative to other chlorophylls is shown in Table I. The spectral absorption

TABLE V

BACTERIOCHLOROPHYLL AND THE CHLOROBIUM CHLOROPHYLLS: WAVELENGTHS OF
THE ABSORPTION MAXIMA (mμ) AND SPECIFIC ABSORPTION
COEFFICIENTS (e) IN ETHER

mμ	$e \times 10^{-3}$	mμ	$e \times 10^{-3}$	Reference
		Bacteriochlorophyll		
358.0	93.7	772.0	105	(65, 175)
358.5	80.5	773.0	100	(65)
357.5	87.6	770.0	102	(139a)
		Chlorobium Chlorophyll 660		
431	143	660.0	95.4	(176)
432	175.5	660.0	112.5	(68, 177)
430	151.1	660.0	98.6	(139a)
		Chlorobium Chlorophyll 650		
425.0	146.0	650.0	113.5	(68, 177)

properties are summarized in Fig. 7 and in Tables I and V. The nuclear magnetic resonance properties and hydrogen (deuterium) exchange have also been reported (74, 172).

9. DEUTERIOBACTERIOCHLOROPHYLL

Deuteriobacteriochlorophyll, $C_{55}D_{74}O_6N_4Mg$, is the major chlorophyll of the purple sulfur bacterium (Rhodospirillum rubrum) grown in heavy

water containing fully deuterated biotin and succinate. The wavelengths
of the spectral absorption maxima and the chromatographic properties
are virtually identical with those of the ordinary Bchl (unpublished observations).

10. CHLOROBIUM CHLOROPHYLLS

Chlorobium chlorophylls, composition uncertain, are the principal
pigments of the green sulfur bacteria (Chlorobacteriaceae). They are
sometimes accompanied by Bchl as already noted (*40, 63, 65, 67, 170–*
173).

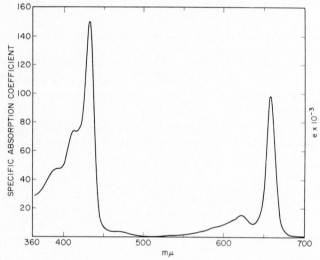

FIG. 8.   Spectral absorption curve for chlorobium chlorophyll 660 in ether.

From the viewpoint of analytical chemistry, the study of the chloro-
bacteriacean chlorophylls is in a very unsatisfactory state. Most green
bacteria yield a chlorophyll or a chlorophyll mixture which has been
called bacterioviridin (*41*) then chlorobium chlorophylls 660 (*67*) and
bacteriochlorophylls *c* (*40, 170*) and which is characterized by a spectral
absorption maximum at 660 mμ, in ether. Two bacterial strains or vari-
ants (*Chlorobium thiosulfatophilum* strain L and *Chlorobium limicola*
CR17), difficult to grow, yield a pigment or mixture which has been
called chlorobium chlorophylls 650 and bacteriochlorophylls *c* and which
is characterized by a spectral absorption maximum at 650 mμ (*67, 68,*
176–183). These two chlorophylls have been reported to yield a variety
of degradation products (detected by partition chromatography) and a
greater variety of oxidation products (detected by gas chromatography).
These results indicate that the chlorobium chlorophylls 650 and 660 are,
respectively, mixtures of six and seven pigments that differ slightly in

chemical composition (*180, 184–186*). Although the detection of several chlorobium chlorophylls by paper chromatography has been reported, the presence of artifacts was not excluded. Our chromatographic studies of chlorobium chlorophyll 660, from *Chloropseudomonas ethylicum* and *Chlorobium thiosulfatophilum*, in long columns of powdered sugar provided no evidence for the separation of this pigment into different fractions. Even the separation of chlorobium chlorophylls 650 and 660 required special solvent mixtures.

The chromatographic behavior of the chlorobium chlorophylls relative to the other chlorophylls is indicated in Table I. Spectral absorption properties are summarized in Fig. 8 and Table V. Those reported by us are based on seven different preparations.

C. Natural Variants of Chlorophylls

If the claims regarding the composition of the chlorobium chlorophylls are correct, one must consider various modifications of the chlorophylls that differ in composition and molecular structure yet are so alike that they cannot be separated by the usual chemical and physical methods. Even with the most sensitive methods, such as nuclear magnetic resonance, no evidence for this kind of variation in the tetrapyrrole units has been observed with Chls *a* and *b* or with the methyl chlorophyllides derived from these chlorophylls.

With isotopes, it is possible to vary the composition and molecular weight of the chlorophylls so that a variety of inseparable functional compounds may be produced, but no variation of the carbon skeleton of the molecules has been observed. Grown autotrophically in heavy water, algae may incorporate into their chlorophyll a little of the residual hydrogen from the traces of ordinary water remaining in the medium. Consequently some chlorophyll molecules with one or a few hydrogen atoms may be formed. As both Chl *a* and Chl *b* contain 37 locations where one or more hydrogen atoms may be attached, the number of possible arrangements or permutations of only a few hydrogen atoms in molecules containing mostly deuterium is enormous. But in media containing 99.8% heavy water, the proportion of these molecules with one or a few hydrogen atoms is far below the sensitivity of the usual analytical techniques. Similarly, algae grown in ordinary water may produce some chlorophyll molecules containing from one to a few of the deuterium atoms normally present in ordinary water. Even though numerous molecules with from one to a few deuterium atoms may be formed, the proportion of these relative to the total chlorophyll must be very small. The average deuterium content of chlorophylls from algae grown in various mixtures of H_2O and D_2O may be estimated by a variety of methods, but there are no analytical methods available for the separation of mixtures of the fully deuterated chlorophylls, partially deuterated chlorophylls, and ordinary chlorophylls.

D. Nomenclature

The nomenclature of the chlorophylls is in a confused state. Variations of the absorption maxima of the pigments in the living cells (65, 91, 95, 96, 187) have been attributed to different "forms" of Chl a (63). But when the pigments are extracted, only one form of Chl a has been found.

The individual chlorophylls isolated from various organisms have been named in different ways. Some have been named alphabetically. Some have been named in relation to the taxonomic name or position of the organism from which the pigment was isolated. Some have been identified by including the wavelength of the absorption maximum in the red region of the spectrum as part of the name (as with the chlorobium chlorophylls). The several "forms" of each chlorophyll within the plant have also been designated in this way (65, 91, 95, 96).

Fully deuterated chlorophylls are called deuteriochlorophylls. This usage distinguishes deuterated compounds from various secondary products that are commonly called deutero compounds, as for example, deuteroporphyrin. It does not distinguish among fully deuterated substances and partially deuterated and specially labeled substances.

E. Criteria of Purity

As noted above, methods for the identification and estimation of each chlorophyll are based upon standards established in relation to the pure pigments. In spite of all the investigations carried out thus far, there is no single technique that will establish the identity and the purity of a particular preparation of chlorophyll.

With Chls a and b, chromatography with mild adsorbents will usually reveal the presence (or absence) of colored contaminants. In fact, this is the only way that Chl a′ may be revealed as a contaminant of Chl a and Chl b′ as contaminant of Chl b. But with the chlorobium chlorophylls, as described by Holt and collaborators (180, 184–186), and with mixtures of the ordinary pigments plus the deuterated pigments, chromatography will not provide a clue to the complexity of the mixture. Chromatography also provides a basis for the identification of chlorophyll preparations, by comparison with authentic samples. Colorless contaminants can be separated from the pigments by chromatography and detected with special reagents (117).

Specific absorption coefficients serve for the identification and estimation of the common chlorophylls (150). They do not provide a reliable basis for the identification or estimation of substances having the same characteristic spectral absorption curves. With preparations of high purity the spectral absorption curves and the specific absorption coefficients should agree with those reported for the individual pigments (see Figs. 4–8 and Tables III–V). Low values are indicative of colorless impurities in the chlorophyll preparations. Low values are also indicative of the fully deuterated pigments (see Table III). With a sensitive microbalance, with a calibrated spectrophotometer, with "balanced" absorption cells, and with great care in the handling of solvents and the control of temperature, specific absorption coefficients of a given preparation are reproducible to ± 1–2%. Great care is required for the isolation of chlorophyll that will provide coefficients within this range of the maximal values (see Section VIII).

The nuclear magnetic resonance spectrum reveals small proportions of some contaminants of low molecular weight in chlorophyll preparations. It provides an indication of the proportions of mixtures of the ordinary chlorophylls with their deuterated counterparts.

The values for the magnesium content of chlorophyll preparations are reduced by colorless impurities. Because the magnesium content of the deuterated pigments is less than that for the ordinary pigments, magnesium analyses are not useful with mixtures of the deuterated pigments.

With "new"chlorophylls (95, 118) the purification and isolation procedures based upon partition, chromatography, and precipitation or crystallization should be repeated until maximal spectral absorption coefficients and constant empirical analyses are obtained. The nuclear magnetic resonance and infrared spectra provide significant supplemental information. With very small quantities of "new" pigments isolated by chromatography, special care should be taken to make certain that the pigments have not arisen from postmortem reactions in the plant (8, 38, 73–77), reactions in the solvent (38, 46, 57, 73, 76, 99, 119), reactions with the adsorbent (109), anomalous zone formation in the chromatographic system (109), or a change in the physical state of the pigment (97).

IV. Desiderata for Estimation of Chlorophylls

Owing to the lability of the chlorophylls, there are many mazes lined with pitfalls in the isolation and determination of these green pigments. Because the chlorophylls occur at high concentration, because their properties are very similar, and because they are associated with the yellow carotenoid pigments and with many kinds of colorless substances within the specialized chloroplasts, there are no widely applicable methods whereby the individual pigments may be determined within the fresh plant material (*188*). The total chlorophyll of some unicellular algae may be examined by fluorescence (*104*) and estimated by the opal-glass absorption method (*63, 189*), by the integrating sphere method (*65*), and by reflectance of red light ("spacelight" spectrometry) (*190–192*).

As a rule, the chlorophylls must be extracted from the plant material before they can be estimated precisely (*38, 193, 194*). With clear plant extracts, the green pigments may be estimated directly by fluorimetry, colorimetry, or spectrophotometry (*63, 65, 122–125, 195*). With turbid extracts containing interfering pigments, the chlorophylls must be transferred to a suitable solvent, such as diethyl ether, thereby separating them from water-soluble pigments and from the turbid substances (*20, 34, 63, 65, 196*). With the extracts of some plant material, it may be necessary to separate the chlorophylls from one another by chromatographic methods as well as from various colored substances, such as the alteration products of the chlorophylls (*9, 76, 110, 111, 116, 122, 124, 127, 134*).

The collection, storage, and treatment of the plant material preceding extraction of the pigments must be considered in relation to the extraction procedure and in relation to the reactions of the chlorophylls that may take place before and during the extraction (*8, 57, 73, 76, 111, 122, 197–199*). Ideally, any method for the estimation of the chlorophylls should be based upon quantitative extraction, and it should reveal the amounts of the individual chlorophylls with an indication of the amounts of their alteration products present in the plant material or formed during the analytical operations.

V. Alteration Products

A. Variation of Alteration Products

There are many more well defined alteration products of the chlorophylls than there are chlorophylls. Many of the conditions that lead to their formation have been carefully described. There are also many alteration products that have not been characterized, and the conditions that lead to their formation have not been defined. It seems inevitable that some of these products will inadvertently be described as "new" chlorophylls (see Section III, E). In view of the large number of these postmortem artifacts, special care should be exercised to make certain that "new" green substances isolated from plants are normal functional pigments and not derived from the common chlorophylls. Unfortunately, there are no methods whereby all the alteration products of each chlorophyll may be converted to a single substance that is easily estimated and which may then serve for estimation of the initial chlorophyll content.

The more common and better characterized alteration products of the chlorophylls fall into two major categories: a large group of green, magnesium-containing compounds and an equally large group of gray or brown, magnesium-free substances. Many of these substances are frequently encountered in analytical work.

B. Well-Defined, Green, Magnesium-Containing Alteration Products

1. CHLOROPHYLL ISOMERS

Isomers of the chlorophylls are formed slowly when the solutions of the natural green pigments are permitted to stand at room temperature. They are formed rapidly when the fresh plants or the chlorophyll solutions are heated even for a short time (15–20 minutes at 100°) yielding an equilibrium mixture (38, 99, 100). Heating in pyridine avoids the allomerization reactions sometimes encountered when the chlorophyll solutions in alcohol are heated (73). Prolonged heating (about 24 hours at 100°) converts all the chlorophyll to the pyrochlorophyll (73). Chlorophylls a, b, and d yield the isomeric chlorophylls a', b', and d', which resemble the parent chlorophylls spectrally, but are slightly less sorbed than the parent chlorophyll in chromatographic columns (38, 84, 99, 100) (see Tables I and II). Chl d yields two additional isomers, isochlorophylls d and d', which exhibit spectral properties like those of Chl a (100). Reports of other isomers in place of Chl a', detected by paper chromatography (85) and by spectral absorption of the eluted pigments in the ultraviolet region (83), could not be confirmed with column chromatography. A pigment found to be more adsorbed than Chl b after it had been heated in propanol was first regarded as an isomer of Chl b, (118). On the basis of the properties reported, this pigment could be an oxidized Chl b, pyrochlorophyll b (73) or various allomerization products (76, 119).

2. CHLOROPHYLLIDES

The chlorophyllides are acidic pigments formed by hydrolysis of the phytyl group from the chlorophyll molecules. This hydrolytic reaction is catalyzed by an enzyme, chlorophyllase, in the plant tissue. It occurs in the presence of nonreactive organic solvents such as acetone and ether. Chlorophyllides a and b, the best known ones, are very strongly adsorbed. They exhibit molecular spectral absorption curves identical with those of the parent chlorophylls (57, 72, 76) (see Table II).

3. Chlorophyllide Isomers

Green, acidic isomers of the chlorophyllides are formed by heating n-propanol solutions of these pigments. The chlorophyllides a′ and b′ have been observed by chromatography (76).

4. Oxidized Chlorophyllides

An acidic, phytol-free, oxidized derivative of Chl a is formed when certain diatoms are shaken with distilled water (77). It is sorbed much more strongly than Chl a, but slightly less strongly than Chl c. Spectroscopically, this product resembles the parent Chl a and oxidized Chl a (73, 76). It appears to be 10-hydroxychlorophyllide a (73, 77).

5. Chlorophyllide Esters

The esters of the chlorophyllides are nonacidic, green pigments that are formed when plant tissues containing the enzyme chlorophyllase are treated with dilute aliphatic alcohols. They form more strongly sorbed green zones than the parent chlorophylls (see Table II). Their spectra are similar to those of the parent chlorophylls, and the molecular absorption coefficients are equal. Common esters are methyl chlorophyllides a and b and ethyl chlorophyllides a and b (72, 73, 200). In plants containing oxidase systems, the alcoholysis reaction must be carried out in an inert gas to prevent oxidation of the green pigments (73).

6. Isomers of Chlorophyllide Esters

When solutions of the chlorophyllide esters are heated, small proportions of the green, nonacidic isomers are formed (73, 76). Like the isomers of the common chlorophylls, these isomers of the chlorophyllides are slightly less sorbed on sugar than the parent pigments (see Table II). They are partially reconvertible to the parent chlorophyllide esters. Their absorption maxima correspond with those of the parent chlorophyllides and with those of the chlorophylls themselves. The methyl and ethyl chlorophyllides a′ and b′ have been observed.

7. Oxidized Chlorophylls

In many plants, oxidative systems, which may involve enzymatic and induced oxidation, promote oxidation of the chlorophylls to green, nonacidic pigments that are usually slightly more sorbed than the parent chlorophylls and somewhat less sorbed than the methyl chlorophyllides in columns of powdered sugar. These products resemble the chlorophylls spectrally but are distinguishable from them (73). In some of the oxidation systems, numerous green products are formed. In a few oxidation systems the chlorophylls are oxidized to colorless substances (75). Some of the green products formed by enzymatically induced oxidation are also formed when the chlorophylls undergo spontaneous oxidation (allomerization) in methanol (76). There is some nuclear magnetic resonance evidence that this oxidation results in the replacement of —H by —OH at the C-10 position (73); hence the oxidized chlorophylls and the oxidized chlorophyllides (77) have the same basic structure. Pigments corresponding to oxidized Chl a and Chl b have been observed (73).

In alkaline solutions (as in the phase test) the chlorophylls are hydrolyzed to green chlorophyllins (102). These reaction products result from splitting of the cyclopentanone ring with the formation of additional carboxyl groups. Their spectra resemble those of the chlorophylls.

8. ALLOMERIZED CHLOROPHYLLS

When dissolved in absolute methanol and exposed to air, the chlorophylls are oxidized rapidly (57). In addition to the oxidized chlorophylls that are formed, the reaction mixture yields various methoxy-containing green pigments called allomerized chlorophylls. Several products, which are more adsorbed than the parent chlorophylls, have been obtained both from Chl a and Chl b (57, 76, 82).

9. PYROCHLOROPHYLLS (DECARBOMETHOXY CHLOROPHYLLS)

When heated in pyridine solution, the chlorophylls lose the carbomethoxy group at C-10 and yield green "pyro" derivatives. The spectral properties are similar to those of the parent chlorophylls. The pyrochlorophylls a and b have been isolated by column chromatography, but isomers of these pigments corresponding to a' and b' were not observed (see Table II) (73).

C. Poorly Defined, Green, Magnesium-Containing Alteration Products

In addition to the well-defined, green pigments, many other green alteration products have been observed. For example, when cocklebur leaves are heated for 1 minute in boiling water, a very strongly sorbed, green, nonacidic pigment different from all those described above was obtained (109). When the leaves of seedlings were killed under mild conditions, as by freezing and thawing, numerous unidentified green pigments were observed. There is reason to believe that various oxidation and pyrolysis products of the chlorophyllide esters may be formed in plant extracts. Decarbomethoxy derivatives of the oxidized and allomerized chlorophylls may also be expected in plant material that has been heated.

The chlorophylls are altered rapidly upon exposure to visible light. This reaction varies with the solvent. It occurs when the pigments are on chromatographic columns. For example, a zone of Chl a formed in a column of powdered sugar with benzene as the solvent and then exposed to sunlight soon yields a gray region next to the glass. When the washing with benzene is continued, the gray zone remains behind the unaltered chlorophyll.

All the chlorophylls are altered rapidly by strong adsorbents yielding various unknown green products. Most of these green substances cannot be removed or eluted from the adsorbents.

D. Well-Defined, Gray-to-Brown, Magnesium-Free Alteration Products

1. PHEOPHYTINS

Chlorophylls exposed to acidic solutions (65, 107) or to acidic resins (106) soon liberate their magnesium atoms and yield gray-brown pheophytins. These gray, nonacidic products are usually somewhat less sorbed in chromatographic columns than the parent chlorophylls (Table II). They exhibit pronounced absorption maxima in the blue-green and red regions of the spectrum. The pheophytins of chlorophylls a, b, c, and d, deuteriochlorophylls a and b, bacteriochlorophyll, and chlorobium chlorophylls have been described (38, 65, 68, 71, 76, 99, 100, 177). The conditions employed to form some of the pheophytins may also produce the pheophorbides (57).

2. ISOMERS OF THE PHEOPHYTINS

Chromatographic evidence for the formation of isomers of the common pheophytins a, b, and d has been obtained (99, 100).

3. PHEOPHORBIDES

These acidic compounds are produced by the removal of magnesium from the chlorophyllides formed in plants by enzymatic hydrolysis of the phytyl group from the chlorophyll. They are also formed by the action of strong acids on the chlorophylls and pheophytins. Typical pheophorbides are those of Chl a and Chl b (9, 57, 73).

4. ISOMERS OF PHEOPHORBIDES

Evidence for isomers of the pheophorbides is based largely upon chromatographic observations.

5. PHEOPHORBIDE ESTERS

The esters of the pheophorbides are formed by way of the chlorophyllides when green plant material containing the enzyme chlorophyllase is treated with dilute aliphatic alcohols and then with acids. The resultant pheophorbide esters in the extracts are easily isolated by chromatography (Table II). These esters may also be prepared by treatment of chlorophylls or pheophytins with acids in alcohols (73).

6. PHEOPHYTINS OF OXIDIZED CHLOROPHYLLS

The oxidized chlorophylls are converted to the corresponding pheophytins when the magnesium is removed with acid (73).

7. PHEOPHYTINS OF ALLOMERIZED CHLOROPHYLLS

With acids, the allomerized chlorophylls yield pheophytins resembling those of the oxidized chlorophylls (57).

8. PHEOPHYTINS OF PYROCHLOROPHYLLS

The pyrochlorophylls yield pyropheophytins through loss of magnesium. Upon pyrolysis, the common pheophytins also yield pyropheophytins by decarbomethoxylation (73).

E. Biological Modifications

In addition to the postmortem alteration products of the chlorophylls, certain green plants sometimes produce tetrapyrrolic substances related to the green pigments. These may result from modification of the genetic composition by irradiation (47, 49, 162) or from chemical inhibition of the metabolism as by the additional of chelating agents (78–81).

VI. Extraction of Chlorophylls

A. Solvents for Extraction

Even though virtually all methods for the estimation of the chlorophylls depend upon their extraction from plants and plant material, there is no single solvent or solvent mixture that will remove the unaltered green pigments rapidly and quantitatively from all kinds of plants (38).

Fresh plant material must be extracted with solvents that remove most of the water before the pigments can be dissolved. Unless an excess of solvent is employed for the extraction, the chlorophylls may be oxidized or hydrolyzed (or both), yielding a variety of the products indicated in Sections V, B and C (73, 76).

The solvents most suitable for the extraction of the chlorophylls from fresh plant material are those miscible with water, such as pyridine, methanol, ethanol, acetone, and acetone plus ethyl acetate (*19, 38, 53, 63, 65, 115, 121, 129, 130, 143, 145, 146, 193, 195, 198*). These solvents also permit the direct estimation of the chlorophylls in the extracts. They allow subsequent transfer of the pigments to diethyl ether for measurement of the spectral absorption properties or to petroleum ether for examination by chromatography. With acidic pigments, however, it may be necessary to buffer the extracts before the pigments can be transferred from the diluted aqueous solution to an organic solvent as described in Section VII, C (*110, 155*). A mixture of methanol plus petroleum ether (2 or 3 to 1) is a widely applicable solvent for the extraction of the chlorophylls. The methanol removes the water from the plant material, and the petroleum ether takes up the pigments before they can undergo secondary reactions such as hydrolysis, allomerization, and oxidation (*38, 194*).

B. Treatment of Plant Material

For the rapid, small-scale, quantitative extraction of the chlorophylls, it is convenient to disintegrate fresh plant material in a blender. A blender with a tight cover reduces loss by splashing and reduces the hazard of ignition of the solvent vapors. The blended mixture is then filtered or centrifuged and the clear extract employed for estimation of the chlorophylls. Alternatively, small portions of the plant material may be ground in a mortar with the solvent and sand and filtered or centrifuged.

Even in a blender, some fresh plant materials, such as the thick rubbery brown algae, will not yield their pigments completely to the organic solvents. Other plant materials do not liberate the pigments without change or alteration of the chlorophylls. With many of these materials, it is desirable to place the samples in a large volume of boiling water for a minute. The water is then cooled quickly with ice or an excess of cold water. The plant material is removed with a brush, a sieve or by centrifugation and then extracted with an organic solvent. Although the heating improves the extractability of the chlorophylls and reduces their enzymatic oxidation and hydrolysis, it accelerates their pyrolysis and isomerization (*38, 73*).

In spite of rapid extraction, plants with strongly acidic cytoplasm yield their chlorophylls as pheophytins. This reaction can usually be prevented by the addition of ammonia or organic bases to the extraction solvents. The addition of calcium or magnesium carbonate, as frequently employed, is no insurance against pheophytin formation. Alternatively the chlorophylls may be estimated as pheophytins (see Section VII, B) (9, 107).

For the large-scale extraction of the chlorophylls from fresh leaves, as in their preparation from spinach, preliminary scalding of the plant material offers many advantages. It reduces the water content of the leaves. It facilitates extraction of the pigments so that blending with the organic solvents is not necessary. It reduces oxidation and hydrolytic reactions (see Section VIII, B) (38, 71). An anomalous effect of heat on the chlorophylls within the leaves of cocklebur was described in Section V, C (109).

Dried or desiccated plant materials often release their green pigments to organic solvents slowly and incompletely. The extraction is sometimes facilitated by rehydration of the material followed by extraction with organic solvents.

Dried, brined, canned, and frozen green plant material often contains various alteration products of the chlorophylls (3, 5, 6, 8). All these products are usually extractable under the conditions employed for extraction of the chlorophylls.

No single extraction procedure is universally applicable to the quantitative removal of the pigments from the leaves of all kinds of plants. Selection of an extraction procedure and extraction solvents remains on an empirical basis (194).

For assurance that the plant extracts contain unaltered chlorophylls, it is essential to examine the pigments by several independent methods, for example, by spectroscopy, colorimetry, and various modifications of chromatography, including paper chromatography and columnar chromatography with mild adsorbents. Conversely, if several pigments are found, it is desirable to establish whether these were normal constituents of the plants or were formed by secondary reactions.

VII. Estimation of Chlorophylls

A. Methods

As indicated in Section IV, there are many methods for the estimation of the chlorophylls in the extracts of green plant material. The most widely applicable and the most direct methods are based upon colorimetry (22, 123–125, 127) and spectrophotometry (9, 53, 63, 65, 98, 122, 123, 125, 138, 140, 143, 145, 146, 148–151, 156, 195, 196, 198, 201, 202). The pigments may also be determined fluorimetrically (105, 121), and they may be converted into other derivatives such as the pheophytins which may then be estimated by spectrophotometry (69, 107, 134).

For the use of spectrophotometric methods it is essential to know the specific absorption coefficients of the pure pigments. The preparation or isolation of relatively pure chlorophylls, described in Section VIII, requires great care and has been accomplished by only a few investigators (71, 72, 113, 117, 143, 150, 154, 177, 203).

The chlorophylls may be separated from one another and from their alteration products by chromatography and recovered by elution. The eluted pigments may then be estimated colorimetrically or spectrophotometrically as noted above for the extracts of leaves. Unfortunately the elution of the separated pigments is rarely quantitative, as discussed in the section on chromatography.

Theoretically, the chlorophylls may be separated from all the other plant constituents by chromatography so that they may be eluted and weighed after evaporation of the solvent. Although chromatography provides great resolving power, it also presents several limitations. Even the mildest adsorbents slowly alter the chlorophyll. Few, if any, chromatographic systems separate the chlorophylls from all the colorless substances in the extracts of plants, even if very long migrations are utilized. Many chromatographic systems liberate organic substances that contaminate the separated and eluted chlorophyll (71). For all these reasons, the chlorophyll eluted after separation by chromatography cannot be determined accurately by evaporation of the elutriate followed by weighing the residue.

Because some of the altered chlorophylls have characteristic spectral absorption curves like those of the parent chlorophylls, extensive alteration of the pigments may not necessarily influence the results calculated from the spectrophotometric observations. On the other hand, other altered chlorophylls may have a pronounced effect; hence the spectrophotometric observations should be supplemented by chromatographic tests to demonstrate that the chlorophylls are unaltered.

B. Estimation in Plant Extracts

The simplest and most direct method for estimation of the individual chlorophylls depends upon measurement of their spectral absorption properties in acetone extracts of plants. For organisms such as those centrifuged from sea water, chlorophylls a, b and c may be estimated by this procedure using a final acetone concentration of 90%. The following equations, derived from the specific extinction coefficients (63), have been recommended (54),

$$\text{Chl } a \text{ (µg/ml)} = 11.64 \, (A_{663}) - 2.16 \, (A_{645}) + 0.10 \, (A_{630})$$
$$\text{Chl } b \text{ (µg/ml)} = -3.94 \, (A_{663}) + 20.97 \, (A_{645}) - 3.66 \, (A_{630})$$
$$\text{Chl } c \text{ (µg/ml)} = -5.53 \, (A_{663}) - 14.81 \, (A_{645}) + 54.22 \, (A_{630})$$

where A is the absorbance at 663, 645, or 630 mμ, less that at 750 mμ (to correct for light scattering due to turbidity). Usually, however, light scattering is greater at shorter wavelengths than at longer wavelengths; hence, if the extinction value at 750 mμ is significant, the values at the shorter wavelengths are certain to be even less dependable. The derivation of the absorption equations from specific extinction coefficients is given in a recent review (63). New equations will be required for each set of specific absorption coefficients (see Tables III–V) and for each solvent or solvent mixture employed for the measurements.

Corresponding equations have been suggested for the determination of Chl a and Chl b in 90% and in 80% acetone extracts, which may be clarified by filtration with a filter aid (47, 198). The equations for 80% acetone extracts (47, 107, 151) are:

$$\text{Chl } a \; (\mu g/ml) = 11.63 \, (A_{665}) - 2.39 \, (A_{649})$$
$$\text{Chl } b \; (\mu g/ml) = 20.11 \, (A_{649}) - 5.18 \, (A_{665})$$
$$\text{Total Chl} \quad (\mu g/ml) = 6.45 \, (A_{665}) + 17.72 \, (A_{649})$$

The chlorophylls in 80% acetone extracts of leaves may also be converted quantitatively into pheophytins with oxalic acid and the pheophytins (and chlorophylls) estimated from the spectral absorption measurements before and after formation of the pheophytins. Combination of the methods, with measurements of the chlorophylls and of the pheophytins, makes possible the estimation of chlorophylls a and b and pheophytins a and b in a given extract (107).

It is scarcely necessary to point out that all these equations are valid only when they have been determined using pure preparations of the pigments, when the only pigments in the solution are the unaltered chlorophylls whose absorption is presumed to be measured, and when there is no turbidity in the solution. But experience has shown that all these conditions are frequently overlooked, leading to invalid conclusions. An example is the untenable result that the organisms in sea water contain as much Chl c as Chl a (19, 53, 146).

C. Estimation after Transference from the Extracts to Diethyl Ether

The turbidity, which is often encountered in acetone extracts of plant material and which frequently resists filtration and centrifugation, may be eliminated by transferring the pigments to diethyl ether (free of peroxides, preferably from newly opened cans) in which the spectrophotometric measurements are then made. The transfer must be made quantitatively, and this may be especially difficult if colloidal solutions are formed when the acetone solutions are diluted with ether and water or aqueous salt solution. With persistent emulsions it may be necessary to repeat the extraction several times, and with very stable emulsions, the extract should be discarded.

Equations recommended for the estimation of Chl a and Chl b in ether (63, 65, 143, 149) are:

$$\text{Chl } a \text{ (}\mu\text{g/ml)} = 9.93 \,(A_{660}) - 0.777 \,(A_{642.5})$$
$$\text{Chl } b \text{ (}\mu\text{g/ml)} = 17.6 \,(A_{642.5}) + 2.81 \,(A_{660})$$
$$\text{Total Chl } \text{ (}\mu\text{g/ml)} = 7.12 \,(A_{660}) + 16.8 \,(A_{642.5})$$

Additional equations for mixtures of Chl a with chlorophylls other than Chl b have also been reported. Some of these are based upon provisional spectral absorption coefficients and are likely to need revision (19, 63).

Ether has significant limitations as a solvent for use in measurements of the spectral properties of the chlorophylls. It is so volatile that the absorption cells must be tightly stoppered to prevent convection currents. Moreover, it readily forms peroxides that destroy the pigments, and it dissolves so much water that the spectra of chlorophylls transferred to the ether solution from aqueous alcohol may exhibit significant variations from the standard solutions prepared with anhydrous, alcohol-free ether (Table VI) (139). Similar effects with water in pyridine have been reported (204).

TABLE VI

EFFECT OF THE TREATMENT OF DIETHYL ETHER ON THE ABSORPTION (e)
IN THE BLUE REGION (b) RELATIVE TO THAT IN THE
RED REGION (r) (139) FOR CHLOROPHYLL a

Ether	$\lambda_{max} \, b$	$\lambda_{max} \, r$	$e_b : e_r$
Sodium dried	426–427	657.5	1.31
"Dry"	427–428	657.5	1.31
U.S.P.	428	660.0	1.21
Plus 6% methanol	429–430	660.0	1.21
Saturated water	428–429	660.0	1.20

D. Estimation after Transference to Nonpolar Solvents and Separation by Chromatography

In quantitative procedures, the chlorophylls in plant extracts may be transferred to petroleum ether, separated by chromatography in columns, paper, or thin layers (see Section II, C), eluted, and determined spectrophotometrically in ether or 80 or 90% acetone. This procedure has the advantage that various alteration products of the chlorophylls do not interfere with the spectrophotometric observations. It has the disadvantage of loss of the chlorophylls during separation, elution, and recovery.

Chromatographic methods are the most effective qualitative procedures for testing the uniformity of the green pigments, except the chlorobium chlorophylls, in the plant extracts. Under the conditions employed to test the homogeneity of the chlorophylls, alteration and multiple zoning may occur. These effects have been observed in paper chromatography and in thin-layer chromatography with diatomaceous earth as the adsorbent. They are less conspicuous in columns of powdered sugar (109). Alteration of the chlorophyll proceeds rapidly in light, especially in benzene solutions as noted already in Section V, C. For definitive results, special care must be employed in the use of the chromatographic methods.

Chemical reactions have been utilized to enhance the dependability of the spectrophotometric observations. Measurements of the green pigments at 450 mμ followed by conversion to the pheophytins with additional measurement at 450 mμ provide a difference value attributable to Chl c (156). Similarly, measurements at 666 mμ before and after formation of the oxime provide a difference value attributable to Chl b (205).

E. Calculation of Quantity of Chlorophyll per Unit of Plant Material

The analytical methods described in this section provide the quantity of the chlorophylls in the extracts. For most practical work it is necessary to calculate the amount of the chlorophylls with reference to some measure of the plant material such as area of leaf surface, weight of fresh or dried material, volume or weight of centrifuged cells, single cells, or single chloroplasts. The variation inherent in these units of reference frequently exceeds by far the error of estimation of the chlorophylls in a standard solution (199).

VIII. Preparation of Chlorophylls

A. Objectives

For analytical estimation of the chlorophylls, it is not essential to isolate the individual pigments provided one is willing to accept the properties described by other workers. On the other hand, the pure pigments may be employed to confirm the chemical composition, the nuclear magnetic resonance, and the spectral absorption properties in the visible and infrared regions. The pure pigments may also be employed to test the reliability of the spectrophotometer which should be calibrated for wavelength and extinction measurements. With new pigments, there is no substitute for determination of the specific extinction coefficients using purified, crystalline preparations.

The most convenient method for the separation and isolation of the chlorophylls in quantities of 0.1–1 gm is chromatography. As elaborated in Section II, C, this procedure separates the chlorophylls from one another, from the carotenoid pigments, and from most of the colorless constituents extracted from plants. The subsequent elution, partition, precipitation, and drying procedures must be directed to removal of residual colorless plant constituents, removal of colorless substances leached from the columns or formed from the solvents and, finally, the removal of the last traces of moisture and solvents (71–73).

B. Chlorophylls *a* and *b*

Fresh spinach is a convenient source of Chls *a* and *b*. It is available at nearly all seasons, and it is relatively cheap. Dried, frozen, and canned spinach are usually unsuitable owing to the presence of large quantities of the chlorophyll alteration products (71). We have used the following procedures to prepare the chlorophylls in a state of high purity.

Fresh spinach leaves free of midribs (200 gm) were dropped into vigorously boiling water (about 2 liters). After 1–2 minutes the water was cooled quickly with an excess of cold water. The water was then decanted, the leaves being held back with a brush. The leaves were washed once with cold water and squeezed by hand, first between paper towels in the hand, then between fresh paper and cloth towels underfoot. These leaves, which weighed 95–100 gm, were separated from one another and placed in methanol (500 ml) plus petroleum ether (125 ml, b.p. 20–40°). After a few minutes most of the chlorophyll was removed. The deep-green extract was decanted through a pad of cotton into a large separatory funnel (2-liter). The leaves were re-extracted twice with the same quantities of methanol and petroleum ether that were used to make the first extraction.

In one partition procedure for transfer of the pigments from methanol to petroleum ether, water (250 ml) was added to the combined leaf extracts, and the solution was swirled gently. If an emulsion was formed, or if the petroleum ether layer was very small, some concentrated sodium chloride solution was added. The lower methanol layer was then drawn into another separatory funnel and shaken vigorously with petroleum ether (60 ml). Sometimes addition of more sodium chloride solution was required. The yellow-green methanol layer, which was about 80% methanol and contained much of the strongly adsorbed, minor xanthophylls, was separated and discarded. The petroleum ether layer was added to the principal petroleum ether solution, then mixed with methanol (200 ml) and water (50 ml). The aqueous methanol was separated and extracted with petroleum ether (60 ml), which was also added to the principal petroleum ether solution. The latter was again extracted with methanol (200 ml) plus water (50 ml). After separation from the petroleum ether layer, the methanol was extracted with petroleum ether (60 ml), which was combined with the principal petroleum ether solution. To prevent separation of the pigments, pure diethyl ether (60 ml), from a freshly opened can, was added to the petroleum ether, from which the residual alcohol was removed by several extractions with water (200 ml each).

In another more direct procedure for transfer of the pigments from methanol to petroleum ether, each methanol plus petroleum ether extract of the leaf material was diluted with about 1 liter of saturated salt solution, thereby transferring most of the pigments to the petroleum ether layer. The aqueous layer was extracted with about 100 ml of fresh petroleum ether. The successive extracts were treated in the same way; all the petroleum ether layers were combined and diluted with a little diethyl ether.

The petroleum ether plus diethyl ether was added to a chromatographic column of powdered sugar (3 by 12 inches) prepared by pressing small portions of the sugar into the chromatographic tube (about 3 pounds of untreated, commercial sugar, containing 3% starch to prevent caking, was used). Percolation of the solution through the sugar was accelerated by slightly reduced pressure (about 45–56 cm of mercury). By the time all the solution had been added to the column, the chlorophylls formed a zone 2–3 inches deep. The solvent containing the nonsorbed carotene then reached nearly to the bottom of the column. At this stage, the column was washed with fresh petroleum ether (about 100 ml), then with petroleum ether containing 0.5% n-propanol. The nonsorbed carotene was carried into the percolate and collected separately. Chl a was next washed into the percolate and collected. Lutein plus zeaxanthin, the quantity of which had been reduced by the methanol extractions of the petroleum ether, was washed into the percolate and discarded. Chl b was now near the bottom of the column. The sugar above the Chl b zone, with very small quantities of the residual minor xanthophylls, was removed and discarded, whereupon the Chl b was washed from the column with petroleum ether containing ethanol (3–5%).

Each chlorophyll elutriate was then extracted with 100-ml portions of 50, 60, 70, 80, and 90% methanol for Chl a and 50, 60, 70, and 80% methanol for Chl b in order to remove colorless (hydrogen-containing) substances eluted from the sugar. The chlorophylls were then isolated in a solid state. To this end, the petroleum ether solutions were placed in separatory funnels and shaken or swirled gently with water several times to remove the residual alcohol. Then the petroleum ether was shaken vigorously with water (150 ml), the chlorophyll collecting on the surface of the water droplets. This mixture was permitted to stand for a few minutes for the petroleum ether to rise; any clear water layer was discarded, and the water plus chlorophyll layer was drained into another separatory funnel. This extraction procedure was repeated until the petroleum ether layer was nearly colorless. Sometimes as many as 8–10 extractions were required for the separation of Chl a leaving almost colorless petroleum ether. Several extractions removed Chl b completely from the petroleum ether elutriate. When each aqueous layer was drained into the second separatory funnel, much of the water separated and could be drained away. Finally there were obtained a little petroleum ether with suspended solid chlorophyll and a little dispersed water. All this material was drained into a flask, the chlorophyll on the walls of the funnel being loosened with a fine stream of water, and the mixture was cooled with solid carbon dioxide. After several hours, the petroleum ether plus chlorophyll was decanted from the frozen water, and the chlorophyll was collected by centrifugation. If much solid chlorophyll remained with the ice, the latter was melted, and more petroleum ether (about 25 ml) was added. This was shaken, the water was frozen again, and the petroleum ether plus solid chlorophyll was poured off and centrifuged. The chlorophyll collected by centrifugation was dried in a vacuum desiccator. From 200 gm of fresh spinach the yields of the separated chlorophylls were: Chl a, 0.12–0.15 gm, and Chl b, 0.05–0.06 gm.

Each chlorophyll, prepared as just described, was purified further by chromatography, by additional partition between methanol and petroleum ether, and by precipitation. As an example, Chl a (0.25 gm) was dissolved in fresh diethyl ether (50 ml), which was then diluted with petroleum ether (200 ml). This solution was adsorbed in a column of powdered sugar (3 by 12 inches, with about 3 pounds of sugar). The adsorbed Chl a, which formed a zone about 1.5 inches deep, was then washed with petroleum ether plus 0.5% n-propanol. Traces of pheophytin a and Chl a' were carried along ahead of the Chl a. When the Chl a zone reached the bottom of the column, the sugar above it was removed, and the pigment was eluted with petroleum ether plus 2–3% ethanol, thus providing a solution of Chl a in about 300 ml of petroleum ether. This petroleum ether solution was treated with methanol (200 ml), and then with water (50 ml), and the aqueous methanol was separated. This treatment was repeated. Then the petroleum ether was washed several times with water by gentle swirling and cooled overnight with solid carbon dioxide. The solid chlorophyll that separated was collected by centrifugation and dried in high vacuum at temperatures up to 100°. It weighed about 0.23 gm. Spectral absorption properties of this chlorophyll agreed well with those reported before, as shown in Table III.

Chl b was purified by the same chromatographic procedure as for Chl a, except that the column was washed with petroleum ether containing from 1 to 1.5% n-propanol. The spectral absorption properties are summarized in Table III.

Two modifications of this preparatory method, that reduce the labor involved in the shaking out procedure, were sometimes employed. The petroleum ether elutriates from the first adsorption columns were extracted with methanol and washed only once or twice with water and evaporated at reduced pressure with a flash evaporator. This provided a sticky chlorophyll preparation that was redissolved and purified by readsorption as just described. Alternatively, the elutriates, washed with methanol and water, were cooled with solid carbon dioxide for about 24 hours. Chl a usually separated incompletely, Chl b almost completely. The solid pigments were then collected by centrifugation and purified by readsorption.

C. Deuteriochlorophylls a and b

These two chlorophylls were prepared from green algae (*Chlorella vulgaris* or *Scenedesmus obliquus*) that had been grown in heavy water using the procedure described above for the ordinary chlorophylls (71). The deuterated algae were treated with boiling D_2O to minimize exchange, but ordinary water and solvents were used in subsequent manipulations. Even with this precaution, hydrogen exchange occurred at the C-10 position as shown by nuclear magnetic resonance. To isolate the completely deuterated pigments, it was necessary to re-exchange this hydrogen atom by reaction of the pigment with deuteriomethanol (CD_3OD) in vacuum. In columns of powdered sugar the deuterated chlorophylls were inseparable from the ordinary chlorophylls (Table I). The spectral absorption properties are summarized in Figs. 4 and 5 and in Table III.

D. Chlorophyll c

Solutions of Chl c free of other pigments have been obtained from diatoms and brown algae by repeated partition of the pigments between aqueous alcohols and petroleum ether. They have also been obtained by chromatography in columns of powdered sugar with petroleum ether plus 7.5% n-propanol as the wash liquid (152, 153).

Crystalline Chl c has been prepared from brown algae by successive chromatographic separations with columns of cellulose, silicic acid, and alumina (117) and by adsorption in paper (155).

By means of adsorption upon powdered sugar alone, thus avoiding the possibly deleterious action of the strong adsorbents, Chl c has now been purified in sufficient quantity so that it can be crystallized. About 35 gm of freshly centrifuged cells of the diatom Nitzschia closterium f. minutissima (from 40 liters of culture) were suspended in a little of the culture medium and added to about 1 liter of the vigorously boiling medium buffered with 20 gm ammonium acetate. The boiling was continued for about 2 minutes. Then the green suspension was cooled with ice and centrifuged in two plastic bottles, 1 liter each. The cells in each bottle were extracted with a mixture of methanol (500 ml), diethyl ether (200 ml), and petroleum ether (100 ml). The cells were collected by centrifugation and extracted with the same quantities of the three solvents. A third extraction was made with methanol (150 ml), diethyl ether (100 ml), and petroleum ether (50 ml), and these two extracts were combined. This procedure provided five solutions, each of which was diluted with aqueous salt solution to remove the methanol and leave most of the pigment in the ether plus petroleum ether. After separation from the ether-petroleum ether layer, each aqueous methanol layer was extracted with carbon tetrachloride (200 ml), which removed all but traces of the pigments. The ether-petroleum ether and carbon tetrachloride extracts were then combined and evaporated with a flash evaporator, and the residue was dried in a high vacuum. This residue was dissolved in ether (80 ml), which was then diluted with petroleum ether (320 ml). Equal portions of this solution (50 ml each) were added to eight columns of powdered sugar (8 by 36 cm); these were washed with petroleum ether and then with this solvent containing 2, 5, and 7.5% n-propanol. This washing carried the yellow-green zone of Chl c, the most strongly adsorbed zone, nearly to the bottom of the columns. It was preceded by a light green zone, presumably an oxidation product of Chl a (77). The Chl c zones from the eight columns were removed with a long spatula and repacked into four clean chromatographic tubes. The Chl c was eluted with a little ethanol followed by washing with ether. The elutriates were washed with water and evaporated to dryness with a flash evaporator. (A portion of this residue dissolved in diethyl ether exhibited a spectral absorption curve like that of Chl c). The gummy residue was taken up in about 15 ml of methanol, transferred to a small flask and diluted with water (about 1.5–2.5 ml added in small portions). After standing overnight in the refrigerator, the preparation was cooled with carbon dioxide for several hours. At this stage, the supernatant was decanted and the residue was shaken with petroleum ether-methanol $(6 + 6$ ml). This solvent mixture dissolved most of the gum and left small black crystals in suspension. The suspension was transferred to a centrifuge tube, and the crystals were collected by centrifugation. The crystals were washed a second time with methanol-petroleum ether $(3.5 + 3.5$ ml) and also a third time with the mixed solvents $(2.5 + 2.5$ ml). They were then washed with petroleum ether and dried. At this stage they were dissolved in tetrahydrofuran (6 ml) which was clarified by centrifugation, and the Chl c was crystallized by the addition of petroleum ether (40 ml). After refrigeration with carbon dioxide for several hours, Chl c was collected by centrifugation and dried (weight, about 16 mg).

This Chl c was virtually insoluble in pertoleum ether, silghtly soluble in diethyl ether, and readily soluble in tetrahydrofuran. The characteristic spectral absorption curve was identical with that reported before (see Fig. 6); the specific coefficients were greater (see Table IV).

E. Chlorophyll d

By means of chromatography in columns of powdered sugar with petroleum ether plus n-propanol as the wash liquid, Chl d is readily separable from the other pigments of marine red algae (38). Thus far, Chl d has not been isolated in solid form (63), but with adequate supplies of a suitable source, such as *Gigartina agardhii*, there should be no difficulty in making such a preparation.

F. Protochlorophyll

Protochlorophyll remains an elusive substance because there are no common rich sources. The inner seed coats from seeds of the pumpkin family are the richest source of the esterified pigment. Etiolated seedlings in which the pigment has been reported in the unesterified form as well as in the esterified form are a rather poor source. Moreover, the green pigment is accompanied by so much fatty material that its isolation in solid form is very difficult (63, 65).

G. Bacteriochlorophyll

The cells from about 12 liters of *Rhodospirillum rubrum* cultures were collected by centrifugation, frozen, and stored in a "deep freeze" refrigerator. These frozen cells were extracted with methanol (500 ml) plus diethyl ether (200 ml) plus petroleum ether (100 ml), and were re-extracted with methanol (200 ml) plus diethyl ether (100 ml). The combined extracts were diluted with water which transferred the pigments to the ether-petroleum ether. This layer was washed with water and evaporated with a flash evaporator. The purplish gray residue was dissolved in diethyl ether (60 ml), diluted with petroleum ether (300 ml), and adsorbed in 3 columns of powdered sugar (8 by 35 cm). These columns were washed with petroleum ether plus 0.5% n-propanol until the blue-gray zone of the Bchl was carried nearly to the bottom. The zones with the Bchl were removed form the three columns, repacked into a fresh tube, and the Bchl was eluted with petroleum ether plus ethanol (3%). The elutriate was diluted with fresh diethyl ether and washed twice with water. Then it was chromatographed in three columns of powdered sugar, as just described, with petroleum ether plus 0.75% n-propanol as the wash liquid. The Bchl was eluted with petroleum ether plus ethanol, and the elutriates were washed with water. Crystals separated almost immediately. After the suspension was cooled with CO_2 overnight, the crystals were collected by centrifugation. They were recrystallized from 5–10 ml of chloroform by the addition of petroleum ether (about 100 ml). The yield was about 90 mg. The chromatographic properties are indicated in Table I, the spectral properties in Fig. 7 and Table V.

H. Deuteriobacteriochlorophyll

The cells from about 10 liters of the culture of *Rhodospirillum rubrum* grown in heavy water with fully deuterated biotin and succinate were utilized for the preparation of deuteriobacteriochlorophyll by means of the procedure just described for ordinary Bchl. The yields of the recrystallized pigment were about 70 mg (see Fig. 7 and Table V).

I. Chlorobium Chlorophyll 660

The cells centrifuged from about 8 liters of the cultures of *Chlorobium thiosulfatophilum 660* or *Chloropseudomonas ethylicum* were treated with boiling water, cooled, and centrifuged. Alternatively, the cells were frozen and preserved frozen until extraction. A mixed solvent, methanol (500 ml), diethyl ether (200 ml), and petroleum ether (100 ml), served for the first and second extractions, and methanol (200 ml), diethyl ether (100 ml) plus petroleum ether (50 ml) served for the third extraction. Each extract was diluted with an excess of aqueous salt solution which transferred the pigments to the petroleum ether-ether layer. This layer was then washed with water and evaporated in a flash evaporator. The resulting green residue was dissolved in ether (20 ml) plus acetone (20 ml) and diluted with petroleum ether (150 ml). Usually a little additional acetone was required to keep the pigments in solution. This solution was added to four columns of powdered sugar (8 by 35 cm) which were washed with a little petroleum ether, then with petroleum ether plus 1, 2, and 2.5% n-propanol. This produced a single, uniform green zone that migrated more than halfway through each column and was preceded by a light purplish-gray zone of Bchl. The green zones from the four columns were removed and repacked into two chromatographic tubes where the green pigment was eluted with ethanol plus petroleum ether. This elutriate was washed twice with water (500 ml) whereupon the chlorophyll crystallized in the petroleum ether. This petroleum ether suspension was cooled with solid carbon dioxide overnight, and the crystals were collected by centrifugation. The pigment was then dissolved in 5–10 ml of chloroform and clarified by centrifugation or filtration. About 100 ml petroleum ether was added to the chloroform solution, from which the chlorophyll crystallized rapidly. Crystals in the cooled suspension were collected in a centrifuge tube and dried in a vacuum. The yield was about 70 mg. Similar results were obtained when the pigments were separated in columns of corn starch. Spectral properties are summarized in Fig. 8 and in Table V.

ACKNOWLEDGMENT

Many of the experimental observations summarized in this report emanated from collaborative investigations in which several of our colleagues have made major contributions. These results would not have been possible without the active leadership of Dr. Joseph J. Katz, particularly in the experiments with heavy water. Dr. Henry L. Crespi, Dr. Robert A. Uphaus, and Mr. Homer F. DaBoll provided many cultures of algae grown under various conditions. Other collaborators are listed as coauthors in papers that are cited.

REFERENCES

(1) D. M. Ramirez and M. L. Tomes, Botan. Gaz. 125, 221 (1964).

(2) W. A. Gortner, J. Food Sci. 30, 30 (1965).

(3) J. L. Hall and D. L. Mackintosh, J. Food Sci. 29, 420 (1964).

(4) W. C. Dietrich, M. M. Boggs, M. D. Nutting, and N. E. Weinstein, Food Technol. 14, 522 (1960).

(5) W. G. C. Forsyth, Ann. Rev. Plant Physiol. 15, 443 (1964).

(6) I. D. Jones, R. C. White, and E. Gibbs, Food Technol. 16, 96 (1962).

(7) V. I. Rogachev, M. L. Frumkin, L. P. Koval'skaya, and K. V. Egorova, Konserv. i Ovoshchesushil. Prom. 15, No. 9, 19 (1960).

(8) S. H. Schanderl, G. L. Marsh, and C. O. Chichester, J. Food Sci. 30, 312 and 317 (1965).

(9) R. C. White, I. D. Jones, and E. Gibbs, J. Food Sci. 28, 431 (1963).

(10) S. C. Lewis, Arch. Environ. Health 9, 308 (1964).

(11) E. Rabinowitch, J. Phys. Chem. 66, 2536 (1962).

(12) B. Rumberg, Nature 204, 860 (1964).
(13) B. Rumberg and H. T. Witt, Z. Naturforsch. 19b, 693 (1964).
(14) G. G. Winberg and T. N. Sivko, Air Water Pollution 6, 267 (1962).
(15) S. Ichimura and H. Fukushima, Botan. Mag. (Tokyo) 76, 395 (1963).
(16) I. Miyata and T. Hosokawa, Ecology 42, 766 (1961).
(17) E. S. Nielsen, Physiol. Plantarum 14, 868 (1961).
(18) E. S. Nielsen, Intern. Rev. Ges. Hydrobiol. Hydrog. 47, 333 (1962).
(19) "Proceedings of the Conference on Primary Productivity Measurement, Marine and Freshwater" (M. S. Doty, ed.), U.S.A.E.C. TID-7633. University of Hawaii, 1961.
(20) J. Seoane-Camba, Compt. Rend. 259, 1432 (1964).
(21) Z. Sestak, Photochem. Photobiol. 2, 101 (1963).
(22) N. D. Tverkina, Chem. Abstr. 59, 13111 (1963).
(23) S. C. Agarwala, C. P. Sharma, and A. Kumar, Plant Physiol. 39, 603 (1964).
(24) G. W. Bryan and L. Bogorad, Studies Microalgae Photosyn. Bacteria, Collection Papers p. 399 (1963).
(25) H. F. Clements, Ann. Rev. Plant Physiol. 15, 409 (1964).
(26) N. N. Dikshit and C. M. Abdul Majid, Agra Univ. J. Res. Sci. 12, Part 3, 115 (1963).
(27) V. K. Girfanov, A. K. Shkurikhina, and N. N. Ryakhovskaya, Fiziol. Obosnovanie Sistemy Pitaniya Rast., Akad. Nauk SSSR, Inst. Fiziol. Rast. p. 46 (1964).
(28) A. Haspelova-Horvatovicova, Biologia 18, 334 (1963).
(29) I. A. Nakaidze and I. E. Andzhaparidze, Agrokhimiya No. 3, p. 55 (1964).
(30) V. G. Nesterov and V. F. Kashlev, Dokl., Mosk. Sel'skokhoz. Akad. 83, 342 (1963).
(31) C. A. Price and E. F. Carell, Plant Physiol. 39, 862 (1964).
(32) G. Rakhimov and T. M. Bushueva, Uzbeksk. Biol. Zh. 8, 31 (1964).
(33) O. Siegel and H. J. Bjarsch, Gartenbauwissenschaft 27, 15 (1962).
(34) W. J. Starnes and H. H. Hadley, Crop Sci. 5, 9 (1965).
(35) M. M. Yakubova and Y. S. Nasyrov, Dokl. Akad. Nauk Tadzh. SSR 7, 36 (1964).
(36) J. Bruinsma, Chem. Abstr. 62, 2191 (1965).
(37) G. M. Smith, "Cryptogamic Botany," 2nd ed., Vol. I. McGraw-Hill, New York, 1955.
(38) H. H. Strain, Ann. Priestley Lectures 32 (1958).
(39) M. B. Allen, L. Fries, T. W. Goodwin, and D. M. Thomas, J. Gen. Microbiol. 34, 259 (1964).
(40) A. Jensen, O. Aasmundrud, and K. E. Eimhjellen, Biochim. Biophys. Acta 88, 466 (1964).
(41) H. H. Strain, in "Photosynthesis in Plants" (J. Franck and W. E. Loomis, eds.), p. 133. Iowa State Coll. Press, Ames, Iowa, 1949.
(42) H. H. Strain, in "Manual of Phycology" (G. M. Smith, ed.), p. 243. Chronica Botanica, Waltham, Massachusetts, 1951.
(43) H. H. Strain, Biol. Bull. 129, 366 (1965).
(44) H. H. Strain, in "Biochemistry of Chloroplasts." NATO Advanced Study Institute, Aberystwyth, Wales, 1965 (in press).
(45) H. H. Strain, Argonne Natl. Lab. Rev. 1, No. 3, 6 (1964).
(46) S. H. Schanderl, C. O. Chichester, and G. L. Marsh, J. Org. Chem. 27, 3865 (1962).

(47) W. Gottschalk and F. Müller, *Planta* **61**, 259 (1964).

(48) G. V. Kasinova, *Radiobiologiya* **4**, 603 (1964).

(49) F. Müller, *Planta* **63**, 65 (1964).

(50) F. Müller, *Planta* **63**, 301 (1964).

(51) G. Röbbelen, *Naturwissenschaften* **44**, 288 (1957).

(52) F. T. Wolf, *Bull. Torrey Botan. Club* **90**, 139 (1963).

(53) G. F. Humphrey, *Australian J. Marine Freshwater Res.* **14**, 24 (1963).

(54) SCOR-UNESCO, "Determination of Photosynthetic Pigments," Mimeo Report of SCOR-UNESCO Working Group 17. Sydney, 1964.

(55) Anonymous, *Chem. Eng. News* **38**, 36 (1960).

(56) S. Aronoff, *in* "Handbuch der Pflanzenphysiologie" (W. Ruhland, ed.), Vol. 5, Part 1, p. 234. Springer, Berlin, 1960.

(57) H. Fischer and A. Stern, "Die Chemie des Pyrrols," Vol. 2, Part II. Akadem. Verlagsges., Leipzig, 1940.

(58) T. N. Godnev, N. K. Akulovich, and R. M. Rotfarb, *Usp. Soverm. Biol.* **55**, 204 (1963).

(59) A. S. Holt and H. V. Morley, *in* "Comparative Biochemistry of Photoreactive Systems" (M. B. Allen, ed.), p. 169. Academic Press, New York, 1960.

(60) R. B. Woodward, *Angew. Chem.* **72**, 651 (1960).

(61) R. B. Woodward, *Proc. 4th Robert A. Welch Found. Conf. Chem. Res., 1960* p. 99. Robert A. Welch Found., Houston, Texas, 1961.

(62) K. Egle, *in* "Handbuch der Pflanzenphysiologie" (W. Ruhland, ed.), Vol. 5, Part 1, p. 323. Springer, Berlin, 1960.

(63) C. S. French, *in* "Handbuch der Pflanzenphysiologie" (W. Ruhland, ed.), Vol. 5, Part 1, p. 252. Springer, Berlin, 1960.

(64) Z. Sestak, *J. Chromatog.* **1**, 293 (1958); *Chromatog. Rev.* **7**, 65 (1965).

(65) J. H. C. Smith and A. Benitez, *in* "Moderne Methoden der Pflanzenanalyse" (K. Paech and M. V. Tracey, eds.), Vol. IV, p. 142. Springer, Berlin, 1955.

(66) H. H. Strain, *in* "Chromatography" (E. Heftmann, ed.), p. 584. Reinhold, New York, 1961.

(67) J. H. C. Smith and C. S. French, *Ann. Rev. Plant Physiol.* **14**, 181 (1963).

(68) R. Y. Stanier, *in* "Comparative Biochemistry of Photoreactive Systems" (M. B. Allen, ed.), p. 69. Academic Press, New York, 1960.

(69) L. P. Vernon and J. Finney, *Photochem. Photobiol.* **4**, 281 (1965).

(70) H. F. DaBoll, H. L. Crespi, and J. J. Katz, *Biotechnol. Bioeng.* **4**, 281 (1962).

(71) H. H. Strain, M. R. Thomas, H. L. Crespi, M. I. Blake, and J. J. Katz, *Ann. N. Y. Acad. Sci.* **84**, 617 (1960).

(72) H. H. Strain, M. R. Thomas, and J. J. Katz, *Biochim. Biophys. Acta* **75**, 306 (1963).

(73) F. C. Pennington, H. H. Strain, W. A. Svec, and J. J. Katz, *J. Am. Chem. Soc.* **86**, 1418 (1964).

(74) J. J. Katz, R. C. Dougherty, W. A. Svec, and H. H. Strain, *J. Am. Chem. Soc.* **86**, 4220 (1964).

(75) H. H. Strain, *Acta Phytochim. (Japan)* **15**, 9 (1949).

(76) H. H. Strain, *Agr. Food Chem.* **2**, 1222 (1954).

(77) J. Barrett and S. W. Jeffrey, *Plant Physiol.* **39**, 44 (1964).

(78) O. T. G. Jones, *Biochem. J.* **88**, 335 (1963).

(79) O. T. G. Jones, *Biochem. J.* **91**, 572 (1964).

(80) V. E. Uspenskaya and E. N. Kondrat'eva, *Dokl. Akad. Nauk SSSR* **157**, 678 (1964).

(81) T. G. Lessie and W. R. Sistrom, *Biochim. Biophys. Acta* **86**, 250 (1964).
(82) L. G. Johnston and W. F. Watson, *J. Chem. Soc.* p. 1203 (1956).
(83) M. R. Michel-Wolwertz and C. Sironval, *Biochim. Biophys. Acta* **94**, 330 (1965).
(84) A. A. Shlyk, V. I. Gaponenko, and T. V. Kukhtenko, *Dokl. Akad. Nauk Belorussk. SSR* **4**, 393 (1960).
(85) V. H. Booth, *Biochem. J.* **84**, 444 (1962); *Chromatog. Rev.* **7**, 98 (1965).
(86) S. Aronoff, *Plant Physiol.* **27**, 413 (1952).
(87) S. Aronoff, *Biochim. Biophys. Acta* **60**, 193 (1962).
(88) D. Frackowiak, *Bull. Acad. Polon. Sci., Ser. Sci., Math., Astron. Phys.* **12**, 357 (1964).
(89) F. F. Litvin and B. A. Gulyaev, *Dokl. Akad. Nauk SSSR* **158**, 460 (1964).
(90) P. S. Stensby and J. L. Rosenberg, *J. Phys. Chem.* **65**, 906 (1961).
(91) Y. F. Frei, *Biochim. Biophys. Acta* **57**, 82 (1962).
(92) S. S. Brody and M. Brody, *Natl. Acad. Sci.—Natl. Res. Council, Misc. Publ.* **1145**, 455 (1963).
(93) W. L. Butler, *Biochem. Biophys. Res. Commun.* **3**, 685 (1960).
(94) R. Govindjee and E. I. Rabinowitch, *Science* **132**, 355 (1960).
(95) C. Sironval, M. R. Michel-Wolwertz, and A. Madsen, *Biochim. Biophys. Acta* **94**, 344 (1965).
(96) J. S. Brown, *Colloq. Intern. Centre Natl. Rech. Sci. (Paris)* **119**, 371 (1963).
(97) H. H. Strain, *Science* **116**, 174 (1952).
(98) W. R. Sistrom, *Biochim. Biophys. Acta* **79**, 419 (1964).
(99) H. H. Strain and W. M. Manning, *J. Biol. Chem.* **146**, 275 (1942).
(100) W. M. Manning and H. H. Strain, *J. Biol. Chem.* **151**, 1 (1943).
(101) B. Ke, *Biochim. Biophys. Acta* **88**, 289 (1964).
(102) G. Oster, S. B. Broyde, and J. S. Bellin, *J. Am. Chem. Soc.* **86**, 1309 (1964).
(103) V. M. Kutyurin and I. Y. Artamkina, *Fiziol. Rast.* **9**, 493 (1962).
(104) R. K. Clayton, *J. Gen. Physiol.* **48**, 633 (1965).
(105) C. S. Yentsch and D. W. Menzel, *Deep-Sea Res.* **10**, 221 (1963).
(106) J. R. Wilson and M. D. Nutting, *Anal. Chem.* **35**, 144 (1963).
(107) L. P. Vernon, *Anal. Chem.* **32**, 1144 (1960).
(108) Z. Sestak, *Preslia* **35**, 123 (1964).
(109) H. H. Strain, J. Sherma, F. L. Benton, and J. J. Katz, *Biochim. Biophys. Acta* **109**, 1, 16, 23 (1965).
(110) L. I. Vlasenok, *Dokl. Akad. Nauk Belorussk. SSR* **6**, 255 (1962).
(111) M. Holden, *Biochim. Biophys. Acta* **56**, 378 (1962).
(112) A. F. H. Anderson and M. Calvin, *Nature* **194**, 285 (1962).
(113) A. F. H. Anderson and M. Calvin, *Arch. Biochem. Biophys.* **107**, 251 (1964).
(114) H. K. Lichtenthaler and M. Calvin, *Biochim. Biophys. Acta* **79**, 30 (1964).
(115) A. A. Shlyk, R. M. Rotfarb, and Y. P. Liyakhnovich, *Byul. Inst. Biol., Akad. Nauk Belorussk. SSR, 1957* No. 3, 115 (1958).
(116) E. M. Stanishevskaya, *Vestsi Akad. Navuk Belarusk. SSR, Ser. Biyal. Navuk* No. 4, 52 (1962).
(117) S. W. Jeffrey, *Biochem. J.* **86**, 313 (1963).
(118) S. Freed, K. M. Sancier, and A. H. Sporer, *J. Am. Chem. Soc.* **76**, 6006 (1954).
(119) H. H. Strain, *J. Am. Chem. Soc.* **77**, 5195 (1955).
(120) J. G. Coniglio and F. T. Wolf, *Phyton (Buenos Aires)* **17**, 189 (1961).
(121) J. B. Weiss and I. Smith, *2nd Chromatog. Symp., Brussels,* p. 293 (1962); *Chem. Abstr.* **60**, 7141 (1964).
(122) Z. Sestak, *Biol. Plant., Acad. Sci. Bohemoslov.* **6**, 132 (1964).

(123) Z. Sestak and J. Ullmann, *Rostlinna Vyroba* **10**, 1196 (1964).
(124) Z. Sestak, *Biol. Plant., Acad. Sci. Bohemoslov.* **1**, 287 (1959).
(125) Z. Sestak, *Rostlinna Vyroba* **10**, 1197 (1964).
(126) D. A. Zakrzhevskii and A. M. Ollykainen, *Fiziol. Rast.* **11**, 1082 (1964).
(127) O. G. Sud'ina and L. A. Sirenko, *Dopovidi Akad. Nauk Ukr. RSR* p. 960 (1961).
(128) E. Heftmann, ed., "Chromatography." Reinhold, New York, 1961.
(129) C. Costes, Thesis, University of Paris, Inst. Natl. Rech. Agron., 1965.
(130) I. A. Popova, *Tr. Botan. Inst., Akad. Nauk SSSR, Ser. 4: Eksperim. Botan.* **16**, 154 (1963).
(131) S. W. Jeffrey, *Biochem. J.* **80**, 336 (1961).
(132) J. M. Anderson, *U.S. At. Energy Comm.* **UCRL-8870** (1959).
(133) J. M. Anderson, *J. Chromatog.* **4**, 93 (1960).
(134) M. F. Bacon, *J. Chromatog.* **17**, 322 (1965).
(135) A. Haspelova-Horvatovicova and F. Fric, *Biologia* **19**, 820 (1964).
(136) U. Kandler and H. Ullrich, *Naturwissenschaften* **51**, 518 (1964).
(137) J. S. Bunt, *Nature* **203**, 1261 (1964).
(138) A. Hager and T. Bertenrath, *Planta* **58**, 564 (1962).
(139) H. J. Perkins and D. W. A. Roberts, *Biochim. Biophys. Acta* **79**, 20 (1964).
(139a) H. H. Strain and W. A. Svec, unpublished data.
(140) A. Fujiwara and T. Mizochi, *Nippon Dojo-Hiryogaku Zasshi* **32**, 356 (1961).
(141) G. Bruchet, *Ann. Univ. Lyon Sci.* **C11-12**, 39 (1959-1960).
(142) R. P. Dales, *J. Marine Biol. Assoc. U.K.* **39**, 693 (1960).
(143) F. P. Zscheile, Jr. and C. L. Comar, *Botan. Gaz.* **102**, 463 (1941).
(144) H. H. Strain, W. M. Manning, and G. Hardin, *Biol. Bull.* **86**, 169 (1944).
(145) T. R. Parsons and J. D. H. Strickland, *J. Marine Res. (Sears Found. Marine Res.)* **21**, 155 (1963).
(146) G. F. Humphrey, *Australian J. Marine Freshwater Res.* **14**, 148 (1963).
(147) L. Felfoldy, E. Szabo, and L. Toth, *Chem. Abstr.* **59**, 914 (1963).
(148) T. R. Parsons, *J. Fisheries Res. Board Can.* **18**, 1017 (1961).
(149) F. P. Zscheile, Jr., C. L. Comar, and G. Mackinney, *Plant Physiol.* **17**, 666 (1942).
(150) G. Mackinney, *J. Biol. Chem.* **132**, 91 (1940).
(151) G. Mackinney, *J. Biol. Chem.* **140**, 315 (1941).
(152) H. H. Strain and W. M. Manning, *J. Biol. Chem.* **144**, 625 (1942).
(153) H. H. Strain, W. M. Manning, and G. Hardin, *J. Biol. Chem.* **148**, 655 (1943).
(154) S. W. Jeffrey, *Nature* **194**, 600 (1962).
(155) S. S. Mel'nikov and V. B. Evstigneev, *Biofizika* **9**, 414 (1964).
(156) T. R. Parsons, *J. Marine Res. (Sears Found. Marine Res.)* **21**, 164 (1963).
(157) A. S. Holt and H. V. Morley, *Can. J. Chem.* **37**, 507 (1959).
(158) A. S. Holt, *Can. J. Botany* **39**, 327 (1961).
(159) H. Sagromsky, *Ber. Deut. Botan. Ges.* **71**, 435 (1958); **73**, 3 and 358 (1960).
(160) N. K. Akulovich, *Byul. Inst. Biol., Akad. Nauk Belorussk. SSR, 1957* No. 3, 99 (1958).
(161) T. N. Godnev and N. K. Akulovich, *Chem. Abstr.* **56**, 2713 (1962).
(162) T. N. Godnev, N. K. Akulovich, and E. V. Khodasevich, *Dokl. Akad. Nauk SSSR* **150**, 920 (1963).
(163) A. Madsen, *Physiol. Plantarum* **16**, 470 (1963).
(164) A. A. Shlyk, L. I. Fradkin, and L. I. Vlasenok, *Vestsi Akad. Navuk Belarusk SSR, Ser. Biyal Navuk* No. 2, 116 (1964).
(165) A. A. Shlyk, V. L. Kaler, and G. M. Podchufarova, *Biokhimiya* **26**, 259 (1961).

(*166*) A. A. Shlyk and I. V. Prudnikova, *Dokl. Akad. Nauk SSSR* **160**, 720 (1965).
(*167*) Z. M. Eidel'man and A. S. Khodzhaev, *Dokl. Akad. Nauk SSSR* **158**, 242 (1964).
(*168*) Y. Inada and K. Shibata, *Plant Cell Physiol.* (*Tokyo*) **1**, 311 (1960).
(*169*) K. E. Eimhjellen, O. Aasmundrud, and A. Jensen, *Biochem. Biophys. Res. Commun.* **10**, 252 (1963).
(*170*) A. Jensen and O. Aasmundrud, *Acta Chem. Scand.* **17**, 907 (1963).
(*171*) I. R. Kaplan and H. Silberman, *Arch. Biochem. Biophys.* **80**, 114 (1959).
(*172*) J. H. Mathewson, W. R. Richards, and H. Rapoport, *Biochem. Biophys. Res. Commun.* **13**, 1 (1963).
(*173*) L. V. Moshentseva and E. N. Kondrat'eva, *Mikrobiologiya* **31**, 199 (1962).
(*174*) K. K. Voĭnovskaya, *Probl. Fotosinteza, Dokl. 2-oi* [*Vtoroi*] *Vses. Konf., Moscow, 1957* p. 67 (1959).
(*175*) J. W. Weigl, *J. Am. Chem. Soc.* **75**, 999 (1953).
(*176*) A. S. Holt and H. V. Morley, *J. Am. Chem. Soc.* **82**, 500 (1960).
(*177*) R. Y. Stanier and J. H. C. Smith, *Biochim. Biophys. Acta* **41**, 478 (1960).
(*178*) R. M. Balitskaya and Y. E. Erokhin, *Dokl. Akad. Nauk SSSR* **153**, 460 (1963).
(*179*) S. F. Conti and W. Vishniac, *Nature* **188**, 489 (1960).
(*180*) A. S. Holt and D. W. Hughes, *J. Am. Chem. Soc.* **83**, 499 (1961).
(*181*) A. S. Holt, D. W. Hughes, H. J. Kende, and J. W. Purdie, *J. Am. Chem. Soc.* **84**, 2835 (1962).
(*182*) E. N. Kondrat'eva and L. V. Moshentseva, *Dokl. Akad. Nauk SSSR* **135**, 460 (1960).
(*183*) H. Rapoport and H. P. Hamlow, *Biochem. Biophys. Res. Commun.* **6**, 134 (1961).
(*184*) A. S. Holt, D. W. Hughes, H. J. Kende, and J. W. Purdie, *Plant Cell Physiol.* (*Tokyo*) **4**, 49 (1963).
(*185*) D. W. Hughes and A. S. Holt, *Can. J. Chem.* **40**, 171 (1962).
(*186*) H. V. Morley and A. S. Holt, *Can. J. Chem.* **39**, 755 (1961).
(*187*) R. A. Moss and W. E. Loomis, *Plant Physiol.* **27**, 370 (1952).
(*188*) W. L. Butler, *Ann. Rev. Plant Physiol.* **15**, 451 (1964).
(*189*) K. Takashima, I. Shihira-Ishikawa, and E. Hase, *Plant Cell Physiol.* (*Tokyo*) **5**, 321 (1964).
(*190*) H. M. Benedict and R. Swidler, *Science* **133**, 2015 (1961).
(*191*) J. E. Tyler, *Proc. Natl. Acad. Sci. U.S.* **51**, 671 (1964).
(*192*) J. E. Tyler, *Proc. Natl. Acad. Sci. U.S.* **47**, 1726 (1961).
(*193*) D. N. Roy, S. N. Ray, and N. C. Ganguli, *Sci. Cult.* (*Calcutta*) **28**, 589 (1962).
(*194*) D. I. Sapozhnikov and S. A. Chernomorskiĭ, *Fiziol. Rast.* **7**, 660 (1960).
(*195*) J. Bruinsma, *Biochim. Biophys. Acta* **52**, 576 (1961).
(*196*) J. Billot, *Physiol. Vegetale* **2**, 195 (1964).
(*197*) M. Holden, *in* "Chemistry and Biochemistry of Plant Pigments" (T. W. Goodwin, ed.), p. 461. Academic Press, New York, 1965.
(*198*) J. Bruinsma, *Photochem. Photobiol.* **2**, 241 (1963).
(*199*) L. V. Kakhnovich and L. A. Khodorenko, *Fiziol. Rast.* **11**, 933 (1964).
(*200*) A. S. Holt and E. E. Jacobs, *Am. J. Botany* **41**, 710 (1954).
(*201*) J. Rácik and V. Mego, *Chem. Zvesti* **15**, 384 (1961).
(*202*) I. A. Shul'gin, V. Z. Podol'nyi, and S. V. Sokolova, *Fiziol. Rast.* **10**, 383 (1963).
(*203*) A. Stoll and E. Wiedemann, *Helv. Chim. Acta* **42**, 679 (1959).
(*204*) P. J. McCartin, *J. Phys. Chem.* **67**, 513 (1963).
(*205*) T. Ogawa and K. Shibata, *Photochem. Photobiol.* **4**, 193 (1965).

—*3*—

The Structure and Chemistry of
Functional Groups*

G. R. SEELY

Charles F. Kettering Research Laboratory, Yellow Springs, Ohio

I. General Aspects

A. Historical Background

After tentative beginnings in the nineteenth century, the systematic study of the chemistry of the chlorophylls was placed on a sound basis by Willstätter and co-workers during the first decade of the present century. They discovered and named pheophytin (*1*), chlorophyllide, pheophorbide, chlorophyllase (*2*), phytol, and allomerization (*3*). They recognized that upon saponification, chlorophyll *a* gave chlorins, and chlorophyll *b*, rhodins (*4*). They applied alkaline degradation (*5*), chromic acid oxidation (*6*), and hydriodic acid reduction (*7*) to the analysis of chlorophyll.

* Contribution No. 216 from the Charles F. Kettering Research Laboratory, Yellow Springs, Ohio. The preparation of this chapter was supported in part by National Science Foundation Grant No. GB-2089.

In 1928 Hans Fischer took up the problem of the structure of chlorophyll, and the early 1930's were years of intensive effort by his group at Munich, by Conant and co-workers at Harvard, and by Stoll and Wiedemann at Zürich, which were to culminate in 1935 with the proposal by Fischer of a structure of chlorophyll a, correct except for the position of the "extra" hydrogens (8) [i.e., the ones now located at the 7 and 8 positions of ring IV (see structure I)].

In 1929, Fischer and Bäumler reported the isomerization of pheophorbide by hydriodic acid to pheoporphyrin a_5 (9), and in 1931, essentially correct structures were announced for this porphyrin and for phylloerythrin (10). Inference of the structure of pheophorbide from that of pheoporphyrin a_5 was delayed, however, by Fischer's acceptance of an empirical formula for chlorophyll containing one more oxygen atom than the five found by Willstätter and Stoll (11).

The assignment in 1935 of the "extra" hydrogens to ring III of chlorophyll rested on arguments concerning the optical activity of chlorophyll derivatives (8, 12); the correct assignment to ring IV in 1939 was based on analysis of the products of chromic acid oxidation (13). These hydrogens, and the hydrogens on ring II of bacteriochlorophyll, were later found to be in the *trans* configuration (14, 15).

The chemistry of bacteriochlorophyll was first studied by Noack and Schneider (16, 17) and the relationship of the bacterial pigment to chlorophyll was established by Fischer et al. (18, 19) by the preparation of common derivatives.

The chemistry of protochlorophyll was first examined by Noack and Kiessling, with pigment extracted from pumpkin seed coats (20, 21). Fischer et al. identified it as a vinyl pheoporphyrin analog of chlorophyll (22, 23).

The structure of phytol was established by F. G. Fischer and Löwenberg (24), who later synthesized it from pseudoionone (25).

By 1939, the emphasis had shifted to those reactions thought to be possible steps in the synthesis of chlorophyll. The total synthesis of chlorophyll a was not attained in Fischer's lifetime, but a synthesis along a path laid out by Fischer was reported by Strell and Kalojanoff in 1962 (26). Two years before, Woodward and co-workers reported synthesis of chlorophyll along quite different lines, discovering in the process the remarkable susceptibility of the porphyrin ring of chlorophyll to electrophilic attack (27, 28). This, and the recent application of nuclear magnetic resonance and infrared techniques to chlorophyll derivatives (29, 30), promise interesting developments in the structure of the chlorophylls and their interaction with other molecules.

Early work on separation, analysis, and degradation of chlorophyll is described in detail by Willstätter and Stoll (*11*). In "Die Chemie des Pyrrols," Vol. 2, Part II, Fischer and Stern summarize knowledge of the chemistry of chlorophyll derivatives to 1939, and present the evidence that established the structures of the chlorophylls (*31*). German work from 1940 to 1947 has been reviewed by Strell, with emphasis on approaches to the synthesis of chlorophyll (*32*). Other reviews dealing in part with the chemistry of the chlorophylls are those of Rothemund (*33*), Stoll and Wiedemann (*34*), Rabinowitch (*35*), Aronoff (*36*), and Holt (*37*).

The intent of this chapter is to review the chemistry of the functional groups of those chlorophylls the structures of which have been known for some time: chlorophylls *a*, *b*, protochlorophyll, and bacteriochlorophyll. The determination of the structures of the chlorobium chlorophylls and chlorophyll *d*, and what is known of the structures of chlorophylls *c* and *e*, and bacteriochlorophyll *b*, are treated in Chapter 4. Recent work on the synthesis of chlorophyll *a* is reviewed in Chapter 5.

B. Structures and Nomenclature

The chlorophylls and their derivatives, the structures of some of which are shown in (I)–(XIII), are regarded as derivatives of the unknown compound phorbin (XIV), and the known compounds chlorin (XV) and porphin (XVI).

Fischer's conventional system of numbering the pyrrole rings, their β-positions, and the methine bridges or *meso* positions (see XVI), is consistent for chlorin and phorbin, at the sacrifice of rigid adherence to I.U.C. rules (*33, 36*). It is, of course, possible to name chlorophyll derivatives systematically as substituted porphins, chlorins, and phorbins, but a large vocabulary of trivial names has arisen, originally out of ignorance of the true structures, and has been retained for convenience. The significance of some of the more frequently encountered terms and symbols in chlorophyll nomenclature will be explained.

Porphyrin: In the broad sense, any macrocyclic tetrapyrrole pigment in which the pyrrole rings are joined by methine bridges and the system of double bonds forms a closed, conjugated loop. Thus, porphins, chlorins, phorbins, and bacteriochlorin (XVII) are porphyrins, but phlorin (XVIII) and corrole (XIX) are not.

In the narrow sense, *porphyrin* designates substituted porphins, as distinct from chlorins, etc. In this chapter, the word "porphyrin" is used only in the broad sense, but is retained in the traditional proper names of porphin derivatives, such as "pheoporphyrin a_5" and "pyrroporphyrin."

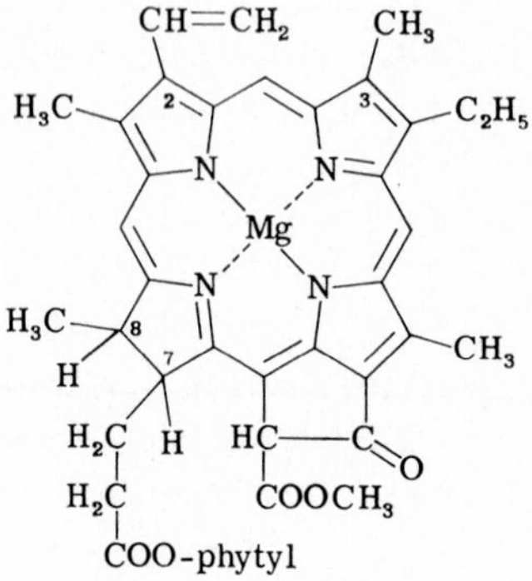

I Chlorophyll a
II Chlorophyll b: 3-CHO
III Chlorophyll d: 2-CHO
IV Protochlorophyll:
    7,8-dehydrochlorophyll

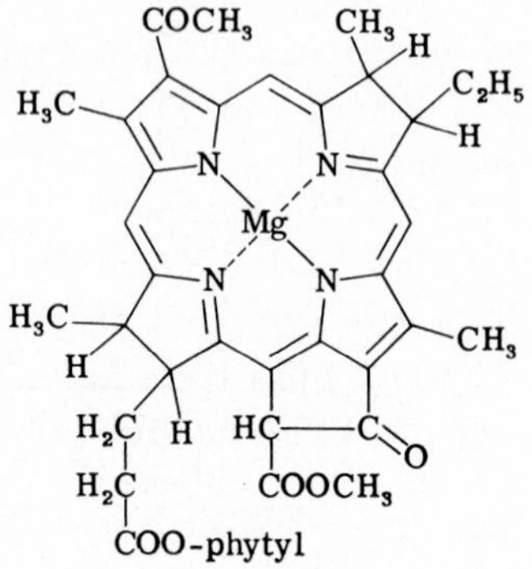

Bacteriochlorophyll

V

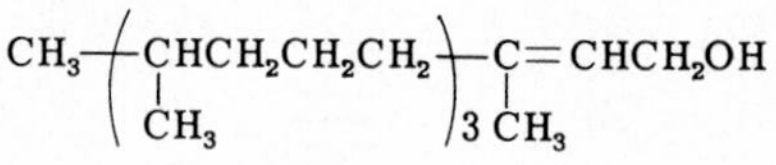

$$CH_3 \left(CHCH_2CH_2CH_2 \atop CH_3 \right)_3 C=CHCH_2OH \atop CH_3$$

Phytol

VI

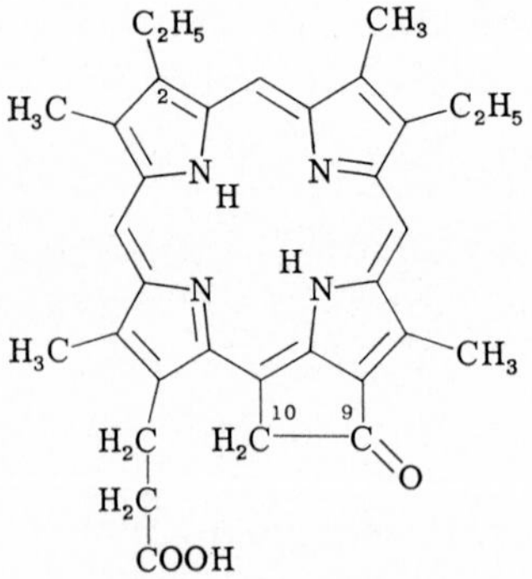

VII Phylloerythrin
VIII Pheoporphyrin a_5:
    10-COOCH$_3$
IX Protopheophorbide:
    2-CH=CH$_2$, 10-COOCH$_3$

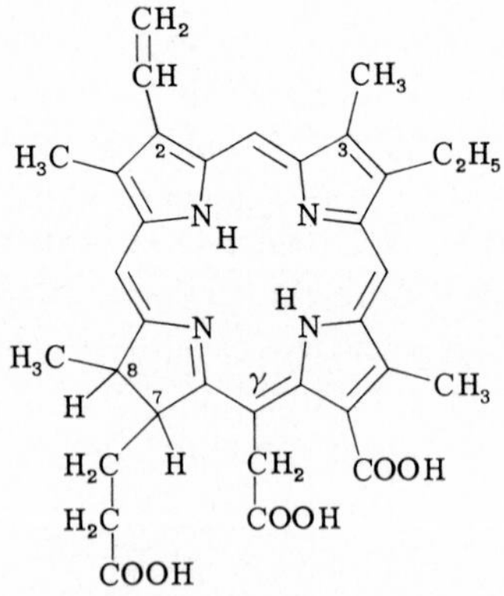

X Chlorin e_6
XI Rhodin g_7: 3-CHO
XII Purpurin 7: γ-COCOOH
XIII Chloroporphyrin e_6:
    2-C$_2$H$_5$, 7,8-dehydrochlorin e_6

Porphin: Term introduced by Fischer as a name for the unsubstituted, fully unsaturated porphyrin (38). The spelling "porphine" has been urged on grounds of the weakly basic properties of compounds of this class (39).

Chlorin in the broad sense means any (7,8-) dihydroporphin. It has been conventional to distinguish four subclasses:

1. *Rhodins,* chlorins of the *b* series
2. *Phorbins* (or *phorbides*), having an isocyclic *ring* V, also called the "*cyclopentanone ring*" or "*carbocyclic ring*"
3. *Purpurins,* having an electron attracting group, such as carbonyl, at the γ-position
4. *Chlorins,* in the narrow sense, not belonging to the former three classes. 7, 8-Dihydroporphins with a *heterocyclic* closure between the 6- and γ-positions (e.g., a lactone) are regarded as chlorins in this sense.

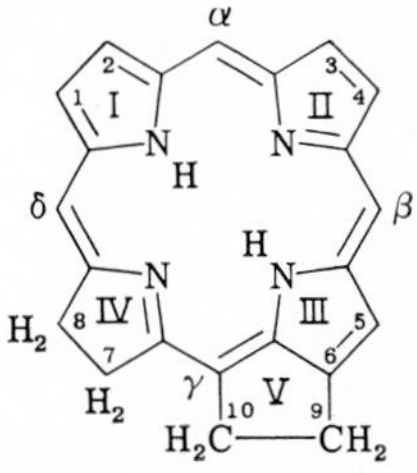

Phorbin
XIV

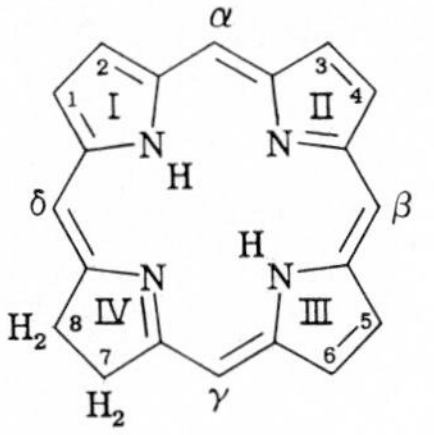

Chlorin
XV

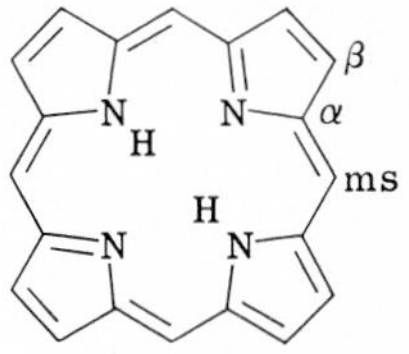

Porphin
XVI

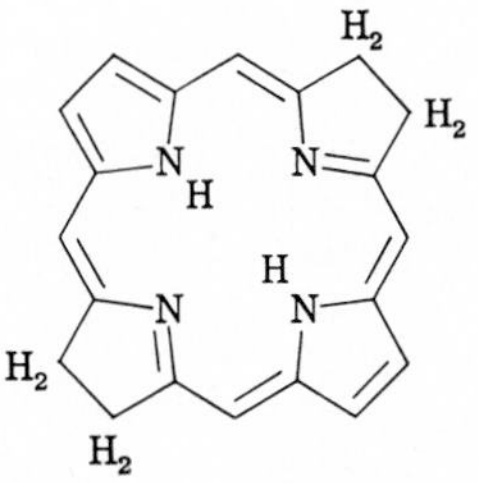

XVII

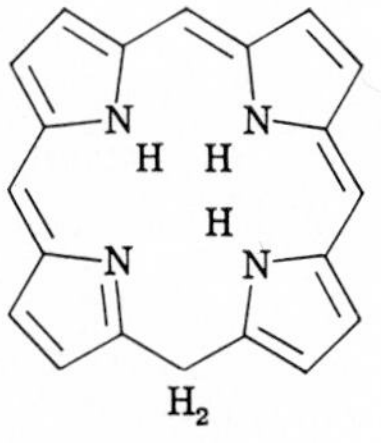

XVIII

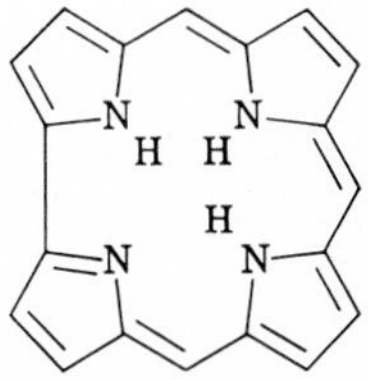

XIX

Corresponding to chlorins (in the narrow sense), rhodins, and phorbins are *chloroporphyrins, rhodinporphyrins,* and *pheoporphyrins.*

The Mg^{++} complexes of chlorins are sometimes called *phyllins;* the $FeCl^{++}$ complexes, *hemins.* Other metal complexes are named as derivatives of pheophytin, chlorin, etc.

Pheophytin, Chlorophyllide, and *Pheophorbide* (a or b) are related to chlorophyll as shown below.

$$\text{chlorophyll} \xrightarrow{\; -\text{phytol} \;} \text{chlorophyllide}$$

$$\Big\downarrow -\text{Mg} \qquad\qquad \Big\downarrow -\text{Mg}$$

$$\text{pheophytin} \xrightarrow{\; -\text{phytol} \;} \text{pheophorbide}$$

In the names of chlorophyll derivatives, a few prefixes have a special significance.

meso indicates reduction of the 2-vinyl group to 2-ethyl. With simpler porphyrins, it indicates substitution at a methine bridge carbon, e.g., *meso* (or *ms*)-tetraphenylporphin.

pyro indicates loss of the C-10 carboxymethyl group.

bacterio indicates derivatives of bacteriochlorin (XVII) or bacterio-chlorophyll (V), i.e., 3, 4, 7, 8-tetrahydroporphins.

proto indicates the more immediate derivatives of protochlorophyll —not to be confused with the "proto" of protoporphyrin, the porphin of heme.

Letters and numbers

a, indicating derivatives of chlorophyll *a,* is often considered understood and omitted in practice, except in pheoporphyrin a_5, etc.

b, appended to the name of a chlorophyll derivative indicates the presence of a 3-formyl group or some recognizable derivative, e.g., chlorophyll *b*-3-methanol or pheoporphyrin b_7 ($=$ pheoporphyrin a_5-3-carboxylic acid).

Chlorin e_6, rhodin g_7: the letters referred to the order of appearance in Willstätter's extractions (*11*). Other letters, similarly introduced, have fallen into disuse. The "e" and "g" are often retained in naming derivatives of these chlorins. The numerical subscripts denote the number of oxygen atoms in the molecule [cf. also pheoporphyrin a_5 (VIII)].

Purpurin 7, purpurin 18: so named by Conant from their HCl numbers (Section II, A, 2); by coincidence, purpurin 7 has 7 oxygen atoms, and Fischer extended the numbering to denote the oxygen content (purpurin 5, purpurin 3).

C. General Chemical Properties

Long years of evolution have selected the chlorophylls as the uniquely successful photosynthetic pigments. This success must be due to the chemical and physical properties of the ground state and of the lower electronic excited states. The properties of the ground state (e.g., charge distribution) determine the mode of aggregation of chlorophylls and their interaction with other components of the photosynthetic apparatus. The properties of the excited states must determine how chlorophyll can act with other molecules to transform light into chemical energy, without being degraded itself.

Chlorophyll can be compared to a tadpole, with a porphyrin "body" and a phytol "tail." The "tail" makes up about one-third of the molecule by weight. The lipophilic phytol group (VI), or farnesol in chlorobium chlorophylls, is present in every functional chlorophyll except, perhaps, Chl c. It is probably responsible for the association of chlorophylls with lipoproteins or carotenoids in nature.

Because of the phytol group, chlorophyll is more soluble than chlorophyllide in hydrocarbon and chlorocarbon solvents, such as benzene and CCl_4. The aggregation that occurs in these solvents is probably through the porphyrin "body," the phytol "tails" projecting into the surrounding solvent. On the other hand, chlorophyll is less soluble than chlorophyllide in polar solvents, such as nitromethane, and mixtures containing water. Here, aggregation is probably assisted by the phytol "tails."

Ring V and the 7-propionic ester group constitute the most hydrophilic part of chlorophyll. In a monolayer over water, chlorophyll sits with these groups against the water, the phytol group and the rest of the porphyrin part folding into a V-shape at an angle to the water surface (see Chapter 8) (40).

All photosynthetically functional chlorophylls are magnesium complexes. The bonding of the Mg^{++} to the nitrogens of the porphyrin ring is probably more nearly ionic than covalent, and the metal retains the capacity to bind to electron donors on either side of the plane of the porphyrin ring. Chlorophyll can thus bind water or polar organic solvents and form aggregates by bonding between the Mg of one ring and the 9-keto group of another (29, 41).

The π-electronic energy of a porphyrin ring is minimal for the planar configuration, and therefore it is unaffected, *to the first order*, by displacement from planarity. If the porphyrin ring is highly substituted, the planar configuration is not necessarily the one of minimum *total* energy, and a certain amount of "puckering" of the ring into the state of lowest total energy might be expected. Such puckering has been found for certain etioporphyrins and tetraphenylporphin (42, 43). It would be of great interest to know to what extent puckering exists in the chlorophylls, and in particular, how far Mg is forced out of the "plane." There is evidence that could be interpreted to mean that the Mg is in an unsymmetrical environment: monosolvates are much more stable against dissociation than disolvates (44, 45), and the formation constants for mixed solvates in binary solvent systems are greater than chance would predict (45).

Corwin et al. have shown that closure of ring V introduces strain into the (planar) chlorin ring, particularly into the bond between ring III and C-γ (46). However, the strain is not such as to prevent easy closure of the ring under basic conditions (Section II, E, 2, e).

Chlorophyll owes its optical activity (8, 47) to the asymmetric carbon atoms C-7, C-8, and C-10. Steric repulsion probably forces the 10-carboxymethyl group to the side of the porphyrin ring opposite the 7-propionic acid phytol ester. The pigments chlorophyll a′ and chlorophyll b′, interconvertible with chlorophylls a and b by heating (48), may be C-10 epimers of the natural pigments (49).

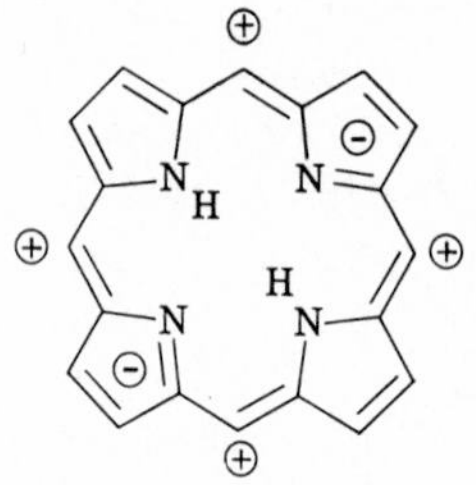

XX 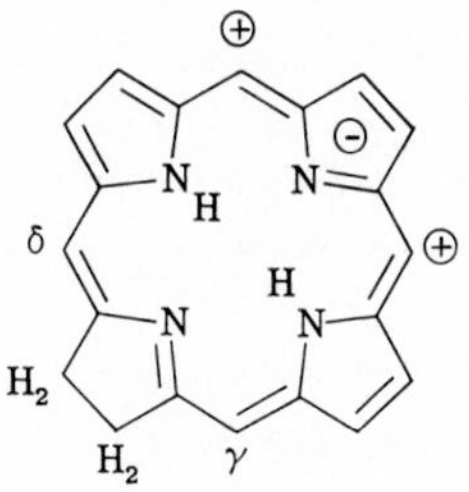   XXI

The application of the rule, that aromatic systems with $4n + 2$ electrons are stable, has led to a new appreciation of the reactivity of the conjugated porphyrin ring. According to Woodward's argument, pyrrolenine rings in porphyrins, such as rings II and IV in (XX) and ring II in (XXI), should attract electrons from neighboring methine bridges, producing the net charge distributions shown (50). Bridge positions not next to a pyrrolenine ring, such as the γ and δ carbons of (XXI), should therefore be more susceptible to electrophilic attack, as nuclear magnetic resonance investigations have proved (50, 51). Bacteriochlorophyll lacks rings of pyrrolenine structure, and all its methine bridge positions are subject to electrophilic attack (51).

D. Characterization of Chlorophyll Derivatives

The main difficulty in characterizing chlorophylls and their derivatives is to distinguish between compounds differing only in the nature of some of the peripheral groups, for example, the thirteen chlorobium chlorophylls. No single technique is capable of this.

Among the chemical characterizations that have been developed by

Willstätter, Fischer, and their colleagues are the following. Their importance has diminished in recent years, relative to that of physical characterization.

Phytol number: The weight per cent phytol content, obtained by saponification, distinguishes chlorophylls from chlorophyllides, pheophytins from pheophorbides (*11*, p. 280).

Silver iodide number: The weight of AgI found in the Zeisel determination of methoxyl or ethoxyl by solvolysis with HI, expressed as percentage of the weight of the chlorophyll, also distinguishes between phytol and methyl or ethyl phorbides (*11*, p. 164).

Zerewitinoff determination of active hydrogen: Measurement of the methane evolved when a chlorophyll derivative is treated with CH_3MgI. The imino hydrogens, unesterified carboxyl groups, and enolizable hydrogen at C-10 count as active (*52*).

Molisch phase test: Appearance of an impermanent brownish color at the interface between concentrated methanolic KOH and an ether solution of chlorophyll derivatives that have activated hydrogen at C-10 (see Section II, E, 1) (*11*, *53*). As a test for allomerization, it is qualitative only.

Hydrochloric acid number: The weight per cent of hydrochloric acid in a solution that will extract two-thirds of a chlorophyll derivative from an equal volume of ether (*11*, p. 243).

Among the physical methods employed, the *melting point* is sharp and characteristic for esters (other than phytol), but derivatives with unesterified acid groups may decompose before melting.

The *crystalline form* is often characteristic, but mixed crystals are easily formed.

The *visible spectrum* is perhaps the most familiar characteristic of chlorophyll derivatives. Fischer and Stern reported band positions in the visible spectrum of most chlorophyll derivatives (*31*), and Stern *et al.* have published and interpreted traces for a large number of chlorophyll (*54–59*) and bacteriochlorophyll (*60*) derivatives.

The effect of substituents on the visible spectrum is very complicated, but it is worth remarking that electron attracting groups at the 2-position ($—CH = CH_2$, —CHO, —COOH, $—COCH_3$), the 6-position (ring V, —COOR), and the γ-position (purpurins), have a bathochromic effect on the spectrum of chlorins, whereas such substituents at the 3-position have a hypsochromic effect. Substituents not directly attached to the conjugated system have little influence on the spectrum; the spectra of chlorophyll *a*, chlorophyllide *a*, 10-hydroxychlorophyll *a*, and pyrochlorophyll *a* are nearly indistinguishable. The spectra are further influenced by the solvent, the bands being shifted toward the red by solvents of high refractive index or dielectric constant (*61*).

The *infrared spectrum* and the *nuclear magnetic resonance spectrum* are proving to be useful tools for structural analysis of chlorophylls, as discussed in Chapter 7.

The *optical activity* has been measured and tabulated for a large number of derivatives of chlorophyll (8, 47, 62) and bacteriochlorophyll (63). The difference in optical activity, measured with red light and with white, has been known for some time (62), and recently attempts have begun to determine the *optical rotatory dispersion* (63a) (see Chapter 13).

The *chromatographic mobility*, or R_f value, is of course of prime importance in the separation and identification of chlorophyll derivatives (see Chapter 2) (64).

II. Chemistry of Functional Groups

A. The Central Metal Ion and Pyrrole Nitrogens

1. PHEOPHYTINIZATION

The central Mg ion of chlorophylls is readily displaced by strong and weak acids (1). In aqueous acetone, the rate of *pheophytinization* is first order in acid concentration (65) and in chlorophyll concentration (66). Schanderl et al. found that specific rate constants for Mg displacement in 20% aqueous acetone decreased in the order: chlorophyllide > methyl chlorophyllide > ethyl chlorophyllide > chlorophyll, in both the *a* and *b* series, but by less than a factor of 2 (67). Rate constants for the *a* series, however, were five or six times as large as for the *b* series. Activation energies in the *a* series were 10.4–10.8 kcal; they were somewhat greater in the *b* series, but curvature of the Arrhenius plot made interpretation of the latter difficult. These values are within the range of those reported earlier by Mackinney and Joslyn (68). Loss of Mg from chlorophyll is 13 times as fast in a monomolecular layer at an air-water interface (pH 4) as in aqueous acetone; the rate in the monolayer is sensitive to surface pressure and the presence of O_2, Ca^{++}, and Mg^{++} (69).

Loss of Mg is rather more rapid from "bacterioviridin" (chlorobium chlorophylls) than from chlorophyll *a*, and the rates for protochlorophyll and bacteriochlorophyll are between those for chlorophylls *a* and *b* (70). Photoreduced chlorophyll loses Mg readily in the presence of weak acids, in aqueous or alcoholic pyridine solutions (71, 72).

2. PROTONATION

The two nitrogens of metal-free porphyrins which do not bear hydrogens are more strongly basic than the other two [cf. structures (XX) and

(XXI)]. The ability of products of degradation of chlorophyll to form water-soluble mono- and diprotonated species enabled Willstätter to separate them by differential extraction out of ether with HCl of increasing strength and to characterize them by their hydrochloric acid number (73). Willstätter and Fischer give the impression that the basicity of the porphyrin is the only factor determining the HCl number (11, 31). But as Conant et al. have pointed out (74), other parts of the molecule are also important in determining the partition of a porphyrin between aqueous and ethereal phases.

TABLE I

HYDROCHLORIC ACID NUMBERS OF VARIOUS CHLOROPHYLL DERIVATIVES

Compound	HCl No.	Reference
Bacteriopheophytin	> 37	(75)
Methyl bacteriopheophorbide	~ 32	(75)
Pheophytin b	35	(11, p. 244)
Methyl pheophorbide b	21	(11, p. 244)
Pheophytin a	29	(11, p. 244)
Methyl pheophorbide a	16	(11, p. 244)
Pheophytin b-3-methanol	25	(76)
Methyl pheophorbide b-3-methanol	14	(76)
Bacteriochlorin e_6 trimethyl ester	16–17	(31, p. 318)
Rhodin g_7	9	(11, p. 244)
Rhodin g_7 trimethyl ester	~ 12–13	(31, p. 266)
Chlorin e_6	3	(11, p. 244)
Chlorin e_6 trimethyl ester	~ 8	(31, p. 98)
γ-Phylloporphyrin methyl ester	0.9	(77)

As shown in Table I, the HCl number increases with esterification and with the size of the esterifying alcohol (phytol > methanol). HCl numbers are larger in the b series than in the a series, and larger yet among bacteriochlorins.

Conant et al. have potentiometrically titrated several porphyrins with perchloric acid, in acetic acid as solvent (74). Their results, a few of which are listed in Table II, show the disparate basicity of the two pyr-

TABLE II

BASICITY OF CHLOROPHYLL DERIVATIVES[a,b]

Compound	pK_1	pK_2	pK_3
Phylloporphyrin methyl ester	$+ 2.3$	$+ 2.3$	$- 2.0$
Methyl pheophorbide a	$+ 1.9$	$- 1.4$	$- 2.3$
Methyl pheophorbide b	$+ 0.3$	$- 1.7$	$- 2.3$
Chlorin e_6	$+ 1.9$	$+ 0.3$	$- 2.2$
Rhodin g_7	$+ 1.9$	0.0	$- 2.0$

[a] After Conant, et al. (74).

[b] $K = \dfrac{[B] [H^+]}{[BH^+]}$, taken from midpoint of titration curve.

rolenine nitrogens (K_1 and K_2) among the chlorins, and the weakening influence of the 3-formyl group in the b series. The HCl number (Table I) correlates qualitatively with the value of K_2.

3. FORMATION OF METAL COMPLEXES

a. Magnesium. Magnesium is easy to remove from the porphyrin ring but hard to replace. Early attempts to introduce Mg into chlorins with Grignard reagents encountered attack on ring V and on the formyl group in the b series (78, 79). Magnesium is introduced more smoothly and in better yield by first treating the Grignard reagent with an alcohol. This method has been applied to pheophytin (80) methyl pheophorbide *a* (81), methyl pheophorbide *b* (82), pheoporphyrin a_5 (80), and ethyl bacteriopheophorbide (19), but only in poor yield with the last. Noack and Kiessling succeeded in replacing Mg in protopheophytin with the aid of the Grignard reagent $p-(CH_3)_2NC_6H_4MgI$ (20, 21). More recently, Mg phenoxide (83), Mg viologen, Mg (pyridine)$_6$ I$_2$, and the Mg complex of 4, 4' dipyridyl (84) have found application in the synthesis of Mg porphins and chlorins.

Magnesium can be introduced into porphins, but not into chlorins, by heating them in pyridine with salts, such as $MgBr_2$(85) or Mg (ClO_4)$_2$ (86). Magnesium complexes of porphins are more stable than those of the corresponding chlorins (83), in keeping with the greater base strength of porphins (74). It is noteworthy that Mg is introduced biochemically into protopheorphorbide before reduction to the chlorin level.

b. Other Metals. Zinc and copper are readily introduced by warming a solution of the chlorin and the acetate of the metal. These metals can be removed only with strong acid (31, p. 332).

Iron is introduced into chlorins by warming with ferrous acetate in acetic acid solution (9, 87). In the absence of air the ferrous complex is formed, but this is rapidly oxidized in the presence of air to the ferric form (88, 89). The spectra of ferrous complexes are normal, but the principal red band in the spectra of ferric complexes lies at unusually short wavelengths, at 620 mµ for Fe^{III} pheophorbide in anhydrous ethanol, as compared with 654 mµ for the ferrous complex (89, 90). The bands of ferric and ferrous chlorin e_6 are at 605 mµ and 630 mµ in ethanol (91). Ferric pheophorbide in solvents containing water forms species absorbing around 675 mµ (90).

Ferric chlorins, like other hemins, form characteristic compounds with organic bases and ions, such as CN$^-$, F$^-$, and CNS$^-$ (31, 92). The ferric complexes of chlorin e_6, pheophorbide *a*, and bacteriochlorin e_6 catalyze the chemiluminescence of luminol (93).

4. SOLVATION

The coordinating power of Mg is not satisfied by bonding to the porphyrin ring, and chlorophylls form solvates with one or two equivalents of Lewis bases such as water, amines, alcohols, ketones, or ethers (44). The formation of solvates, or change from one solvate to another, can be detected by changes in fluorescence yield (94), or in the visible (45, 94), infrared (30, 95), and nuclear magnetic resonance (29, 95, 96) spectra.

Chlorophyll as normally prepared retains 0.5–2 moles of water (97), probably bound to the Mg. Dissolved in solvents without distinctly basic properties, such as benzene, cyclohexane, and CCl_4, chlorophyll retains 1 mole of water. If the water is removed by rigorous drying, chlorophyll forms dimers or higher aggregates in which the Mg of one molecule is "solvated" by the 9-carbonyl of another (29, 30).

Water is bound more firmly to chlorophyll b than to chlorophyll a. Whereas chlorophyll a lost all water on heating for 1.5 hour at 60° at 3×10^{-5} mm pressure, chlorophyll b retained 3.7 moles of water under slightly more severe conditions (98).

Chlorophyll and pheophytin form molecular complexes with $MgCl_2$, $BeCl_2$, $AlCl_3$, $BiCl_3$, $NbCl_5$, $AgClO_4$ (99), $FeCl_3$ (100), and $SnCl_2$ (101). These may be similar to the "sitting-atop" complexes found for protoporphyrin by Fleischer and Wang (102). Their structure is uncertain, but perhaps involves bonding of the metal to the imino nitrogens of porphyrin. These complexes may be intermediates in the insertion of metal ions into pheophytin (99) and protoporphyrin (102) and the displacement of Mg by $FeCl_3$ in chlorophyll (100).

B. The 7-Propionic Acid Group

1. CHLOROPHYLLASE

The only direct way to convert chlorophyll to chlorophyllide, or to an ester of chlorophyllide other than phytyl, is with the aid of an enzyme, chlorophyllase. Enzymes of this class, apparently universal in green plants and photosynthetic bacteria, have the natural function of attaching phytol to chlorophyllides. The enzyme occurs in especially high concentrations in the leaves of certain plants, among them *Heraclium spondylium*, *Datura stramonium*, and *Beta vulgaris*. Chlorophyllase was first identified by Willstätter and Stoll (2, 103); the scope of its activity has been studied by Fischer and Lambrecht (104), and its properties have been reviewed recently by Holden (105) and by Egle (106).

In aqueous acetone or ether, chlorophyllase catalyzes the hydrolysis of 7-propionic acid esters; in aqueous methanol or ethanol, it catalyzes transesterification. Chlorophyllides *a* and *b*, methyl chlorophyllides, or ethyl chlorophyllides can be isolated directly from plants rich in chlorophyllase by first allowing the leaf meal to digest with aqueous acetone, methanol, or ethanol (*2*). The leaf meal of plants rich in the enzyme can be used repeatedly to catalyze hydrolysis or esterification of chlorophyll derivatives. Willstätter and Stoll reported that chlorophyllide could be esterified with phytol to regenerate chlorophyll *in vitro* (103), but Klein and Vishniac were unable to repeat this under somewhat different conditions (*107*).

Chlorophyllase is active with pheophorbides and chlorophyllides of the *a* and *b* series (*2*), and with bacteriochlorophyll (*19*), but not with chlorins and pheoporphyrin a_5 (*104*). Purpurin 7 trimethyl ester is hydrolyzed in acetone to the dimethyl ester. The only definitely established structural requirement is that the 7, 8-positions be reduced. The 10-carboxymethyl group may be necessary for activity of the chlorophyllase from higher plants, because this enzyme is inactive with pyropheophorbide (*104*) and with chlorobium chlorophylls.

There has been some question whether the 7-propionic acid group is esterified before or after biosynthetic reduction of protochlorophyll (see Chapter 14). The inability of chlorophyllase to act on porphins supports the position that reduction precedes esterification.

2. NONENZYMATIC HYDROLYSIS AND ESTERIFICATION

Esters of the 7-propionic acid group in chlorophyll derivatives are rapidly hydrolyzed by cold concentrated HCl, more rapidly than esters of other acid groups (*108*). Thus, pheophorbides *a* and *b* can be prepared from the pheophytins (*11*, p. 249), and the chlorin e_6 dimethyl ester from the trimethyl ester (109). The 7-propionic acid group is also the one most easily esterified by cold methanolic HCl (*110*). Pheophytin in thus converted almost quantitatively into methyl pheophorbide (*111*). These reactions are obviously not applicable to Mg-containing derivatives.

Pheophorbide *a* has been esterified with higher alcohols, including phytol, by phosgene in cold pyridine solution (*112*). "Protopheophytin," prepared by this method from vinylpheoporphyrin a_5 and phytol, was indistinguishable from a product of oxidation of pheophytin (see Section II, G, 1) (*23*).

C. The Vinyl Group

Recognition that chlorophylls *a* and *b* have a vinyl substituent at the 2-position came rather late, largely because it is converted to ethyl during the pheoporphyrin reaction (Section II, F, 2), the route favored by

Fischer for conversion of chlorophyll derivatives into porphins. Only in 1935 was it reported that pheophorbide *a* reacts with diazoacetic ester like other vinyl porphyrins, and that after chromic acid oxidation, one of the products was the same as (XXII) obtained from oxidation of the adduct of protoporphyrin and diazoacetic ester (*113*).

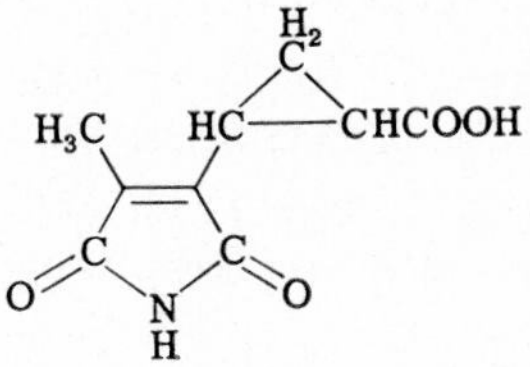

XXII

Under neutral or alkaline conditions the vinyl group reacts as though it were independent of the conjugated porphyrin ring, but under acid conditions, it often reacts as an extension of the conjugated system.

1. REDUCTION

Upon cautious hydrogenation of chlorophyll *a* derivatives with Pd or PtO$_2$ in acetone or dioxane, the vinyl group can be saturated without excessive formation of leuco compounds (*114, 115*). Some reduction of the 3-formyl group also occurs with chlorophyll *b* derivatives (*116*). Spectra of the *meso*-compounds prepared this way resemble those of the parent compound, but the bands are displaced 5–15 mμ to the blue (*58*).

The vinyl group is also reduced to ethyl by hydrazine hydrate at 60°, but side reactions are common. For example, pheophorbide *a* is converted to mesochlorin e$_6$ 6-hydrazide (*117*).

2. OXIDATION

The vinyl group of chlorins and phorbins is oxidized by KMnO$_4$ in aqueous pyridine to give successively glycol, formyl, and carboxylic acid substituents (*118*). Chlorophyllides and pheophorbides are also attacked

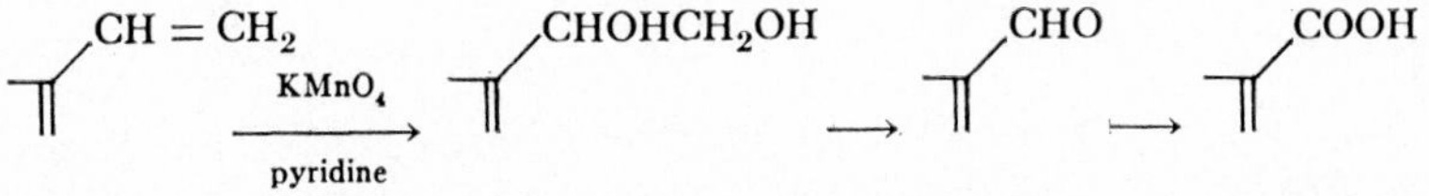

at ring V under these conditions, but in acetone, not containing base, it is possible to confine oxidation to the vinyl group (*119*). 2-Devinyl 2-formyl chlorophyll *a* prepared this way is spectrally identical with chlorophyll *d*, a pigment found in certain red algae (*119*).

3. Addition

Hydrobromic acid adds to the vinyl group to give the α-bromoethyl derivative; this is converted by hydrolysis in 15% HCl to the α-hydroxyethyl derivative, and this, by oxidation with $KMnO_4$ in pyridine, to the 2-acetyl derivative (120–122). 2-α-Hydroxyethyl derivatives revert to vinyl by loss of water on heating in high vacuum (121, 123).

Vinyl pheo- or chloroporphyrins react with dilute HI in acetic acid at room temperature to give 2-α-hydroxyethyl derivatives (23); vinyl chlorins, however, give the 2-acetyl derivative of the corresponding porphin (124, 125). The latter reaction, called the *oxo reaction* by Fischer, presumably takes place by addition of HI, hydrolysis, and oxidation, but it is not clear at what stage dehydrogenation to porphin occurs. Examples are the conversions of pheophorbide *a* to oxopheoporphyrin a_5 monomethyl ester, and chlorin e_6 trimethyl ester to oxochloroporphyrin e_6 trimethyl ester (125). The reaction provided an

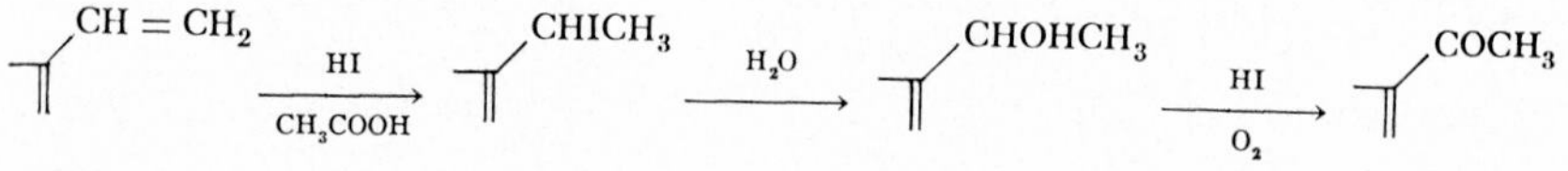

important clue to the presence of an unsaturated side group in chlorophyll (126).

Chlorophyll *b* derivatives are apparently not susceptible to the oxo reaction, but the 2-acetyl derivatives of rhodin, for example, can be obtained by the HBr-addition path above (127).

4. Elimination

The vinyl group is lost when the chloroferric complex of a chlorin or a phorbin is heated in a resorcinol melt. Oxidation to porphin occurs as a side reaction. Examples are the conversion of methyl pheophorbide *a* hemin to 2-devinyl pyropheophorbide *a* hemin methyl ester, and of pyropheophorbide *b* hemin methyl ester to the 2-devinyl 3-deformyl derivative (87).

D. Carbonyl Groups

The carbonyl substituents of naturally occurring chlorophylls are the 9-keto group common to all, the 3-formyl group of chlorophyll *b*, the 2-formyl group of chlorophyll *d*, and the 2-acetyl group of bacteriochlorophylls. Also of interest are carbonyl substituents of some chlorophyll derivatives, such as that of 6-formyl mesoisochlorin e_4 [cf. structure (XLIX)], the γ-formyl group of purpurins 3 and 5, and the γ-COCOOH of purpurin 7. Carbonyl substituents provide points for attachments of various groups to chlorophylls without grossly disrupting the molecule, and have been used in attempts at synthesis (see Chapter 5).

1. CONDENSATION

2-Devinyl 2-formyl chlorin e_6 trimethyl ester (*118, 119*) and methyl pheophorbide *b* (*128*) form 2- and 3-monoximes readily at room temperature. At 100°, methyl pheophorbide *b* forms the 3, 9-dioxime and methyl pheophorbide *a* the 9-monoxime (*125, 129, 130*). Acid hydrolysis of methyl pheophorbide *b* dioxime releases the 3-formyl group first, leaving the 9-monoxime of pheophorbide *b* (*130*).

Although oxime formation is general for chlorophyll *b* derivatives, reports of formation of phenylhydrazones, semicarbazones, etc., are rarer. However, the formation of the semicarbazone and the phenylhydrazone of chlorophyll *b* has recently been described, and the 3-oxime has been applied to analysis of mixtures of chlorophylls *a* and *b* (*131*). The ability of bases, such as phenylhydrazine, to open ring V (*80*) is a complicating factor. Girard's Reagent "T," $[NH_2NHCOCH_2N(CH_3)_3]^+$ Cl^-, forms water-soluble compounds with chlorophyll *b* derivatives, but not with chlorophyll *a* derivatives or purpurin 7. This has been suggested as a basis for separating chlorophyll derivatives of the *a* and *b* series (*132*).

Fischer *et al.* describe a condensation product of rhodin g_7 and *o*-phenylenediamine (*133*).

Neopurpurin reaction. The γ-formyl group of purpurin 3-monomethyl ester (XXV) or purpurin 5-dimethyl ester (XXIII) condenses under alkaline conditions with the 7-propionic acid group (*134*).

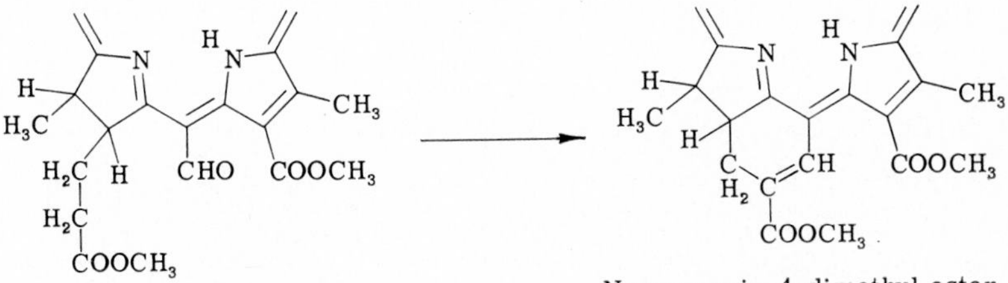

XXIII   →   Neopurpurin-4-dimethyl ester

XXIV

XXV   CH₃NO₂, pyridine + C₂H₅NH₂   XXVI

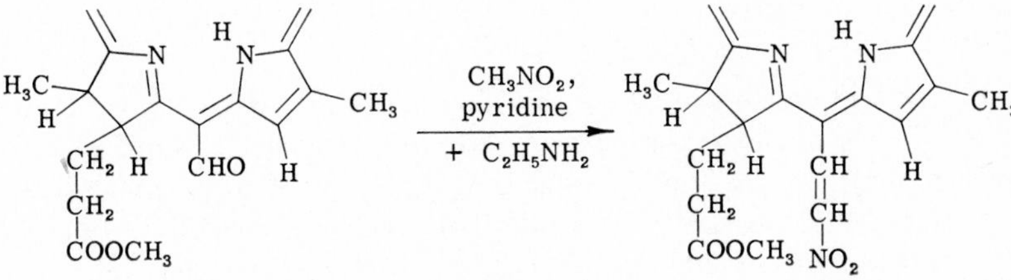

Purpurin 3-methyl ester (XXV) condenses with nitromethane under basic conditions to the γ-vinylnitro derivative (XXVI) (135). Strell has reviewed these and other condensation and addition reactions of γ- and 6-formyl porphins and chlorins (32).

2. ADDITIONS

a. Acetals. The apparently uncatalyzed addition of methanol to chlorophyll *d* and pheophytin *d* to form acetals (136) was instrumental in identifying them as 2-devinyl 2-formyl chlorophyll *a* derivatives (119). Formation of an acetal with methyl pheophorbide *b* is more difficult, but is accomplished with methyl orthoformate/HCl in methanol (128). Formation of acetal is marked by reversion of the spectrum to the type (59) characteristic of the *a* series.

b. Other Additions. Formyl groups in the 3-, 6-, and γ-positions readily add HCN, again with reversion of the spectra to the *a* type (32, 128).

Cysteine reacts with methyl pheophorbide *b* in pyridine to form an adduct, presumably with the structure (XXVII) (137). The adduct reverts to starting materials in 5% HCl, and has an *a*-type spectrum. It is oxidized by O_2 in the light with evolution of CO_2.

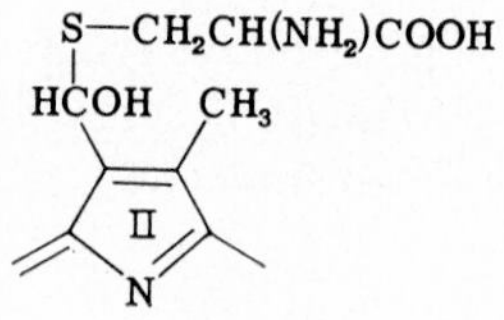

$$S{-}CH_2CH(NH_2)COOH$$

$$H\overset{|}{C}OH \quad CH_3$$

XXVII

3. REDUCTION

a. Sodium Borohydride. Methyl chlorophyllide *b* and other derivatives of the *b* series are reduced in two distinct steps, first to the 3-methanol derivative, then to the 9-hydroxy-3-methanol compound (XXVIII) (138). Chlorophyll *a* derivatives are reduced to the 9-hydroxy derivative by borohydride (138), and chlorophyll *d* is reduced to the 2-methanol derivative (119). The method appears to be of general utility.

b. Aluminum Isopropoxide. Fischer and Mittenzwei have reduced pheophorbide *a* and *b* derivatives with this reagent in isopropanol, including purpurin 7 trimethyl ester and 2-devinyl 2-acetyl-pheophorbide *a*, the last in connection with studies on bacteriochlorophyll (63, 123). The course of the reaction is like that of borohydride reduction.

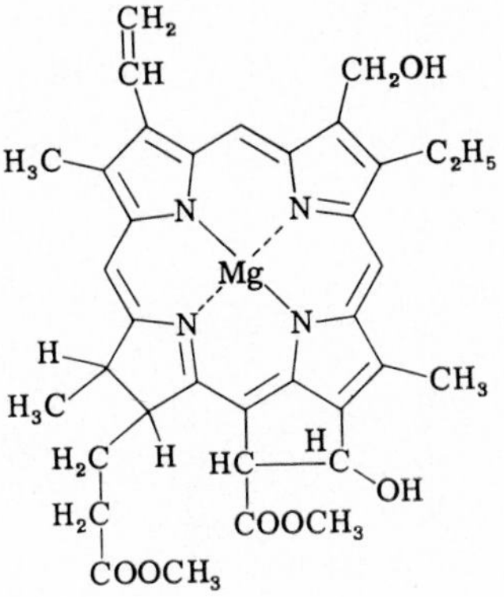

XXVIII

c. Wolff-Kishner. Fischer and Gibian have applied the Wolff-Kishner reduction (hydrazine hydrate + $NaOCH_3$, 120°, 8 hours) to pyropheophorbides. Groups are reduced in the order: vinyl, 3-formyl, 9-keto. Thus, pyropheophorbide *b* (XXIX) gave first mesopyropheophorbide *a*, then mesodeoxopyropheophorbide *a* (XXX) (*139*). Fischer and Gibian also noted that optically active pheophorbides were racemized by this treatment (*140*).

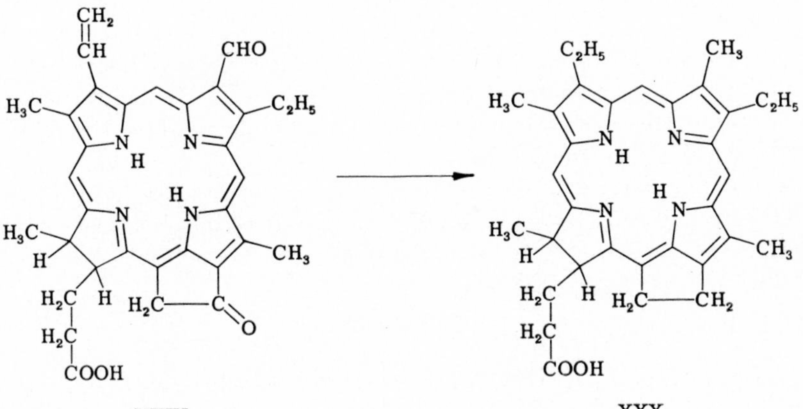

XXIX   XXX

4. OXIDATION

Pheophorbide *b* is oxidized to pheophorbide b_7 (XXXI) by HI/O_2 in acetic acid at 0° (*116*), conditions similar to those under which the oxo reaction occurs with pheophorbide *a*.

As mentioned earlier, 2-formylchlorins are oxidized to 2-carboxylic acids by $KMnO_4$ in pyridine (*118*).

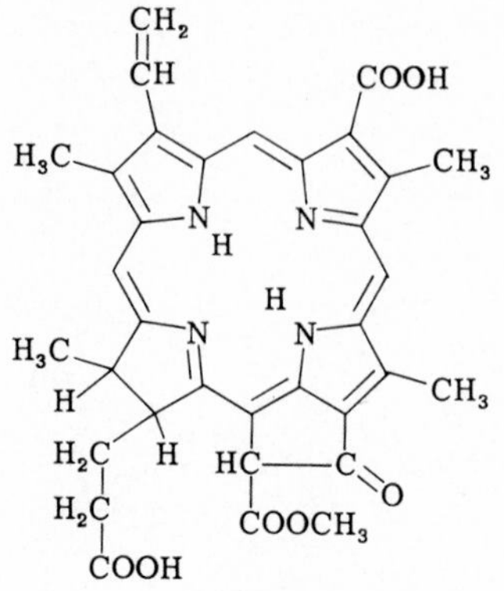

XXXI

E. Ring V

So far as the reactions of ring V are concerned, chlorophyll is a chlorin-substituted β-keto ester, with an activated hydrogen at C-10. It is therefore subject to enolization and other reactions under basic conditions.

The extent to which ring V exists in the enol form has been much debated in the past, but according to recent interpretations of infrared spectra, it is ordinarily almost entirely in the keto form (30). Closure of ring V introduces strain into the chlorin ring, according to the analog models of Corwin et al. (46, 83), and the further strain that would be entailed by a double bond between C-9 and C-10 probably militates against the presence of much enol under nonbasic conditions. The C-10 hydrogen does exchange with solvent methanol, even in neutral solution, and rate constants have been measured by a nuclear magnetic resonance technique (51). Exchange is much accelerated in the presence of base, and in pyridine, even the C-10 hydrogens of pyrochlorophyll are exchanged at an appreciable rate.

1. ENOLIZATION

Strong base (e.g., KOH) abstracts the C-10 hydrogen of chlorophylls and pheophorbides, leaving a yellow to reddish brown species which is presumably the enolate ion (141). The color change is the basis of the *phase test*, which has been traditionally used to determine whether or not a sample of chlorophyll had become allomerized (142).

The phase test is commonly run by layering 30% KOH in methanol under an ether solution of the chlorophyll (31, p. 331). The color of the enolate ion or "phase test intermediate," appears at the boundary between the two liquids. Under these conditions, the "phase test intermediate" is soon saponified, and the green chlorin or phyllin is extracted into the alcohol layer. In pyridine or dimethyl formamide, the intermediate is much more stable and can be kept indefinitely in the absence of oxygen (*138, 143*).

Phase test intermediates have been generated from chlorophylls *a* and *b* by reaction with isopropylamine at − 113° (*144*) and as a secondary product of electrolytic reduction in dimethyl sulfoxide (*145*).

A positive phase test, or better, quantitative conversion to the intermediate in an aprotic solvent, is considered proof of the presence of the 9-keto, the 10-hydrogen, and the 10-carboxymethyl groups. Pheophorbides *a* and *b*, protochlorophyll, bacteriochlorophyll (*146*) and pheoporphyrin a$_5$ (*147*) give the phase test, but pyrochlorophylls, 9-oximes, chlorins, and allomerized chlorophylls do not. Conant and Moyer, however, noted that chlorin e$_6$ trimethyl ester slowly developed the phase test color, owing to base catalyzed recyclization to pheophorbide (*148*).

The first recorded spectra of phase test intermediates were obtained by a flow technique (*149*). Weller and Holt have reported spectra of the intermediates of chlorophyll *a* (*138, 143*) and chlorophyll *b* (*143*) in pyridine, and of pheophytin *a* in dimethyl formamide (*138*). Holt remarked that the spectrum of the phase test intermediate of pheophytin *a* depended on the base [KOH or Mg(OCH$_3$)$_2$] used. Spectra of the phase test intermediates of chlorophyll *a*, chlorophyll *b*, pheophytin *a*, and pheophytin *b*, prepared by adding tetrabutylammonium hydroxide (1 *M* in methanol) to a solution of the pigment in pyridine, are shown in Figs. 1–4.

The profound alteration in the spectrum on making the phase test intermediate suggests that its negative charge is not confined to the oxygen of the enolate ion but is distributed over the entire conjugated system (*143*). As the pyrrolenine rings in porphyrin systems tend to attract electrons (*50*), it is possible that the resonance structure (XXXII)

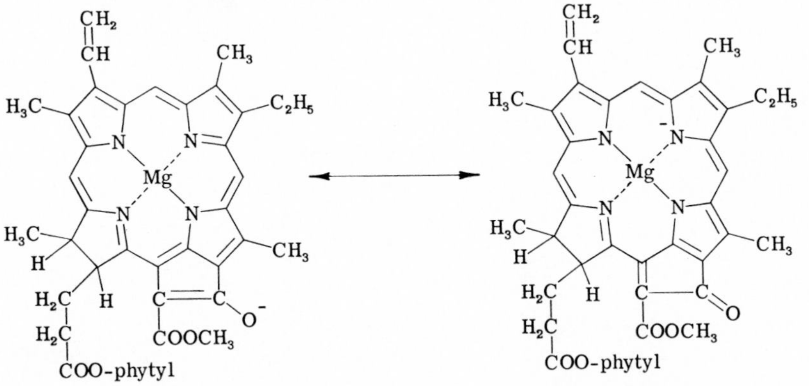

XXXII

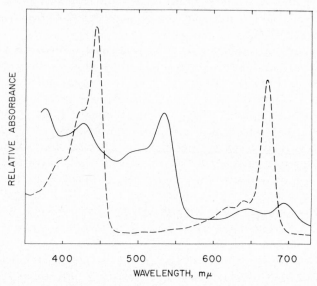

Fig. 1.   Spectra of chlorophyll *a* (broken line) and its phase test intermediate (solid line) in pyridine.

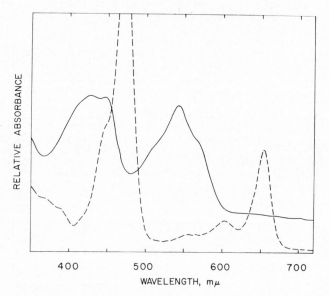

Fig. 2.   Spectra of chlorophyll *b* (broken line) and its phase test intermediate (solid line) in pyridine.

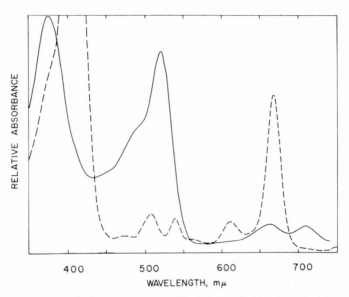

FIG. 3. Spectra of pheophytin *a* (broken line) and its phase test intermediate (solid line) in pyridine.

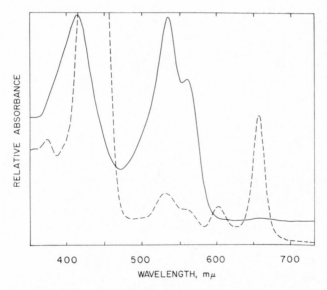

FIG. 4. Spectra of pheophytin *b* (broken line) and its phase test intermediate (solid line) in pyridine.

makes an important contribution to the state of the ion. In this structure, the conjugated system no longer makes a closed loop, and an altered spectrum might be expected.

2. SOLVOLYSIS

a. Saponification. Alkaline hydrolysis of chlorophyll produces Mg chlorin e_6, also known as phyllin e_6[°] and chlorophyllin (5, 73, 138, 150). Chlorophyllin is soluble in water, but tends to form aggregates (150).

Saponification of pheophorbides *a* and *b* by KOH/CH$_3$OH gives chlorin e_6 and rhodin g_7, respectively (4, 9). These compounds have proved of central importance to studies of degradation and synthesis of chlorophyll. Saponification is often accompanied by allomerization, decarboxylation, and dehydrogenation to porphins. Most favorable for hydrolysis is a brief reaction in hot, dilute KOH, dissolved in methanol rather than in a higher alcohol (9, 79, 148).

b. Methanolysis by Diazomethane. A solution of diazomethane in methanol containing pyridine converts pheophorbide *a* into chlorin e_6 trimethyl ester, on standing for a day under nitrogen (151, 152). Pheophorbide *b* is similarly converted to rhodin g_7 trimethyl ester (153), pheoporphyrin a_5 to chloroporphyrin e_6 trimethyl ester (154), and methyl bacteriopheophorbide to bacteriochlorin e_6 trimethyl ester (155). As with the phase test, this reaction requires an activated hydrogen, and pyropheophorbide does not react (152).

c. Aminolysis. In reactions similar to hydrolysis, chlorophyllides and pheophorbides react with ammonia, primary, amines, and secondary amines to form 6-carboxamides (80, 127, 140). Spectra of the amides

resemble those of the corresponding acids and esters. Pheoporphyrin a_5 also reacts with piperidine (151), ammonia, and phenylhydrazine (80), to form chloroporphyrin 6-carboxamides.

Weller and Livingston measured rates of aminolysis of chlorophylls *a* and *b* with piperidine, isoamylamine, isobutylamine, *sec*-butylamine, and phenylhydrazine (156). The reaction was first order in chlorophyll, and chlorophyll *b* reacted more rapidly than chlorophyll *a*. There was a good correlation between reaction constant and base strength, but steric effects were also important.

[°] The spectrum cited by Fischer (31, p. 93) under this name is for an improperly prepared compound; for the correct spectrum, cf. Holt (138).

d. Esterification. Chlorins are stabilized against decarboxylation by esterification, usually effected with diazomethane (*31*, p. 329). Esterification with diazomethane is rapid and complete, but stepwise esterification is possible with HCl/CH_3OH at 0°. Thus, chlorin e_6 gave the dimethyl ester with the 6-carboxylic acid group still free after 5 hours (*157*). Carboxylic acid groups are esterified in the order: 7-propionic acid, γ-acetic acid, 6-carboxylic acid (*110*).

e. Ring Closure. Hydrolysis may be reversed and ring V re-formed by heating esters under anhydrous basic conditions. Thus, chlorin e_6 trimethyl ester and rhodin g_7 trimethyl ester gave methyl pheophorbides *a* and *b* on heating with KOH in pyridine and methanol (*141, 158*), and chloroporphyrin e_6 trimethyl ester gave pheoporphyrin a_5 dimethyl ester on boiling in pyridine with soda (*10, 22*). Re-formation of ring V by this reaction makes synthesis of chlorin e_6 tantamount to synthesis of chlorophyll, as introduction of phytyl and magnesium are feasible steps.

3. Allomerization

Willstätter *et al.* noted that, after standing in alcohol solution in contact with air, chlorophyll lost the ability to give the phase test (*3, 142*). Allomerization, as they called the process, has continued to plague researchers of chlorophyll, because some of the products are almost indistinguishable spectrally from chlorophyll, and even a positive phase test does not mean that a large part of the chlorophyll is not allomerized. The products of allomerization vary with the circumstances, but they all have oxygen, instead of hydrogen, bonded to C-10.

Conant *et al.* characterized "unstable" chlorins and purpurins as products of KOH-catalyzed allomerization of pheophorbides in propanol, or in methanol at − 10° (*148, 159*). They found that 1 mole of O_2 was consumed per mole of chlorophyll allomerized; they also introduced potassium molybdate as an allomerizing agent (*160, 161*). Fischer *et al.* studied allomerization in air, and by benzoquinone, I_2, and $KMnO_4$ in the absence of air (*81, 82, 85, 152, 154, 162–169*). The bleached product of the oxidation of chlorophyll by $FeCl_3$ in methanol is allomerized in the presence of air (*170*). Pheophytin is much more resistant to allomerization by O_2 in alcohol than is chlorophyll (*169*).

Allomerization during extraction of chlorophyll is inhibited by making the solvent slightly acidic.

a. By Oxygen. According to Conant *et al.* (*148, 159*) the first products of allomerization of methyl pheophorbide *a* (XXXIII) in the presence of hydroxide are "unstable" chlorin acid (XXXIV) and ester (XXXV); these, on standing, go over to the purpurin 18 (XXXVI) and purpurin 7 trimethyl ester (XXXVII). Both "unstable" chlorin and its monomethyl ester are converted to purpurin 7 trimethyl ester by esterifi-

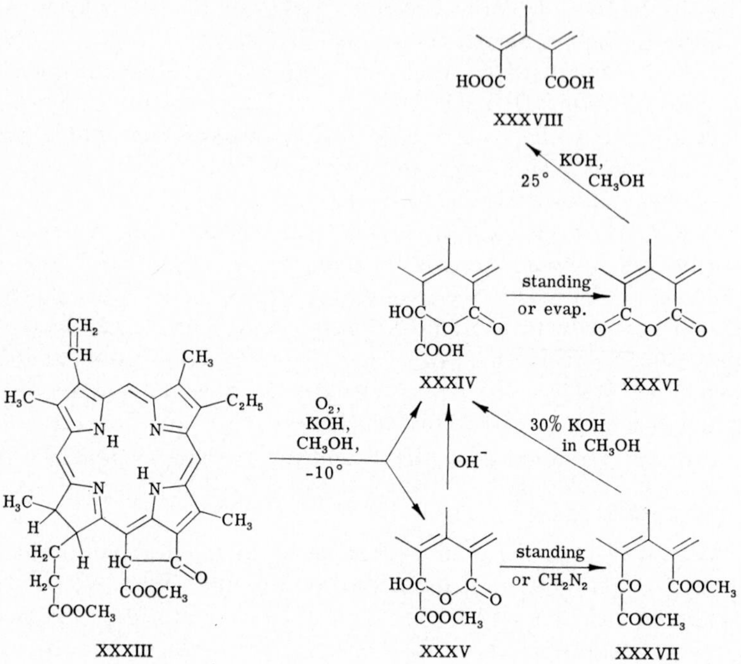
HOOC COOH
XXXVIII
KOH, 25° CH3OH
standing or evap.
HOC O C O
COOH
XXXIV
O C O C O
XXXVI
CH2
CH
CH3
H3C
C2H5
NH
N
O2, KOH, CH3OH, -10°
30% KOH in CH3OH
OH⁻
H N N
H3C
H
CH3
H2C H HC C O
H2C COOCH3
COOCH3
XXXIII
standing or CH2N2
HOC O C O
COOCH3
XXXV
CO COOCH3
COOCH3
XXXVII

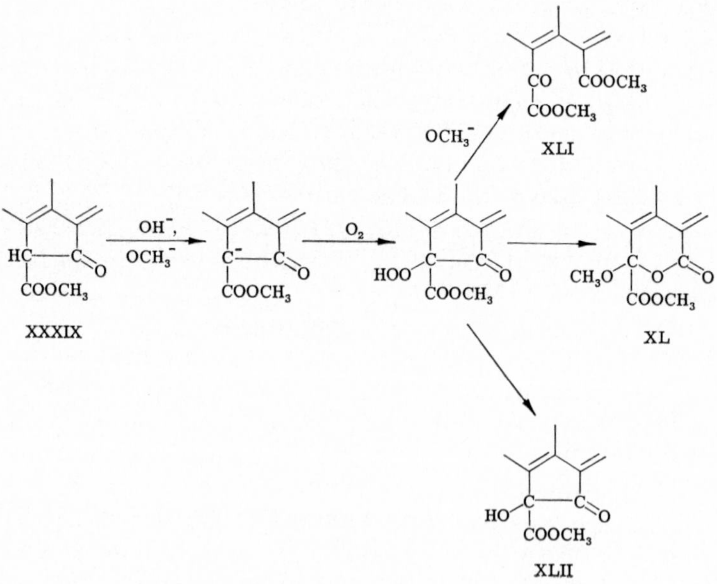
CO COOCH3
COOCH3
OCH3⁻
XLI
OH⁻, OCH3⁻
O2
HC C O
COOCH3
XXXIX
C⁻ C O
COOCH3
HOO C C O
COOCH3
CH3O C O C O
COOCH3
XL
HO C C O
COOCH3
XLII

cation with diazomethane. Protracted treatment with base increases the amounts of purpurin 18 and chlorin p_6 (XXXVIII) at the expense of purpurin 7. Fischer and Pfeiffer identified purpurin 7–lactone ethyl (or methyl) ether diester as a product of allomerization of ethyl chlorophyllide a in ethanol (or methanol) (169). They postulated that an initially formed C-10 hydroperoxide was cleaved solvolytically, to give the final product after considerable rearrangement.

After allomerization of methyl chlorophyllide a (XXXIX) for 12 hours in methanol which may have contained an adventitious base impurity, Holt isolated Mg purpurin 7–lactone methyl ether dimethyl ester (XL) and two other products, Mg purpurin 7 trimethyl ester (XLI) and 10-hydroxymethyl chlorophyllide a (XLII) (138). None of these was a precursor of another. Holt concluded that allomerization in alcohols involved oxidation of traces of the phase test intermediate.

The rate of allomerization is much increased in the presence of metal salts, especially those of Mg^{++} and La^{3+}; although at low chlorophyll a concentration ($< 10^{-4} M$), the rate of allomerization is first order in chlorophyll and only one product is formed initially, at higher concentrations the rate becomes greater than first order and two additional products are formed (170a).

b. By Other Oxidizing Agents. Allomerization by strong oxidizing agents in the absence of air apparently need not go through the phase test intermediate, because reaction in acetic acid has been reported (154, 163, 171). The products characteristically retain an isocyclic ring V, but the substituent at C-10 depends upon the solvent, as the following examples show.

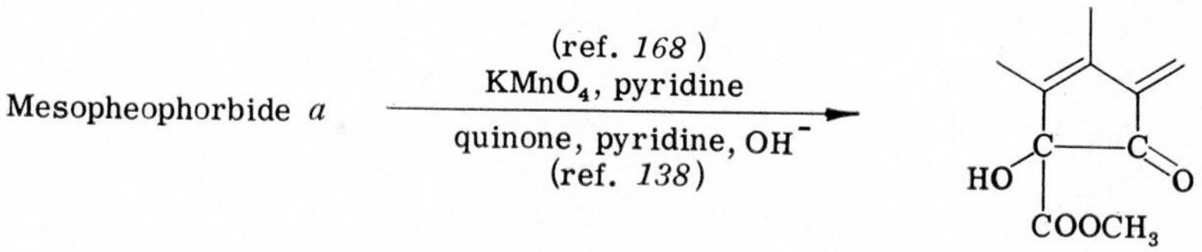

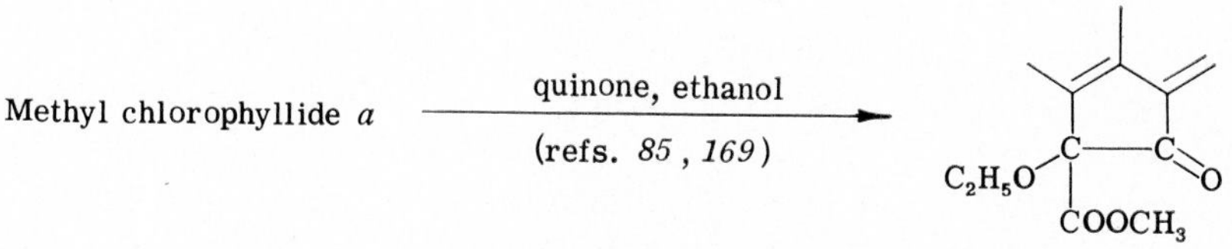

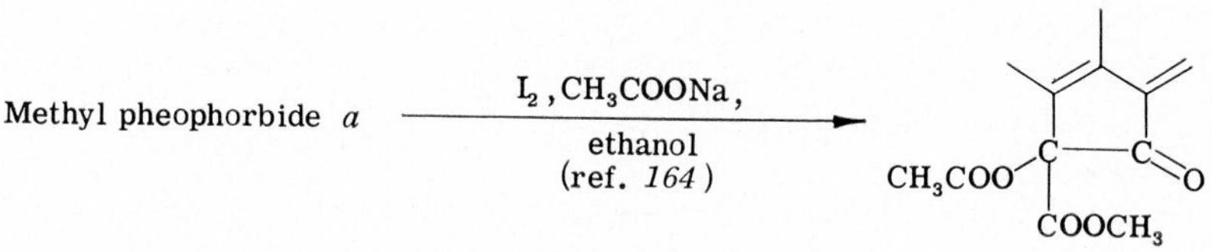

Mesopheophorbide a → (ref. 168) KMnO$_4$, pyridine / quinone, pyridine, OH$^-$ (ref. 138) → [structure: HO—C, C=O, COOCH$_3$]

Methyl chlorophyllide a → quinone, ethanol (refs. 85, 169) → [structure: C$_2$H$_5$O—C, C=O, COOCH$_3$]

Methyl pheophorbide a → I$_2$, CH$_3$COONa, ethanol (ref. 164) → [structure: CH$_3$COO—C, C=O, COOCH$_3$]

The oxidation potentials of the effective allomerizing agents ($FeCl_3$: $E_0 = + 0.77$ v; $Mo(CN)_8{}^{3-}$: $+ 0.73$; benzoquinone: $+ 0.71$; $MnO_4{}^-$: $+ 0.56$; I_2: $+ 0.54$) are close enough to that reported for chlorophyll ($+ 0.64$) (*172*) to suggest that the reaction is initiated by electron abstraction from chlorophyll.

It has been suggested that quinone catalyzed allomerization goes through solvolysis of an intermediate hydroquinone ether (XLIII) (*169*), and that I_2-catalyzed allomerization goes via a 10-iodo intermediate (*138*). In neither case has the intermediate yet been detected.

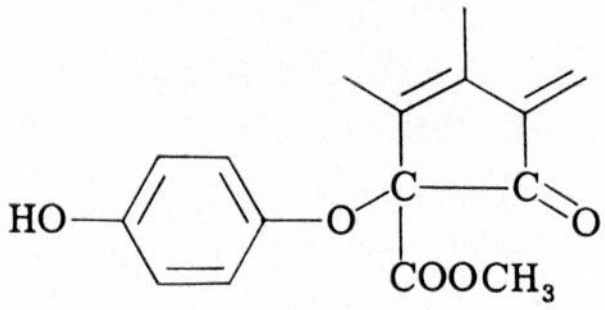

XLIII

c. *Other Chlorophylls.* The allomerization of chlorophyll *b* proceeds similarly to that of chlorophyll *a*, but the yields, as usual, are poorer (*171, 173–175*). Methyl bacteriopheophorbide was allomerized by O_2 in KOH/propanol/ether at $- 10°$ and esterified to bacteriopurpurin 7 trimethyl ester (*19*). Pheoporphyrin a_5 is allomerized by 25% KOH/propanol to rhodoporphyrin-γ-carboxylic acid anhydride, the porphin analog of purpurin 18 (*162*).

Pyro derivatives do not give the phase test, and they are therefore not subject to allomerization in the original sense (*3*). They are, however, oxidized to 9,10-diketo compounds by O_2 in alkaline dimethyl formamide (*176, 177*). Ring V of methyl pyropheophorbide *a* (*178*) or phylloerythrin (*179*) is opened by alcoholic KOH/O_2 to give chlorin p_6 and rhodoporphyrin-γ-carboxylic acid, respectively, and other compounds.

4. DECARBOXYLATION

a. *Pyrolysis.* Chlorophylls *a* and *b*, chlorophyllides, and pheophorbides lose the 10-carboxymethyl group when heated to 180°–250° in biphenyl (*180*) or boiled for a day in pyridine (*95, 114, 115, 181*). The *pyro* compounds so formed closely resemble the parent compounds spectrally (*95*) but have lost the ability to give the phase test and to allomerize. Because of their stability they are useful in chemical applications that do not require the presence of an activated hydrogen at C-10. The chlorobium chlorophylls of green bacteria are derivatives of pyrochlorophyllide.

Loss of the 10-carboxymethyl group also occurs in acid strong enough to hydrolyze the ester. Thus, pheophorbide a is converted to phylloery-thrin (VII) on heating for 16 hours in 20% HCl (178, 182).

b. *Degradation.* Decarboxylation of chlorophyllin (5, 183) and pheophorbide (9, 184–186) was investigated rather thoroughly in search of simple derivatives of chlorophyll. Decarboxylation is accompanied by isomerization or oxidation to 2-devinyl-2-ethyl porphins, and by much general degradation, so that complex mixtures of products are obtained and the yield of any one is poor. For details concerning decarboxylation, Willstätter and Stoll (11) and Fischer and Stern (31) should be consulted, but the general situation is somewhat as follows.

Under alkaline conditions pheophorbide is partly saponified to chlorin e_6 and partly oxidized by air to purpurin 7. On pyrolysis, chlorin e_6 gives

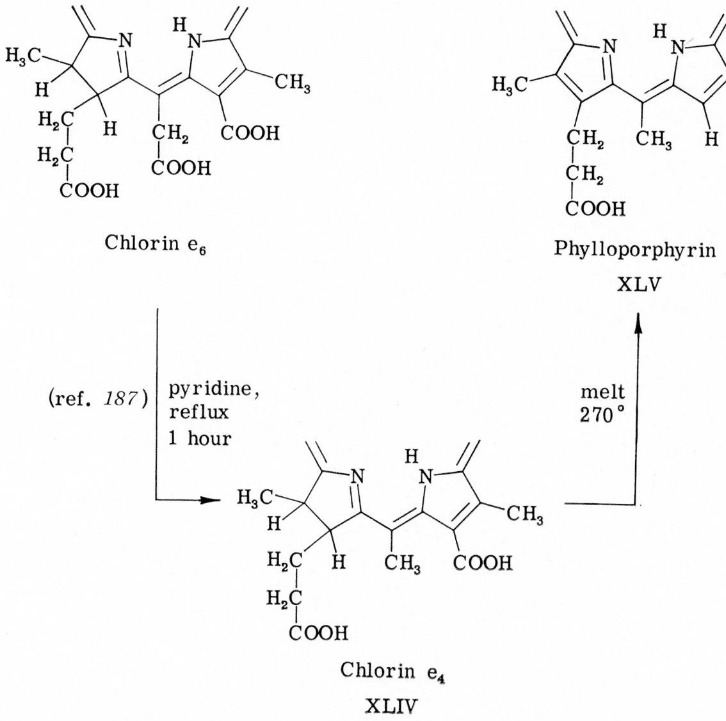

Chlorin e_6 Phylloporphyrin

                                                    XLV

(ref. *187*)   pyridine, reflux 1 hour   melt 270°

Chlorin e_4

XLIV

rise to chlorin e_4 (XLIV) then to phylloporphyrin (XLV). Purpurin 7 is decarboxylated to rhodochlorin (XLVI), which then rearranges to rhodoporphyrin* and is decarboxylated to pyrroporphyrin (XLVII).

Under acid conditions, loss of the 10-$COOCH_3$ group competes with

---

* Unless otherwise specified, trivially named chlorins derived from chlorophyll are by convention assumed to be 2-vinyl derivatives, and trivially named porphins, 2-ethyl derivatives.

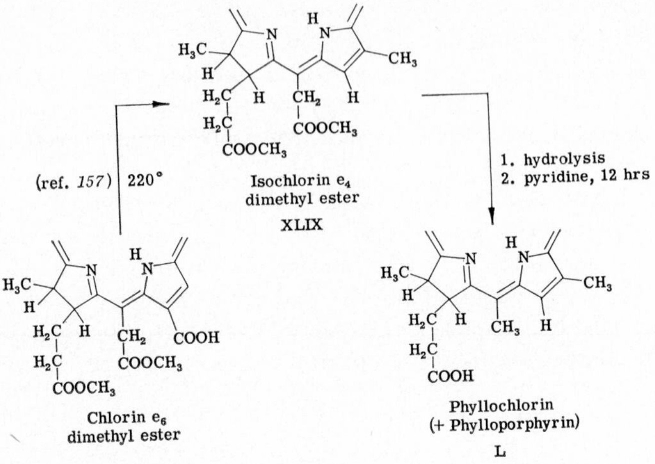

(ref. 157) 220°

Isochlorin e₄
dimethyl ester
XLIX

1. hydrolysis
2. pyridine, 12 hrs

Chlorin e₆
dimethyl ester

Phyllochlorin
(+ Phylloporphyrin)
L

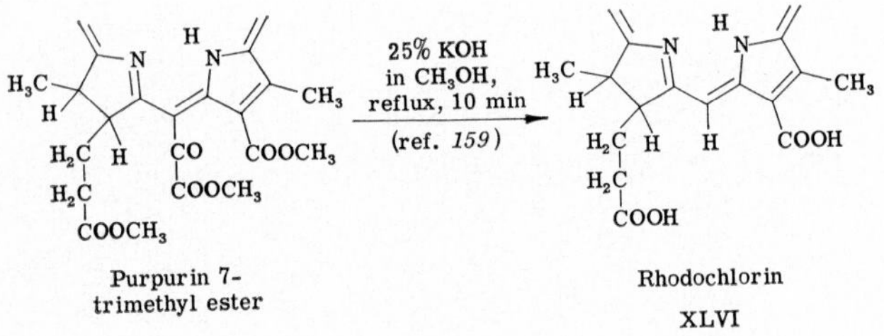

25% KOH
in CH₃OH,
reflux, 10 min
(ref. 159)

Purpurin 7-
trimethyl ester

Rhodochlorin
XLVI

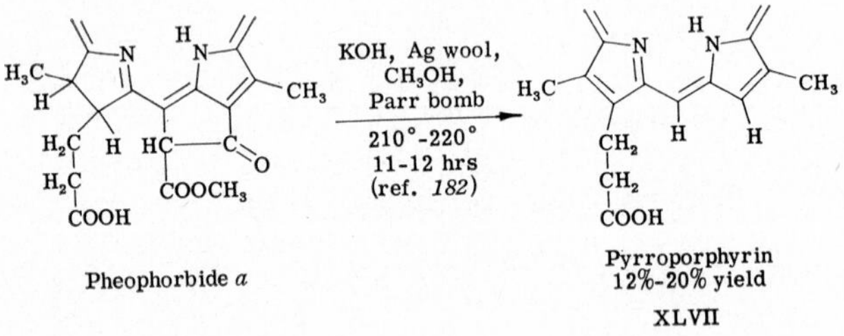

KOH, Ag wool,
CH₃OH,
Parr bomb
210°-220°
11-12 hrs
(ref. 182)

Pheophorbide a

Pyrroporphyrin
12%-20% yield
XLVII

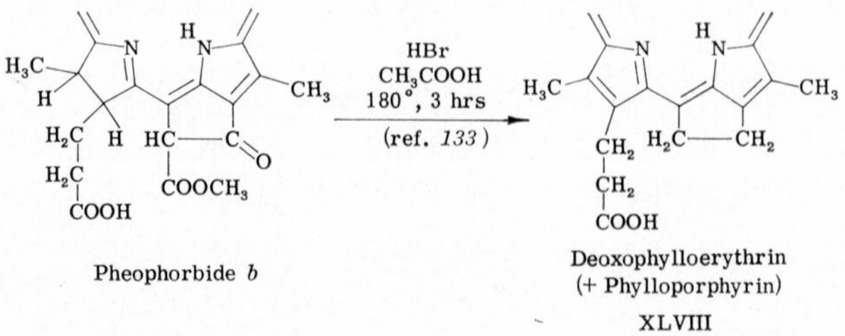

HBr
CH₃COOH
180°, 3 hrs
(ref. 133)

Pheophorbide b

Deoxophylloerythrin
(+ Phylloporphyrin)
XLVIII

opening of ring V, so that phylloerythrin (VII) and deoxophylloery-thrin (XLVIII) are included among the products.

The γ-CH$_2$COOH is most easily decarboxylated, the 6-COOH group is next, and the 7-CH$_2$ CH$_2$ COOH is almost impossible to decarboxylate (184). The examples on pp. 95 and 96 will illustrate these generalizations.

F. The Conjugated Double Bond System

1. CHLORINATION

In 1937 Fischer and Lautsch reported the preparation of compounds they described as "7,8-dioxychlorins" (LI) by O$_2$/Ag$_2$O oxidation of chlorin e$_6$ trimethyl ester, mesochlorin e$_6$ trimethyl ester, pyropheophor-bide, and chlorin p$_6$ trimethyl ester (188). Under the same conditions,

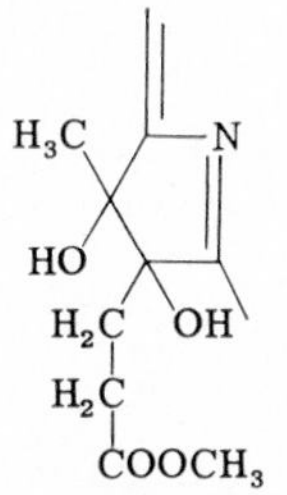

LI

pheophorbide was allomerized. Woodward and Skaric, however, found by nuclear magnetic resonance studies that the δ-bridge carbon of chlorins is remarkably sensitive to electrophilic substitution, and that the "dioxychlorins" were really δ-chlorochlorins, the chlorine having been introduced during extraction of the "product" by HCl in the presence of oxidizing agents (132, 189). The chlorine is easily displaced by reducing agents under acid conditions (hydroquinone/CH$_3$COOH) (50).

Chlorins have been chlorinated intentionally with acid chlorides/AlCl$_3$ (190) and HCl/peroxide (50, 190–192), the chlorine apparently going to the δ-position.

Bromination of isochlorin e$_4$ derivatives also leads to δ-substitution preferentially (190, 193–195).

2. THE PHEOPORPHYRIN REACTION

Treatment of pheophorbide with HI in acetic acid at 60° for a few minutes converts it into an isomer, pheoporphyrin a$_5$ (VIII). This reaction, reported in 1929 (9), provided Fischer with a direct link between chlorophyll and the more familiar porphin system. Indeed, degradation of pheoporphyrin a$_5$ (186, 196) led to an essentially correct assignment of its structure in 1931 (*10*); four years of intensive effort followed before the nearly correct 1935 structure was assigned to pheophorbide.

The reaction is general for phorbins, chlorins, and purpurins. *Meso* derivatives are also converted to porphins, but exposure to air is

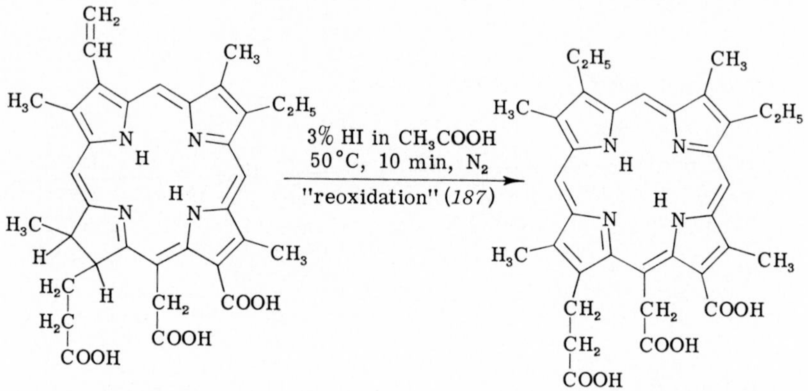

Chlorin e$_6$ 3% HI in CH$_3$COOH   50°C, 10 min, N$_2$ "reoxidation" (*187*)   Chloroporphyrin e$_6$

X   XIII

necessary to effect dehydrogenation (*114*). Apparently the only structural requirement is for an electron-attracting group in the 6-position; 9-hydroxypheophorbides, mesodeoxopyropheophorbide (XXX), and isochlorin e$_4$ (cf. XLIX) are not readily converted to porphins (*76, 140, 157, 190*). Decarboxylation may accompany the reaction (*197*), as may oxidation of the formyl group of chlorophyll *b* derivatives (*116, 133*), but these side reactions can be reduced by operating under nitrogen (*128, 187*).

How the pheoporphyrin reaction occurs is still unclear. It is often regarded as a "reduction" by HI to a "leuco" compound, followed by reoxidation of porphin (*23*). But conditions are hardly suitable for reduction of pheophorbide, because I$_2$ is generated if the reaction is run in the air. Furthermore, similar isomerizations can be achieved with HBr and HCl at higher temperatures (*151, 178*). It is more likely that HI adds reversibly across the conjugated double bond system of chlorins, and that relocation of hydrogen takes place within this adduct (*178*). We have noted that HI converts (protonated) pheophorbide in acetic acid to a compound, presumably an adduct, with a principal absorption band around 515 mμ. The spectrum resembles that of the acid form of photo-reduced pheophytin (*198*) and other compounds in which the closed conjugated system of double bonds is interrupted, but, unlike reduced pheophytin, it is not rapidly oxidized by air. Pheoporphyrin a$_5$ in acetic

acid undergoes color changes on addition of HI, similar to those undergone by pheophorbide (178).

The exact mechanistic relationship of the pheoporphyrin reaction
to the oxo reaction (125) (which replaces it at room temperature), and
the HBr-catalyzed hydration of the vinyl group (121), remains obscure.
In this connection, it is interesting that HI in CH_3COOH at 65° converts
methyl bacteriopheophorbide into two porphins, one of which is oxopheoporphyrin a_5; that is, the 2-acetyl group does not necessarily react
(155).

Other evidence of the reactivity of the conjugated system under
acidic conditions is the exchange of the δ-hydrogen of chlorin e_6 trimethyl
ester in CH_3COOD at 80° (50) and certain rearrangements of porphins
found by Woodward et al. in the course of their work on the synthesis of
chlorophyll (27, 28, 129).

The reported addition of thiolacetic acid or methyl mercaptan to a
porphin (189) is perhaps analogous to the postulated addition of HI to
chlorins.

G. Oxidation

1. DEHYDROGENATION

a. By Oxygen. Simple chlorins are dehydrogenated by oxygen on
refluxing in pyridine to which 30% KOH in methanol has been added.
The 2-vinyl group is retained, if originally present in the chlorin (157,
199). A similar reaction is the oxidation of rhodochlorin (XLVI) to
vinylrhodoporphyrin by $Fe(CN)_6^{3-}$ in alkali (200). As already noted,
considerable conversion to porphins occurs when chlorins are degraded,
but the vinyl group is not usually retained.

Phorbins can be dehydrogenated with retention of the 2-vinyl group
by boiling for 3½ minutes in formic acid with a little iron powder. In
this way, 2-vinyl pheoporphyrin a_5 was prepared from methyl pheophorbide a, and found to be identical with protopheophorbide (20, 22,
23, 122). Pyropheophorbide is oxidized to vinyl phylloerythrin. Phorbides
of the b series must be converted to oximes before dehydrogenation (22).
The dehydrogenation of methyl bacteriopheophorbide and bacteriochlorin e_6 trimethyl ester by oxygen to 2-devinyl 2-acetyl methyl pheophorbide a and 2-devinyl 2-acetyl chlorin e_6 trimethyl ester is catalyzed
by cupric acetate in acetic acid (19).

Phlorins, which are isomeric with chlorins, are rapidly dehydrogenated by oxygen (189).

b. By Quinones. Simple chlorins are dehydrogenated by 2, 3-

dichloro-5, 6-dicyanobenzoquinone in benzene at room temperature, and by less powerfully oxidizing quinones at higher temperatures (*201, 202*). Photooxidation at room temperature is also effective (*203, 204*). Phorbins, however, resist both chemical and photochemical dehydrogenation by quinones. The greater difficulty encountered in oxidizing phorbins and heavily substituted chlorins may be due to crowding of substituents at 6, γ, and 7 positions (*15, 28*).

Bacteriochlorophyll is oxidized to 2-devinyl-2-acetyl chlorophyll *a* by $FeCl_3$ in methanol (*75*), and by other oxidants (*17*). 2-Deacetyl 2-vinyl bacteriochlorin e_6 trimethyl ester is dehydrogenated to chlorin e_6 trimethyl ester by quinone in alcohol acidified with HCl (*63*). Bacteriochlorin e_6 trimethyl ester (LII) is dehydrogenated almost quantitatively to 2-devinyl 2-acetyl chlorin e_6 trimethyl ester (LIII) at room temperature,

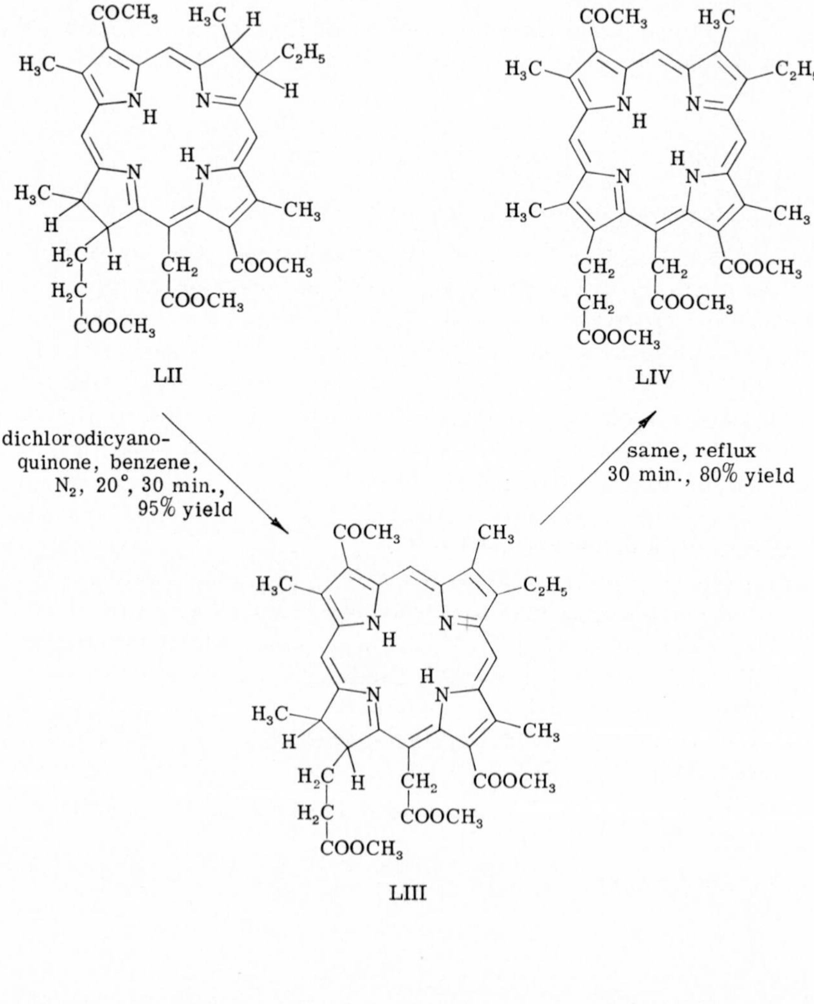

LII

LIV

dichlorodicyano-
quinone, benzene,
N_2, 20°, 30 min.,
95% yield

same, reflux
30 min., 80% yield

LIII

and at higher temperatures to 2-deethyl 2-acetyl chloroporphyrin e_6 trimethyl ester (LIV).

2. OSMIUM TETROXIDE

Simple porphins (pyrro-, rhodo-, phylloporphyrins), form adducts with OsO_4, which are hydrolyzed in sodium sulfite solution to dihydroxyporphins (175, 205). These have chlorin-type spectra, but differ from the "dioxychlorins" prepared earlier by Fischer and Lautsch (188), now believed to be chlorochlorins.[*]

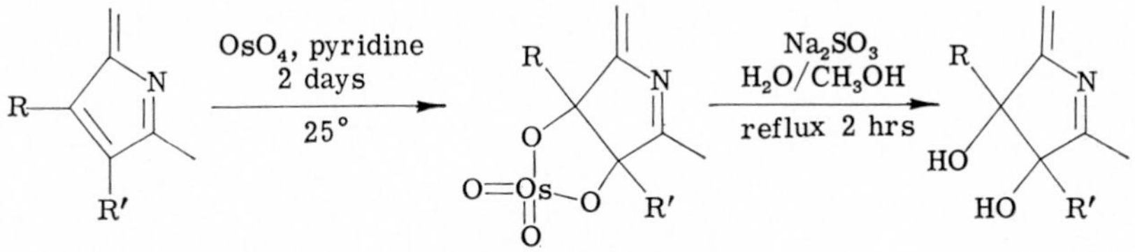

Chlorins also reacted with OsO_4, but the products could not be fully identified. The 2-vinyl group, when present, is oxidized by OsO_4 to 2-glycol.

3. CHROMIC ACID

The maleimides and succinimides produced by destructive oxidation of chlorophylls have provided fundamental structural evidence. On chromic acid oxidation, pyrroporphyrin, rhodoporphyrin, and phylloporphyrin give methyl ethyl maleimide (LV) from ring II and hematinic acid imide from ring IV (6). Chlorins of the a series give (LV) from ring II (13), and dihydrohematinic acid imide (LVII) from ring IV (206). Ring II is degraded on oxidation of chlorins of the b series. Ring I in 2-vinyl chlorins is degraded, but in meso chlorins it is oxidized to (LV) (207), and in adducts with diazoacetic ester it is oxidized to (XXII) (113). Recovery of citraconimide (LVIII) from oxidation of phyllochlorin (L) (13), and dihydrohematinic acid imide from oxidation of pheophorbide a (206), established that the "extra" hydrogens of chlorophyll were on ring IV, not on ring III. Ficken et al. determined that these hydrogens were in the trans configuration (14).

Mittenzwei isolated (LVII) from ring IV and perhaps methyl ethyl succinic anhydride from ring II after oxidation of bacteriochlorin e_6 and

---

[*] It might at first seem odd that products of oxidation of porphins should have spectra resembling those of products of reduction. But spectra of porphyrins are determined largely by the symmetry and the topology of the conjugated system, which are the same for dihydroxyporphins as for chlorins.

methyl bacteriopheophorbide (*63*). Golden *et al.* isolated (LVII), (LV), and *trans*-methyl ethyl succinimide from bacteriochlorin e_6 trimethyl ester, confirming the reduced state of rings II and IV in bacterio compounds and establishing the configuration of the hydrogens on ring II (*15*).

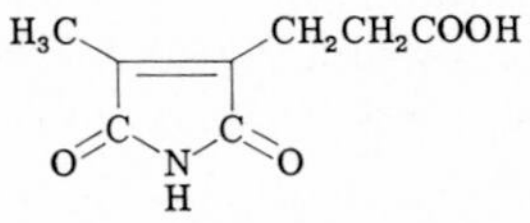

LVI

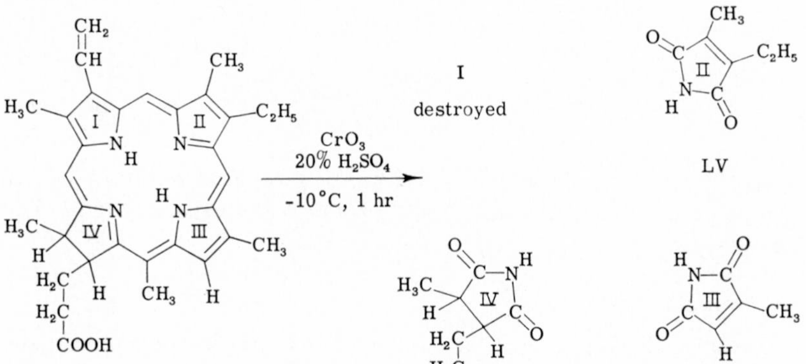

L   destroyed

CrO₃
20% H₂SO₄
-10°C, 1 hr

LVII

LV

LVIII

H. Reduction

1. HYDROGENATION

On hydrogenation with $H_2/(Pd$ or $PtO_2)$ in acetic acid, pheophorbides and chlorins take up 2–3 moles of H_2 beyond the one needed to reduce the vinyl group; porphins are recovered by air oxidation (*114, 147, 208, 209*). The *leuco compounds* from pheophorbide, methyl pheophorbide, and pyropheophorbide have been crystallized. They probably have a porphyrinogen structure, i.e., with the four bridge positions hydrogenated (*210*). They are stable to air when dry, but are reoxidized in acetone solution to pheoporphyrin a_5 and phylloerythrin.

Fischer and Bub found that optical activity was lost when the leuco compounds formed by hydrogenation of pheophorbides in acetone were reoxidized by air to mesopheophorbides (*210*). It may be that partial reduction of the porphyrin ring disposes it to acid-catalyzed isomerizations like those of the pheoporphyrin reaction.

2. Chemical Reduction

Numerous attempts have been made to prepare chlorins from porphins by chemical reduction. The leucoporphyrins formed under acidic conditions revert to porphins on reoxidation, but under alkaline conditions (sodium in isoamyl alcohol) chlorins may be formed. The state of knowledge of chemical reduction of porphyrins has been reviewed briefly by Corwin and Collins (211). They were able to reduce tin 2-vinyl chloroporphyrin e_6 to tin mesochlorin e_6 with Na/isoamyl alcohol.

Photochemical reduction of the chlorophylls has been more fruitful; see Chapter 18.

3. Exhaustive Reduction

Pheophorbide a and chlorin e_6 are reduced by heating at 100° for 2 hours in $HI/PH_4I/CH_3COOH$ to a mixture of pyrroles, including hemopyrrole (LIX), phyllopyrrole (LX), and opsopyrrole carboxylic acid (LXI) (7, 212). These reductions were important in demonstrating the relationship between chlorophyll and the porphyrins derived from heme.

LIX   LX   LXI

I. Conclusion

The aspects of chlorophyll chemistry that have received the most attention in the past are those connected with structural determination and synthesis. Although a very great amount of information has accumulated about these reactions of chlorophyll, there are still reactions of biochemical interest that have not been reproduced in vitro: for example, the introduction of Mg into porphyrins under aqueous conditions, the conversion of protochlorophyllide to chlorophyllide and to bacteriochlorophyllide, the conversion of chlorophyll a or some precursor to chlorophyll b (or a precursor), and the nonenzymatic transesterification of the 7-propionic acid group of Mg-containing chlorophyll derivatives. These examples show that our understanding of the essential biological chemistry of chlorophyll is still very incomplete.

References

(1) R. Willstätter and F. Hocheder, Ann. Chem. 354, 205 (1907).
(2) R. Willstätter and A. Stoll, Ann. Chem. 378, 18 (1910).
(3) R. Willstätter and A. Stoll, Ann. Chem. 387, 317 (1911).

(4) R. Willstätter and M. Isler, *Ann. Chem.* **390**, 269 (1912).

(5) R. Willstätter and H. Fritzsche, *Ann. Chem.* **371**, 33 (1909).

(6) R. Willstätter and Y. Asahina, *Ann. Chem.* **373**, 227 (1910).

(7) R. Willstätter and Y. Asahina, *Ann. Chem.* **385**, 188 (1911).

(8) H. Fischer and A. Stern, *Ann. Chem.* **520**, 88 (1935).

(9) H. Fischer and R. Bäumler, *Ann. Chem.* **474**, 65 (1929).

(10) H. Fischer, O. Moldenhauer, and O. Süs, *Ann. Chem.* **486**, 107 (1931).

(11) R. Willstätter and A. Stoll, "Investigations on Chlorophyll" (translated by F. M. Schertz and A. R. Merz), Science Printing Press, Lancaster, Pennsylvania, 1928.

(12) H. Fischer and K. Kahr, *Ann. Chem.* **524**, 251 (1936).

(13) H. Fischer and H. Wenderoth, *Ann. Chem.* **537**, 170 (1939).

(14) G. E. Ficken, R. B. Johns, and R. P. Linstead, *J. Chem. Soc.* p. 2272 (1956).

(15) J. H. Golden, R. P. Linstead, and G. H. Witham, *J. Chem. Soc.* p. 1725 (1958).

(16) K. Noack and E. Schneider, *Naturwissenschaften* **21**, 835 (1933).

(17) E. Schneider, *Z. Physiol. Chem.* **226**, 221 (1934).

(18) H. Fischer and R. Lambrecht, *Z. Physiol. Chem.* **249**, I (1937).

(19) H. Fischer, R. Lambrecht, and H. Mittenzwei, *Z. Physiol. Chem.* **253**, 1 (1938).

(20) K. Noack and W. Kiessling, *Z. Physiol. Chem.* **182**, 13 (1929).

(21) K. Noack and W. Kiessling, *Z. Physiol. Chem.* **193**, 97 (1930).

(22) H. Fischer, H. Mittenzwei, and A. Oestreicher, *Z. Physiol. Chem.* **257**, IV (1939).

(23) H. Fischer and A. Oestreicher, *Z. Physiol. Chem.* **262**, 243 (1940).

(24) F. G. Fischer and K. Löwenberg, *Ann. Chem.* **464**, 69 (1928).

(25) F. G. Fischer and K. Löwenberg, *Ann. Chem.* **475**, 183 (1929).

(26) M. Strell and A. Kalojanoff, *Ann. Chem.* **652**, 218 (1962).

(27) R. B. Woodward, W. A. Ayer, J. M. Beaton, F. Bickelhaupt, R. Bonnett, P. Buchschacher, G. L. Closs, H. Dutler, J. Hannah, F. P. Hauck, S. Itô, A. Langemann, E. LeGoff, W. Leimgruber, W. Lwowski, J. Sauer, Z. Valenta, and H. Volz, *J. Am. Chem. Soc.* **82**, 3800 (1960).

(28) R. B. Woodward, *Pure Appl. Chem.* **2**, 383 (1961); *Angew. Chem.* **72**, 651 (1960).

(29) G. L. Closs, J. J. Katz, F. C. Pennington, M. R. Thomas, and H. H. Strain, *J. Am. Chem. Soc.* **85**, 3809 (1963).

(30) J. J. Katz, G. L. Closs, F. C. Pennington, M. R. Thomas, and H. H. Strain, *J. Am. Chem. Soc.* **85**, 3801 (1963).

(31) H. Fischer and A. Stern, "Die Chemie des Pyrrols," Vol. II, Part 2. Akad. Verlagsges., Leipzig, 1940.

(32) H. Fischer and M. Strell, *FIAT Rev. Ger. Sci.* Part 1, p. 142 (1947).

(33) H. Rothemund, *in* "Medical Physics" (O. Glasser, ed.), p. 154. Year Book Publ., Chicago, Illinois, 1944.

(34) A. Stoll and E. Wiedemann, *Fortschr. Chem. Forsch.* **2**, 538 (1952).

(35) E. I. Rabinowitch, "Photosynthesis and Related Processes," Vol. 1, Chapter 16. Wiley (Interscience), New York, 1945; also Vol. 2, Part 2, Chapter 37B, 1956.

(36) S. Aronoff, *in* "Handbuch der Pflanzenphysiologie" (W. Ruhland, ed.), Vol. 5, Part 1, p. 234. Springer, Berlin, 1960.

(37) A. S. Holt, *in* "Chemistry and Biochemistry of Plant Pigments" (T. W. Goodwin, ed.), p. 3. Academic Press, New York, 1965.

(38) H. Fischer and P. Halbig, *Ann. Chem.* **448**, 193 (1926).

(39) H. V. Knorr and V. M. Albers, *J. Chem. Phys.* **9**, 197 (1941).

(40) W. D. Bellamy, G. L. Gaines, Jr., and A. G. Tweet, *J. Chem. Phys.* **39**, 2528 (1963).

(41) A. F. H. Anderson and M. Calvin, *Arch. Biochem. Biophys.* **107**, 251 (1964).

(42) E. B. Fleischer, *J. Am. Chem. Soc.* **85**, 146 (1963).

(43) E. B. Fleischer, C. K. Miller, and L. E. Webb, *J. Am. Chem. Soc.* **86**, 2342 (1964).

(44) S. Freed and K. M. Sancier, *J. Am. Chem. Soc.* **76**, 198 (1954).

(45) G. R. Seely, *Spectrochim. Acta* **21**, 1847 (1965).

(46) A. H. Corwin, J. A. Walter, and R. Singh, *J. Org. Chem.* **27**, 4280 (1962).

(47) A. Stoll and E. Wiedemann, *Helv. Chim. Acta* **16**, 307 (1933).

(48) H. H. Strain and W. M. Manning, *J. Biol. Chem.* **146**, 275 (1942).

(49) H. H. Strain, *J. Agr. Food Chem.* **2**, 1222 (1954).

(50) R. B. Woodward and V. Škaric, *J. Am. Chem. Soc.* **83**, 4676 (1961).

(51) R. C. Dougherty, H. H. Strain, and J. J. Katz, *J. Am. Chem. Soc.* **87**, 104 (1965).

(52) H. Fischer and S. Goebel, *Ann. Chem.* **522**, 168 (1936).

(53) H. Molisch, *Ber. Deut. Botan. Ges.* **14**, 16 (1896).

(54) A. Stern and H. Wenderlein, *Z. Physik. Chem.* **A174**, 81 (1935).

(55) A. Stern and H. Wenderlein, *Z. Physik. Chem.* **A174**, 321 (1935).

(56) A. Stern and H. Wenderlein, *Z. Physik. Chem.* **A176**, 81 (1936).

(57) A. Stern and H. Wenderlein, *Z. Physik. Chem.* **A177**, 165 (1936).

(58) A. Stern and H. Molvig, *Z. Physik. Chem.* **A178**, 161 (1937).

(59) A. Stern and F. Pruckner, *Z. Physik. Chem.* **A180**, 321 (1937).

(60) A. Stern and F. Pruckner, *Z. Physik. Chem.* **A185**, 140 (1939).

(61) G. R. Seely and R. G. Jensen, *Spectrochim. Acta* **21**, 1835 (1965).

(62) F. Pruckner, A. Oestreicher, and H. Fischer, *Ann. Chem.* **546**, 41 (1941).

(63) H. Mittenzwei, *Z. Physiol. Chem.* **275**, 93 (1942).

(63a) K. Sauer, *Proc. Natl. Acad. Sci. U.S.* **53**, 716 (1965).

(64) M. J. Hendrickson, R. R. Berueffy, and A. R. McIntyre, *Anal. Chem.* **29**, 1810 (1957).

(65) M. A. Joslyn and G. Mackinney, *J. Am. Chem. Soc.* **60**, 1132 (1938).

(66) G. Mackinney and M. A. Joslyn, *J. Am. Chem. Soc.* **62**, 231 (1940).

(67) S. H. Schanderl, C. O. Chichester, and G. L. Marsh, *J. Org. Chem.* **27**, 3865 (1962).

(68) G. Mackinney and M. A. Joslyn, *J. Am. Chem. Soc.* **63**, 2530 (1941).

(69) M. Rosoff and C. Aron, *J. Phys. Chem.* **69**, 21 (1965).

(70) A. A. Krasnovskii and E. V. Pakshina, *Dokl. Akad. Nauk SSSR* **148**, 935 (1963).

(71) V. B. Evstigneev and V. A. Gavrilova, *Dokl. Akad. Nauk SSSR* **91**, 899 (1953).

(72) T. T. Bannister, *Plant Physiol.* **34**, 246 (1959).

(73) R. Willstätter and W. Meig, *Ann. Chem.* **350**, 1 (1906).

(74) J. B. Conant, B. F. Chow, and E. M. Dietz, *J. Am. Chem. Soc.* **56**, 2185 (1934).

(75) A. S. Holt and E. E. Jacobs, *Am. J. Botany* **41**, 718 (1954).

(76) A. S. Holt, *Plant Physiol.* **34**, 310 (1959).

(77) H. Fischer and H. Orth, "Die Chemie des Pyrrols," Vol. II, Part 1, p. 359. Akad. Verlagsges., Leipzig, 1937.

(78) R. Willstätter and L. Forsén, *Ann. Chem.* **396**, 180 (1913).

(79) H. Fischer and H. Siebel, *Ann. Chem.* **499**, 84 (1932).

(80) H. Fischer and S. Goebel, *Ann. Chem.* **524**, 269 (1936).

(81) H. Fischer and G. Spielberger, *Ann. Chem.* **510**, 156 (1934).

(82) H. Fischer and G. Spielberger, *Ann. Chem.* **515**, 130 (1935).

(83) A. H. Corwin and P. E. Wei, *J. Org. Chem.* **27**, 4285 (1962).

(84) P. E. Wei, A. H. Corwin, and R. Arellano, *J. Org. Chem.* **27**, 3344 (1962).

(85) H. Fischer, L. Filser, and E. Plötz, *Ann. Chem.* **495**, 1 (1932).

(86) S. J. Baum, B. F. Burnham, and R. A. Plane, *Proc. Natl. Acad. Sci. U.S.* **52**, 1439 (1964).

(87) H. Fischer and A. Wunderer, *Ann. Chem.* **533**, 230 (1938).

(88) M. S. Ashkinazi and B. Ya. Dain, *Dokl. Akad. Nauk SSSR* **80**, 385 (1951).

(89) M. S. Ashkinazi, I. P. Gerasimova, and B. Ya. Dain, *Dokl. Akad. Nauk SSSR* **102**, 767 (1955).

(90) M. S. Ashkinazi, I. P. Gerasimova, and B. Ya. Dain, *Dokl. Akad. Nauk SSSR* **108**, 655 (1956).

(91) M. S. Ashkinazi and A. I. Kryukov, *Ukr. Khim. Zh.* **23**, 448 (1957).

(92) M. S. Ashkinazi and A. I. Kryukov, *Dopovidi Akad. Nauk Ukr. RSR* p. 490 (1960).

(93) E. Schneider, *J. Am. Chem. Soc.* **63**, 1477 (1941).

(94) R. Livingston, W. F. Watson, and J. McArdle, *J. Am. Chem. Soc.* **71**, 1542 (1949).

(95) F. C. Pennington, H. H. Strain, W. A. Svec, and J. J. Katz, *J. Am. Chem. Soc.* **86**, 1418 (1964).

(96) C. B. Storm and A. H. Corwin, *J. Org. Chem.* **29**, 3700 (1964).

(97) A. S. Holt and E. E. Jacobs, *Am. J. Botany* **41**, 710 (1954).

(98) V. M. Kutyurin and V. P. Knyazev, *Dokl. Akad. Nauk SSSR* **149**, 456 (1963).

(99) I. I. Dilung and S. S. Butsko, *Dokl. Akad. Nauk SSSR* **131**, 312 (1959).

(100) S. S. Butsko and B. Ya. Dain, *Zh. Obshch. Khim.* **28**, 2603 (1958).

(101) I. I. Dilung and B. Ya. Dain, *Zh. Fiz. Khim.* **33**, 2740 (1959).

(102) E. B. Fleischer and J. H. Wang, *J. Am. Chem. Soc.* **82**, 3498 (1960).

(103) R. Willstätter and A. Stoll, *Ann. Chem.* **380**, 148 (1911).

(104) H. Fischer and R. Lambrecht, *Z. Physiol. Chem.* **253**, 253 (1938).

(105) M. Holden, *Photochem. Photobiol.* **2**, 175 (1963).

(106) K. Egle, *in* "Handbuch der Pflanzenphysiologie" (W. Ruhland, ed.), Vol. 5, Part 1, p. 387. Springer, Berlin, 1960.

(107) A. Klein and W. Vishniac, *J. Biol. Chem.* **236**, 2544 (1961).

(108) M. Strell and E. Iscimenler, *Ann. Chem.* **557**, 175 (1947).

(109) M. Strell, *Ann. Chem.* **546**, 252 (1941).

(110) M. Strell and E. Iscimenler, *Ann. Chem.* **557**, 186 (1947).

(111) A. Stoll and E. Wiedemann, *Helv. Chim. Acta* **16**, 183 (1933).

(112) H. Fischer and W. Schmidt, *Ann. Chem.* **519**, 244 (1935).

(113) H. Fischer and H. Medick, *Ann. Chem.* **517**, 245 (1935).

(114) H. Fischer and E. Lakatos, *Ann. Chem.* **506**, 123 (1933).

(115) H. Fischer, E. Lakatos, and J. Schnell, *Ann. Chem.* **509**, 201 (1934).

(116) H. Fischer and W. Lautenschlager, *Ann. Chem.* **528**, 9 (1937).

(117) H. Fischer and H. Gibian, *Ann. Chem.* **548**, 183 (1941).

(118) H. Fischer and H. Walter, *Ann. Chem.* **549**, 44 (1941).

(*119*) A. S. Holt and H. V. Morley, *Can. J. Chem.* **37**, 507 (1959).
(*120*) H. Fischer and J. Hasenkamp, *Ann. Chem.* **519**, 42 (1935).
(*121*) H. Fischer, W. Lautsch, and K. H. Lin, *Ann. Chem.* **534**, 1 (1938).
(*122*) H. Fischer, A. Oestreicher, and A. Albert, *Ann. Chem.* **538**, 128 (1939).
(*123*) H. Fischer, H. Mittenzwei, and D. B. Hever, *Ann. Chem.* **545**, 154 (1940).
(*124*) H. Fischer and J. Riedmair, *Ann. Chem.* **505**, 87 (1933).
(*125*) H. Fischer, J. Riedmair, and J. Hasenkamp, *Ann. Chem.* **508**, 224 (1934).
(*126*) H. Fischer and J. Hasenkamp, *Ann. Chem.* **513**, 107 (1934).
(*127*) H. Fischer and M. Conrad, *Ann. Chem.* **538**, 143 (1939).
(*128*) H. Fischer, S. Breitner, A. Hendschel, and L. Nüssler, *Ann. Chem.* **503**, (1933).
(*129*) A. Stoll and E. Wiedemann, *Helv. Chim. Acta* **17**, 163 (1934).
(*130*) A. Stoll and E. Wiedemann, *Helv. Chim. Acta* **17**, 456 (1934).
(*131*) K. Ogawa and K. Shibata, *Photochem. Photobiol.* **4**, 193 (1965).
(*132*) H. R. Wetherell and M. J. Hendrickson, *J. Org. Chem.* **24**, 710 (1959).
(*133*) H. Fischer, F. Broich, S. Breitner, and L. Nüssler, *Ann. Chem.* **498**, 228 (1932).
(*134*) H. Fischer and M. Strell, *Ann. Chem.* **538**, 157 (1939).
(*135*) H. Fischer and F. Gerner, *Ann. Chem.* **553**, 67 (1942).
(*136*) W. M. Manning and H. H. Strain, *J. Biol. Chem.* **151**, 1 (1943).
(*137*) E. Tyray, *Ann. Chem.* **556**, 171 (1944).
(*138*) A. S. Holt, *Can. J. Biochem. Physiol.* **36**, 439 (1958).
(*139*) H. Fischer and H. Gibian, *Ann. Chem.* **552**, 153 (1942).
(*140*) H. Fischer and H. Gibian, *Ann. Chem.* **550**, 208 (1942).
(*141*) H. Fischer and A. Oestreicher, *Ann. Chem.* **546**, 49 (1941).
(*142*) R. Willstätter and M. Utzinger, *Ann. Chem.* **382**, 129 (1911).
(*143*) A. Weller, *J. Am. Chem. Soc.* **76**, 5819 (1954).
(*144*) S. Freed and K. M. Sancier, *Science* **117**, 655 (1953).
(*145*) R. Felton, G. M. Sherman, and H. Linschitz, *Nature* **203**, 637 (1964).
(*146*) J. H. C. Smith and A. Benitez, *in* "Moderne Methoden der Pflanzenanalyse" (K. Paech and M. V. Tracey, eds.), Vol. IV, p. 142. Springer, Berlin, 1955.
(*147*) A. Stoll and E. Wiedemann, *Helv. Chim. Acta* **16**, 739 (1933).
(*148*) J. B. Conant and W. W. Moyer, *J. Am. Chem. Soc.* **52**, 3013 (1930).
(*149*) B. Dunicz, T. Thomas, M. V. Pee, and R. Livingston, *J. Am. Chem. Soc.* **73**, 3388 (1951).
(*150*) G. Oster, S. B. Broyde, and J. S. Bellin, *J. Am. Chem. Soc.* **86**, 1309 (1964).
(*151*) H. Fischer, W. Gottschaldt, and G. Klebs, *Ann. Chem.* **498**, 194 (1932).
(*152*) H. Fischer and J. Riedmair, *Ann. Chem.* **506**, 107 (1933).
(*153*) H. Fischer, A. Hendschel, and L. Nüssler, *Ann. Chem.* **506**, 83 (1933).
(*154*) H. Fischer, J. Heckmaier, and W. Hagert, *Ann. Chem.* **505**, 209 (1933).
(*155*) H. Fischer and J. Hasenkamp, *Ann. Chem.* **515**, 148 (1935).
(*156*) A. Weller and R. Livingston, *J. Am. Chem. Soc.* **76**, 1575 (1954).
(*157*) H. Fischer and H. Kellermann, *Ann. Chem.* **519**, 209 (1935).
(*158*) H. Fischer and W. Lautsch, *Ann. Chem.* **528**, 265 (1937).
(*159*) J. B. Conant, J. F. Hyde, W. W. Moyer, and E. M. Dietz, *J. Am. Chem. Soc.* **53**, 359 (1931).
(*160*) J. B. Conant, S. E. Kamerling, and C. C. Steele, *J. Am. Chem. Soc.* **53**, 1615 (1931).
(*161*) J. B. Conant, E. M. Dietz, C. F. Bailey, and S. E. Kamerling, *J. Am. Chem. Soc.* **53**, 2382 (1931).

(*162*) H. Fischer, O. Süs, and G. Klebs, *Ann. Chem.* **490**, 38 (1931).

(*163*) H. Fischer and W. Hagert, *Ann. Chem.* **502**, 41 (1933).

(*164*) H. Fischer and J. Heckmaier, *Ann. Chem.* **508**, 250 (1934).

(*165*) H. Fischer, J. Heckmaier, and T. Scherer, *Ann. Chem.* **510**, 169 (1934).

(*166*) H. Fischer and T. Scherer, *Ann. Chem.* **519**, 234 (1935).

(*167*) H. Fischer and K. Kahr, *Ann. Chem.* **531**, 209 (1937).

(*168*) M. Strell, *Ann. Chem.* **550**, 50 (1942).

(*169*) H. Fischer and H. Pfeiffer, *Ann. Chem.* **555**, 94 (1944).

(*170*) W. F. Watson, *J. Am. Chem. Soc.* **75**, 2522 (1953).

(*170a*) L. G. Johnston and W. F. Watson, *J. Chem. Soc.* p. 1203 (1956).

(*171*) H. Fischer and A. Albert, *Ann. Chem.* **599**, 203 (1956).

(*172*) J. C. Goedheer, G. H. Horreus de Haas, and P. Schuller, *Biochim. Biophys. Acta* **28**, 278 (1958).

(*173*) J. B. Conant, E. M. Dietz, and T. H. Werner, *J. Am. Chem. Soc.* **53**, 4436 (1931).

(*174*) H. Fischer and K. Bauer, *Ann. Chem.* **523**, 235 (1936).

(*175*) H. Fischer and H. Pfeiffer, *Ann. Chem.* **556**, 154 (1944).

(*176*) A. S. Holt, D. W. Hughes, H. J. Kende, and J. W. Purdie, *J. Am. Chem. Soc.* **84**, 2835 (1962).

(*177*) A. S. Holt, D. W. Hughes, H. J. Kende, and J. W. Purdie, *Plant Cell Physiol. (Tokyo)* **4**, 49 (1963).

(*178*) H. Fischer, L. Filser, W. Hagert, and O. Moldenhauer, *Ann. Chem.* **490**, 1 (1931).

(*179*) K. Noack and W. Kiessling, *Z. Angew. Chem.* **44**, 93 (1931).

(*180*) J. B. Conant and J. F. Hyde, *J. Am. Chem. Soc.* **51**, 3668 (1929).

(*181*) H. Fischer and H. Siebel, *Ann. Chem.* **494**, 73 (1932).

(*182*) J. L. Wickliff and S. Aronoff, *Anal. Biochem.* **6**, 39 (1963).

(*183*) R. Willstätter, M. Fischer, and L. Forsén, *Ann. Chem.* **400**, 147 (1913).

(*184*) H. Fischer and A. Treibs, *Ann. Chem.* **466**, 188 (1928).

(*185*) A. Treibs and E. Wiedemann, *Ann. Chem.* **471**, 146 (1929).

(*186*) H. Fischer and R. Bäumler, *Ann. Chem.* **480**, 197 (1930).

(*187*) H. Fischer, J. Heckmaier, and E. Plötz, *Ann. Chem.* **500**, 215 (1933).

(*188*) H. Fischer and W. Lautsch, *Ann. Chem.* **528**, 247 (1937).

(*189*) R. B. Woodward, *Ind. Chim. Belge* **27**, 1293 (1962).

(*190*) H. Fischer and F. Gerner, *Ann. Chem.* **559**, 77 (1947).

(*191*) H. Fischer and W. Klendauer, *Ann. Chem.* **547**, 123 (1941).

(*192*) H. Fischer and E. Dietl, *Ann. Chem.* **547**, 234 (1941).

(*193*) H. Fischer and J. M. Ortiz-Velez, *Ann. Chem.* **540**, 224 (1939).

(*194*) H. Fischer, H. Kellermann, and F. Balaz, *Chem. Ber.* **75**, 1778 (1942).

(*195*) H. Fischer and F. Baláz, *Ann. Chem.* **555**, 81 (1944).

(*196*) H. Fischer and O. Süs, *Ann. Chem.* **482**, 225 (1930).

(*197*) H. Fischer and O. Moldenhauer, *Ann. Chem.* **478**, 54 (1930).

(*198*) M. S. Ashkinazi, I. A. Dolidze, and V. E. Karpitskaya, *Biofizika* **6**, 294 (1961).

(*199*) H. Fischer, K. Herrle, and H. Kellermann, *Ann. Chem.* **524**, 222 (1936).

(*200*) J. B. Conant and C. F. Bailey, *J. Am. Chem. Soc.* **55**, 795 (1933).

(*201*) U. Eisner and R. P. Linstead, *J. Chem. Soc.* p. 3749 (1955).

(*202*) U. Eisner, A. Lichtarowicz, and R. P. Linstead, *J. Chem. Soc.* p. 733 (1957).

(*203*) M. Calvin and G. D. Dorough, *J. Am. Chem. Soc.* **70**, 699 (1948).

(*204*) F. M. Huennekens and M. Calvin, *J. Am. Chem. Soc.* **71**, 4024 (1949).

(*205*) H. Fischer and H. Eckoldt, *Ann. Chem.* **544**, 138 (1940).

(*206*) H. Fischer and H. Wenderoth, *Ann. Chem.* **545**, 140 (1940).
(*207*) H. Fischer and S. Breitner, *Ann. Chem.* **522**, 151 (1936).
(*208*) J. B. Conant and J. F. Hyde, *J. Am. Chem. Soc.* **52**, 1233 (1930).
(*209*) E. M. Dietz and T. H. Werner, *J. Am. Chem. Soc.* **56**, 2180 (1934).
(*210*) H. Fischer and K. Bub, *Ann. Chem.* **530**, 213 (1937).
(*211*) A. H. Corwin and O. D. Collins, III, *J. Org. Chem.* **27**, 3060 (1962).
(*212*) H. Fischer, A. Merka, and E. Plötz, *Ann. Chem.* **478**, 283 (1930).

—4—

Recently Characterized Chlorophylls*

A. S. HOLT

Division of Biosciences, National Research Council Ottawa, Canada

I. Introduction

The structures and properties of chlorophylls *a* and *b*, bacteriochlorophyll *a*, and protochlorophyll and their derivatives are outlined in detail in Chapter 3. In this chapter we will consider chlorophyll *d* and chlorobium chlorophylls 650 and 660, whose structures have been determined within the past several years. Information available concerning chlorophyllide(s) *c*, bacteriochlorophyll *b*, and other chlorophylls or chlorophyll-like pigments is also presented.

II. Chlorophyll *d*

This pigment (Mg-2-devinyl 2-formylpheophytin *a*) was discovered by Manning and Strain (*1*) in extracts of Rhodophyceae; Smith and Benitez determined its specific absorption coefficient (*2*). Later, Holt and Morley (*3*) converted chlorophyll *a* to its 2-formyl derivative by oxidation of the 2-vinyl group with permanganate in acetone and found its absorption spectrum to be identical with that of chlorophyll *d*. In common with chlorophyll *d* it reacted reversibly with methanol to form a dimethyl acetal derivative, which previously had been referred to as "isochlorophyll *d*" (*1*).

---

* Issued as N.R.C. No. 8756.

Attempts to isolate chlorophyll *d* from fresh samples of red algae collected near Halifax, Nova Scotia, were unsuccessful. It was obtained later from fresh samples collected at Moss Beach, San Mateo County, California (*4*). When the natural and synthetic samples were compared, they yielded: (a) identical absorption spectra; (b) one zone when co-chromatographed on a sucrose column; (c) products with identical absorption spectra when treated with borohydride, i.e., after reduction of the formyl group to a hydroxymethyl group; (d) a red-colored phase-test intermediate. The absorption spectrum of chlorophyll *d* is given in Fig. 1.

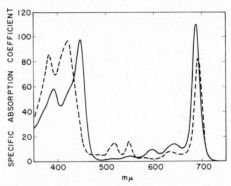

Fig. 1. Absorption spectra of chlorophyll *d* (solid line) and pheophytin *d* (broken line) in ether (*2*).

$$\alpha_{sp} = \frac{1}{Cl} \log_{10} \frac{I_0}{I} = \frac{\alpha_M}{\text{Mol. wt.}}$$

where C = concentration of chlorophyll in grams per liter and l = centimeters path length.

The problem of whether chlorophyll *d* exists *in vivo* and is not an artifact of isolation remains to be solved. Certain thalli of *Gigartina papillata* gave relatively good yields of chlorophyll *d* while others gave none. Sagromsky (*5, 6*) has claimed to have isolated chlorophyll *d* and to have found that more was formed in strongly illuminated plants than in shaded ones; also that it can be produced on standing in alkaline methanol. No chemical evidence for the presence of a 2-formyl group was presented, however. Recently trace amounts of a chlorophyll *d*-like pigment in *Chlorella* extracts were reported (*7*).

III. Chlorobium Chlorophylls 650 and 660

These pigments (farnesyl esters of 2-α-hydroxyethyl 2-devinyl pyro-chlorophyllide *a*) constitute the bulk of the chlorophylls of two different groups of Chlorobacteriaceae. They are divided according to the wavelengths (mμ) of the "red" absorption maxima of their ether solutions; *in vivo* they absorb at 725 and 746 mμ, respectively. Neither group gives a positive phase test nor contains a methoxyl group; both contain magnesium and a *trans-trans*-farnesyl ester group (8–10).

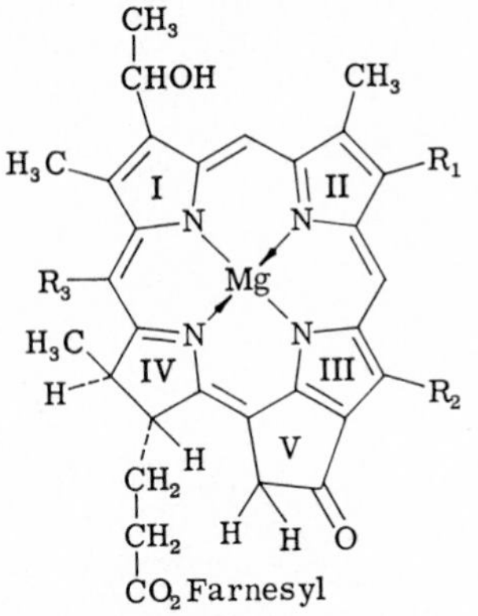

650 Series	R_1	R_2	R_3
Fraction 1	isobutyl	ethyl	H
2	n-propyl	ethyl	H
3	isobutyl	methyl	H
4	ethyl	ethyl	H
5	n-propyl	methyl	H
6	ethyl	methyl	H
660 Series	R_1	R_2	R_3
Fraction 1	isobutyl	ethyl	ethyl
2	isobutyl	ethyl	methyl
3	n-propyl	ethyl	ethyl
4	n-propyl	ethyl	methyl
5	ethyl	ethyl	methyl
6	ethyl	methyl	methyl

FIG. 2.   Structures of fractions 1–6 chlorobium chlorophyll 650 and 660.

Partition chromatography between hydrochloric acid and ether on Celite of the free acids obtained by alkaline hydrolysis revealed that each class of pigment contains six different compounds. The fractions obtained from each were numbered 1–6 in the order of their elution from the column (11). The chief difference between the two classes of compounds is the presence of a δ-alkyl substituent in the "660" series.

114   A. S. HOLT

The β-substituents and their order on each of the four pyrrole nuclei were established by a combination of (a) conversion to various derivatives and correlation with spectral shifts of the absorption bands; (b) chromic acid oxidation and comparison of the resulting imides with known imides using gas–liquid–partition chromatography; and (c) com-

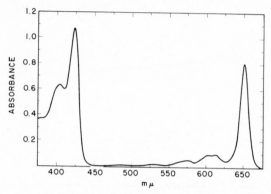

Fig. 3. Absorption spectrum of chlorobium chlorophylls 650 in ether (23) α_{sp} at $650 = 113.5$ (10). See Fig. 1 for definition of α_{sp}.

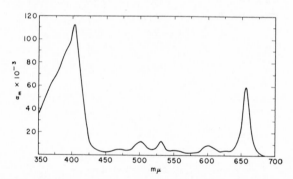

Fig. 4. Absorption spectrum of chlorobium pheophorbide 650 methylester (fraction 2) in ether (23).

parison of porphyrin derivatives with synthetic samples of known structure (11–19).

The presence of ring V in the "650" series as it occurs in pyropheophorbide a was indicated by (a) the "rhodo" type spectrum of the porphyrins obtained by treatment of the "650" fractions with hydriodic acid in acetic acid (65°); (b) the fact that fraction 6 could be dehydrated to yield pyropheophorbide a, and (c) the identity of the visible absorption spectra of all six fractions.

Treatment of the "660" free acids with hydriodic acid yielded in each case a porphyrin with an "etio" instead of a "rhodo" spectrum despite the presence of the cyclopentanone ring. This type of spectrum resulted from the presence of a "meso" substituent. The presence of the δ-substituent was shown by proton magnetic resonance studies (16, cf. 20). Confirmation was provided by the fact that the δ-phylloporphyrins from fractions 5 and 6 were identical with synthetically prepared samples (21).

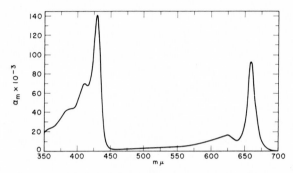

FIG. 5. Absorption spectrum of chlorobium chlorophylls 660 in ether (14).

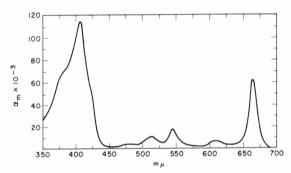

FIG. 6. Absorption spectrum of chlorobium pheophytins 660 in ether (14).

Recent determinations of the molecular weights by mass spectrometry (22) of the methyl esters of fractions 1, 2, 4, and 6 and of the pyrroporphyrin from fraction 3 of the "650" series are in agreement with those predicted from previously published data (9, 18). The molecular weights of fractions 2 and 4 of the "660" series indicated the presence of a δ-methyl instead of an ethyl group as previously published (23). We now assign a δ-ethyl group to fractions 1 and 3. The structures of the different chlorophylls based on these results are given in Fig. 2.

Pigments with absorption spectra corresponding to those of the "660" and "650" pheophorbides have recently been reported from extracts of

Rhodopseudomonas spheroides grown in media containing 8-hydroxy-quinoline (*24*). These pigments, however, gave a positive phase test.

The absorption spectra of the chlorophylls and magnesium-free derivatives are shown in Figs. 3–6.

IV. Chlorophyllide(s) *c*

The presence of this pigment was noted as early as 1864 (*25, 26*) and was called "chlorofucine." It was believed by Willstätter and Page to be an artifact (*27*), but shown not to be by Strain *et al.* (*28, 29*). *In vivo* it absorbs at 640 mμ (*30*).

Recent work (*31*) in the author's laboratory has shown the presence of at least three different chlorophyllides *c* in extracts from *Phaeodactylum tricornutum*. All three contain a free carboxyl group. The wavelengths of the absorption maxima of the different fractions are identical, but the ratios of their absorbancies differ. Molecular weight calculations based on methoxyl content of mono- and dimethyl esters indicate values of 550 and 600. All three fractions give an orange phase test (*29*). The information obtained from the spectra of the "pheophorbides" indicate that they are similar to protopheophorbide. Exhaustive catalytic reduction with hydrogen and subsequent degradation yields pyrroporphyrin XV from two of the fractions.

V. Bacteriochlorophyll *b*

This pigment was recently discovered in two strains of *Rhodospeudomonas* sp. It differs spectroscopically from bacteriochlorophyll *a* by the fact that the "blue" and "red" maxima are shifted 10 and 20 mμ, respectively, toward longer wavelengths in acetone solution (*32, 33*). *In vivo* it absorbs at 1012 mμ (*34*). It gives a positive phase test, and from the spectral changes on reduction with borohydride appears to possess the 2-carbonyl group.

VI. Seed-Coat Protochlorophyll

Extraction of seed coats of *Cucurbita pepo* with acetone yielded two fractions (*35*). One was identical with protochlorophyll, while the other possessed the properties of Mg 2,4-divinylpheoporphyrin a_5, which has been identified as bacterioprotochlorophyll *a* (*36*).

VII. "P750" of Blue-Green Algae

The presence of a pronounced absorption band at 750 mμ in *Anacystis nidulans* was first reported by Govindjee *et al.* (*37*). Gassner also found a similar band in *Synechococcus cedrorum* (*38*) and noted a similarity to bacteriopheophytin *a*. In acetone solution it absorbed at 730 mμ. No photosynthetic function has yet been ascribed to it.

VIII. "F698"

Broyde and Brody (*39*) reported the presence of traces of a chlorophyll-like pigment in samples of chlorophyll *a* or chlorophyll *b*. It was characterized by a fluorescence maximum at 698 mμ. Difference spectroscopy indicated absorption maxima at about 415 and 675 mμ in acetone solution.

Interest in this pigment comes from the fact that the temperature dependence and the quenching characteristics of its fluorescence parallel those of a pigment which fluoresces at 698 mμ in whole cells of *Chlorella pyrenoidosa*. There is speculation that the pigment responsible for the fluorescence at 698 mμ in whole cells is connected with the photochemistry of system II in photosynthesis.

IX. Chlorophyll *e*

Strain (*40*) reported the presence in small amounts of a pigment which absorbs at 415 and 654 mμ (methanol solution) in extracts from *Tribonema bombycinum*. No further information is available.

References

(*1*) W. M. Manning and H. H. Strain, *J. Biol. Chem.* **151**, 1 (1943).
(*2*) J. H. C. Smith and A. Benitez, *in* "Moderne Methoden der Pflanzenanalyse" (K. Paech and M. V. Tracey, eds.), Vol. IV, pp. 142-196. Springer, Berlin, 1955.
(*3*) A. S. Holt and H. V. Morley, *Can. J. Chem.* **37**, 507 (1959).
(*4*) A. S. Holt, *Can. J. Botany* **39**, 327 (1961).
(*5*) H. Sagromsky, *Ber. Deut. Botan. Ges.* **73**, 3 (1960).
(*6*) H. Sagromsky, *Ber. Deut. Botan. Ges.* **73**, 358 (1960).
(*7*) M. R. Michel-Wolwertz, C. Sironval, and J. C. Goedheer, *Biochim. Biophys. Acta* **94**, 584 (1965).
(*8*) H. Rapoport and H. P. Hamlow, *Biochem. Biophys. Res. Commun.* **6**, 134 (1961).
(*9*) A. S. Holt, D. W. Hughes, H. J. Kende, and J. W. Purdie, *Plant Cell Physiol.* (*Tokyo*) **4**, 49 (1963).

(*10*)   R. Y. Stanier and J. H. C. Smith, *Biochim. Biophys. Acta* **41**, 478 (1960).

(*11*)   D. W. Hughes and A. S. Holt, *Can. J. Chem.* **40**, 171 (1962).

(*12*)   H. V. Morley, F. P. Cooper, and A. S. Holt, *Chem. & Ind.* (*London*) p. 1018 (1959).

(*13*)   A. S. Holt and H. V. Morley, *J. Am. Chem. Soc.* **82**, 500 (1960).

(*14*)   A. S. Holt and H. V. Morley, *in* "Comparative Biochemistry of Photoreactive Systems" (M. B. Allen, ed.), p. 169. Academic Press, New York, 1960.

(*15*)   A. S. Holt and D. W. Hughes, *J. Am. Chem. Soc.* **83**, 499 (1961).

(*16*)   A. S. Holt, D. W. Hughes, H. J. Kende, and J. W. Purdie, *J. Am. Chem. Soc.* **84**, 2835 (1962).

(*17*)   J. L. Archibald, S. F. MacDonald, and K. B. Shaw, *J. Am. Chem. Soc.* **85**, 644 (1963).

(*18*)   J. W. Purdie and A. S. Holt, *Can. J. Chem.* **43**, 3347 (1965).

(*19*)   A. S. Holt, J. W. Purdie, and J. W. F. Wasley, *Can. J. Chem.* **44**, 88 (1966).

(*20*)   R. B. Woodward and V. Škarić, *J. Am. Chem. Soc.* **83**, 4676 (1961).

(*21*)   J. L. Archibald, D. M. Walker, K. B. Shaw, A. Markovac, and S. F. MacDonald. *Can. J. Chem.* **44**, 345 (1966).

(*22*)   A. H. Jackson, personal communication (1965).

(*23*)   A. S. Holt, *in* "Biochemistry of Plant Pigments" (T. W. Goodwin, ed.), Part I, p. 3. Academic Press, New York, 1965.

(*24*)   O. T. G. Jones, *Biochem. J.* **88**, 335 (1963).

(*25*)   G. G. Stokes, *Proc. Roy. Soc.* **13**, 144 (1864).

(*26*)   H. C. Sorby, *Proc. Roy. Soc.* **21**, 442 (1873).

(*27*)   R. Willstätter and H. J. Page, *Ann. Chem.* **404**, 237 (1914).

(*28*)   H. H. Strain and W. M. Manning, *J. Biol. Chem.* **144**, 625 (1942).

(*29*)   H. Strain, W. M. Manning, and G. Hardin, *J. Biol. Chem.* **148**, 655 (1943).

(*30*)   C. S. French and R. F. Elliot, *Carnegie Inst. Wash. Year Book* **57**, 278 (1958).

(*31*)   J. W. F. Wasley, J. W. Purdie, and A. S. Holt, unpublished data (1964).

(*32*)   K. E. Eimhjellen, O. Aasmundrud, and A. Jensen, *Biochem. Biophys. Res. Commun.* **10**, 232 (1963).

(*33*)   K. E. Eimhjellen, A. Jensen, and O. Aasmundrud, *Biochim. Biophys. Acta* **88**, 466 (1964).

(*34*)   A. S. Holt and R. K. Clayton, *Photochem. Photobiol.* **4**, 829 (1965).

(*35*)   O. T. G. Jones, *Biochem. J.* **96**, 6P (1965).

(*36*)   O. T. G. Jones, *Biochem. J.* **89**, 182 (1963).

(*37*)   Govindjee, C. Cederstrand, and E. Rabinowitch, *Science* **134**, 391 (1961).

(*38*)   E. B. Gassner, *Plant Physiol.* **37**, 637 (1962).

(*39*)   S. B. Broyde and S. S. Brody, *Biochem. Biophys. Res. Commun.* **19**, 444 (1965).

(*40*)   H. H. Strain, *in* "Manual of Phycology" (G. M. Smith, ed.), p. 243. Chronica Botanica, Waltham, Massachusetts, 1951.

—5—

The Synthesis of Chlorophyll a

WALTER LWOWSKI

Department of Chemistry, Yale University, New Haven, Connecticut

I. Introduction

Work on the synthesis of chlorophyll *a* began in Hans Fischer's laboratory even before all the structural details of chlorophyll were known. Fischer outlined a number of partial synthetic sequences based on conversions discovered in the course of his structural work. Since Fischer's death, these sequences have been supplemented by Strell, and more recently by Inhoffen. The yields in many of these conversions, all of which have been studied only through the use of numerous different relay substances—obtained by degradation of natural chlorophyll—are very low; some of the structures of the intermediates involved have not been rigorously established, by modern standards, and some connections between various partial syntheses have yet to be substantiated. None the less, there is reason to suppose that Fischer's scheme for chlorophyll synthesis can be brought to a successful conclusion, if the effort is continued.

R. B. Woodward's (*1*) synthesis consists of a continuous sequence of fully synthetic reactions giving yields high enough to proceed from simple pyrroles to racemic purpurin 5 and allowing the resolution of the latter into enantiomers. One of these was shown to be identical with material from natural sources. The short subsequent path from purpurin 5 to chlorin e_6 is followed by the simple steps, known for a long time, from

119

chlorin e_6 to chlorophyll *a* itself. Thus, Woodward's synthesis makes it possible actually to prepare chlorophyll *a* from simple pyrroles.

In the following sections, Fischer's synthetic work will be portrayed by sketching those of his synthetic attempts that show best his concepts for a chlorophyll synthesis. This will be followed by a discussion of M. Strell's attempts to complete Fischer's work, and finally by a description of Woodward's synthesis.

II. Hans Fischer's Work on the Synthesis of Chlorophyll *a*

A. Introduction

After he had synthesized hemin in 1930, Fischer began work on the synthesis of chlorophylls, even before he had established all the details of their structures. He and his collaborators investigated numerous substances related to chlorophylls *a* and *b* and discovered many interconversions of these compounds. The basic approach to the synthesis involved making suitably substituted porphyrins and then introducing the still-missing substituents as well as the two "extra" hydrogens in positions 7 and 8. Fischer has discussed various methods for synthesizing porphyrins (2–4). All these reactions involve condensations of two pyrromethene "halves" under drastic conditions and usually give poor yields. Whenever the two components are different from one another, mixtures of isomeric porphyrins are produced. In addition to isomeric porphyrins, transformations of the starting materials under the extreme reaction conditions lead to the formation of other, more extensively modified porphyrins, and very formidable separation problems are created (4).

The synthesis of phylloporphyrin in 3.7% yield, in admixture with numerous other porphyrins, for example, is accomplished by melting the two pyrromethene halves with pyruvic acid (5):

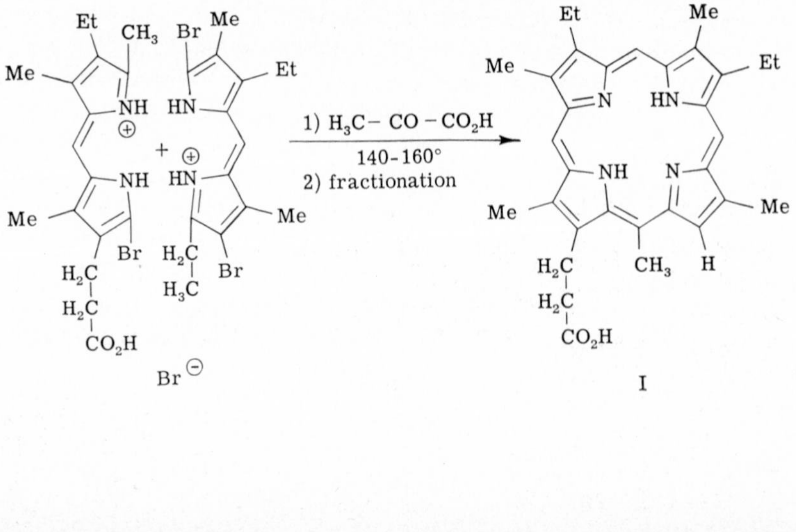

The reactions discovered during the broad-scale investigation of chlorophyll chemistry, Fischer then incorporated in his synthetic plan whenever appropriate. A number of such approaches were unsuccessful and modifications in the plan were made according to the difficulties encountered. For example, the special problems associated with the reactivity of the vinyl group in position 2 led Fischer to attempt first the synthesis of mesochlorophyll *a* (containing ethyl instead of vinyl in the 2-position) (6).

B. Attempted Synthesis of Mesochlorophyll *a*

After having experienced difficulties in a variety of conversions of compounds containing a vinyl group in the 2-position, Fischer decided (6) to attempt the synthesis of mesochlorophyll *a* (II), which lacks the reactive olefinic grouping.

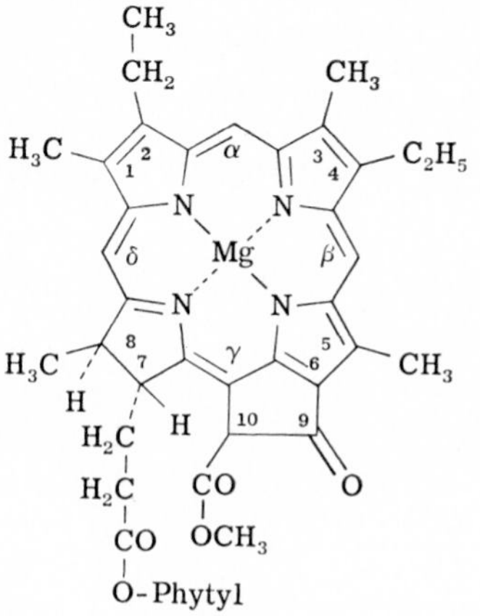

II

Heating of phylloporphyrin (I) to 185° with sodium ethoxide in ethanol for 8 hours led to a substance which was regarded as racemic phyllochlorin (III), in 20% yield (7). The nature of the process is not known in detail, but the similarity of the qualitative visible spectrum of the product to that of optically active phylloporphyrin, obtained by degradation of chlorophyll, provides support for the assumed structure.

Oxidation of III, derived from natural sources, with potassium permanganate gave a 10% yield of the γ-aldehyde, mesopurpurin 3 (IV) (8), identical with material prepared by degradation.

A parallel oxidation of the racemic material, presumably of the structure (III), obtained by synthesis, gave an inactive aldehyde, assumed to be the racemic form of (IV).

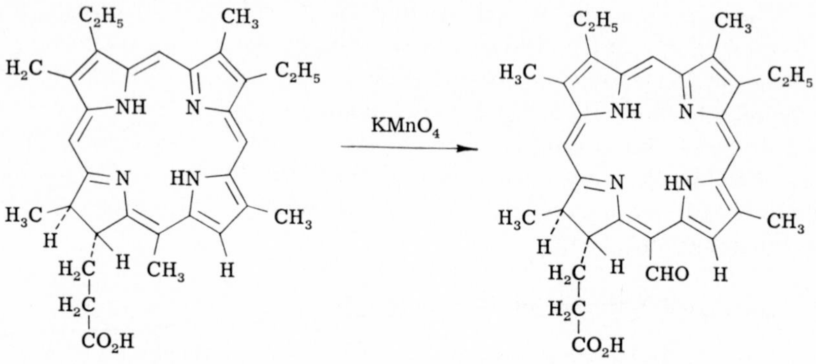

III

IV

Mesopurpurin 3 (IV) from natural sources was treated with HCN and then with methanolic HCl (6) to give a substance, chlorin 0.5,* to which the structure (Va) was assigned. However, the mode of formation and the properties of this substance suggest strongly that it should be reformulated as (Vb). When chlorin 0.5 was hydrogenated to a leuco compound and reoxidized mesopyrrochlorin-γ-glycolic acid dimethyl ester (VI) was obtained (6).

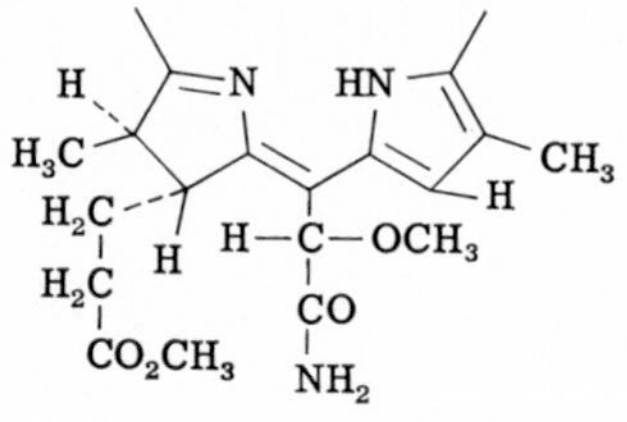

Va

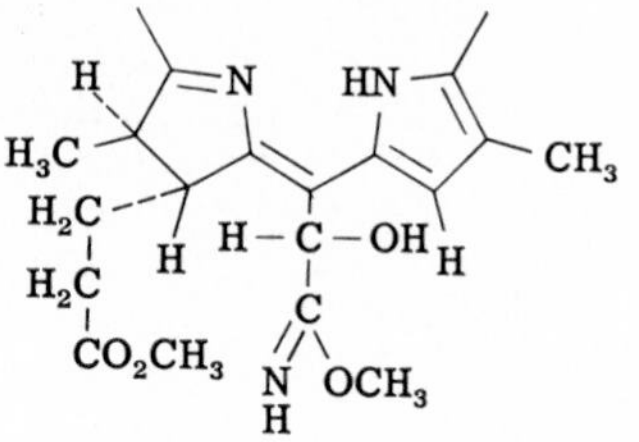

Vb

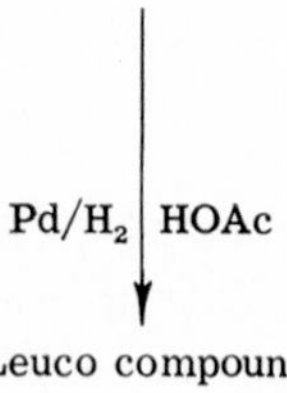

Pd/H₂ | HOAc

Leuco compound

Treatment of (VI) with hydriodic acid in glacial acetic acid gave principally isochloroporphyrin e₄ (VII) and a little pyrroporphyrin, plus a small quantity of a chlorin, not characterized in any way, of which it was stated (6) variously that (a) "it is still not clear whether starting material (VI) or mesoisochlorin e₄ (VIII) is in hand"; (b) "with great probability it is mesoisochlorin e₄ (VIII)." For details of the reaction of chlorins with hydriodic acid see Chapter 3, Section II, F.

---

* So called from its HCl number (see Chapter 3).

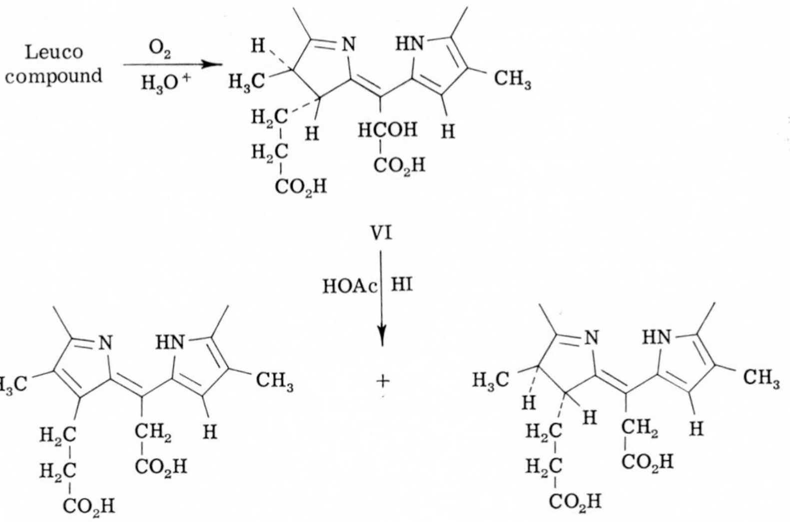

VI

HOAc | HI

VII   +   VIII

Accepting the tentative identification of (VIII), the synthetic scheme continues with the conversion of (natural) mesoisochlorin e_4 to its ferric complex (IX). (IX) was treated (9) with Fischer's "dichloro ether" reagent, a mixture of dichloromethyl ethyl ether, phosphorous oxychloride, and ethyl formate (10). This reaction introduced a carbon bridge (C-9) between position 6 and the γ-side chain. Removal of the iron and re-esterification with diazomethane gave a product regarded as 9-hydroxydeoxo-meso-pheophorbide *a* dimethyl ester (X), in spite of its not having the predicted qualitative visible absorption spectrum (11, 12). Oxidation of this substance with *p*-quinone in acetic acid gave in 2% yield a product regarded as methyl *meso*pheophorbide (XI), although

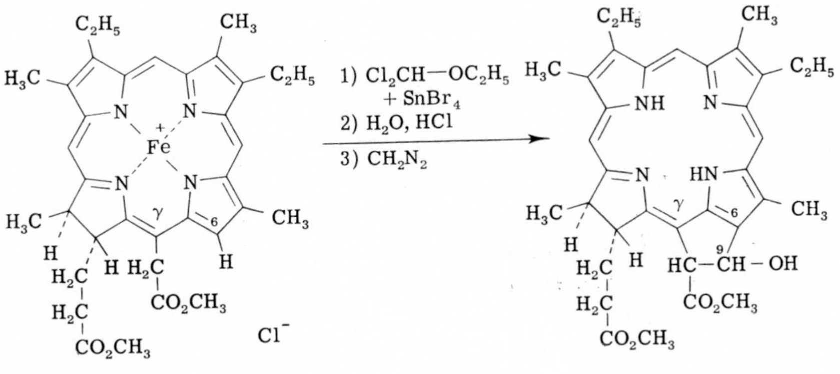

IX   1) $Cl_2CH—OC_2H_5$ + $SnBr_4$ 2) H_2O, HCl 3) CH_2N_2 → X

it could not be crystallized and was characterized only by qualitative
visible absorption spectra.

The conversion of mesopheophorbide to mesochlorophyll *a* should be
possible using the paths already established for the conversion of pheophorbide to chlorophyll *a* itself (*13, 14*) (also see Chapter 2, Section II,
A, 3). Fischer believed that repetition on a larger scale of the reactions

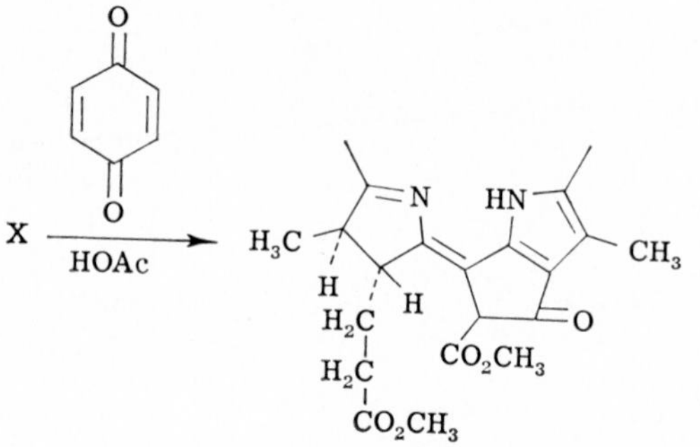

XI

which had not been unambiguously established would permit proper
characterization of the presumed intermediates. Had this been possible,
it would indeed have required only the resolution of one of the synthetic
intermediates to bring his plan for the synthesis of mesochlorophyll *a* to a
successful conclusion.

For the synthesis of chlorophyll *a* itself Fischer laid down a plan
closely paralleling that described above for mesochlorophyll *a*. Since
it was clear that the crucial vinyl group at position 2 could not be carried
through the often severe operations, it was decided to use the corresponding 2-unsubstituted, and at later stages 2-acetyl, analogs, which
afford possibilities for conversion to vinyl derivatives at a late stage of
the route. Fischer was unable to make much progress with this plan
before his death in 1945, but his collaborators carried on the work.

C. The Continuation of Fischer's Work on Synthesis

After World War II, A. Treibs, W. Siedel, and M. Strell set out to
complete Fischer's chlorophyll synthesis (*15*). Following the plan of
avoiding, as long as possible, intermediates containing a vinyl group at
C-2 (*16*), 2-deethylphylloporphyrin (XII) was synthesized in 0.6% yield,
along with more than ten other porphyrins, from the appropriate two
pyrromethene halves (*17*). A reaction sequence patterned after Fischer's
synthesis of pheoporphyrin a_5 dimethyl ester (XIII) (*18*) was to lead
to 2-devinylisochlorin e_4 monomethyl ester (XIX). (XII) had been re-

duced earlier (*19*) to a substance regarded as the racemic phyllochlorin (XIV) in unspecified yield; as in the case of the 2-ethyl analog (III), the assignment of structure rests solely upon comparisons of qualitative spectroscopic observations in the visible region. Authentic, optically active 2-devinylphyllochlorin, obtained by degrading chlorophyll, was oxidized to 2-devinylpurpurin 3 (XV) in 5% yield (*16*).

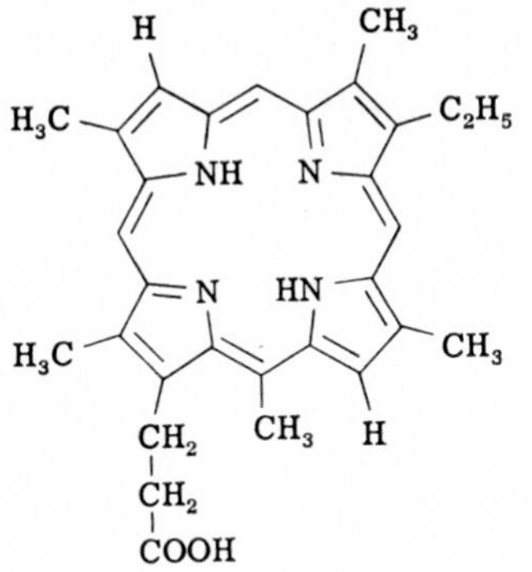

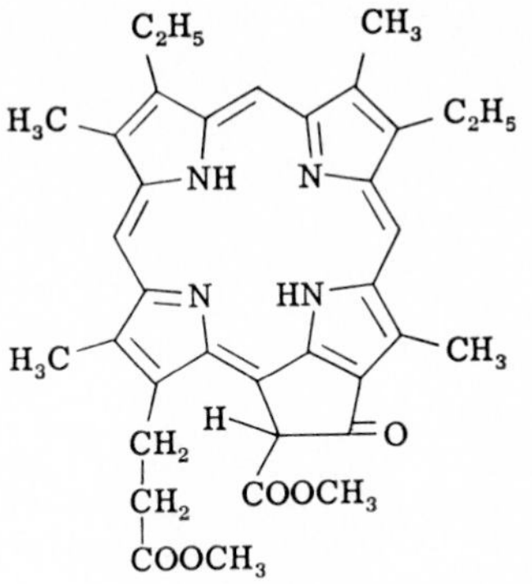

The subsequent conversions (XV through XVIII) (*20*) seem to be analogs to Fischer's (*18*) reactions in the porphyrin series, but Strell's (*20*) sequence has only been announced in a lecture, and no details have been given.

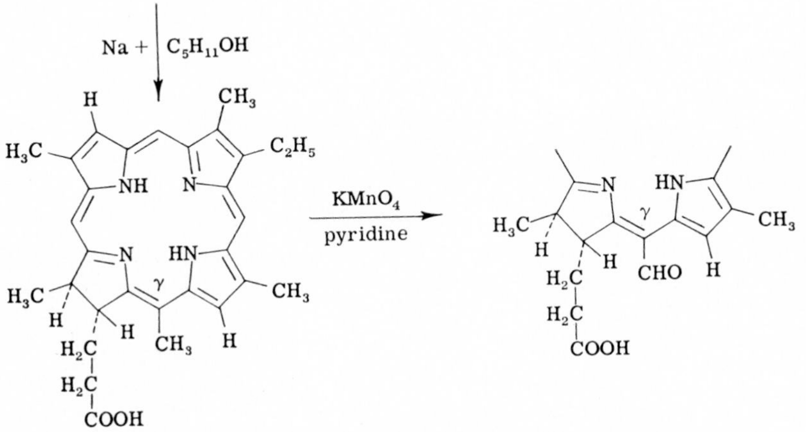

The next link in the synthetic sequence is a remarkable report by Fischer (*21*) that (XIX) can be acetylated in the 2- and 6-positions, then partially deacetylated by benzoyl chloride and pyridine to 2-mono-

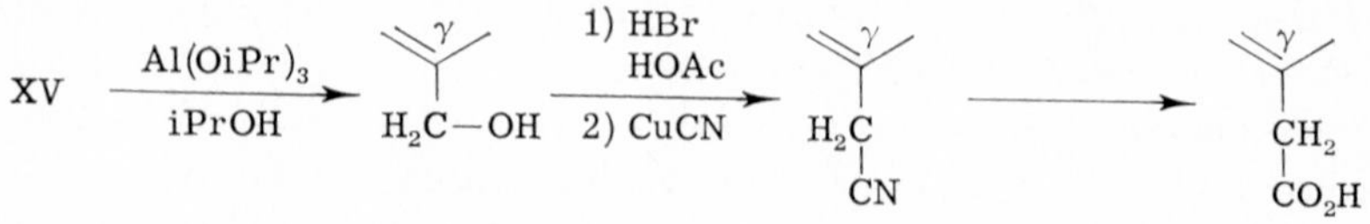

XV $\xrightarrow[\text{iPrOH}]{\text{Al(OiPr)}_3}$ (XVI) $\text{H}_2\text{C}-\text{OH}$ $\xrightarrow[\text{2) CuCN}]{\begin{array}{c}\text{1) HBr}\\\text{HOAc}\end{array}}$ (XVII) H_2C / CN $\longrightarrow$ (XVIII) CH_2 CO_2H

XVI	XVII	XVIII

acetyl-2-devinylisochlorin e_4. Inhoffen (22) has shown this claim to be in error but has recently developed (22) a method for the direct conversion of optically active (XIX), prepared by degradative methods, to the monoacetyl compound (XX).

XVIII $\xrightarrow[\text{fication}]{\text{esteri-}}$ (XIX) $\xrightarrow[\text{H}_2\text{O}]{\text{Ac}_2\text{O}}$ (XX)

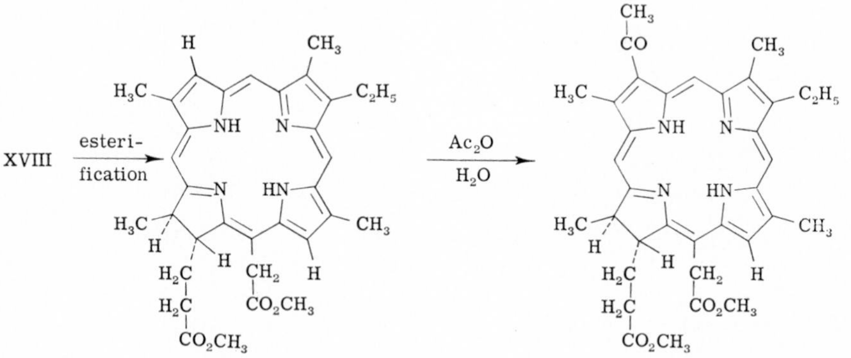

XIX	XX

Reduction of the 2-monoacetyl compound (XX), available at the time only by oxidation of naturally derived 2-α-hydroxyethylisochlorin e_4 (XXI), with sodium borohydride (23, 24) gave a 90% yield of crystalline 2-α-hydroxyethylisochlorin e_4 (XXI)—presumably a mixture of diastereomers differing in the configuration at C_α in the side chain at C-2. From (XXI), two routes were given. The Fe complex of (XXI) was treated (23) with Fischer's dichloro ether reagent (10) and then with alkali to give 2-(α-hydroxymeso)-9-hydroxydeoxopheophorbide *a* dimethyl ester (XXII), which was oxidized with platinum dioxide (23, 24) to the 9-oxo compound (XXIII). The structures thus obtained in unspecified yield by this partial synthetic route were characterized only by qualitative visible spectroscopic measurements; they were not obtained in the crystalline state, even though the corresponding authentic structures are described as crystalline compounds. Fischer (25) had reported that heating (XXIII) to 180° gave pheophorbide *a* (XXIV) in minute yield, identified only by its visible absorption spectrum. In 1960 the utterly premature claim was put forward that the transformations just described constituted the completion of Fischer's plan for the synthesis of chlorophyll.

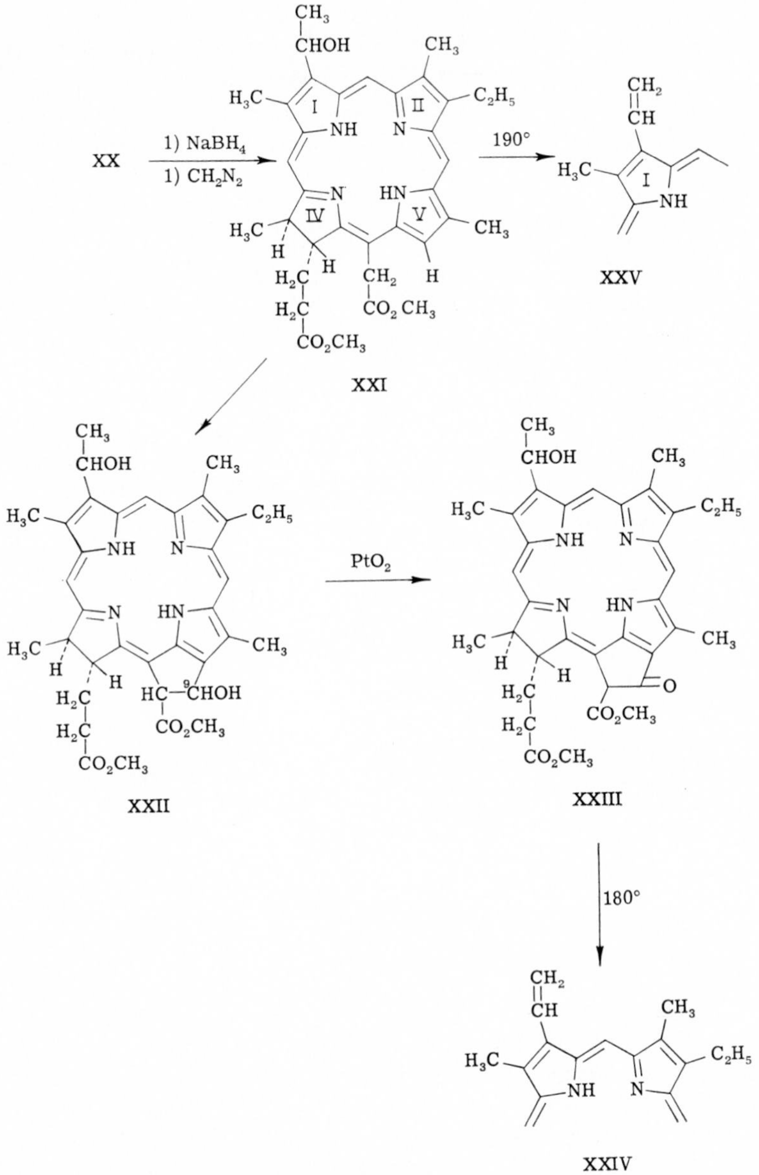

XX $\xrightarrow[\text{1) CH}_2\text{N}_2]{\text{1) NaBH}_4}$ (XXI) $\xrightarrow{190°}$ (XXV)

XXI

XXV

XXII

PtO$_2$

XXIII

180°

XXIV

In 1962, Strell (*24*) described a different route from (XXI) to (XXIV), which avoids the not definitively established pyrolysis step from (XXIII) to (XXIV). Vacuum pyrolysis of (XXI) gave an unspecified yield of isochlorin e$_4$ dimethyl ester (XXV), identified by its qualitative spectrum and mixture melting point with authentic material. The Fe

complex of (XXV), on treatment with the dichloro ether reagent and
tin tetrabromide, followed by hydrolysis and re-esterification, gave a 12%
yield of the desired 9-hydroxydeoxopheophorbide *a* dimethyl ester
(XXVI). Oxidation of (XXVI) with quinone in acetic acid produced
pheophorbide *a* dimethyl ester (XXIV) in 30% yield. (XXIV) and

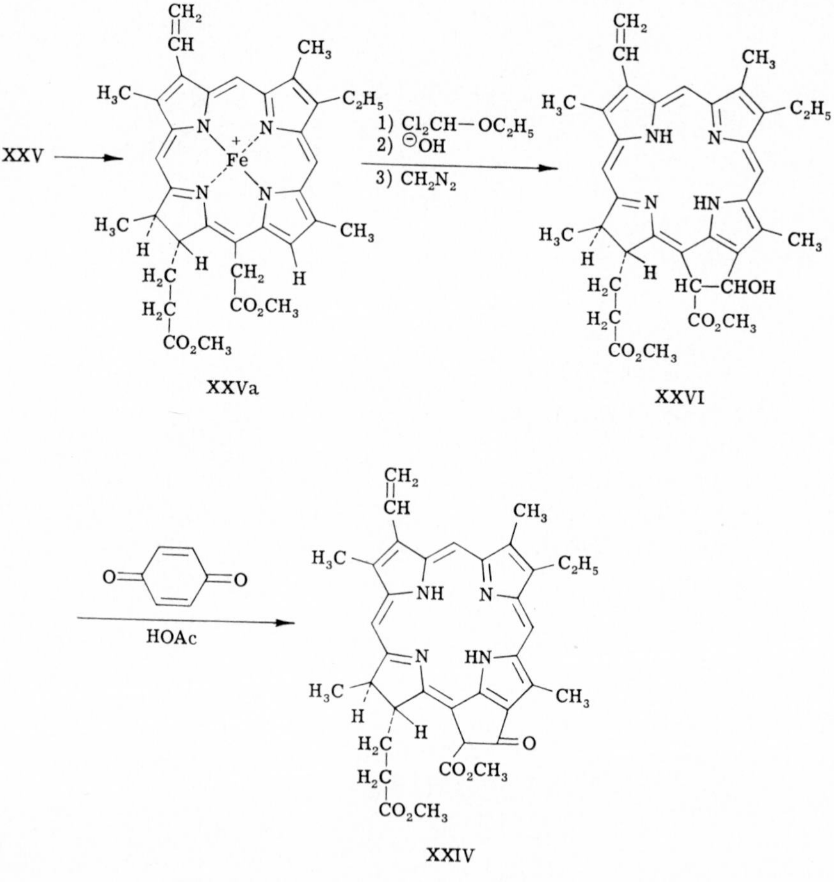

XXVa

XXVI

XXIV

(XXVI) were identified by qualitative absorption spectra, elemental
analyses, and mixture melting points.

The conversion of pheophorbide *a* to chlorophyll *a* had been carried
out earlier: Willstätter (26) had shown how to prepare the phytol ester
at the propionic acid side chain on position 7 and how to introduce
magnesium into the center of the molecule (27), and Fischer (13, 14)
subsequently improved these methods. Phytol had also been synthesized
earlier (28). The most complex of these reactions, the formation of the
magnesium complex, is discussed in Chapter 3, Section A, 3.

Thus, at the present writing (1965) the final realization of Fischer's plan for establishing many sequences of partial synthetic operations, which, taken together, would constitute in a formal sense a synthesis of chlorophyll, still requires (a) the resolution of the presumed synthetic 2-devinylphyllochlorin or 2-devinylpurpurin 3 into enantiomers and rigorous establishment of structure through comparison with authentic material from natural sources, and (b) provision of experimental evidence in support of the transformations (XV) to (XVIII). Such a scheme, conceived a generation ago, should not be expected to meet the synthesis standards of today, and it is of course inconceivable that it could actually be used as a basis for the preparation of chlorophyll by synthesis. Nevertheless, its proper realization—with strict adherence to the careful experimental standards Fischer himself would certainly have demanded—would constitute an appealing tribute to the great pioneer of chlorophyll chemistry.

III. Woodward's Synthesis of Chlorophyll *a*

A. Introduction

In 1956, R. B. Woodward began work on a synthesis designed to provide a continuous sequence of unambiguous conversions, leading from simple pyrroles to chlorophyll *a*. Careful study of the chemistry of pyr-

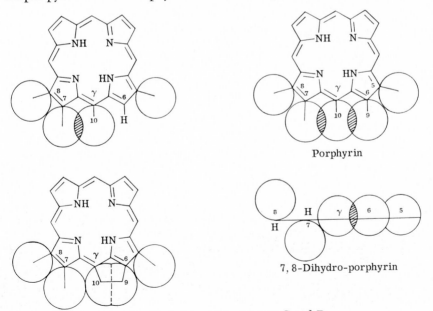

Porphyrin

7, 8-Dihydro-porphyrin

FIG. 1.   Steric interactions in rings C and D.

roles, porphyrins, and chlorins resulted in a plan for synthesis based on a number of premises: (a) A basic porphyrin should be synthesized in high yield, and free from isomers, so that the long sequence of further transformations could be carried out on a reasonable scale. (b) The introduction of the "extra" hydrogens exclusively in positions 7 and 8 and with the required *trans* stereochemistry, was expected to be facilitated by the relief of steric compression between the groups at 7, 8, and γ. Indeed, it appeared that the *trans*-7,8-dihydroporphyrins are the most stable of the tautomers having the substitution pattern of chlorophyll. Thus, thermodynamically controlled product formation was expected to lead to the right dihydroporphyrins once the sterically important substituents had been placed in their proper positions. Steric factors were also expected to aid the closure of the carbocyclic ring bridging positions 6 and γ. Figure 1 illustrates these steric interactions. (c) Although an acetic acid side chain in γ-position appears desirable for the formation of the carbocyclic ring, this side chain should be introduced late in the synthesis because it is easily extruded from dihydroporphyrins, e.g.:

(d) The 2-vinyl group should be formed only at a late stage in the synthesis because of its high reactivity, which is similar to that of the vinyl group in styrene.

B. Woodward's Porphyrin Synthesis

The porphyrin (XXVII) was chosen as the first synthetic goal. In the 2-position, it bears a β-aminoethyl group, later to be converted to the vinyl function, and in the γ-position (XXVII) contains a propionic acid side chain, later to be degraded to an aldehyde group and then to be built up to the desired acetic acid side chain.

The synthesis (*1, 29*) started with four pyrroles corresponding to the rings A, B, C, and D in (XXVII). The pyrroles that were to form rings C and D were already known (*30, 31*). The pyrrole for ring A was made from 2,5-dicarboxy-3-methyl-4-formylpyrrole (*32*).

The progenitor of ring B was obtained by chlorination of 2-(β,β-diacyanovinyl)-3,5-dimethyl-4-ethylpyrrole (*33*).

XXVII

XXVIII

XXIX

The component containing rings A and D was prepared by condensing (XXVIII) with (XXX) (*31*) and reducing the pyrromethene with sodium borohydride. Although (XXVIII) has two free α-positions, only one dipyrromethene (XXXI) was formed. The lack of reactivity of the 5-position in (XXVIII) is due presumably to steric hindrance and perhaps to the electron-withdrawing effect of the β-ethylammonium side chain in position 4. The recrystallized dihydrobromide of the pyrromethene (XXXI) was obtained in 71% yield.

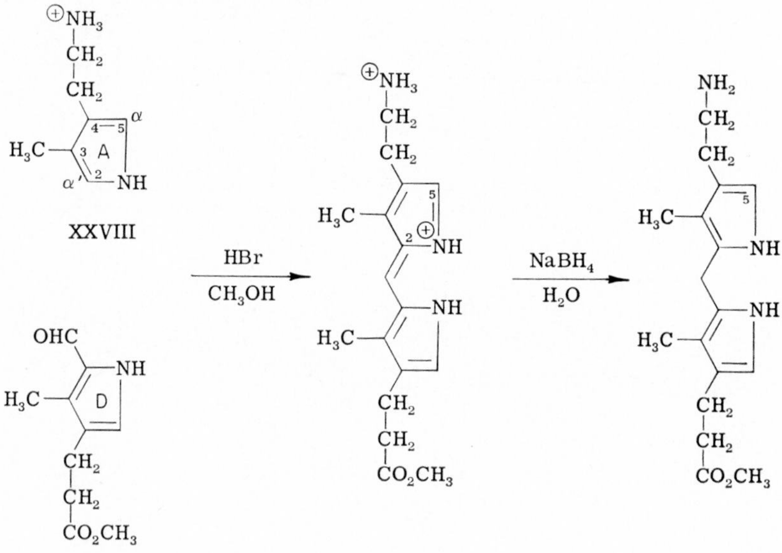

To build up the other half of the porphyrin ring, (XXIX) was condensed with 3-carbethoxy-4-methylpyrrole (XXXIII) (*30*) to give 3′,4-dimethyl-3-ethyl-4′-carbethoxy-5-(β,β-dicyanovinyl)dipyrrylmethane (XXXIV). An undesirable isomer of (XXXIV) (with methyl and carbethoxy reversed on ring C) might have been formed because (XXXIII) has two free α-positions. However, the carbethoxy group deactivates the adjacent α-position through its electron-withdrawing effect and also sterically hinders attack on the α-position more than the smaller methyl group discourages attack on the α′-position. Moreover, the α′-position is activated electronically by the adjacent methyl group. (XXXIV) was obtained in 61% yield, with respect to (XXIX), and 44%, based on (XXXIII).

While the α-position in (XXXIII) is relatively unreactive, it can be made to react in (XXXIV), where the α′-position is already occupied.

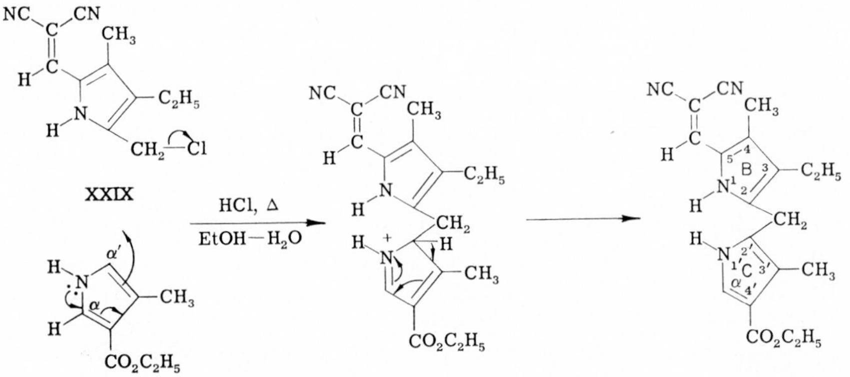

XXIX

XXXIII

HCl, Δ
EtOH—H₂O →

XXXIV

In the presence of anhydrous zinc chloride, (XXXIV) was condensed with β-carbomethoxypropionyl chloride to give (XXXV). Once the condensation steps had been carried out, the dicyanovinyl group was converted to the aldehyde group in the 5-position which was needed for the assembly of the porphyrin ring. Vigorous hydrolysis, followed by re-esterification of the carboxyl groups, gave (XXXVI). (XXXIV) was obtained in 76% yield, (XXXVI) in 80% yield. (XXXVI) contains an

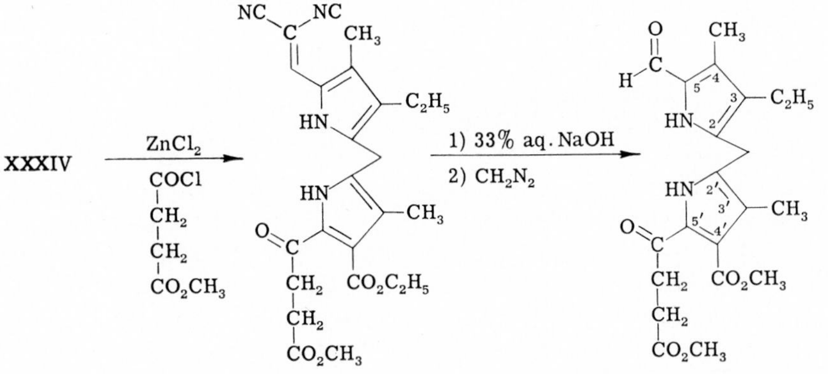

XXXIV

ZnCl₂
COCl
CH₂
CH₂
CO₂CH₃ →

XXXV

1) 33% aq. NaOH
2) CH₂N₂ →

XXXVI

aldehyde function in position 5 and a ketone function in position 5′ and is thus suitable for condensation with dipyrrylmethanes having free α-positions in both rings. However, an unsymmetrical dipyrrylmethane, such as (XXXII), can form two isomeric condensation products: the desired one in which ring A is connected to ring B, and ring C to ring D; and another condensation product in which rings A and C, and B and D, are linked together. Such a reaction would waste much of the starting

materials and the product mixture would be hard to separate. To overcome this difficulty, it was planned to link the two halves, (XXXII) and
(XXXVI), together by forming a Schiff base from the aldehyde function
on ring B and the amino group on the side chain of ring A. Such a
link would not interfere with the desired condensation; indeed, a protonated azomethine function on ring B was expected to react even more
readily than an aldehyde group. The "wrong" isomer, however, can no
longer be formed, for rings B and D can no longer come close to each
other in an intramolecular reaction. In practice, a new problem was encountered: the aldehyde group proved not reactive enough to form the
Schiff base under conditions that would not also destroy the dipyrrylmethanes. The aldehyde carbon is only weakly electrophilic, due to its
resonance with the pyrrole ring, while the dipyrrylmethanes are very
sensitive toward acid (34) which otherwise might have been used to
catalyze the Schiff base formation. The problem finally was solved by
converting the aldehyde (XXXVI) to the corresponding thioaldehyde
(XXXVII) in 79% overall yield, by way of a simple azomethine.

XXXVI   XXXVII

The thioaldehyde (XXXVII) reacts rapidly with amines—including
(XXXII)—without acid catalysis. After considerable experimentation,
conditions were found under which the porphyrin (XLII) was obtained
in 50% overall yield from (XXXI) and (XXXVII). The dipyrrylmethane
(XXXII) was not isolated in this procedure but was prepared in aqueous solution by reducing (XXXI), and then extracted into dichloromethane. The dichloromethane solution was passed through a layer of anhydrous sodium sulfate into a flask containing an equivalent amount of
the thioaldehyde (XXXVII). Then the solution was quickly evaporated
in vacuo; the residue was dissolved in a 1:2 mixture of benzene and
methanol and immediately injected into a stirred, ice-cold 12 N solution
of hydrogen chloride in methanol. Weaker acid did not cause condensation on both ends of the system rapidly enough to prevent extensive
decomposition of an intermediate containing four pyrrole rings in a chain.

A dihydroporphyrin (XL) is the first product in the reaction sequence containing an uninterrupted sequence of conjugated double
bonds. Acids no longer destroy it but establish an equilibrium of tauto-

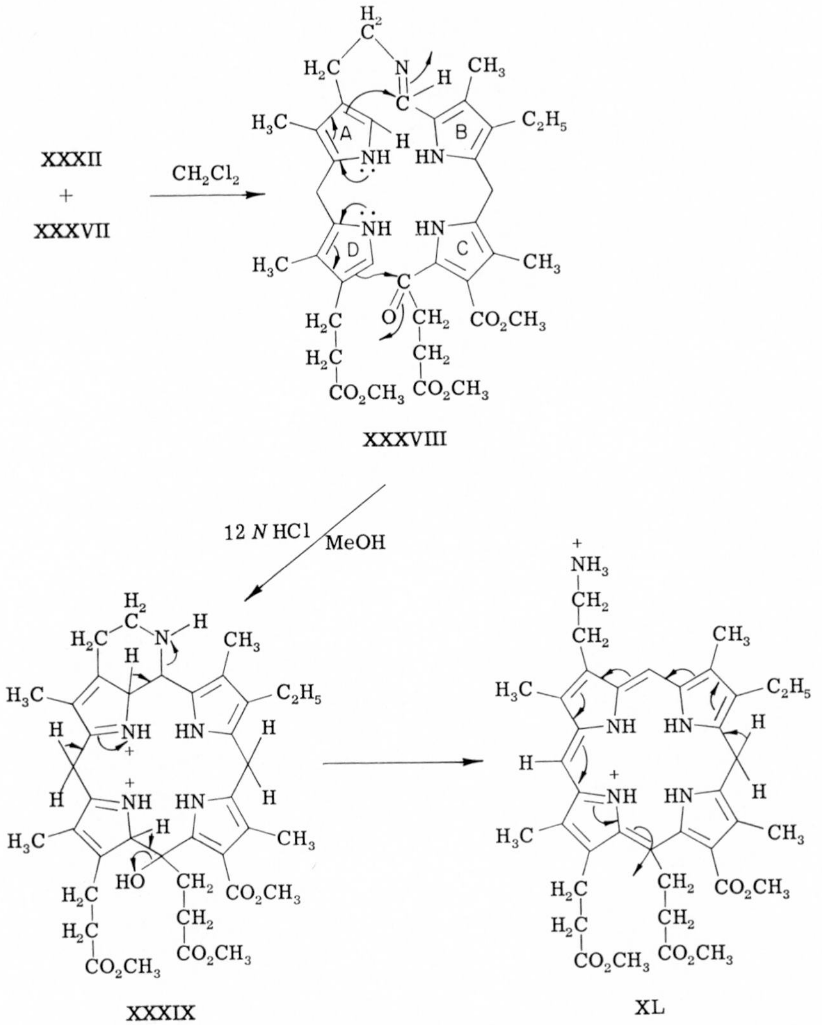

mers containing a saturated carbon at the positions α, β, γ, or δ. (XLI) is the stablest of these tautomers because its propionic acid chain in the γ-position has moved out of the plane of the 16-membered ring and no longer interferes with the groups in positions 6 and 7. Thus (XLI) is the first isolable product of the condensation reaction.

The resonance contributors (XLIa) and (XLIb) illustrate the de-localization which occurs in this new class of dihydroporphyrins, called phlorins. Their salts give deep green solutions, while the free phlorins dissolve with a pure blue color in organic solvents. Phlorins are rather stable substances and can be dissolved in concentrated sulfuric acid and

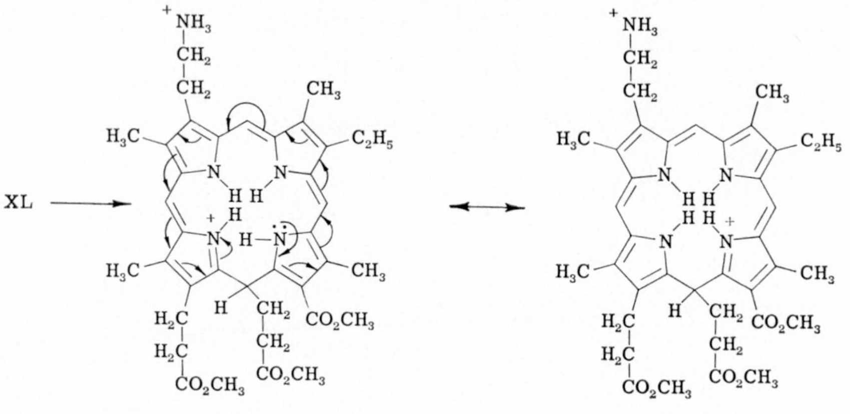

XLIa   XLIb

recovered without change. They are, however, easily oxidized to porphyrins. In the synthesis of (XXVII), this was done by adding iodine to the reaction mixture after cyclization. The mixture was kept at room temperature for an hour, the solvents were evaporated *in vacuo*, and acetic acid and pyridine were added to the residue in order to acetylate the amino group on the side chain on ring A. Chromatography gave the acetylated porphyrin (XLII), free from other porphyrins and in 50% overall yield from (XXXI) and (XXXVII).

Had the synthesis been carried out with an acetic acid side chain in the γ-position, rather than with a propionic acid side chain, loss of acetic acid and formation of porphyrin (XLIII) might have consumed much or all of the phlorin.

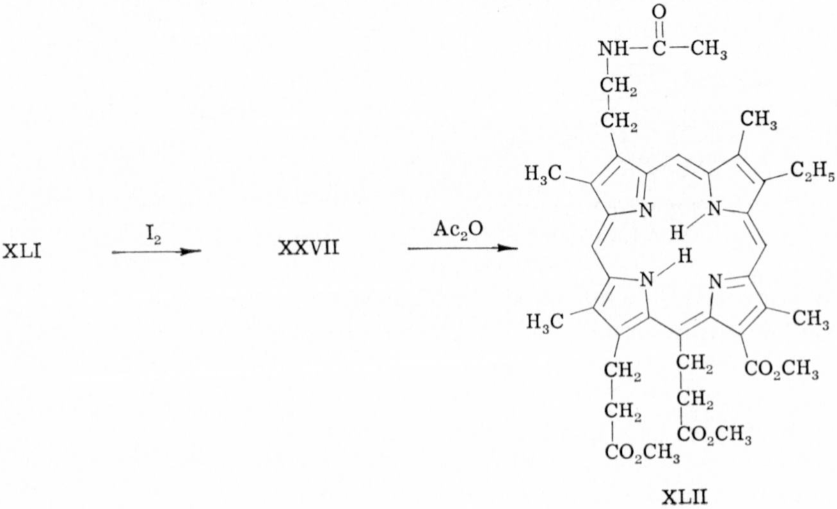

XLI $\xrightarrow{\text{I}_2}$ XXVII $\xrightarrow{\text{Ac}_2\text{O}}$

XLII

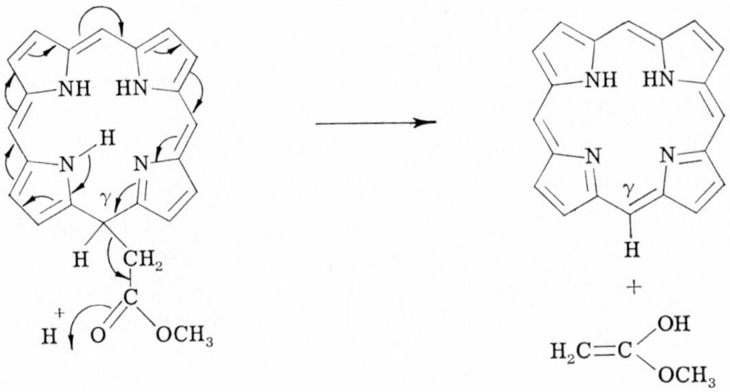

XLIII

C. Woodward's Route from Porphyrin to Chlorophyll

In the porphyrin (XLII), full conjugation around the 16-membered ring provides resonance stabilization greater than that in the phlorins. However, much more steric strain is present, since the side chains in positions 6, γ, 7, and 8 are all in the same plane. According to premise b of the general synthetic plan (Section III, A), such steric crowding should facilitate isomerization. Indeed, heating (XLII) in acetic acid under nitrogen led to tautomerization, but the product was not a chlorin (in which the groups at 7 and 8 would have left the plane of the 16-membered ring), but a new phlorin (XLIV) in which the γ-carbon again had become tetrahedral. The phlorin (XLIV) was readily oxidized by air, and heating (XLII) in the presence of air gave an almost quantitative yield of the porphyrin (XLV).

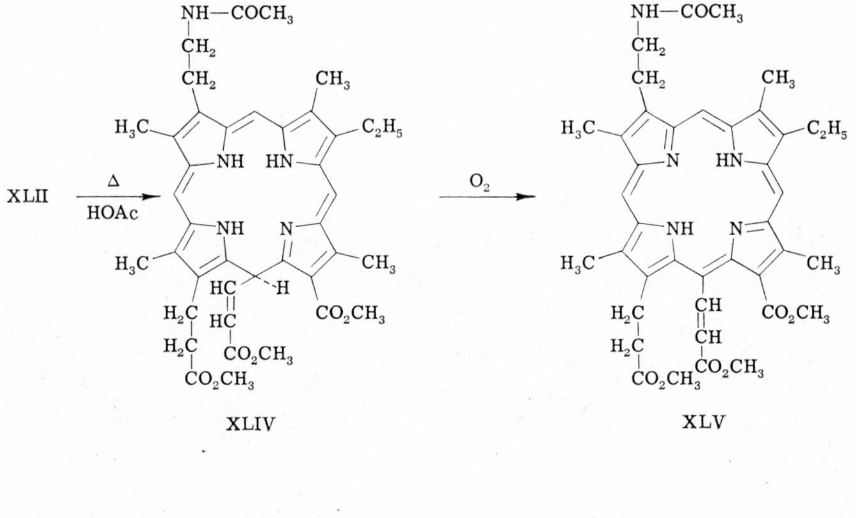

XLIV

XLV

The porphyrin (XLV) could be formed in quantitative yield only because of the irreversibility of the oxidation step. The steric compression of the side chains attached to rings C and D again is severe. Consequently, tautomerization occurs again on heating of (XLV) in acetic acid to 110° for 30 hours under nitrogen. An equilibrium mixture containing 63% (XLVI), a purpurin, is produced. (Purpurins form a subclass of chlorins; their visible absorption spectra are shifted by an unsaturated substituent in γ-position, resulting in a purple rather than green color.) The equilibration of (XLV) with (XLVI) is the first observed example of a reversible interconversion between a porphyrin and a chlorin.

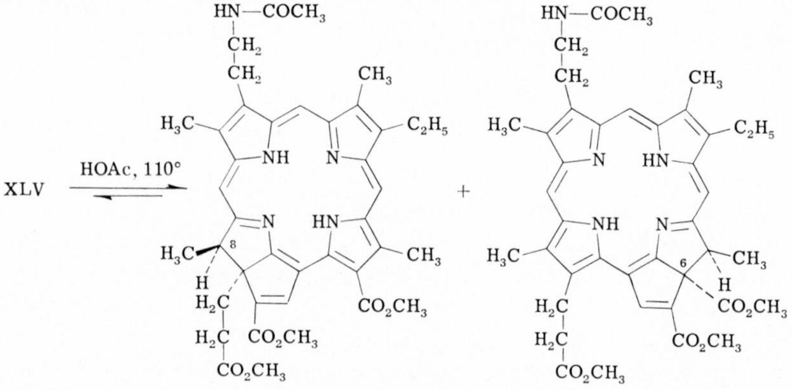

XLVI   XLVII

Apart from (XLVI), the equilibrium mixture consists principally of the starting material (XLV). A small component (XLVII) was identified as an isomer of (XLVI) in which the five-membered carbocyclic ring had been closed not to position 7, but to 6. In the formation of (XLVII), the conjugation of the 6-carbomethoxy group with the 16-membered ring is destroyed. This, together with the greater effective bulk of the carbomethoxy group relative to a propionic acid chain, appears to be why (XLVII) is a minor component in the equilibrium mixture. Since (XLV) can be recovered from the mixture, it can be converted to (XLVI) in 70% yield.

The stereochemistry at positions 7 and 8 must correspond to the more stable arrangement, since (XLVI) is formed in an equilibrium reaction. Thus (XLVI) must contain the side chains at 7 and 8 *trans* to each other.

Before further conversions on rings C and D of (XLVI) were undertaken, the β-acetylaminoethyl side chain in the 2-position was converted to a vinyl group in 84% yield. Hydrolysis to the free amine, followed by exhaustive methylation with methyl sulfate and treatment with methanolic sodium hydroxide led to Hofmann elimination and formation of the 2-vinyl compound (XLVIII).

(XLVIII) was found to undergo a very useful photooxidation reaction. Oxygen, in the presence of visible light, attacked exclusively the double bond in the five-membered ring, opening it to the dicarbonyl compound (XLIX), formed in 74% yield. Neither the vinyl group nor the chlorin ring system was attacked. The photooxidation product (XLIX) contains an aldehyde group in γ-position, as indicated by its nuclear magnetic resonance spectrum, in which a singlet at −1.22 τ appears. This signal is caused by the aldehyde proton under the influence of the ring current of the conjugated 16-membered ring. The nuclear magnetic resonance spectrum of purpurin 5 (L) exhibits a similar signal at −1.48 τ.

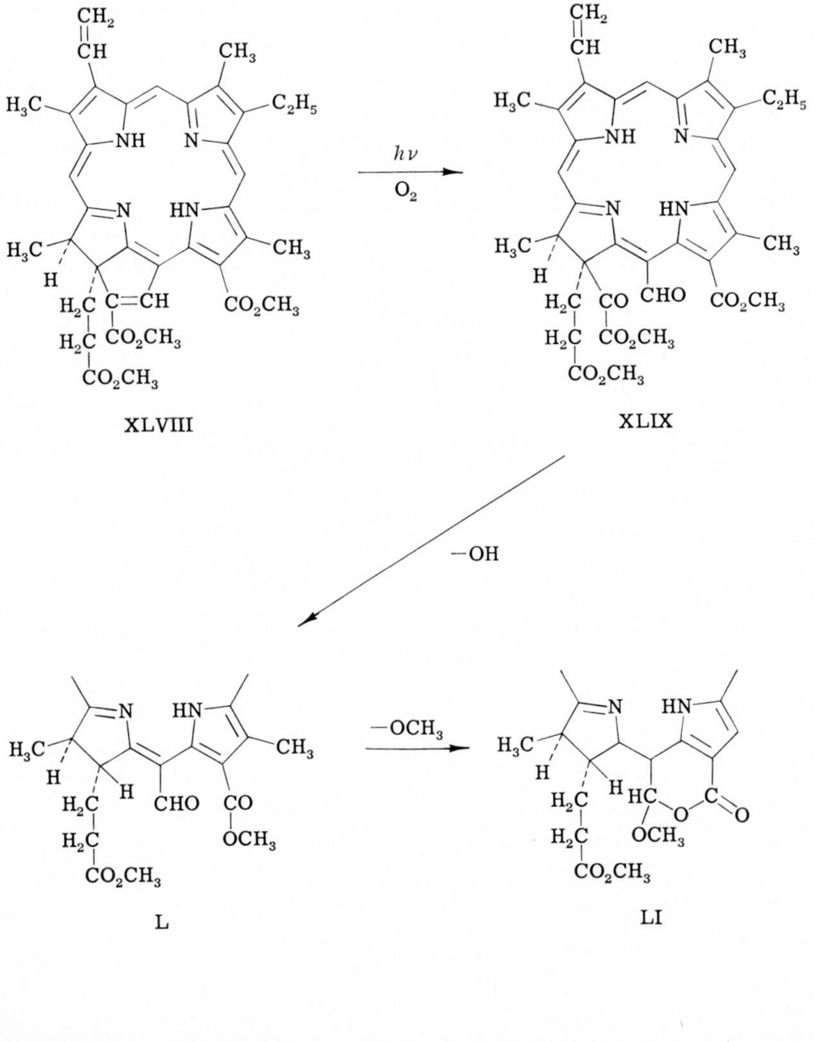

The methoxalyl side chain in (XLIX) could be removed by treatment with dilute methanolic potassium hydroxide. Position 7 is well suited to bearing negative charge, which can be distributed into the carbonyl groups attached to the γ- and 6-positions. Protonation of the 7-anion results in the introduction of the second of the "extra" hydrogens. The stereochemistry of this reaction is again controlled by steric repulsions between the side chains, and the *trans*-7-β-carbomethoxyethyl-8-methyl compound (L) results. The primary product (L) reacts further with methoxide to give the methoxy lactone (LI), the racemic form of a known substance, isopurpurin 5 methyl ester. This compound had been prepared by Fischer (35), who, however, did not recognize its true structure. (LI) was obtained from (XLIX) in 23% yield.

Fischer's product had been obtained by degrading natural chlorophyll, making the synthetic substance (LI) the first compound in the synthetic sequence that could be compared with material of natural origin. The infrared and visible absorption spectra of solutions of synthetic (racemic) (LI) and (optically active) natural isopurpurin 5 methyl ester were identical in every respect, thus confirming the *trans* disposition of the hydrogen atoms at C-7 and C-8. Hydrolysis of the synthetic isopurpurin 5 methyl ester (LI) with very dilute sodium hydroxide in a mixture of dioxane and water gave the corresponding hydroxy lactone (LII), the racemic counterpart to the chlorin 5 obtained from natural sources. Crystalline quinine salts of racemic chlorin 5 (LII) were obtained, and fractional crystallization of the mixture of diastereomeric salts led to material of $[\alpha]_{546}^{23} = +1236°$ optical rotation. The quinine salt obtained from natural chlorin 5 showed a rotation of $[\alpha]_{546}^{23} = +1220°$. The free chlorin 5 from the resolution had a rotation of $[\alpha]_{546}^{23} = +1810°$ while chlorin 5 from natural sources exhibited a rotation of $[\alpha]_{546}^{23} = +1823°$. Measuring the optical rotation was greatly aided by the chlorin 5 absorption spectrum having a minimum near 546 mμ and by the occurrence of a mercury arc emission line at the same wavelength. Reliable measurements thus were possible, using a high-pressure mercury lamp, a prism monochromator, and a photoelectric polarimeter. At other wavelengths, light absorption by the chlorin 5 and emission of the available light sources combined unfavorably.

When the resolved chlorin 5 (LII) was treated with diazomethane, totally synthetic, optically active purpurin 5-dimethyl ester (L) was obtained, which was identical in all respects with a sample of natural origin (*36*). Melting point, mixture melting point, and quantitative infrared and visible absorption spectra were compared. The conversion of the quinine salt of (LII) to (L) gave a 90% yield. Purpurin 5 can crystallize in two dimorphic modifications, one appearing as long thin

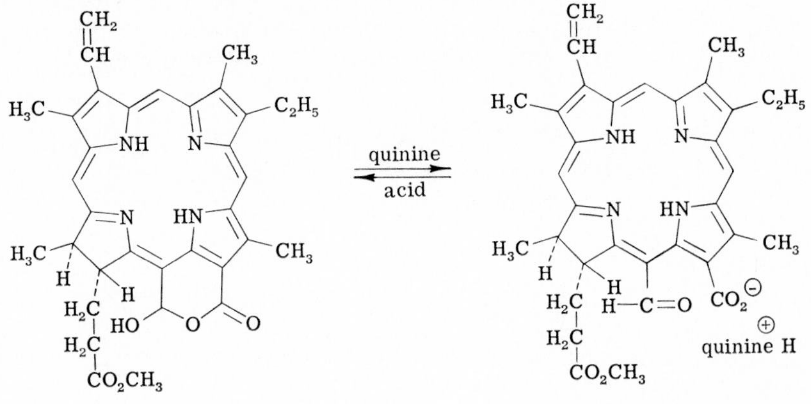

LII

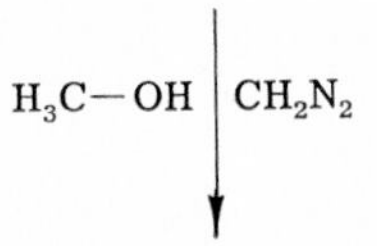

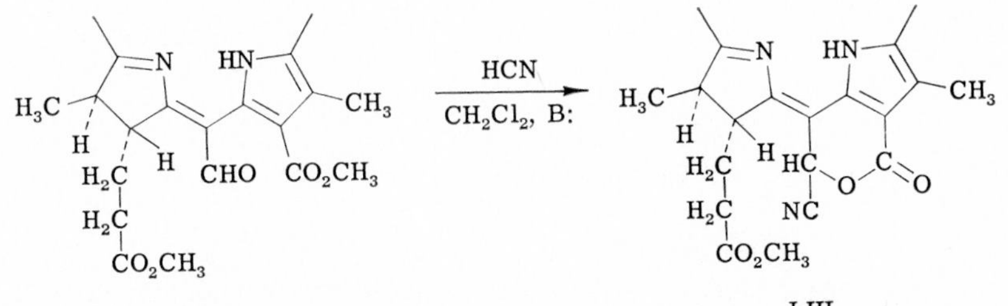

prisms and the other as relatively large hexagonal plates. These crystal forms are not simply different habits of the same lattice, for their infrared spectra are drastically different. This gave rise to some consternation when the material of natural provenance was first compared with the synthetic (L). The modifications can be interconverted by seeding solutions with the desired dimorph.

Treating purpurin 5-dimethyl ester (L) with hydrocyanic acid and triethylamine in dichloromethane solution gave a 84% yield of the cyanolactone (LIII) (37). This material was reduced with zinc and acetic acid to the γ-cyanomethyl compound (LIV), which gave the ester (LV) upon treatment with diazomethane. (LIII) was converted to (LV) in 10% yield. Action of methanolic hydrogen chloride on (LV) produced a 40% yield of chlorin e_6 trimethyl ester (LVI) whose melting point, mixture melting point, and quantitative visible and infrared absorption spectra were identical with those of an authentic sample.

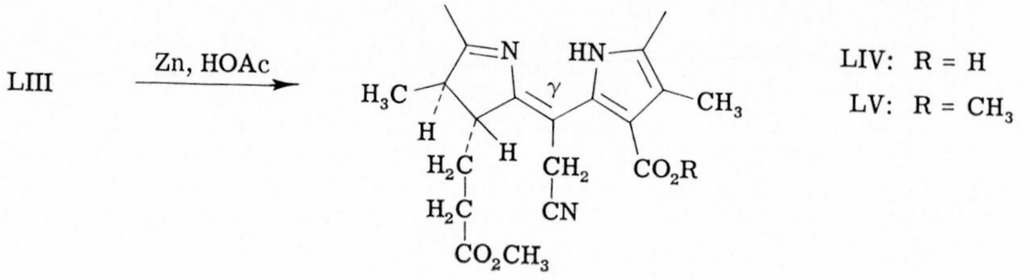

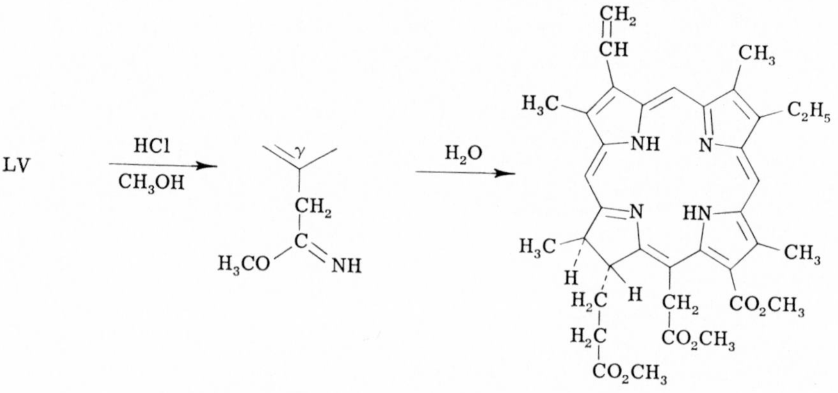

The steps leading from the four pyrroles (XXVIII), (XXIX), (XXX), and (XXXIII) to chlorin e_6 (LVI) prove the structure of chlorin e_6 in all its details. The route from (LVI) to chlorophyll a itself had already been laid out by Willstätter and Fischer (26–28, 38). Dieckmann condensation with sodium methoxide closes the ring between C-10 and the carbomethoxy group at C-6 of chlorin e_6, to give methyl pheophorbide a (XXIV). The route from (XXIV) to chlorophyll a has been outlined on page 128. Optically active phytol of known stereochemistry had been synthesized in 1959 (39). Thus, all the components needed to make chlorophyll a are accessible by total synthesis.

REFERENCES

(1) R. B. Woodward, *Pure Appl. Chem.* **2**, 383 (1961); *Angew. Chem.* **72**, 651 (1960).
(2) H. Fischer and J. Klarer, *Ann. Chem.* **448**, 178 (1926).
(3) H. Fischer and H. Orth, "Die Chemie des Pyrrols," Vol. II, Part 1, pp. 163 ff. Akad. Verlagsges., Leipzig, 1937.
(4) H. Fischer and A. Stern, "Die Chemie des Pyrrols," Vol. II, Part 2, pp. 7 ff. Akad. Verlagsges., Leipzig, 1940.
(5) H. Fischer and H. Helberger, *Ann. Chem.* **480**, 235 (1930).
(6) H. Fischer and M. Strell, *Ann. Chem.* **556**, 224 (1944).
(7) A. Treibs and E. Wiedemann, *Ann. Chem.* **471**, 146 (1929).
(8) H. Fischer and F. Gerner, *Ann. Chem.* **553**, 67 (1942).
(9) H. Fischer and F. Gerner, *Ann. Chem.* **559**, 77 (1947).
(10) H. Fischer and G. Wecker, *Z. Physiol. Chem.* **272**, 1 (1942).
(11) H. Fischer and F. Gerner, *Ann. Chem.* **559**, 80, 81 (1947).
(12) H. Fischer, H. Mittenzwei, and D. B. Hevér, *Ann. Chem.* **545**, 154 (1940).
(13) H. Fischer and G. Spielberger, *Ann. Chem.* **510**, 156 (1934).
(14) H. Fischer and W. Schmidt, *Ann. Chem.* **519**, 244 (1935).
(15) H. Wieland, *Angew. Chem.* **62**, 1 (1950).
(16) M. Strell and A. Kalojanoff, *Ann. Chem.* **577**, 97 (1952).
(17) A. Treibs and R. Schmidt, *Ann. Chem.* **577**, 105 (1952).
(18) H. Fischer, E. Stier, and W. Kanngiesser, *Ann. Chem.* **543**, 258 (1940).
(19) H. Fischer and F. Baláz, *Ann. Chem.* **553**, 166 (1942).
(20) M. Strell and A. Kalojanoff, *Angew. Chem.* **66**, 445 (1954).
(21) H. Fischer, F. Gerner, W. Schmelz, and F. Baláz, *Ann. Chem.* **557**, 134 (1947).
(22) H. H. Inhoffen, *Angew. Chem.* **76**, 383 (1964).
(23) M. Strell, A. Kalojanoff, and H. Koller, *Angew. Chem.* **72**, 169 (1960).
(24) M. Strell and A. Kalojanoff, *Ann. Chem.* **652**, 218 (1962).
(25) H. Fischer, H. Mittenzwei, and D. B. Hevér, *Ann. Chem.* **545**, 154 (1940).
(26) R. Willstätter and A. Stoll, *Ann. Chem.* **378**, 18 (1911).
(27) R. Willstätter and L. Forsén, *Ann. Chem.* **396**, 180 (1913).
(28) F. G. Fischer and K. Löwenberg, *Ann. Chem.* **475**, 183 (1929).
(29) R. B. Woodward, W. A. Ayer, J. M. Beaton, F. Bickelhaupt, R. Bonnett, P. Buchschacher, G. L. Closs, H. Dutler, J. Hannah, F. P. Hauck, S. Itô, A. Langemann, E. LeGoff, W. Leimgruber, W. Lwowski, J. Sauer, Z. Valenta, and H. Volz, *J. Am. Chem. Soc.* **82**, 3800 (1960).
(30) A. Treibs, R. Schmidt, and R. Zinsmeister, *Chem. Ber.* **90**, 79 (1957).
(31) H. Fischer and Z. Czukás, *Ann. Chem.* **508**, 167 (1934).
(32) H. Fischer and K. Zeile, *Ann. Chem.* **483**, 251 (1930).
(33) H. Fischer and M. Neber, *Ann. Chem.* **496**, 1 (1932).
(34) cf. G. F. Smith, *Advan. Heterocyclic Chem.* **2**, 287 (1963).
(35) H. Fischer and M. Strell, *Ann. Chem.* **538**, 157 (1939).
(36) cf. H. Fischer and A. Stern, "Die Chemie des Pyrrols," Vol. II, Part 2, p. 115. Akad. Verlagsges., Leipzig, 1940.
(37) cf. H. Fischer and M. Strell, *Ann. Chem.* **543**, 143 (1940).
(38) cf. A. Stoll and E. Wiedemann, *Fortschr. Chem. Forsch.* **2**, 538 (1952).
(39) J. W. K. Burrell, L. M. Jackman, and B. C. L. Weedon, *Proc. Chem. Soc.* p. 263 (1959).

Section II
Physical Properties in Solution and in Aggregates

—6—

Visible Absorption and Fluorescence of Chlorophyll and Its Aggregates in Solution

J. C. GOEDHEER

Biophysical Research Group, Physics Institute,
University of Utrecht, Utrecht, the Netherlands

I. Absorption and Fluorescence Spectra of Monodisperse Chlorophylls and Pheophytins

A. Relation between Absorption Spectra and Electron Transitions: Theoretical Remarks

Absorption spectra of pigments derived from the basic porphin structure are characterized by a number of relatively sharp bands in the yellow, red, or near infrared region, and in the violet or near ultraviolet (the "Soret bands").

The porphin skeleton is built up of pyrrole nuclei and consists of carbon, hydrogen, and nitrogen atoms only. Several attempts have been made to estimate the wavelength, intensity, and polarization of the absorption bands by calculation of the electronic levels in the π-electron system of the closed circuit of conjugated bonds. Of these we mention the free-electron method (*1, 2*), the molecular orbital method (L.C.A.O.) (*3, 4*), and the calculations of Gouterman (*5, 6*), based on electronegativity. It is beyond the scope of this chapter to consider in detail the merits and imperfections of the various methods. A general survey of principles related to light absorption and fluorescence emission of chlorophylls has been given by Rabinowitch (*7*) and, more recently, by Kamen (*8*).

However, some remarks relative to the quantum mechanical considerations underlying the methods of calculation will be made here and used in the interpretation of absorption spectra in terms of electronic transitions or, where a classical picture is helpful in visualization of events, in terms of a vibrating harmonic oscillator (dipole).

The dipole moments associated with the π-electronic transitions of large planar, conjugated molecules like porphin or coronene lie in the plane of the molecule. If a threefold or higher axis of rotational symmetry is present, the transitions are doubly degenerate, and their intensities are strongly influenced by configuration interaction (*4*). The conjugated double bond systems of the chlorophylls generally lack this high degree of symmetry; their absorption bands are therefore associated with a definite transition direction in the molecule, and their energies and intensities are less affected by configuration interaction and by the presence of substituents with conjugated bonds.

The different electronic transitions leading to the various absorption bands characterizing the absorption spectrum of the chlorophylls, pheophytins, and other porphin derivatives, can be detected experimentally with the help of polarization optics and fluorescence spectra. Additional information can be obtained from a consideration of the influence of interaction with solvent molecules on the shape and position of the absorption bands.

Whether an absorption band corresponds to a transition parallel or perpendicular to the plane of the molecule may be determined by measuring the dichroism of the pigment dissolved in an artificially oriented system, such as a "liquid crystal" or a stretched polyvinyl film. A certain degree of orientation may also occur in a laminar flow system or in a strong electric field. Such measurements distinguish between π-π transitions, which are parallel to the plane of the molecule, and others such as n-π transitions, which may be perpendicular. N-π transitions, involving excitation of nitrogen or oxygen lone pair electrons into the conjugated double bond system, have not yet been identified for any of the chlorophylls with certainty.

The orientation of an electronic transition with respect to the one emitting fluorescence can be determined by measuring the degree of fluorescence polarization (p) as a function of the wavelength of the exciting light. ($I_\parallel$ and $I_\perp$ are the intensities of fluorescence with polariza-

$$p = \frac{I_\parallel - I_\perp}{I_\parallel + I_\perp}$$

tion directions parallel and perpendicular to that of the exciting beam.) When the absorbing and emitting oscillator are the same, $p = +0.50$, provided Brownian rotation is negligible during the lifetime of the excited state, and the concentration is so low that no energy transfer occurs (9, 10). If the absorbing oscillator is perpendicular to the emitting one, $p = -0.33$. In practice values of $p = 0.40$ are typical if the absorbing and fluorescing oscillator are identical, which makes the maximum obtainable negative polarization about -0.26.

Within the absorption band system of a single transition (main band and vibrational levels), the polarization value should be constant. If overlap occurs with differently oriented oscillators the degree of fluorescence polarization should vary within the absorption band region. However, the method of fluorescence polarization does not yield direct information concerning the direction of the transition of lowest energy in the plane of the molecule.

The fluorescence spectrum is mirror symmetrical relative to the absorption spectrum belonging to the oscillator of lowest energy. Hence, absorption bands ascribed to vibrational levels belonging to this electron transition will appear in the fluorescence spectrum, while bands belonging to a different transition will not. According to Förster (11) this mirror symmetry between the fluorescence (quantum) spectrum and the absorption spectrum (on an energy scale) holds only in a narrow region about the center of gravity of fluorescence and absorption bands. If the assumption is made that the frequencies of vibrations in the ground and the first excited state are equal, then the measured fluorescence spectrum should be corrected by the factor γ_a/γ_e^3 (γ_a and γ_e are the corresponding frequencies of the absorption and the fluorescence bands) to obtain mirror symmetry. This correction is of the order of 0.8 for the first and second absorption bands of chlorophylls.

Interaction with solvent molecules results in a red shift of the absorption bands as compared with their estimated position in vacuum. The extent of this interaction is mainly determined by the dielectric properties of the solvent, such as the dielectric constant and the refractive index. Especially in polar solvents, the interaction is strong, and there is experimental evidence that different transitions are affected differently.

Interaction with other pigment molecules affects the shape and the position of the absorption bands, usually broadening and shifting them to the red. It seems likely that such shifts are mainly determined by dipole-dipole interactions of the π-electron systems of the pigments. The strengths of such interactions are, in first approximation, proportional to the oscillator strengths of the transitions [cf. Heller and Marcus (12)]. However, if aggregates are formed, the absorption spectrum may be changed more drastically (13).

Interaction of pigments with proteins or other "carrier" molecules will also influence the position and shape of absorption bands, and, as in the case of aggregation or crystallization, fluorescence may be affected strongly. Study of these phenomena and comparison of both *in vivo* and in-solution properties yield information about the state of binding of the chlorophylls and their function in photosynthesis.

In the above no mention has been made of an important constituent of the chlorophylls: the central magnesium atom. Its presence influences the spectroscopic properties of each transition in a different way. Therefore, for each chlorophyll the corresponding pheophytin is also considered. Also, the transitions considered above are only the formally allowed singlet-singlet ones. With flash spectroscopy evidence for triplet-triplet absorption has been found, but little or no phosphorescence of the chlorophylls has been detected. The knowledge of optical properties of these spectra, therefore, is scanty.

B. Experimental Results: Absorption, Fluorescence, and Fluorescence Polarization Spectra

1. BACTERIOCHLOROPHYLLS

a. Bacteriochlorophyll a. In most purple bacteria, both Thiorhodaceae and Athiorhodaceae, only a single chlorophyll pigment is present. The absorption of this pigment, and the absorption and fluorescence spectra of bacteriopheophytin dissolved in ether, are given in Fig. 1. Absorption spectra in the visible region were taken from Smith and Benitez (14), and in the ultraviolet from Holt and Jacobs (15), and the fluorescence spectrum was determined in this laboratory. There are slight differences in the shapes of the absorption spectra—with respect to band heights of absorption maxima and shoulders—between the spectra determined by Smith and Benitez (14), Holt and Jacobs (15), and Jensen et al. (16), but the locations of the absorption maxima do not differ by more than 1 or 2 mμ.

Purple bacteria have no accessory pigments like the chlorophylls *b* and *c* of green and brown algae or the phycobilins of blue and red algae, but Bchl *in vivo* occurs in spectroscopically and photochemically different forms. These are discussed in Chapters 12 and 13.

In the fluorescence spectrum of Bchl no symmetrical mirror image

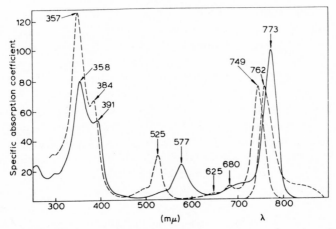

FIG. 1. Absorption spectrum of bacteriochlorophyll a (————), and absorption and fluorescence spectra of bacteriopheophytin a (- - - - -) dissolved in ether. The specific absorption coefficient has the dimensions liter gram^{-1} cm^{-1}.

of the 577 mµ absorption band is visible. Hence this band is assumed to correspond to a second electronic transition which differs from the fluorescing one. This assumption agrees well with fluorescence polarization data obtained by Goedheer (*17, 18*). These measurements show that a negative polarization ($p = -0.18$) obtains for Bchl at wavelengths around 580 mµ, indicating that the transition responsible for the yellow absorption band is perpendicular to the long wavelength one (Fig. 2). The measurements also show changes in the degree of polarization within the Soret band suggesting that the 391 mµ shoulder belongs to a transition perpendicular to, and the 358 mµ band to a transition parallel to, the fluorescing one. The relative intensities of the two components

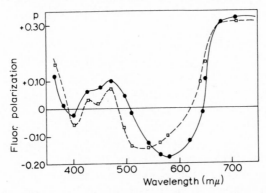

FIG. 2. Fluorescence polarization spectra of bacteriochlorophyll a (————), and bacteriopheophytin a (- - - - -) dissolved in cyclohexanol.

can be calculated from the absorption spectrum and the polarization degree, using the formula given by Stupp and Kuhn (19).

Orientation experiments with Bchl in ammonium oleate (18), indicate that both the 773 mμ and the 577 mμ absorption bands correspond to transitions located in the plane of the molecule. Hence they are likely to correspond with π-π transitions. As the Soret band of Chl a is also probably oriented in this plane, it is assumed that the overlapping transitions of the Soret band in the chlorophylls in general are π-π transitions.

The second electronic transition of Bchl has some remarkable properties. Its absorption band is located at around 580 mμ in most nonpolar solvents, moving to longer wavelengths with increasing refractive index of the solvent. In methanol, however, the band is located at 606 mμ (20), a shift of about 800 cm⁻¹ from its position in ether. In the higher alcohols it is located at slightly shorter wavelengths than in methanol (21). In pyridine, the band is at 612 mμ (22). The shift observed in polar solvents is much less marked for the main band near 770 mμ. The marked effect of polar solvents on the position of the band due to the second transition is not observed for bacteriopheophytin (21). The influence of strongly polar solvent molecules thus seems to be exerted primarily via the magnesium atom.

b. Bacteriochlorophyll b. Jensen *et al.* (16), in an investigation of 18 species of photosynthetic bacteria, found one species of *Rhodopseudomonas* containing a Bchl different from Bchl a. The general shapes of the absorption spectra were similar, but the maxima in acetone solution for the new Bchl were located at longer wavelengths. The values of band positions and relative band heights given by these authors are respectively 795 (96), 677 (15), 582 (30), 407 (87), 368 (100) (see Chapter 12 for more details). As no Bchl a was found in this species, this new type of Bchl is most probably the photochemically active pigment. The authors mentioned that Lascelles also has isolated a similar Bchl compound.

c. Bacteriopheophytin a. Removal of the magnesium atom affects the absorption bands of Bchl in different ways (Fig. 1). The transition of lowest energy is shifted 24 mμ toward the blue in ether. The second transition is influenced more, its band shifting from 577 to 525 mμ. No marked new bands of absorption appear. The Soret band is only slightly shifted. All bands are somewhat sharpened, while the vibrational structure is more clearly visible. The fluorescence polarization spectrum is similar in shape to that of Bchl (Fig. 2). The region where the value is negative is shifted, corresponding to its absorption. Since bacteriopheophytin has a conjugated system similar to that of tetrahydroporphin and, having a "long field" type of spectrum, is likely to be influenced less than porphin by side-chain effects, a comparison can be made between the values of wavelengths, relative intensity and polarization calculated by the simplified molecule orbital theory (3) and the measured ones. As the values thus calculated give only the center of gravity of singlet and triplet pairs of one transition, a rough estimation of singlet-triplet energy gaps can be obtained in this way. In view of the various assumptions and approximations which have to be made in such calculations (value of overlap integral, influence of extra electrons of nitrogen atoms, etc.) the results do not show too wide a discrepancy [cf. Goedheer (17, 18)].

d. *Chlorobium Chlorophyll 660.* Green sulfur bacteria usually contain the dihydroporphin, chlorobium chlorophyll 660, as the main chlorophyll pigment (23–25). The "660" designation indicates the absorption maximum of this chlorophyll in ether, as described in Chapter 4. The shape of the absorption spectrum is generally similar to that of Chl *a*, the main difference being the absence of a band near 580 mμ in chlorobium chlorophyll (Fig. 3). As will be shown, this weak band for Chl *a* most probably corresponds to the second electronic transition. The spectrum of fluorescence polarization indicates that the 580 mμ band is displaced to the red for chlorobium chlorophyll 660, and is hidden by the vibrational level of the first electronic transition.

e. *Chlorobium Chlorophyll 650.* Stanier and Smith (26) found that some strains of green photosynthetic bacteria contain a different chlorobium chlorophyll with an absorption maximum at 650 mμ in ether. Jensen *et al.* (16) have investigated a number of species with respect to their pigment content. The absorption spectrum of chlorobium chlorophyll 650 shows a measurable band at 575 mμ, and fine structure in the band near 620 mμ.

f. *Chlorobium Pheophytins 660 and 650.* The changes in the absorption spectra resulting from elimination of magnesium from the chlorobium chlorophylls are more marked than in the case of Bchl *a*, and resemble more closely those for Chl *a* (Figs. 3 and 4). The fluorescence polarization spectra are also similar to that of pheophytin *a*, and will be discussed with this pigment.

g. *Bacteriochlorophyll a in Green Bacteria.* Besides the chlorobium chlorophylls, a pigment with absorption spectrum similar to that of Bchl *a* is found in green bacteria (27, 28). Jensen *et al.* (16) chromatographed both pigments together and could not detect any separation. It thus seems likely that Bchl *a* occurs also in green bacteria (see also Chapter 12).

2. PLANT CHLOROPHYLLS

a. *Chlorophyll a.* The absorption and fluorescence spectra of Chl *a*, dissolved in ether, are given in Fig. 4. The bands are sharp in this solvent,

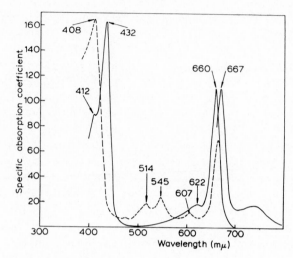

FIG. 3. Absorption and fluorescence spectra of chlorobium chlorophyll 660 (————), and absorption spectrum of chlorobium pheophytin 660 (- - - - -) dissolved in ether.

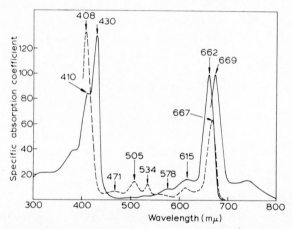

FIG. 4. Absorption and fluorescence spectra of chlorophyll *a* (————) and absorption spectrum of pheophytin *a* (- - - - -), dissolved in ether.

indicating a relatively weak interaction between solvent molecules and pigment, but the presence of some water or polar solvents affects the absorption spectrum. Perkins and Roberts (29) found the intensity ratio of the Soret band to the red band to be 1.11–1.20 in ether containing some water or methanol, and 1.31 in dry ethyl ether.

Elaborate spectral data of chlorophylls in various solvents have been presented by Rabinowitch (30), French (31), and Smith and Benitez

(*14*). Seely and Jensen (*32*) studied the effect of solvent on the absorption spectrum of Chl *a* in about 40 different solvents. They found that spectral parameters such as band position, band half-width, and Soret band:red band intensity ratio depended on the chemical nature of the solvent as well as on the dielectric constant and the refractive index.

In the fluorescence spectrum of Chl *a* there is no distinct mirror image of the 578 mμ band. This suggests that this band belongs to a transition different from the fluorescing one. This conclusion is supported by the fluorescence polarization spectrum (*17, 33, 35*). The degree of fluorescence polarization drops steeply around 580 mμ to a slightly negative value, but it is less negative than with Bchl. The difference between the two pigments can be explained by the fact that for Chl *a* the position of the red band entails considerably more overlap between the absorption bands of the first transition (comprising a very weak second vibrational level) and the perpendicular transition of the 578 mμ band.

From the mirror symmetrical shape of the fluorescence spectrum of Chl *a* it can be estimated that about 35% of the absorption at 578 mμ is due to the second electronic transition, a value which corresponds well with the one calculated from the degree of polarization, assuming the occurrence of two perpendicular transitions (*19*). Thus, upon conversion of the conjugated ring system from the tetrahydroporphin type of Bchl into the dihydroporphin type of Chl *a*, the band related to the second electronic transition is not shifted, whereas the maximum related to the first transition moves from 770 to 660 mμ. This might indicate that the second transition is oriented in the unchanged dimension of the conjugated ring system of these pigments. The transition probability ("oscillator strength") of the second transition, however, is markedly decreased.

Analogously to Bchl, the 578 mμ band in Chl *a* shifts appreciably toward the red in polar solvents. In methanol it is no longer visible as a separate band but is masked by the 620 mμ vibrational level of the main red band.

The fluorescence polarization spectrum (Fig. 5) determined with a monochromator shows a minimum (*32*) or a shoulder (*31*) at about 620 mμ. This could indicate that at about 620 mμ also a different transition is present, overlapping the first one. It might possibly be an n-π transition [cf. Fernandez and Becker (*36*)]. In some cases the absorption band at about 620 mμ is doubled (*32, 37*), which might be due to the presence of such a band.

The degree of fluorescence polarization changes markedly in the region of the Soret band. Hence this band probably also corresponds to at least two perpendicular transitions. Since the polarization spectrum is slightly negative at 400–440 mμ, the main component at 430 mμ is supposed to be perpendicular, and the shoulder at 390 mμ parallel, to the fluorescence emitting transition. Both components are differently affected by polar solvents [cf. Seely and Jensen (32).] Thus increased interaction in polar solvents results in a broadening and shifting of bands which differ for the various electronic transitions and in a marked increase in the "Stokes shift" (wavelength difference between absorption and fluorescence maximum).

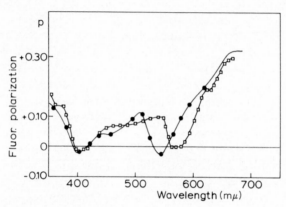

Fig. 5. Fluorescence polarization spectrum of chlorophyll *a* (□) [reproduced from Bär *et al.* (35)] and pheophytin *a* (●) dissolved in castor oil.

Measurements of dichroism of chlorophyll in liquid crystals, such as ammonium oleate and lecithin, indicate that for chlorophyll, absorption occurs in the plane of the molecule for all bands in the visible part of the spectrum (11). All these bands, then, are most probably caused by π-π transitions, at least for the major fraction of absorption. Zocher (38) reported that a dichroism in the blue is seen which is different in sign from the dichroism in the red. It is not impossible that this is caused by pheophytinization of part of chlorophyll by the oleic acid.

In completely dry hydrocarbons, e.g., benzene, the shape of the absorption spectrum, compared with that in solvents containing some water, is slightly altered. The bands are sharpened, while a shoulder on the long wave side of the red band appears. Chl *a* dissolved in dry hydrocarbons is only weakly fluorescent or nonfluorescent (34). In dry pyridine a new band is found at about 638 mμ, while the shape of the Soret band is markedly different from its shape in the presence of water (39, 40). Changes in the absorption spectrum may also occur after cooling to low temperatures. With Chl *a* all bands are shifted slightly toward the red and sharpened, while a new maximum is found between the red band and its first vibrational level (41). Brody and Broyde (42), measured the

of the intensity of the main maximum to that of the first vibrational level is 4.5, as compared with 6.7 for Chl a.

The spectrum of Chl b changes more in completely dry hydrocarbons than is the case with Chl a. Also, the spectrum measured at low temperatures is quite different from the room temperature one. Cooling to $-196°$ makes the red band shift to 662 mμ, while marked changes occur also in the Soret band [cf. Freed and Sancier (41)].

d. *Pheophytin b*. Removal of magnesium from Chl b results in a slight red shift of those bands which are assumed to belong to the first transition to 655 and 599 mμ, while new bands appear at 550 and 520 mμ. The maximum of the Soret band is shifted from 453 to 434 mμ and sharpened. The fluorescence polarization spectrum has minima at about 530 and 450 mμ, though the value of the polarization does not become negative. The fluorescence spectrum is similar in shape to that of Chl b (Fig. 6).

e. *Chlorophyll c*. Chl c, the accessory pigment of brown algae, is a pigment with a porphin-type system of conjugated bonds. The absorption spectrum also is of the "round field" type (4). The red band is much weaker than in the other chlorophylls and is at 626 mμ in ether. The Soret band is high and does not show a marked composite structure. The Soret band:red band intensity ratio is about 10. No fluorescence polarization spectra have been made with purified preparations of this pigment. The shape of the fluorescence spectrum is similar to that of the other chlorophylls, but the intensity ratio of the main band to the vibrational level is only 2.7. As the absorption band at 577 mμ is of about the same height as the red band at 626 mμ, there is no approximate mirror symmetry between the fluorescence and absorption spectra. This is an indication that the 577 mμ band is caused by more than one electron transition.

f. *Pheophytin c*. In the porphins, elimination of the central magnesium atom results in a much more drastic change of the absorption and fluorescence spectra than in the dihydroporphins (Fig. 8). The visible part of the absorption spectrum shows four bands at 641, 589, 564, and 521 mμ, similar to the absorption spectra of many porphyrins. Apparently the central Mg atom distorts the symmetry of the "round field" spectrum in such a way that the absorption spectrum of Chl c resembles that of a dihydroporphin like Chl b. However, the absorption spectra of pheophytin b and c are markedly different. The low value of the band of longest wavelength is characteristic of the porphin type of spectrum. Probably this transition suffers more from symmetry selection rules than the vibrational levels. The fluorescence spectrum also is changed markedly as compared to that of the other pheophytins. Two bands of approximately equal height occur at 651 and 717 mμ, while no further bands occur in the near infrared. Hence no mirror symmetry occurs between any part of the spectrum, indicating the overlap of several transitions in the absorption spectrum. The absorption bands of pheophytin *c* are appreciably broader than those usually encountered with porphin derivatives.

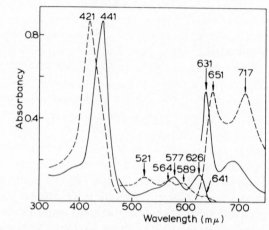

Fig. 8. Absorption and fluorescence spectra of chlorophyll *c* (————) and pheophytin *c* (- - - - -), dissolved in ether. Absorbancies are in relative units.

g. Chlorophyll d. Chl *d* has an absorption spectrum of a shape similar to that of Chl *a*, but its main bands are located at longer wavelengths, 688 and 447 mμ in ether solution (Fig. 9). There is a difference in the shape of the Soret band. It thus shows a dihydroporphin type of spectrum. The fluorescence spectrum also is of a similar shape to that of Chl

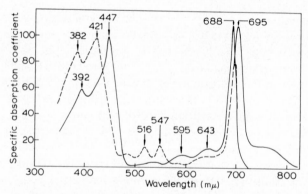

Fig. 9. Absorption and fluorescence spectra of chlorophyll *d* (————) and absorption spectrum of pheophytin *d* (- - - - -), dissolved in ether.

a, though the vibrational level is seen as a shoulder instead of as a separate maximum (*44*).

By oxidation of chlorophyll with potassium permanganate, Holt and Morley (*45*) obtained a pigment which was spectroscopically identical to Chl *d*. This pigment was oxidized in the 2-vinyl side chain. Sagromsky (*46*), measured the Chl *d* content of freshly collected red algae, and of algae which had been illuminated in the absence of CO_2. The latter condition favored the formation of Chl *d*. It thus seems likely that Chl *d* is formed by oxidation of Chl *a*. An indication of the occurrence of a pigment like Chl *d* in green algae was found by Michel-Wolwertz *et al.* (*47*). Absorption and fluorescence measurements of chromatographed extracts from *Chlorella* showed the presence of a small amount of a pigment with absorption bands at about 690 and 445 mμ. The possibility that this pigment occurs in a low percentage in the living cell cannot be excluded.

h. Pheophytin d. Pheophytinization of Chl *d* results in spectral changes which are similar to those for Chl *a;* the band at 595 mμ is shifted to 547 mμ, while a new band arises at 516 mμ. The red band is shifted somewhat toward longer wavelength. In contrast to pheophytin *a*, the fluorescence spectrum of pheophytin *d* shows less pronounced vibrational structure than the corresponding chlorophyll.

i. Protochlorophyll. This pigment can be obtained from etiolated plants or from squash or pumpkin seeds. The absorption spectra of protochlorophyll derived from different sources are nearly identical (*48*) (Fig. 10). The pigment has a spectrum generally similar to that of Chl *c*, but the bands are sharper and show more detail. The Soret band:red band intensity ratio is about 7.5. The ratio of the bands at 623 and 571 mμ is appreciably higher than for Chl *c*. The fluorescence spectrum shows two bands at 630 and 690 mμ in ether solution, with an intensity ratio of about 5. The fluorescence spectrum is not mirror symmetrical to the red part of the absorption spectrum, though it deviates less than for Chl *c*. This indicates that the red bands in protochlorophyll are due to more than one electronic transition, which might be expected in a "round field" type of absorption spectrum.

The fluorescence polarization spectrum does not show marked maxima and minima. The maximum value in the red is $p = 0.14$, while in the Soret band it is $p = 0.12$. A similar spectrum is found with Zn-tetraphenylporphin (*35*).

j. Protopheophytin. As with Chl *c*, removal of the central Mg atom results in drastic changes in the absorption and fluorescence spectra. The absorption spectrum of protopheophytin is characteristic of a metal-free porphyrin. In the fluorescence spectrum the second band is of higher intensity than the first. Mirror symmetry is absent. The fluorescence polarization spectrum, obtained with interference filters, does not show marked maxima and minima. The relative polarizations of individual bands cannot therefore be concluded. Low-temperature spectra made with narrow monochromatic light beams may show more structure. Only the second fluorescence band has been used for the determination of the polarization spectrum. It may be that the polarization spectrum

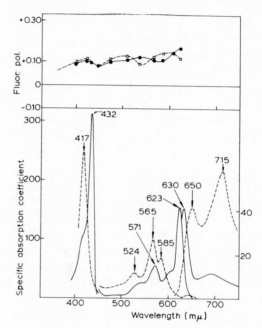

Fig. 10. *Lower:* Absorption and fluorescence spectra of protochlorophyll (——) and protopheophytin (- - - -) from squash seeds, dissolved in ether. *Upper:* Fluorescence polarization spectra of protochlorophyll (●—●) and protopheophytin (□- -□) dissolved in castor oil.

differs for the two fluorescence bands, as is the case for tetraphenylporphin (35).

3. CHLOROPHYLL DERIVATIVES

The shapes of the absorption spectra of chlorophylls and their derivatives are determined primarily by their porphin, dihydroporphin or tetrahydroporphin skeleton, but changes in the side chains or the cyclopentanone ring can affect the position and the shape of absorption bands.

Chl *a* in organic solvents may be changed, during storage in the dark or during purification, into a number of compounds which move separately during chromatography, but which may have very similar absorption spectra (see also Chapter 2). It is generally assumed that in the plant only one type of Chl *a* is present, while the other compounds obtained after chromatography are decomposition products. Michel-Wolwertz and Sironval (*37*), however, using fast and nondestructive techniques for purification of carefully extracted chlorophyll, concluded that probably several modified forms of Chl *a* occur in the chloroplasts. The absorption spectra of these compounds show slight differences, especially in the region of the Soret band and the minor absorption bands.

With most photosynthetic purple bacteria only one type of Bchl is found, though the *in vivo* absorption spectra show large differences. These differences must be due to different environments for the same Bchl molecule.

The absorption spectra of some chlorophyll derivatives differing in the side chains or in the structure of the cyclopentanone ring will be considered in investigating the influence of structural variations on the spectrum.

a. Chlorophyllides. Removal of the long phytol chain from the chlorophylls, e.g., by the enzyme chlorophyllase, does not appreciably affect either the band shape or the position of the absorption bands.

If Chl *a* is stored in alcoholic solvents in the dark for a prolonged time, it is slowly converted into a product with a slightly different absorption spectrum: allomerized chlorophyll. This change is thought to be caused by an oxidation in the cyclopentanone ring (*30*). Holt and Jacobs (*49*) separated allomerized Chl *a* into at least three fractions with slightly different absorption spectra. The main fraction, when measured in methanol, had a red band shifted somewhat to the blue (from 663 to 650 mμ) and a Soret band without a shoulder. Allomerization is also observed for Chl *b*. With the bacterial pigments, however, no absorption changes corresponding to those induced by allomerization can be observed after prolonged storage in the dark in alcoholic solvent.

b. Chlorophyllin. Addition of concentrated alkali to a methanolic chlorophyll solution results in cleavage of the cyclopentanone ring and removal of the phytol chain. The saponification products of Chl *a* thus obtained are soluble in water and alcohol. When dissolved in buffer solution at pH 8, fluorescence can be measured. The fluorescence yield is lower than for the chlorophylls. The spectral properties of chlorophyllin *a* have been studied by Oster *et al.* (*50*) and by Savkina and Evstigneev (*51*). Oster *et al.* (*50*) also measured dichroism of chlorophyllin *a* in stretched films of polyvinylalcohol. According to their figure, dichroism was measurable throughout the visible part of the spectrum. The dichroic ratio was somewhat lower in the Soret band than in the red band. As the molecular planes are probably preferentially slanted toward the direction of stress, the experiment indicates that absorption occurs mainly in the plane of the molecule.

In Fig. 11 the absorption spectra of chlorophyllin *a*, bacteriochlorophyllin, and chlorobium chlorophyllin 660 are given, measured in a mixture of 10% methanol and 90% phosphate buffer of pH 7.8. The spectrum of chlorophyllin *a* is similar to that given by Oster *et al.* (50). The bands are broadened, and the ratio of the red band to the Soret band is markedly decreased. The broadening is probably due to strong interaction of the polar molecules with the saponified side chains of the molecule.

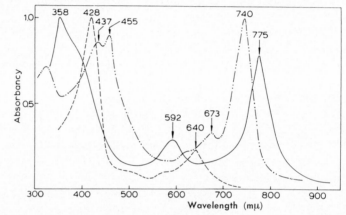

Fig. 11. Absorption spectra of chlorophyllin *a* (- - - - -), chlorobium chlorophyllin 660 (—··—), and bacteriochlorophyllin (———) in phosphate buffer (pH 7.8) containing 10% methanol.

The position of the yellow maximum of bacteriochlorophyllin is at about 592 mμ, in methanol as well as in phosphate buffer. This indicates that the strong influence of polar solvents on this band is exerted only when the cyclopentanone ring is intact.

The absorption spectrum of chlorobium chlorophyllin 660 in 10% methanol, 90% phosphate buffer is markedly different from that in methanol. In the buffer the red band is at 740 mμ and the Soret band is at 455 mμ. The general shape thus still is that of dihydroporphin. The intensity ratio of the Soret band to the red band is 0.9. The fluorescence spectrum shows two bands at 680 and 770 mμ, indicating the presence of two components, perhaps a monomer and a dimer.

4. Oxidized and Reduced Chlorophylls

A change in the electron distribution in the conjugated ring system of a chlorophyll will result in the appearance of an absorption spectrum which differs strongly from that of the chlorophyll. Oxidation or reduction of the pigment molecule as a whole can bring about such a change. These processes may be caused by light, by addition of chemicals, or by a combination of both. The products are often unstable, and their spectra are seen only as transient phenomena. The same is true of the absorption spectra of chlorophylls in their first excited triplet states. The absorption spectra of various derivatives will be considered in relation to the pigments from which they are derived.

a. *Bacteriochlorophyll*. If oxidants such as iodine, potassium permanganate, or ferric chloride are added in low concentration ($10^{-4} M$)

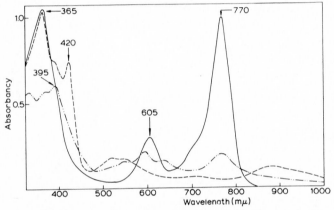

Fig. 12. Absorption spectra of bacteriochlorophyll: (———) in methanol; (- - - - -) immediately after addition of $10^{-4} M$ iodine; (—··—··) after photobleaching in methanol. A similar spectrum occurs after addition of iodine and storage.

to a methanolic Bchl solution, the absorption spectrum changes momentarily into a spectrum with weak bands at about 900, 520–540 mμ, a marked band at about 420 mμ, and a shoulder at 390 mμ (Fig. 12). The Soret band at 365 mμ remains unchanged. The original Bchl spectrum is instantaneously restored by addition of excess reductant. If no reductant is added, the spectrum gradually changes into one with weak bands at 680, 630, 580, and 530 mμ, and a broad "Soret band" around 400 mμ. The last-mentioned spectrum is also obtained after photobleaching. In this case the 680 mμ band is absent immediately after photobleaching, and appears only gradually. It belongs to a pigment with a dihydroporphin type of spectrum [main bands at 675 and 435 mμ, cf. Holt and Jacobs (15)], formed by secondary processes. The spectral identity of the products obtained by photooxidation and chemical oxidation suggests that both are produced by reaction with oxygen of a photochemically produced Bchl$^+$ compound. This suggestion seems confirmed by the observation that bacteriochlorophyllin (buffered at pH 8) is also bleached rapidly by light, and the resulting absorption spectrum is similar to that obtained immediately after addition of $10^{-4} M$ iodine. Partial reversion can be obtained by ferrous salts in both cases (*51a*). Apparently the cleavage of the cyclopentanone ring prevents the reaction with oxygen.

In nonpolar solvents addition of oxidants to Bchl does not result in the formation of the second type of spectrum. After addition of oxidants, the Bchl$^+$ intermediate in acetone lacks a marked 520 mμ band, but retains the band at 900 mμ.

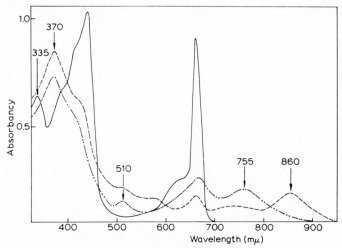

FIG. 13. Absorption spectra of chlorobium chlorophyll 660: (————) in methanol; (- - - - -) immediately after addition of $10^{-4} M$ iodine; (—··—··) after addition of iodine and 15 minutes' storage in the dark. A similar spectrum occurs after photobleaching of chlorobium chlorophyll 660 in methanol.

Besides being oxidized, Bchl can be reduced in air-free pyridine in the presence of a suitable electron donor (e.g., ascorbic acid) and light (*52*). The reduced compound shows a band at about 660 mμ (*53*). No mention is made of a band in the reduced compound on the long wave side of the 770 mμ Bchl band.

Pekkarinen and Linschitz (*22*) measured the absorption spectrum of a metastable, probably triplet, state of Bchl obtained with brief flashes (10^{-5} sec) of high intensity light. The absorption spectrum (in air-free pyridine) shows a Soret band at about 400 mμ and a broad plateau from about 500 to 650 mμ. No measurable band was found in the 900 mμ region.

b. Chlorobium Chlorophyll 660. Though the absorption spectrum of chlorobium chlorophyll 660 is of the dihydroporphin type and thus similar to that of Chl *a*, its redox properties are similar to those of Bchl (54). The absorption spectrum of chlorobium chlorophyll+ 660 intermediate, obtained after addition of $10^{-4}\,M$ ferric chloride or iodine, is generally similar to that of the Bchl intermediate, with bands at about 860, 660, 540, and 370 mμ (Fig. 13). Only the 420 mμ band of Bchl+ is missing. This spectrum soon changes into one with bands at 755, 660, 510, and 370 mμ. A similar spectrum is obtained when chlorobium chlorophyll 660 is photobleached, but in this case no chlorobium chlorophyll+ intermediate is seen. The "bleached" products do not fluoresce.

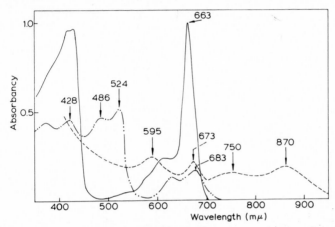

FIG. 14.   Absorption spectra of chlorophyll *a*: (————) in methanol; (- - - - -) after addition of $10^{-4}\,M$ ferric chloride; (—··—··—) the brown intermediate in the Molisch phase reaction in pyridine solution. Reproduced from Weller (58).

This effect of oxidation on the absorption spectrum is opposite to what is found with Bchl and Chl *a*. The maximum at 755 mμ of oxidized chlorobium chlorophyll disappears completely after addition of alkali.

Reversible photoreduction of chlorobium chlorophyll was observed by Krasnovsky and Pakshina (55). The spectrum of the reduced compound resembles that of reduced Chl *a*.

c. Chlorophyll a. Chl *a* dissolved in methanol can be oxidized reversibly to a metastable Chl+ by ferric chloride (56), but not by iodine or potassium permanganate. Chl *a* is less easily oxidized than the bacterial pigments (54, 57). The Chl+ form is rather unstable in solutions containing oxygen. It changes in a few seconds into a second intermediate, and reversibility is lost. As shown in Fig. 14, the Chl+ *a* compound has weak absorption bands at about 870, 750, 595, and, probably, 430 mμ in methanol. The weak, broad bands in far red and blue resemble the broad bands of oxidized cytochromes. No fluorescence can be detected.

Contrary to chlorobium chlorophyll and Bchl, the original type of spectrum returns slowly upon standing in methanol, though the red band of the resulting spectrum is shifted from 663 to 673 mμ.

A product with probably similar absorption bands as Chl^+ a can be obtained by irradiation of chlorophyll in glassy solvents at liquid nitrogen temperature and in the absence of oxygen (59). The light effect is enhanced and stabilized by the presence of a quinone. This suggests that the product formed by addition of oxidants can also be produced by absorption of light.

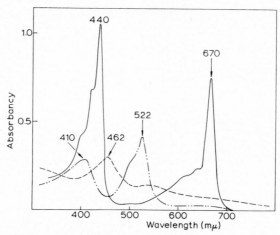

FIG. 15. Absorption spectra of chlorophyll a: (————) in pyridine; (—··—··—) photoreduced chlorophyll a in the presence of ascorbic acid [reproduced from Zieger and Witt (62)]; (-----) the triplet state of chlorophyll a [reproduced from Linschitz and Sarkanen (63)].

Photoreduction of Chl a in air-free pyridine containing traces of water and electron donors of sufficient reducing potential results in a spectrum with a marked band at 522 mμ, but little or no absorption in and beyond the red band of Chl (60–62) (Fig. 15). The fluorescence spectrum of the reduced compound Chl a, measured by Gachkovsky (64), shows maxima or shoulders at 620, 651 and 731 mμ. Pheophytin a is reduced as effectively as Chl a.

Flash illumination of Chl a solutions from which oxygen has been eliminated produces an absorption spectrum assumed to be due to Chl a in its first excited triplet state (63, 65). This occurs in various solvents, such as pyridine, benzene, and methanol. The absorption spectrum comprises a broad band at about 460 mμ, a weak band at 520 mμ and a gradual decrease in absorption toward the red. No absorption band beyond the red chlorophyll band is visible. Flash illumination of pheophytin a also results in the appearance of the triplet state spectrum.

Irradiation of oxygen-free methanolic Chl a solutions with high-intensity continuous light results in reversible absorption changes (66, 67). The absorption spectrum of the changed fraction has not been measured. An absorption spectrum of generally the same appearance as that found with reversible photochemical reduction is observed for the intermediate of the Molisch "phase test" (reaction of chlorophyll with alcoholic alkali) (Fig. 14) (58, 68, 69).

More experimental data, including those obtainable with polarization optics and with related but more simple porphin structures, are needed in order to attempt a fruitful interpretation of absorption spectra of oxidized and reduced chlorophylls and intermediates obtained by illumination of chlorophyll solutions.

d. Chlorophyll b. Chl b also can be oxidized with ferric chloride, but less easily than Chl a. The complete absorption spectrum of the oxidized compound has not been measured, but an increase in absorption beyond the red absorption band is observed, as with the other chlorophylls.

Chl b can be photoreduced in air-free pyridine with ascorbic acid (70). Reoxidation in the dark gives a pigment with bands at 693 and 432 mμ (71). Flash illumination converts Chl b into its lowest triplet state in higher yield than with Chl a. The absorption spectrum of the triplet state shows weak bands at about 490, 450, and 550 mμ. The absorption declines gradually toward the red, as with Chl a (63, 72).

e. Protochlorophyll. No reversible chemical oxidation with ferric chloride could be measured with protochlorophyll isolated from squash seeds, but reversible photoreduction has been measured with protochlorophyll from pumpkin seeds (52).

5. RELATED METAL PORPHYRINS

a. Zinc Tetraphenylporphin. The Zn and Mg complexes of the porphin compounds have many common characteristics. Both types show fluorescence, and triplet state absorption spectra can be determined after flash illumination. Also, photoreduction of these pigments with ascorbic acid can be measured. Complexes with the paramagnetic Fe, Cu, and Co show no fluorescence, and no triplet state absorption spectra could be obtained (65).

The absorption spectrum of tetraphenylporphin shows four bands at about 660, 590, 550, and 520 mμ with increasing intensity, while the Zn complex has two bands at 600 mμ and 560 mμ (the intensity ratio 600:560 is about 0.5 in pyridine solution), and a minor band at 525 mμ, besides the Soret band. According to Seely and Calvin (73), the ratio of the first two bands is strongly dependent on the type of solvent. This applies also to the two bands in the fluorescence spectrum. According to Bär *et al.* (35) the fluorescence polarization spectrum consists of a fairly straight line with a value of about $p = 0.10$, resembling the protochlorophyll polarization spectrum. The triplet state absorption spectra of zinc tetraphenylporphin and tetraphenylporphin are of a similar shape, with relative broad bands at about 840, 730, and 460 mμ (22). The influence of the presence of the metal thus is much more marked in the singlet ground state absorption spectrum than it is in the first excited triplet state one. A similar absorption spectrum is shown by zinc porphin in methanol. In this spectrum, bands at 575 and 530 mμ are found, of which the second one dominates. Comparison with Mg tetraphenylporphin and the porphin-type chlorophylls shows that the second band in the Zn complex is more pronounced than it is in the Mg complexes. This is probably due to the different electronegativities of the metals.

b. Magnesium Phthalocyanine. Phthalocyanine differs from tetrabenzporphin by the presence of nitrogen atoms instead of carbons at the sites connecting the pyrrole nuclei. The absorption bands of the metal-free compound, though generally similar to those of porphyrins, are shifted toward longer wavelength and have a different intensity ratio. The presence of extra electrons of the nitrogen atoms may affect π-π transitions, and bands corresponding to n-π transitions might be anticipated.

The spectrum of the Mg complex is generally similar to those of the chlorophylls, with bands at 670, 610, and 645 mμ and a Soret band below 400 mμ. The absorption spectrum of the pigment in its triplet state shows bands at 400, 470, and possibly beyond 700 mμ (65). The fluorescence spectrum also resembles those of chlorophylls, and no marked maxima and minima are visible. With the Mg complex, as with Chl *a*, a bright chemiluminescence can be measured after heating the pigment in tetralin.

c. Magnesium Tetrabenzporphin (Mg TBP). This porphyrin, prepared by Barrett *et al.* (74), has the general metal porphin type of spectrum. The sharp bands indicate weak pigment-solvent interaction, even in polar solvents. The fluorescence spectrum also is similar to that of the porphins protochlorophyll and Chl *c*.

Addition of oxidants (iodine, ferric chloride, potassium permanganate) produces a spectrum generally similar to that of the oxidized chlorophylls, the redox potential being of the same order of magnitude as that of bacterial pigments. The spectrum of the oxidized compound shows weak bands at 778, 714, 675, 610, about 500 and a Soret band at 418 mμ (Fig. 16). Complete and immediate restoration of the original spectrum is obtained after addition of excess ferrous sulfate, ascorbic acid, or NADH.

A second state of oxidation can be found after addition of ceric chloride. With this strong oxidant the bleached product has no absorption bands in the visible or near infrared spectrum, while after addition of reductants the original spectrum is about 80% restored.

Illumination of Mg TBP dissolved in methanol results in the disappearance of the Mg TBP spectrum and the appearance of a compound with an absorption spectrum similar to that of Mg TBP+, oxidized by iodine. Addition of reductants results in the reappearance of the original absorption spectrum. As the photooxidized product does not fluoresce, bleaching and restoration can be followed by measuring fluorescence.

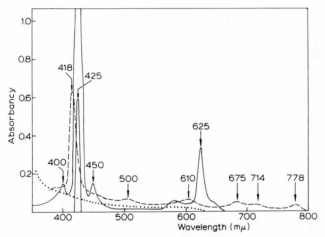

Fig. 16. Absorption spectrum of Mg-tetrabenzporphin: (———) in methanol; (- - - - -) after addition of 10^{-4} M iodine; (. . . . .) after oxidation with ceric chloride.

The behavior of bacteriochlorophyllin and Mg TBP toward photo-oxidation and chemical oxidation appears to be similar. It seems likely that this property is common for all Mg and Zn porphyrins. The absence of oxidizable side chains, such as the cyclopentanone ring and vinyl group in these compounds probably favors the stability of the intermediates in polar solvents containing oxygen. The reversible absorption changes of Chl a in oxygen-free polar solvents (59, 65, 66) are perhaps also caused by photooxidation, as secondary reactions of Chl+ with oxygen are prevented [cf. also Tollin and Green (75)]. The absorption spectrum of the intermediate, however, has not been measured.

An indication of the occurrence of a second oxidized state without marked absorption bands in the visible spectrum may also be obtained with some chlorophylls.

The redox potential in methanol of the system Mg TBP/Mg TBP$^+$ can be influenced markedly by the addition of alkali or acid, without the occurrence of side reactions. After addition of $10^{-2}\,M$ HCl the Mg TBP$^+$ spectrum appears, possibly produced by the oxygen in solution. The Mg TBP spectrum is restored either upon addition of alkali, or upon addition of reductants, even acid ones such as ascorbic acid. A change in the redox potential by addition of alkali was also observed with Bchl and chlorobium chlorophyll. The observation of Rabinowitch (30), that the first step of pheophytinization of chlorophyll a is reversible and affected by light, may well be related to this phenomenon.

II. Absorption and Fluorescence Spectra of Chlorophyll Aggregates

The high concentration of chlorophyll in the photosynthetic apparatus, and the difference in shape between the absorption spectra of chlorophylls in dilute solution and *in vivo*, have motivated investigations on effect of high pigment concentration on spectral properties.

A. Colloidal Suspensions

If a solution of Bchl or chlorophyll in methanol or acetone is diluted with water, an opalescent, nonfluorescent colloid forms. The absorption spectrum differs from the spectrum in solution, especially with Bchl. According to Krasnovsky *et al.* (76), freshly prepared colloidal solutions show bands at 785 and 848 mμ, instead of the single band at 770 mμ (Fig. 17). They assumed these bands corresponded to different states of aggregation in the colloidal particles. Similar spectra were found by Komen (77) for a 45:55 mixture of acetone and water. In a 10:90 mixture the two bands were located at 850 and 920 mμ. Comparison with Bchl crystals led Rabinowitch (30) to suggest that the 800 mμ band corresponds to a nonordered colloidal coagulation of pigment molecules, while the 840–860 mμ band corresponds to a "crystallized" fraction of the colloid.

The nature of the 920 mμ band is then obscure. As this band is appreciably further in the infrared than the band of relative large Bchl crystals, it seems plausible that factors other than just pigment-pigment interaction are affecting the spectral position.

With Chl a no formation of a second long wavelength band is usually found, but the absorption spectrum of a colloidal Chl a suspension may vary to a certain extent with conditions of preparation, purity, and time

(76, 78). The red absorption band is usually broadened and shifted to 672 mμ from its 662 mμ position in acetone.

Other chlorophyll colloids can be prepared in aqueous dioxane (77–80) or in aqueous carbitol (81). In dioxane the red band of Chl a is shifted farther than in aqueous colloids (to 685 mμ), while the band width is less and scattering is decreased. The minor bands between the red band and Soret band are much less pronounced in the colloids than in solution. Pheophytin does not form colloids in dioxane. Consequently this phenomenon requires the presence of the central Mg atom. Of other metal derivatives, only the Zn complex of pheophytin is able to form a colloid in aqueous dioxane.

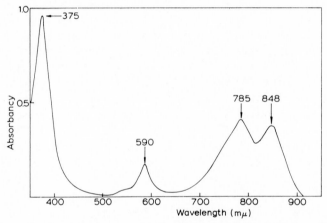

FIG. 17. Absorption spectrum of colloidal bacteriochlorophyll in a 80% water/ acetone mixture.

Addition of water to a solution of Chl a in butyl carbitol also results in a shift of the red band to longer wavelengths. Here fluorescence can be measured with a water percentage as high as 70–80%. If butyl carbitol is added to dioxane colloids, band shifts to 690 and 695 mμ can occur.

The mechanism resulting in the spectral shifts related to the water and dioxane colloids might be a dispersion of unordered particles into more regular aggregates (82). The decrease in scattering and apparent increase in absorption coefficient may be caused by a decrease of the "sieve effect" (83), due to splitting up of large particles. Dioxane probably effects polymerization via the central Mg or Zn atom.

B. Aggregates Formed at High Pigment Concentration

Rodrigo determined molecular weights by light scattering in concentrated chlorophyll solutions in acetone and methanol (84). Using a formula for light scattering derived by Debye (85), he determined a molecular weight for chlorophyll in acetone of about 2700 for concentrations exceeding $10^{-3} M$. This was interpreted to indicate the formation of a chlorophyll trimer. He found no detectable shift of the absorption maximum, though his curves indicated an approximately 15% increase in half-width value of the red absorption band. In methanol the results were complicated, probably owing to allomerization of chlorophyll in this solvent. Brody and Brody (86) measured the absorption spectrum of concentrated chlorophyll ($3 \times 10^{-2} M$) in methanol at room temperature. This spectrum showed a broadening and red shift of the main red absorption maximum. By subtraction from a spectrum of a $10^{-5} M$ solution they found a difference spectrum with bands at 648 and 682 mμ. In analogy with the theory of McRae and Kasha (13), they assumed this to be the absorption spectrum of a chlorophyll dimer. According to McRae and Kasha the absorption bands in an aggregated pigment molecule complex split into components numerically dependent on the number of constituent molecules. In the case of a dimer the absorption band of the monomer is split into two components. The distance between these components depends on the angle between the two molecules. The ability of the dimer to fluoresce is usually decreased, while its ability to phosphoresce is enhanced. Brody and Brody calculated that the two chlorophyll molecules should be placed at an angle of about 138°. They also reported the occurrence of a fluorescence band at 720 mμ in concentrated Chl a solutions after cooling to liquid air temperature (87), and ascribed it to chlorophyll-dimer emission.

The interpretation of these phenomena, as being brought about by a chlorophyll dimer, meets with serious objections. Subtraction of a sharp band from a broad band always results in the occurrence of a difference spectrum containing a double band. Moreover there is no reason why dimer fluorescence should be present at low temperatures and not at room temperature; according to McRae and Kasha (13), the triplet state at room temperature effectively drains the energy from the lowest dimer singlet state. It is not likely that this process is reversed at liquid nitrogen temperature. It seems more probable that the 720 mμ low temperature band represents the vibrational level in the chlorophyll fluorescence spectrum, which is sharpened and enhanced as a result of cooling. The main band, though enhanced at low concentration or in thin layers at high concentration, may be decreased in moderately thick layers due to increased scattering and reabsorption [cf. Goedheer (88)].

None of the various colloidal chlorophyll suspensions exhibit a splitting of the red band into two components, though they are assumed to represent various states of aggregation. The electron distribution in the π systems of large planar molecules such as the chlorophylls does not seem to be favorable for the occurrence of dimers, oriented at a fixed angle with each other, as a first state of aggregation. Higher aggregated forms, in which the molecules are oriented in more or less parallel stacks, seem to be more likely. In such polymers, a red shift and a band broadening are to be expected, but these need not be related, as the latter is caused primarily by the three-dimensional structure of the oriented aggregate.

The band shifts are supposed to be maximal, if the transition directions as well as the porphyrin planes of the chlorophyll molecules are oriented. Such orientation can be brought about if the chlorophyll molecules are attached by one of their side chains to an oriented structure, e.g., a detergent micelle.

Marcus and Haugen (89) measured chlorophyll absorption and fluorescence spectra in aqueous hexane solution. Though the method of preparation was not given, it seems likely that these solutions contain a few per cent of methanol, the solubility of water in pure hexane being slight. A difference spectrum between 5×10^{-5} M (1 cm) and a 5×10^{-6} M (10 cm) solution showed a double band (maxima at 660 and 678 mμ) in the red, which the authors ascribed to pigment aggregation according to McRae and Kasha [cf. also Brody and Brody (86)].

The fluorescence spectra of methanolic and aqueous hexane solutions of Chl a were generally similar, though some difference in spectral position occurred with different wavelength of excitation. The authors conclude from the absence of the "Stokes shift" that this fluorescence corresponds to resonance emission in the aggregate. This, however, is quite unlikely. It is more plausible to assume that more than one chlorophyll-solvent form is present in the methanol–hexane–water mixture, and the aggregated form with somewhat broadened absorption spectrum is non-fluorescent.

III. Fluorescence and Other Luminescence Properties of Chlorophylls and Analogs

A. Fluorescence

1. QUANTUM YIELDS AND LIFETIMES

The quantum yield of fluorescence of Chl a at low concentration in organic solvents was measured by Förster and Livingston (90). They found a value of 0.22 for Chl a dissolved in ether, while in other solvents the values were but little different. For Chl b they determined a value of 0.09 in ether and of 0.05 in methanol. The quantum yield of pheophytin *a* fluorescence was 0.14. Latimer *et al.* (*91*) measured values of 0.33 for Chl *a* and 0.16 for Chl *b*, both in ether solution.

The quantum yields do not differ markedly for light absorbed in different parts of the spectrum. Förster and Livingston (*90*) found, contrary to earlier measurements, the same quantum yield for light absorbed in either the Soret band or in the orange-red region. The independence of quantum yield with wavelength of excitation indicates an effective coupling between the mutually parallel and perpendicular transitions within the molecule. This is confirmed by measurements of fluorescence action spectra of bacteriopheophytin and pheophytin *a*. In these spectra the bands at 545 mμ, which are due to a transition perpendicular to the fluorescing one, are as pronounced as in the absorption spectra.

The wavelength-independence of quantum yield does not hold on the long-wave side of the red absorption band. According to Förster and Livingston the quantum yield of Chl *a* fluorescence excited at 689 mμ is less than half that excited at 662 mμ. Such a long wavelength decline of quantum yield of fluorescence in organic solution can be due either to the presence of nonfluorescing, but nevertheless absorbing, molecules (e.g., chlorophyll dimers, polymers, or oxidation products), or to some property of the pigment itself (e.g., conversion of the low-energy light quantum directly to the triplet state, or the presence of an n-π level with a weak coupling to the general π-π system).

Fluorescence of allomerized Chl *a* is reported to have a yield of about half that of Chl *a*, while the yield of allomerized Chl *b* was twice that of Chl *b* (*92*). The products of photooxidation of Chl *a* usually do not show fluorescence. Also the products formed by oxidation with ferric chloride or other oxidants are nonfluorescent. The photoreduced forms of chlorophyll, however, are reported to emit fluorescence (*64*). The quantum yields of fluorescence of Bchl, chlorophylls *c* and *d,* protochlorophyll, and their corresponding pheophytins have not been measured.

Cooling a solution of Chl *a* to liquid nitrogen temperature yields an increase in fluorescence of about 75% in both the main peak and the vibrational level. Since the bands are sharpened as a result of cooling, this does not represent an increase in quantum yield of the same amount.

In completely dry hydrocarbons chlorophyll is nonfluorescent, or only very weakly fluorescent (*34*). Addition of a few per cent of water or polar solvents restores the fluorescence yield to its "normal" value. Pheophytin, however, when dissolved in completely dry hydrocarbons fluoresces with its "normal" yield. A similar phenomenon is found with phthalocyanine and its magnesium complex. Thus it is not related to the presence of the cyclopentanone ring.

Actual fluorescence lifetimes, measured by flash excitation and detected by a fast oscilloscope (93), were found to be approximately 3.9 $\times 10^{-9}$ seconds for Chl a in ethyl ether and 5.1×10^{-9} seconds for Chl b. Butler and Norris (94), using a phase fluorometer, found a value of 2.5×10^{-9} seconds for Chl a [cf. also Terenin (95)]; after cooling to liquid nitrogen temperature the value was increased to 4.4×10^{-9} seconds.

The natural lifetime τ_r is defined as that lifetime which would result if all absorbed light energy were emitted as fluorescence; that is, without the occurrence of competing dark processes, such as internal conversion or interaction with special quenching molecules in the solvent. The quantum yield of emission from the natural lifetime should equal unity. The natural lifetime depends on the transition probability between first excited state and ground state ("oscillator strength") and can be determined by measurement of the area under the absorption bands, insofar these bands are due to the first transition (96).

The actual fluorescence lifetime determined by radiationless processes, and the ratio $\tau : \tau_r$ equals the quantum yield of fluorescence of the pigment.

The actual fluorescence lifetime can also be determined, in an indirect way, by measuring fluorescence polarization as a function of viscosity of the solvent. This may be seen as follows. If a molecule can rotate freely during the time elapsing between light absorption and emission of fluorescence (the time the molecule is in its excited state), fluorescence resulting from polarized incident light is completely depolarized. The Brownian rotation of the pigment molecule depends on its molecular volume and shape, viscosity of solvent, and temperature. If these parameters are known, the lifetime of the excited state can be calculated (9–11, 17).

2. ENERGY TRANSFER BETWEEN PIGMENTS IN SOLUTION

Several mechanisms exist by which excitation energy is transferred from one pigment molecule to another. Which mechanism acts under given conditions depends mainly on the average distance between the molecules. If the pigment concentration is such that the average distance between the molecule centers is larger than the molecular dimension but smaller than the wavelength of light, excited pigment molecules can be coupled by inductive resonance with neighboring ones. The efficiency of excitation energy transfer depends on pigment concentration, mutual orientation, quantum yield of fluorescence, overlap integral of absorption and fluorescence spectrum, and the refractive index of the solvent. In solution, energy transfer by inductive resonance results in such phenomena as concentration depolarization and concentration quenching. If the solution consists of a mixture of two or more kinds of chlorophyll, inductive resonance leads to sensitized fluorescence.

a. Concentration Depolarization. Irradiation with plane polarized light results in a fluorescence polarization of about $p = 0.40$ in a randomly oriented pigment solution if: (a) the "oscillators" for absorption and emission are parallel; (b) no rotation occurs during the fluorescence lifetime; (c) the pigment concentration is low.

If the last condition does not hold, energy transfer by inductive coupling of two differently oriented molecules results in depolarization of fluorescence. At a certain concentration, called the "critical concentration," the fluorescence will be half depolarized. This critical concentration was found for Chl *a* in castor oil to be $0.6 \times 10^{-3} M$; for Chl *b*, $3.5 \times 10^{-3} M$ (17). There is reasonable agreement with values calculated from equations of energy transfer.

b. Fluorescence Concentration Quenching. At concentrations appreciably exceeding the critical ones for fluorescence depolarization, energy absorbed by a pigment molecule can excite quite a number of other molecules, some of which may be nonfluorescent and act as energy "sinks." According to Förster and Livingston (90) concentration quenching indeed occurs with Chl *a*. The energy "sinks" may be chlorophyll dimers or polymers [cf. Lavorel (97)], pigment molecules associated with foreign quenching molecules, or pigment molecules oxidized by the illumination used in the experiments, but still showing some absorption in the red.

That the presence of chlorophyll dimers should account for the long wavelength decline does not seem very likely because the effect occurs even at very low pigment concentrations (90). Measurement of polarization of fluorescence in the long wavelength region can show whether the long wavelength decline is due to a decrease in fluorescence lifetime of all chlorophyll molecules, or to the absence of fluorescence in a small fraction of chlorophyll molecules which have their absorption maximum at a slightly longer wavelength.

c. Sensitized Fluorescence. If a mixture of chlorophylls *a* and *b* is excited with light absorbed mainly by Chl *b*, and the pigment concentration is sufficiently high, fluorescence will be emitted mainly by Chl *a* [cf. Duysens (98)]. Because of the positions of the absorption and fluorescence maxima, energy will be transferred mainly from *b* to *a*, while the reverse is improbable.

A different mechanism for energy transfer can occur if the concentration is so high that the average distance between the molecule is of the order of the molecular dimensions. In this case the π-electron clouds may overlap to such a degree that an electron can migrate easily from the excited state of one pigment molecule to another, while an electron of the ground state may move from the second molecule to the first. In solution, however, aggregation probably occurs before such high concentrations are reached. Within such aggregates these "solid state" phenomena can occur, but the concentration is reduced so that between the aggregates no electron transport occurs.

At all concentrations, energy transfer can occur by secondary fluorescence. Light emitted by one molecule is absorbed by another, as absorption and fluorescence spectra overlap partly. This process is inefficient, but it can interfere with measurements of fluorescence polarization and fluorescence quantum yields.

B. Phosphorescence, Afterglow, and Chemiluminescence

1. Reports of Phosphorescence

Though the possibility of exciting a triplet state of Chl a has been made plausible by flash photolysis experiments, measurements of phosphorescence or slow fluorescence from it are scanty. The lifetime of the triplet state, according to flash absorption spectrophotometry, is approximately 3×10^{-4} seconds for Chl a, and it depends strongly on pigment concentration (63). It is apparently mainly determined by nonradiative processes such as collisional quenching and internal conversion. In oxygen-free benzene, Chl b can be converted into the triplet state by flash excitation with about 100% yield (72).

Measurements of phosphorescence of chlorophylls a and b have been reported by Becker et al. (36, 99, 100). Fernandez and Becker (36) measured an extremely weak afterglow of chlorophylls a and b in completely dry solvents. The emission maximum was located at about 755 mμ for Chl a and 733 mμ for Chl b. They assumed this emission to correspond to π-n phosphorescence. Singh and Becker (100) recorded afterglows from chlorophylls a and b in rigid E.P.A. (ether, pentane, alcohol) glasses, by excitation with a 1000-W Xe-Hg arc and measurement with sensitized photographic plates at 77°K. They found poorly resolved maxima at 885 ± 5 and 925 ± 5 mμ for Chl a, while Chl b showed bands at 875 ± 5 and 915 ± 5 mμ. They assumed that these emissions represented π-π phosphorescence.

However, as a variety of compounds show afterglow at 77°K, it is possible that impurities present in trace amounts could give rise to afterglow with such high intensity light sources.

2. Afterglows

Afterglows from solid or dissolved chlorophylls may also be chemiluminescence instead of phosphorescence. In this case oxygen does not need to be excluded rigorously. Kautsky et al. (101) reported an afterglow of chlorophyll when investigating phosphorescence of various dyestuffs, but only when isoamylamine was used as a solvent. Albrecht et al. (102) found an afterglow of chlorophyll in glassy cellulose butyrate lacquers, with a lifetime of several minutes. Goedheer (103) measured a weak afterglow of chlorophyll in air-saturated methanol, with a half-life of about 1 minute. No such afterglow could be observed with pheophytin at a similar concentration.

A different type of afterglow can be demonstrated after photobleaching of Bchl. This pigment, when dissolved in alcohol is photooxidized rapidly, and oxygen is taken up. A peroxide test (104) indicated that peroxides were formed. The original spectrum was largely restored by adding ascorbic acid or hydrogen sulfide. This restoration of the spectrum coincided with the emission of light in the spectral region of Bchl fluorescence (103). Illumination of Bchl-ascorbic acid mixtures resulted in a rapid Bchl-sensitized decomposition of ascorbic acid and emission of afterglow. It thus seems likely that all these afterglow phenomena are chemiluminescence caused by the production of organic peroxides.

3. Chemiluminescence

The chlorophylls, as well as various other metal porphyrins, such as magnesium tetrabenzporphin, zinc tetraphenyl porphin, and magnesium phthalocyanine, emit a bright red chemiluminescence when heated in tetralin to about 150°C (105). The mechanism was found to involve a metal porphyrin-catalyzed decomposition of tetralin hydroperoxide (106, 107). The explanation of this chemiluminescence lies in the capacity of peroxides to undergo oxidizing as well as reducing reactions. If the redox potential of the metal porphyrin is of the right order of magnitude, this pigment may act as an electron donor for the oxidizing reaction, and as an electron acceptor for the reducing reaction. The average energy difference between the two reactions (with H_2O_2 about 1.1 ev) appears as heat, unless sufficient thermal energy is added to excite the first singlet state of the pigment. In this case luminescence appears with the spectrum of fluorescence. The pheophytins, which are not easily oxidizable, do not emit chemiluminescence when heated in tetralin.

In view of its general appearance with Mg and Zn porphyrins, the phenomenon depends on the redox potential of the pigment. No chemiluminescence is found with the paramagnetic Fe and Cu complexes, probably because of the lack of fluorescence of these compounds. The thermoluminescence measured by heating chloroplasts (108) could also be explained as chemiluminescence of chlorophyll and peroxidized lipids [cf. Bellamy (109)].

A chemiluminescence with a much lower efficiency is measured when methanolic Chl a solutions are oxidized by addition of ferric salts and subsequently reduced by excess ferrous salts or ascorbate (103). Again this reaction does not occur with the pheophytins. The appearance of luminescence coincides with the restoration of the absorption spectrum. The energy emitted by this low yield reaction (it is at least a factor of 10^8 lower intensity than the reaction mentioned above) may possibly originate in the energy difference between ferric and ferrous ions being separated and being mixed. It may, however, also be caused by other reactions such as allomerization.

REFERENCES

(1) H. Kuhn, *J. Chem. Phys.* **17**, 1198 (1949).
(2) W. T. Simpson, *J. Chem. Phys.* **17**, 1218 (1949).
(3) H. C. Longuet-Higgins, C. W. Rector, and J. R. Platt, *J. Chem. Phys.* **18**, 1174 (1950).
(4) J. R. Platt, *in* "Radiation Biology" (A. Hollaender, ed.), p. 71. McGraw-Hill, New York, 1956.
(5) M. Gouterman, *J. Chem. Phys.* **30**, 1139 (1959).
(6) M. Gouterman, *J. Mol. Spectry.* **6**, 138 (1961).
(7) E. I. Rabinowitch, "Photosynthesis and Related Processes," Vol. II, Part 1, p. 620. Wiley (Interscience), New York, 1951.
(8) M. D. Kamen, "Primary Processes in Photosynthesis." Academic Press, New York, 1963.
(9) F. Perrin, *Ann. Phys. (Paris)* **12**, 160 (1929).
(10) G. Weber, *in* "Advances in Protein Chemistry" (M. L. Anson and J. T. Edsall, eds.), Vol. 8, p. 415. Academic Press, New York, 1953.
(11) T. Förster, "Fluoreszenz Organischer Verbindungen." Van den Hoeck and Ruprecht, Göttingen, 1951.
(12) W. Heller and A. Marcus, *Phys. Rev.* **84**, 809 (1951).
(13) E. G. McRae and M. Kasha, *J. Chem. Phys.* **28**, 721 (1958).
(14) J. H. C. Smith and A. Benitez, *in* "Modern Methods of Plant Analysis" (K. Paech and M. V. Tracey, eds.), p. 143. Springer, Berlin, 1955.
(15) A. S. Holt and E. E. Jacobs, *Am. J. Botany* **41**, 710 (1954).
(16) A. Jensen, O. Aasmundred, and K. E. Eimhjellen, *Biochim. Biophys. Acta* **88**, 466 (1964).
(17) J. C. Goedheer, Thesis, University of Utrecht (1957).
(18) J. C. Goedheer, *Nature* **176**, 928 (1955).
(19) R. Stupp and H. Kuhn, *Helv. Chim. Acta* **35**, 2469 (1952).
(20) J. W. Weigl, *J. Am. Chem. Soc.* **75**, 999 (1952).
(21) J. C. Goedheer, *Biochim. Biophys. Acta* **27**, 478 (1958).
(22) L. Pekkarinen and H. Linschitz, *J. Am. Chem. Soc.* **82**, 2407 (1960).
(23) P. Metzner, *Ber. Deut. Botan. Ges.* **40**, 171 (1962).

(24) H. Larsen, Kgl. Norske Videnskab. Selskabs, Skrifter No. 1, p. 147 (1953).
(25) T. W. Goodwin, Biochim. Biophys. Acta 18, 309 (1955).
(26) R. Y. Stanier and J. H. C. Smith, Biochim. Biophys. Acta 41, 478 (1960).
(27) J. M. Olson and C. A. Romano, Biochim. Biophys. Acta 59, 726 (1962).
(28) J. M. Olson, D. Folmer, R. Radloff, C. A. Romano, and C. Sybesma, in "Bacterial Photosynthesis" (H. Gest, A. San Pietro, and L. P. Vernon, eds.), p. 113. Antioch Press, Yellow Springs, Ohio, 1963.
(29) H. J. Perkins and D. W. A. Roberts, Biochim. Biophys. Acta 79, 20 (1964).
(30) E. I. Rabinowitch, "Photosynthesis and Related Processes," Vol. 1, p. 638. Wiley (Interscience), New York, 1945.
(31) C. S. French, in "Handbuch der Pflanzenphysiologie" (W. Ruhland, ed.), Vol. 5, p. 252. Springer, Berlin, 1960.
(32) G. R. Seely and R. G. Jensen, Spectrochim. Acta 21, 1835 (1965).
(33) M. Gouterman and L. Stryer, J. Chem. Phys. 37, 2260 (1962).
(34) R. Livingston, W. F. Watson, and J. McArdle, J. Am. Chem. Soc. 71, 1542 (1949).
(35) F. Bär, H. Lang, E. Schnabel, and H. Kuhn, Z. Elektrochem. 65, 346 (1961).
(36) J. Fernandez and R. Becker, J. Chem. Phys. 31, 476 (1959).
(37) M. R. Michel-Wolwertz and C. Sironval, Biochim. Biophys. Acta 94, 330 (1965).
(38) M. Zocher, Trans. Faraday Soc. 35, 34 (1939).
(39) A. A. Krasnovsky and G. P. Brin, Dokl. Akad. Nauk SSSR 89, 527 (1953).
(40) P. J. McCartin, J. Phys. Chem. 67, 513 (1963).
(41) S. Freed and K. M. Sancier, Science 114, 275 (1951).
(42) S. S. Brody and S. B. Broyde, Nature 199, 1097 (1963).
(43) H. H. Strain, M. R. Thomas, and J. J. Katz, Biochim. Biophys. Acta 75, 306 (1963).
(44) C. S. French, J. H. C. Smith, H. I. Virgin, and R. L. Airth, Plant Physiol. 31, 369 (1956).
(45) A. S. Holt, and H. V. Morley, Can. J. Chem. 37, 507 (1959).
(46) H. Sagromsky, Ber. Deut. Botan. Ges. 77, 323 (1964).
(47) M. R. Michel-Wolwertz, C. Sironval, and J. C. Goedheer, Biochim. Biophys. Acta 94, 584 (1965).
(48) V. M. Koski and J. H. C. Smith, J. Am. Chem. Soc. 70, 3558 (1948).
(49) A. S. Holt and E. E. Jacobs, Am. J. Botany 41, 463 (1954).
(50) G. Oster, S. B. Broyde, and J. S. Bellin, J. Am. Chem. Soc. 86, 1309 (1964).
(51) I. G. Savkina and V. B. Evstigneev, Biofizika 8, 335 (1963).
(51a) J. C. Goedheer, unpublished data (1965).
(52) A. A. Krasnovsky and K. K. Voinovskaya, Dokl. Akad. Nauk SSSR 66, 603 (1949).
(53) A. A. Krasnovsky and E. V. Pakshina, Dokl. Akad. Nauk SSSR 135, 1258 (1960).
(54) J. C. Goedheer, G. H. Horreus de Haas, and P. Schuller, Biochim. Biophys. Acta 28, 278 (1958).
(55) A. A. Krasnovsky and E. V. Pakshina, Dokl. Akad. Nauk SSSR 127, 913 (1959).
(56) E. Rabinowitch and J. Weiss, Proc. Roy. Soc. A162, 251 (1937).
(57) A. Stanienda, Naturwissenschaften 50, 731 (1963).
(58) A. Weller, J. Am. Chem. Soc. 76, 1575 (1954).
(59) H. Linschitz and J. Rennert, Nature 169, 193 (1952).

(60) A. A. Krasnovsky, *Ann. Rev. Plant. Physiol.* 11, 363 (1960).
(61) T. T. Bannister, *Plant Physiol.* 34, 246 (1959).
(62) G. Zieger and H. T. Witt, *Z. Physik. Chem.* Neue Folge 28, 286 (1961).
(63) H. Linschitz and K. Sarkanen, *J. Am. Chem. Soc.* 80, 4026 (1958).
(64) V. I. Gachkovsky, *Biofizika* 4, 16 (1959).
(65) R. Livingston and E. Fujimori, *J. Am. Chem. Soc.* 80, 5610 (1958).
(66) D. Porret and E. I. Rabinowitch, *Nature* 140, 321 (1937).
(67) J. J. McBrady and R. Livingston, *J. Phys. Colloid Chem.* 51, 775 (1947).
(68) B. Dunicz, T. Thomas, M. Van Pee, and R. Livingston, *J. Am. Chem. Soc.* 73, 3388 (1951).
(69) S. Freed and K. M. Sancier, *Science* 116, 175 (1952).
(70) V. B. Evstigneev and V. A. Gavrilova, *Dokl. Akad. Nauk SSSR* 74, 781 (1950).
(71) A. A. Krasnovsky, G. P. Brin, and K. K. Voinovskaya, *Dokl. Akad. Nauk SSSR* 69, 393 (1949).
(72) S. Claesson, L. Lundquist, and B. Holmström, *Nature* 183, 661 (1959).
(73) G. R. Seely and M. Calvin, *J. Chem. Phys.* 23, 1068 (1955).
(74) P. A. Barrett, R. P. Linstead, F. G. Rundall, and G. A. D. Tuey, *J. Chem. Soc.* p. 1097 (1940).
(75) G. Tollin and G. Green, *Biochim. Biophys. Acta* 60, 524 (1962).
(76) A. A. Krasnovsky, K. K. Voinovskaya, and L. M. Kosobutskaya, *Dokl. Akad. Nauk SSSR* 85, 389 (1952).
(77) J. G. Komen, *Biochim. Biophys. Acta* 22, 9 (1956).
(78) K. P. Meyer, *Helv. Phys. Acta* 12, 349 (1939).
(79) B. B. Love and T. T. Bannister, *Biophys. J.* 3, 99 (1963).
(80) A. A. Krasnovsky and L. Kosobutskaja, *Dokl. Akad. Nauk SSSR* 91, 343 (1953).
(81) B. B. Love, *Biochim. Biophys. Acta* 64, 318 (1962).
(82) M. Belavtseva and A. A. Krasnovsky, *Biofizika* 4, 521 (1959).
(83) L. N. M. Duysens, *Biochim. Biophys. Acta* 19, 1 (1956).
(84) F. A. Rodrigo, Thesis, University of Utrecht (1955).
(85) P. Debye, *J. Appl. Phys.* 15, 338 (1944).
(86) S. S. Brody and M. Brody, *Nature* 189, 547 (1961).
(87) S. S. Brody and M. Brody, *Trans. Faraday Soc.* 58, 416 (1962).
(88) J. C. Goedheer, *Biochim. Biophys. Acta* 88, 304 (1964).
(89) R. J. Marcus and G. R. Haugen, *Photochem. Photobiol.* 4, 183 (1965).
(90) F. S. Förster and R. Livingston, *J. Chem. Phys.* 20, 1315 (1952).
(91) P. Latimer, T. T. Bannister, and E. Rabinowitch, *Science* 124, 585 (1956).
(92) R. Livingston, in "Photosynthesis in Plants" (J. Franck and W. E. Loomis, eds.), p. 179. Iowa State Coll. Press, Cedar Falls, Iowa, 1949.
(93) S. S. Brody and E. Rabinowitch, *Science* 125, 555 (1957).
(94) W. L. Butler and K. H. Norris, *Biochim. Biophys. Acta* 66, 73 (1963).
(95) A. N. Terenin, *2nd All-Union Conf. Photosynthesis, Moscow, 1957* (*English Transl.*) p. 9. Consultants Bureau, New York.
(96) J. A. Prins, *Nature* 134, 457 (1934).
(97) J. Lavorel, *J. Phys. Chem.* 61, 1600 (1957).
(98) L. N. M. Duysens, Thesis, University of Utrecht (1952).
(99) R. S. Becker and M. Kasha, *J. Am. Chem. Soc.* 77, 3669 (1955).
(100) I. S. Singh and R. S. Becker, *J. Am. Chem. Soc.* 82, 2083 (1960).

(*101*) H. Kautsky, H. Hirsch, and W. Flesch, *Ber. Deut. Chem. Ges.* **68**, 152 (1935).

(*102*) H. O. Albrecht, W. C. Dennison, L. G. Livingston, and G. E. Mandeville, *J. Franklin Inst.* **268**, 278 (1959).

(*103*) J. C. Goedheer, *Nature* **193**, 875 (1962).

(*104*) A. A. Krasnovsky and K. K. Voinovskaya, *Dokl. Akad. Nauk SSSR* **81**, 879 (1951).

(*105*) A. Stewart, H. V. Knorr, and V. M. Albers, *Phys. Rev.* **61**, 730 (1942).

(*106*) H. Linschitz and E. W. Abrahamson, *Nature* **172**, 909 (1953).

(*107*) G. D. Dorough, J. R. Miller, and F. M. Huennekens, *J. Am. Chem. Soc.* **73**, 4315 (1951).

(*108*) W. Arnold and H. Sherwood, *Proc. Natl. Acad. Sci. U.S.* **43**, 105 (1957).

(*109*) W. D. Bellamy, *Photochem. Photobiol.* **3**, 259 (1964).

—7—

Infrared and Nuclear Magnetic Resonance Spectroscopy of Chlorophyll*

J. J. KATZ, R. C. DOUGHERTY, AND L. J. BOUCHER

Argonne National Laboratory, Argonne, Illinois

* Based on work performed under the auspices of the U.S. Atomic Energy Commission.

185

I. Introduction

The application of physical methods to organic chemistry has resulted in notable advances in our understanding of the structure and function of chemical compounds. Classical absorption spectroscopy in the visible and ultraviolet has provided much of what is now known about the photosynthetic pigments. The extension in recent years of absorption spectroscopy to other regions of the electromagnetic spectrum has furnished new and widely applicable tools for the study of complex organic compounds, and these methods promise to make possible new insights into the behavior and properties of chlorophyll and the other complex compounds that constitute the photosynthetic apparatus. In this chapter we shall be concerned with two of the most powerful of the newer techniques, infrared and nuclear magnetic resonance spectroscopy, particularly as applied to the chlorophylls and some closely related compounds.

The absorption of visible or ultraviolet light by organic molecules is accompanied by the excitation of (valency) electrons to higher energy levels. Electronic excitation requires quite high-energy quanta, corresponding to light of relatively short wavelength, i.e., in the ultraviolet and visible regions. Electronic transition spectra give information about the presence and nature of unsaturation, particularly conjugated double bonds and aromatic ring systems. Certain functional groups, such as ketonic or aldehydic carbonyl groups have characteristic absorption spectra, and, because of this, electronic spectra can be usefully employed for structural studies. Since so many of the compounds involved in photosynthesis have conjugated double-bond systems, it is easy to see why electronic spectra have played such a large role in the study of photosynthetic pigments. Absorption spectroscopy in this region is highly sensitive, excellent spectrometers are available, and this makes possible many analytical applications. Nevertheless, absorption spectroscopy in the visible and ultraviolet has some serious limitations, not the least of which is the relative insensitivity of spectra originating in conjugated systems to chemical changes at sites not coupled to the conjugated systems. This is particularly important in the case of the chlorophylls, where major chemical changes in groups not conjugated with the macrocyclic ring may make only trivial changes in the visible absorption spectrum.

The infrared region of the electromagnetic spectrum extends from just above $1\,\mu$ ($10,000\,\text{Å}$ or $10,000\,\text{cm}^{-1}$) to about $500\,\mu$ ($20\,\text{cm}^{-1}$), where the microwave region begins. It is only the region from 10,000 to $200\,\text{cm}^{-1}$ in which molecular vibrations occur that concerns us here. The vibrational frequencies of molecules lie in the infrared region, and the absorption of radiation in this frequency range excites higher vibrational states. Infrared spectra thus yield information about stretching and bending motions of the atoms in a molecule. In theory infrared absorption bands arise from molecular motions of the molecule as a whole, but in practice the absorption frequencies of many structural units are sufficiently independent of the rest of the molecule in which they occur to have characteristic *group frequencies*. Group frequencies can be used to establish the presence or absence of many functional groups in organic compounds. Group frequencies of functional groups are often sensitive to their environment, and consequently, it is often possible to obtain valuable information from infrared spectra on hydrogen bonding, chelation, conformation, and *cis-trans* isomerism, in addition to purely structural details.

Nuclear magnetic resonance (NMR) spectroscopy is still another form of absorption spectroscopy. Atomic nuclei with an odd mass number, or nuclei with even mass but odd charge, possess a mechanical spin, and because atomic nuclei are also associated with an electric charge, the spin endows these nuclei with a magnetic moment. In a uniform magnetic field, the magnetic dipoles of such nuclei can assume a small number of quantum-restricted orientations relative to the field. Each spin orientation corresponds to an energy state, the energy of which is a function of the spin number (I), the magnetic moment (μ) and the strength of the applied magnetic field. Transitions between adjacent energy states are possible, and can be made to occur if the nuclei are irradiated with electromagnetic radiation of a frequency comparable to the energy differences between the states. For the magnetic fields now available, the energy differences between spin states fall in the radiofrequency region, and NMR spectra for various nuclei can be observed in the radiofrequency range of 5–100 megacycles. Nuclear magnetic resonance signals can be observed, for example, with $^{1}_{1}H$, $^{2}_{1}H$, $^{19}_{9}F$, $^{31}_{15}P$, $^{14}_{7}N$, $^{15}_{7}N$, $^{13}_{6}C$ and $^{17}_{8}O$, but not with $^{12}_{6}C$ or $^{16}_{8}O$. Proton magnetic resonance spectra are by all odds the most commonly studied.

Proton magnetic resonance (PMR) spectroscopy derives its great importance from the facts that the resonance frequency of a proton at constant magnetic field is strongly affected by its molecular environment, and hydrogen is omnipresent in organic compounds.* The applied magnetic field induces circulation of extranuclear electrons, and these circulations generate small magnetic fields which may either oppose or augment the imposed external magnetic field. Thus, depending on the structure of the molecule, chemically distinct protons will come into

---

* The deuteron resonance spectra of fully deuterated methyl pheophorbide *a* and chlorophyll *a* have been successfully recorded [R. C. Dougherty, G. Norman, and J. J. Katz, *J. Am. Chem. Soc.* **87**, 5801 (1965)].

resonance at different frequencies (at constant field), and these *chemical shifts* are important diagnostic criteria in the interpretation of NMR spectra. Because the area under the resonance curve is directly proportional to the number of protons involved in a particular absorption, chemical shift data coupled with information about the relative areas allow the classification of protons in an organic compound by kind and by number.

Nuclear magnetic resonance techniques are useful in the study of many problems. NMR data can be used for the unambiguous detection of certain functional groups and for structure determinations of organic compounds. They provide information about stereochemical configuration and the conformation of molecules. Molecular motion and hindered rotation about single or partial double bonds can be investigated. Kinetic processes involving hydrogen exchange, introduction, or abstraction can be readily followed by NMR methods. Because the formation of intermolecular aggregates or coordination with other compounds can change the magnetic environment of a given nucleus, NMR spectroscopy is a valuable tool for the study of coordination and self-aggregation. These are only some of the more important ways in which this powerful technique can be used, and it is, therefore, not surprising that PMR spectroscopy should be brought to bear on the many difficult problems that still remain to be unraveled in the field of chlorophyll chemistry.

II. Infrared Spectra

A. General Aspects of the Interpretation of Infrared Spectra

As already hinted at, the interpretation of the infrared (IR) spectra of complex compounds is essentially an empirical procedure. To be sure, a rigorous mathematical analysis of the spectra of simple molecules in the gas phase can be carried out, but even for compounds far less complicated than chlorophyll a completely rigorous analysis of an IR is not yet possible. However, it is practical to assign absorption bands to the vibration of particular molecular groupings by empirical spectra-structure correlations. Even empirical interpretations are found to yield much useful structural information by way of functional group identification.

The theory and practice of IR spectroscopy are admirably described in a number of recent books (*1–3*). The books of Bellamy (*4*) and Nakanishi (*5*) are particularly useful for spectra-structure correlations.

B. Methods

The experimental techniques for securing IR spectra of chlorophyll and related compounds in the 4000–600 cm^{-1} region are quite straightforward and conventional (6). Solid-state spectra obtained with mulls or potassium bromide discs are free of background solvent absorption. The waxy consistency of the chlorophylls and many of their derivatives makes proper preparation of mulls or potassium bromide discs difficult. Solid-state spectra, moreover, are generally not as highly resolved as are solution spectra, and for the magnesium-containing chlorophylls are of limited utility. We shall see that the IR spectra of magnesium-containing chlorophyll compounds are strongly dependent on aggregation state. Solid-state spectra are characteristic of the highly aggregated or partially crystalline state, and thus it is difficult to observe changes in aggregation with solid-state spectra. On occasion, however, KBr discs can provide useful data, particularly for magnesium-free compounds, and they are sometimes the only practical way of securing spectra, especially where solubility is limiting.

Solution spectra give the most precise information, even if they are not always the most easily obtained. More precise control can be exercised over concentration, sample thickness, and the environment of the solute molecules. Because of the necessity to conserve the difficulty obtainable pure chlorophylls, microtechniques for securing solution spectra are almost essential. Type D Irtran-2 microcavity cells with a 0.050-mm light path length make it possible to secure useful spectra with approximately 0.5–1 mg of chlorophyll dissolved in 10 μl of solvent. Irtran-2 cells are made of microcrystalline zinc sulfide and are transparent through the rock salt range, 4000–600 cm^{-1}. They are mechanically strong and are impervious to water, alcohols, and most organic solvents. Microcells require focusing of the light beam. Commercial instruments (Beckman IR-7 or IR-12, Perkin-Elmer 521) are easily fitted with beam-condensing systems that are available as accessories. All the IR spectra in the rock-salt region shown here are uncompensated and were recorded on a Beckman IR-7. The instrument was fitted with a beam condenser and an attenuator inserted in the reference beam. Type D Irtran-2 cells with a 0.05-mm light path were used in our solution measurements.

Solvent selection is an important consideration in obtaining solution spectra. No solvent is fully transparent over the entire IR range. Weak solvent absorption bands may be compensated for by the introduction of solvent in a variable path length cell into the reference beam. Generally, several solvents are required to observe the full spectrum. For routine identification and characterization, carbon tetrachloride is useful, but it is important that IR spectra in CCl$_4$ be supplemented by measurement in tetrahydrofuran (or other polar solvents) in the 1800–1600 cm^{-1} region.

Relatively few aspects in the middle and far infrared (650–200 cm^{-1}) have been as yet recorded for substances of biological importance. The far-infrared is particularly important for compounds such as the chlorophylls that contain coordinately bound metal ions, because metal-ligand vibrations occur in this region (7).

The interpretation of spectra in the far-infrared is more difficult than in the much more intensively studied rock-salt region, and only a few spectra-structure correlations are available (2). Several excellent commercial far-infrared spectrophotomers (Beckman IR-11, IR-12, Perkin-Elmer 301) are now available, and satisfactory spectra in this region are no longer difficult to obtain. Sample preparation presents some special problems. Mulls in mineral oil can be used, but relatively large amounts (20–50 mg) of chlorophyll are required because absorption bands in this region are normally weak. Only a limited number of solvents are free of strong absorptions in the far-infrared (8). At 5-mm light path lengths, cyclohexane is transparent from 650 to 100 cm^{-1}, except for a strong absorption at 540–500 cm^{-1}. Benzene, chloroform, and carbon tetrachloride are also useful solvents, but must be used at shorter path lengths and over a smaller spectral range than cyclohexane. Cells made of high density polyethylene are suitable (9). Our 5.00-mm cells require approximately 15 mg of chlorophyll dissolved in 2 ml of solvent.

C. Historical Survey

The first examination of the IR spectrum of chlorophyll and the important related compounds ethyl chlorophyllide, pheophytin, and phytol was made by the pioneer investigator of infrared phenomena, Coblentz (10), in 1933. It was not until twenty years later that these compounds were critically re-examined, with very much improved instrumentation, by Weigl and Livingston (11). This study, and the subsequent work of Holt and Jacobs (12), amply demonstrated the utility and applicability of IR spectroscopy to structural problems in chlorophyll chemistry.

A considerable number of chlorophylls and related compounds have now been examined in the infrared, and Table I provides a key to the literature. Spectral region and sample phase are shown.

Wetherell et al. (18) have measured the IR spectra of chlorin-e$_4$, mesochlorin-e$_4$, rhodochlorin, and mesorhodochlorin from 1500 to 3000 cm^{-1}. The spectra of the divalent metal (Ni, Zn, Ag, Cu, Cd, Hg) derivatives of pheophytin a and pyropheophytin a were examined in the rock salt region by Holt (19) and by Sidorov and Terenin (20).

In 1960, fully deuterated chlorophylls became available by isolation from algae grown autotrophically in 99.8% D$_2$O (15, 21). In the course of isotope effect studies on these unusual substances, it became apparent that the then current interpretation of the IR spectra of chlorophyll con-

TABLE I

INFRARED SPECTRA OF THE CHLOROPHYLLS REPORTED
PREVIOUSLY IN THE LITERATURE[a]

Compound	Region (cm^{-1})	Medium[b]	Reference
Chlorophyll a	650–3500	Film	(11)
Chlorophyll a	650–3800	Nujol mull	(12)
Chlorophyll a	600–4000	KBr disc	(13)
Chlorophyll a	650–3800	$CHCl_3$, CCl_4, pyridine	(12)
Chlorophyll a	1600–3800	CS_2, Et_2O	(12)
Ethyl chlorophyllide a	650–3800	$CHCl_3$	(12)
Pheophytin a	650–3800	CCl_4, pyridine	(12)
Pheophytin a	650–3500	Film	(11)
Methyl pheophorbide a	650–3800	$CHCl_3$	(11)
Methyl pheophorbide a	650–4000	Nujol mull	(14)
Ethyl pheophorbide a	650–3800	$CHCl_3$	(12)
Chlorophyll b	650–3500	Film	(11)
Chlorophyll b	650–3800	Nujol mull	(12)
Chlorophyll b	600–4000	KBr disc	(13)
Chlorophyll b	650–3800	$CHCl_3$, CCl_4, CS_2	(12)
Ethyl chlorophyllide b	650–3800	$CHCl_3$	(12)
Pheophytin b	650–3800	CCl_4	(12)
Methyl pheophorbide b	650–4000	Nujol mull	(14)
Ethyl pheophorbide b	650–3800	$CHCl_3$	(12)
Bacteriochlorophyll	650–3500	Film	(11)
Bacteriochlorophyll	650–3800	CCl_4	(12)
Methyl bacteriochlorophyllide	650–3800	$CHCl_3$	(12)
Deuteriochlorophyll a	600–4000	CCl_4	(15)
Deuteriochlorophyll b	600–4000	Film	(15)
Pyrochlorophyll a	1600–1800	$CHCl_3$, THF	(16)
Methyl pyropheophorbide a	650–3800	$CHCl_3$	(11)
Chlorobium chlorophyll (660)	1550–1850	CCl_4	(17)
Chlorobium pheophytin (660)	1550–1850	CCl_4	(17)

[a] This review includes some of our previously unreported data on the infrared spectra of chlorobium chlorophyll (660), bacteriochlorophyll, and on the far-infrared spectra of magnesium-containing chlorins (17a). Table I is intended to provide a key to the published literature to mid-1965.

[b] Spectra taken as films, Nujol mulls, or KBr discs are solid-state spectra; all others are solution spectra in the solvents indicated.

tained serious inconsistencies. Investigation then led to a reinterpretation that appears to account for the data more adequately (22).

Thus, a considerable body of information is now available on the IR spectroscopy of chlorophyll and its cognates. Not only can structural information be derived, but the Russian school of investigators have used IR spectroscopy to derive information about the state of water associated with chlorophyll, and to study the hydrogen exchange properties of chlorophyll (23, 23a,b,c, 24). The IR spectra of the porphyrins have been discussed critically by Schwartz et al. (25).

D. Interpretation of Infrared Spectra of Chlorophyll: Solvent Effects

In the course of their work, Holt and Jacobs (12) discovered that the IR spectra of chlorophyll a are solvent dependent in the 1750–1600 cm^{-1} region. These authors, as well as others (23b), attributed this phenomenon to keto-enol tautomerism involving the β-keto-ester in ring V. Subsequent work by Katz et al. (22) has shown that an alternate explanation in terms of the state of aggregation of the chlorophylls is more acceptable. The IR spectra of chlorophyll a and b in nonpolar solvents in the 1750–1600 cm^{-1} region are best interpreted on the basis of intermolecular aggregation involving coordination of ketone and aldehyde carbonyl oxygen atoms of one molecule with the central magnesium atom of another. In basic or polar solvents, the coordination unsaturation of the magnesium is satisfied by the solvent, and the chlorophylls exist predominantly in monomeric form. The same conclusions were also reached by Anderson and Calvin (26). A recent study of the far-infrared spectra (17a) of the chlorophylls and their derivatives lends additional support to the aggregation hypothesis.

Typical IR spectra (4000–600 cm^{-1}) of the chlorophylls and certain of their derivatives are displayed in Figs. 1–6. For the ensuing discussion these spectra are divided into five regions: 4000–2700 cm^{-1}, 1750–1600 cm^{-1}, 1600–1300 cm^{-1}, 1300–650 cm^{-1}, and 650–200 cm^{-1}. (The ordinary hydrogen-containing chlorophylls are transparent in the region 2700–1750 cm^{-1}.) Absorption peaks in the first two regions are primarily associated with the stretching vibrations of a few atoms and thus assignments are relatively easily made. Absorptions in the other regions and in the far-infrared, however, are associated with both stretching and bending fundamentals, and the complexity of the spectra makes definite assignments in these regions difficult and correlations tentative.

E. The 4000–2700 cm^{-1} Region

Absorption bands found in the high-frequency region of the spectrum are usually associated with O-H, N-H, and C-H stretching modes. Some of the spectra shown here, as well as some of those reported in the literature, show a broad, medium intensity band centered at ~3400 cm^{-1}. This absorption band can be logically assigned to an O-H stretching vibration of water. Adventitious water can arise from incompletely dried solvents or, in the case of the solid state spectra, from traces of water in the potassium bromide. More importantly, water can be an integral com-

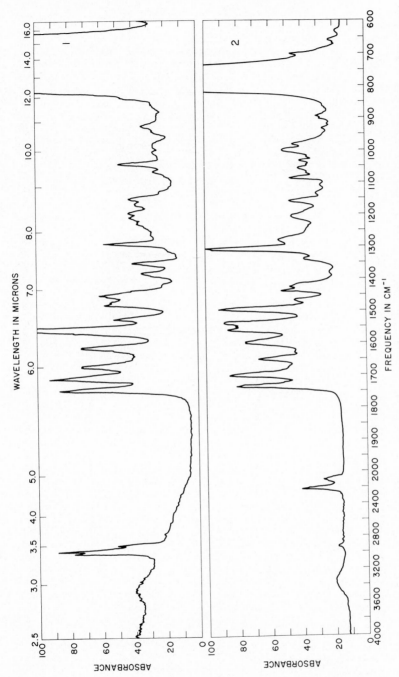

FIG. 1. Infrared spectra of chlorophyll *a* (curve *1*) and deuteriochlorophyll *a* (curve *2*) in CCl_4 solution. Concentration, 10% (w/v). (CCl_4 absorption peaks are present at 1570–1530 and 850–650 cm^{-1}.)

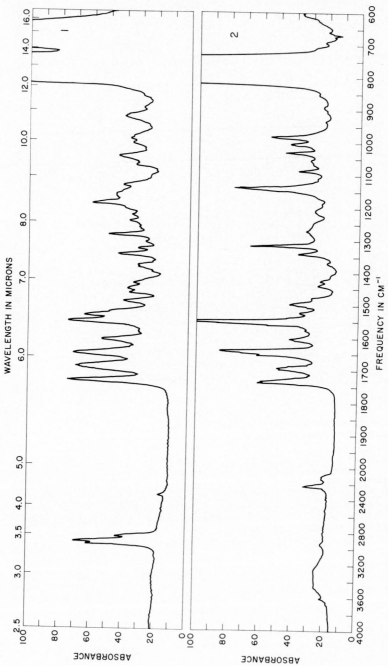

Fig. 2. Infrared spectra of cholorophyll *b* (curve *1*) and deuteriochlorophyll *b* (curve *2*) in CCl₄ (10% w/v).

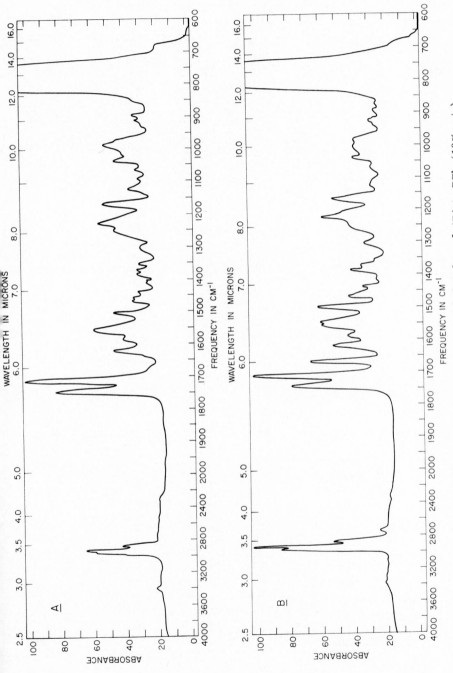

FIG. 3. Infrared spectra of pheophytin *a* (A) and pheophytin *b* (B) in CCl₄ (10% w/v).

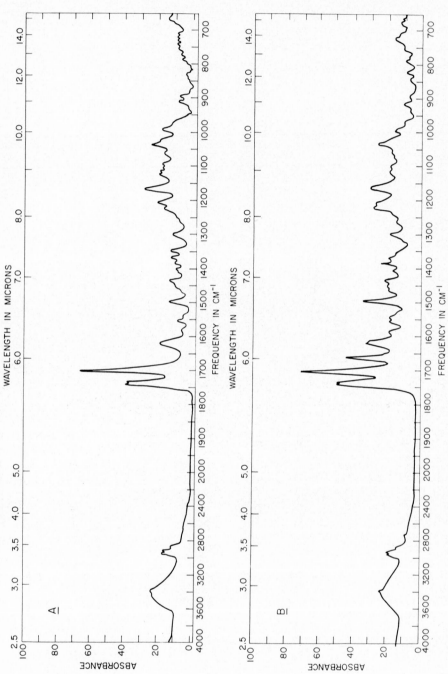

Fig. 4. Infrared spectra of solid methyl pheophorbide a (A) and methyl pheophorbide b (B) in KBr discs (0.5% w/w).

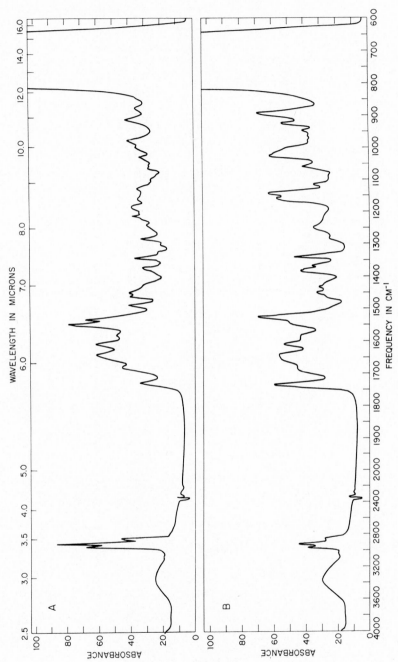

FIG. 5.   Infrared spectra of pyrochlorophyll *a* (A) and bacteriochlorophyll (B) in CCl_4 (10% w/v).

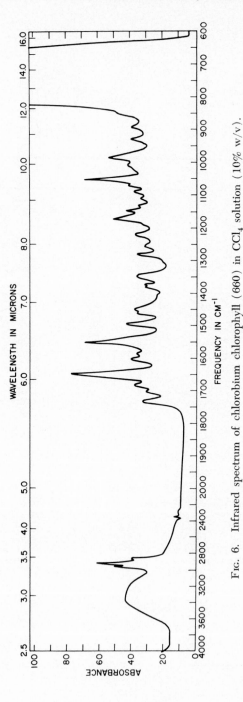

Fig. 6. Infrared spectrum of chlorobium chlorophyll (660) in CCl_4 solution (10% w/v).

ponent of the chlorophyll samples themselves. The chlorophylls and pheophytins are known to crystallize with an indeterminate number of water molecules (27). After routine vacuum drying (10^{-2} mm) films of these materials still show the broad water absorption band. Exhaustive drying at 50–60°C at 10^{-5} mm Hg pressure, however, removes the water from chlorophyll a (27), and the band disappears (23). Codistillation with carbon tetrachloride on a vacuum line also is effective in the removal of water. Nevertheless, under the same conditions that yield anhydrous chlorophyll a, chlorophyll b (23) and pheophytins a and b still appear to retain some water (23a).

After removal of the 3400 cm^{-1} band by drying of chlorophyll a, a narrow, weak band is observed at 3495 cm^{-1} (23) in the solid and at 3525 cm^{-1} in CCl$_4$ and CHCl$_3$ solution (22). The divalent metal derivatives of pheophytin a also show this weak band at 3525 cm^{-1} in CCl$_4$ solution (20). These data have been interpreted to indicate the presence of an observable quantity of the enolic tautomer, with the band at 3525–3495 cm^{-1} arising from the stretching mode of the resulting O-H group. Other evidence, however, indicates that chlorophyll a is almost totally in the keto form (22, 28) (see Section IV, C). The observation that the $\sim$3500 cm^{-1} band persists even after prolonged exposure to D$_2$O vapor, whereas enolic O-H should exchange rapidly, makes it difficult to designate this absorption band as an enolic O-H stretch. An alternate assignment might be to the first overtone of the ester carbonyl vibration, as aliphatic esters usually show a weak overtone band in this region (29). However, the observed absorption peak is somewhat higher in frequency than would be predicted for the first overtone ($1740 \times 2 = 3480$ cm^{-1}), and generally a lowering of the frequency of the first overtone (by anharmonicity effects) is observed. The possibility exists that the band is a combination tone, or an impurity. The nature of this absorption mode thus remains unclear.

The broad band of sample water in the spectrum of chlorobium chlorophyll 660 shows an overlying absorption at $\sim$3370 cm^{-1}. This band can be assigned to the O-H stretch of the hydroxy ethyl group which has been shifted from its normal position at 3580 cm^{-1} by hydrogen bonding with the sample water.

Metal-free derivatives of chlorophyll, the pheophytins and the methyl pheophorbides, show a weak band at 3370–3400 cm^{-1}, which is assigned to a N-H stretching vibration. This band is partially obscured in KBr disc spectra by the water band. Hydrogen bonding to adjacent pyrrolic nitrogen atoms could account for the downward shift of this absorption peak from its position in the vapor phase spectrum of free pyrrole (3535 cm^{-1}) (30), but a decrease in N-H bond strength because of electron delocalization in porphyrins may also contribute. Exposure of films of the pheophytins to D_2O vapor results in hydrogen-deuterium exchange; the resulting N-D stretch absorbs at 2520 cm^{-1} (23a).

The spectra of the chlorophylls and pheophytins show a number of strong absorptions in the C-H stretching region, 3000–2700 cm^{-1}. The bands are found at 2965–2955, 2940–2925, and 2865–2880 cm^{-1}. The C-H absorptions of the methyl pheophorbides are markedly decreased in intensity, but the absorptions appear at the same positions with different relative intensities. This shows that the phytyl (or farnesyl) group contributes a large part of the intensity of these bands in the spectra of the chlorophylls. The 2965–2955 cm^{-1} absorption is assigned to an asymmetric C-H stretching mode of the —CH$_3$ group, while the 2940–2925 cm^{-1} band is assigned to the antisymmetric C-H stretching mode of the —CH$_2$ group. Finally, the 2865–2880 cm^{-1} band (some fine structure is evident) represents the overlapping of the symmetric —CH$_3$ and —CH$_2$ stretching absorptions. The methine and vinyl C-H stretching vibrations are too weak to be observed, while the C-8, C-9, and C-10 hydrogen absorption peaks are most likely masked by the much more numerous methyl and methylene bands. On the other hand, the formyl C-H absorption is observed at 2720 cm^{-1} in chlorophyll b and its derivatives.

In the spectra of the fully deuterated chlorophylls, the absorption peaks associated with the C-D stretching modes appear at 2220, 2130, and 2110 cm^{-1}. The downward shift of the frequency is the expected isotope effect.

F. The 1750–1600 cm^{-1} (Carbonyl) Region

The positions of the absorption bands in the 1750–1600 cm^{-1} (carbonyl) region for the chlorophylls and their derivatives are given in Table II. Measurements were made in nonpolar (CCl$_4$, C$_2$Cl$_4$, or CHCl$_3$) and polar (tetrahydrofuran, THF) solvents, or in the solid state (KBr disc). The spectra in polar solvents are taken to be typical of disaggregated monomeric compounds, while those in nonpolar solvents and in the solid state are taken to be typical of the aggregated state. In this region of the spectrum, the presence or absence of magnesium is decisive, since coordination aggregation occurs only with metal-containing chlorins. Interpretation of the spectra in this region follows Katz et al. (22).

In nonpolar solvents, chlorophyll a shows four absorption bands in the carbonyl region, whereas in polar solvents the spectra contain three peaks (Fig. 7). The same solvent dependence is noted for methyl chlorophyllide a, indicating that the phenomenon is independent of the phytyl group. On the other hand, the spectra of the magnesium-free pheophytin a and methyl pheophorbide a show only three absorption bands, at 1743–1732, 1708–1701, and 1619–1616 cm^{-1}, in both polar and nonpolar solvents. The first (high-frequency) strong band is assigned to the ester at C-7 and C-10 carbonyl absorptions. The two overlapping bands are not completely resolved, but there is a definite splitting in the ester ab-

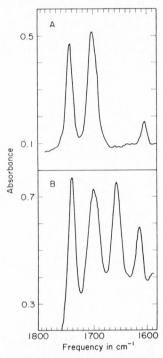

Fig. 7.   Infrared spectra of chlorophyll a in the carbonyl region in a polar and nonpolar solvent: (A) in tetrahydrofuran; (B) in CCl$_4$.

sorptions that is apparent in hydrogen-bonding solvents such as methanol and ethanol. The carbomethoxy (at C-10) absorption then appears near 1715 cm^{-1}, and the propionic acid ester near 1735 cm^{-1}. The second strong absorption at 1708–1701 cm^{-1} can be assigned to the stretching mode of the ketone carbonyl group. This assignment has been shown to be plausible by the absence of this absorption peak in the spectrum of the oxime and 2,4-dinitrophenyl hydrazine derivatives of methyl and ethyl pheophorbide a (12); these reagents are carbonyl specific.

A number of possible assignments for the third relatively weak band at 1619–1616 cm^{-1} exist. The possibility that it arises from the vibration

TABLE II

INFRARED ABSORPTION BANDS OF THE CHLOROPHYLLS IN THE 1750–1600 CM^{-1} (CARBONYL) REGION

Compound	Solvent[a]	Ester carbonyl	Ketone carbonyl	Aggregated ketone carbonyl	Aldehyde carbonyl	Aggregated aldehyde carbonyl	Skeletal $\diagup C = C \diagdown$
Chlorophyll a	CCl₄	1736	1695	1653	—	—	1608
	THF	1739	1696	—	—	—	1597
Methyl chlorophyllide a	CHCl₃	1732	1680	1648	—	—	1610
	THF	1743	1682	—	—	—	1597
Pheophytin a	CCl₄	1743	1708	—	—	—	1618
	THF	1743	1705	—	—	—	1616
Methyl pheophorbide a	KBr	1732	1701	—	—	—	1618
	THF	1739	1703	—	—	—	1619
Chlorophyll b	CCl₄	1735	1695	1652	1652	1608	1608
	THF	1740	1702	—	1663	—	1597
Methyl chlorophyllide b	CHCl₃	1738	1698	1655	1655	1609	1609
	THF	1744	1706	—	1669	—	1599
Pheophytin b	CCl₄	1741	1712	—	1667	—	1617
	THF	1744	1710	—	1667	—	1618
Methyl pheophorbide b	KBr	1739	1705	—	1663	—	1620
	THF	1741	1707	—	1665	—	1617
Pyrochlorophyll a	CCl₄	1735	1686	1643	—	—	1616
	THF	1737	1688	—	—	—	1598
Pyropheophytin a	CCl₄	1733	1694	—	—	—	1620
	THF	1728	1695	—	—	—	1619

	Solvent						
Methyl pyropheophorbide a	THF	1736	1695	—	—	—	1617
Bacteriochlorophyll	C_2Cl_4	1738	1692	1638	1656[b]	1612[c]	1575
Bacteriopheophytin	THF	1743	1693	—	1665[b]	—	1571
	CCl_4	1745	1709	—	1675[b]	—	1620
	THF	1743	1702	—	1675[b]	—	1624
Methyl bacteriopheophorbide	KBr	1740	1696	—	1667[b]	—	1620
	THF	1743	1703	—	1668[b]	—	1623
Chlorobium chlorophyll (660)	CCl_4	1734	1690	1655	—	—	1604
		1704	—	—	—	—	—
	THF	1735	1685	—	—	—	1594
		1714	—	—	—	—	—
Methyl chlorobium pheophorbide (660)	THF	1738	1698	—	—	—	1613

[a] THF, tetrahydrofuran.
[b] Carbonyl absorption from $CH_3\!\!-\!\!\overset{\displaystyle \|}{\underset{O}{C}}\!\!-$ group at position 2.
[c] Aggregation peak from acetyl carbonyl group interaction.
[d] Includes coupled C=C, C=N ring vibrations.

of the vinyl substituent at position 2 in chlorophyll a is ruled out by its presence in the spectra of the chlorins that do not have a vinyl group, and its absence in the spectra of those porphyrins that do (18). Apparently, the vinyl absorption is weak and is masked by other stronger bands. The observation that the ~ 1620 cm^{-1} band is present in the spectrum of chlorin e_4 indicates that the absorption is not dependent on the presence of an intact ring V. Further, this absorption peak is not observed in the spectra of the porphyrins or tetrahydroporphyrins (for example, monomeric bacteriochlorophyll). The assignment of the ~ 1620 cm^{-1} absorption to the vibration of the semi-isolated $C(3)=C(4)$ in ring II is thus not unreasonable. In a highly conjugated system such as the tetrapyrrole system, however, all $C=C$ (and $C=N$) vibrations are not isolated, but are undoubtedly coupled to each other to a greater or lesser extent. It is, therefore, probably more correct to assign the 1620 cm^{-1} peak to a general skeletal vibration characteristic of the dihydroporphyrin (chlorin) ring system.

In nonpolar solvents, the spectra of both chlorophyll a and methyl chlorophyllide a have, in addition to the three absorption peaks seen for the magnesium-free derivatives, a strong band at around 1650 cm^{-1}. This peak is assigned to a vibration of ketone carbonyl coordinated to magnesium in another molecule. Such coordination leads to intermolecular aggregation, and the carbonyl absorption peak that results may be designated an *aggregation peak* (22).

Both pheophytin b and methyl pheophorbide b show four absorption bands in the carbonyl region, at 1744–1739, 1712–1705, 1667–1663, and 1620–1617 cm^{-1}, in both polar and nonpolar solvents. The medium intensity band at ~ 1665 cm^{-1} can be assigned to the C-3 aldehyde carbonyl stretching vibration. This attribution is supported by the absence of this peak in the spectra of pheophytin a and methyl pheophorbide a, as well as by the disappearance of this peak upon formation of the phenyl hydrazine derivative of pheophorbide b (12). The other three bands can be assigned in the same way as those for chlorophyll a. The spectra of chlorophyll b (Fig. 8) and methyl chlorophyllide b, in polar and nonpolar solvents show four absorption peaks in the 1750–1600 cm^{-1} region. The four bands observed in polar solvents are assigned in the same way as those of the magnesium-free derivatives.

The interpretation of the spectra of chlorophyll b in nonpolar solvents is much less definitive than for chlorophyll a. Both the aldehyde and ketone carbonyl oxygens can participate in aggregate formation by coordination to magnesium, and there are thus two sources of aggregation peaks. Further, the aggregation peak originating from the ketone carbonyl oxygen–magnesium coordination appears to be close in wavelength to the absorption arising from the free aldehyde carbonyl. It appears likely that the spectral consequences of (aldehyde carbonyl) oxygen–magnesium coordination are to be found very near the 1610–1620 cm^{-1} C=C absorption. Thus, the 1660 and 1610 cm^{-1} peaks are probably composite in nonpolar solvents: (a) the peak near 1660 cm^{-1} is partially due to the aggregation (ketone) peak and partially to free aldehyde; (b) the peak near 1610 cm^{-1} is partially due to the C=C skeletal vibration,

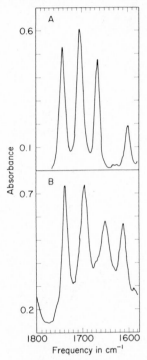

FIG. 8.   Infrared spectra of chlorophyll b: (A) in tetrahydrofuran solution; (B) in CCl$_4$ solution.

but also contains a contribution from an aldehyde aggregation peak. The relative intensities of the four bands in polar and nonpolar solvents lends support to this viewpoint: The relative intensity of the ketonic absorption is increased, and the peak near 1610 cm^{-1} is diminished in polar solvents as compared to nonpolar solvents. Further, the intensity of the 1660 cm^{-1} band is never greater than the ester absorption, while the absorption near 1610 cm^{-1} is more intense than the ester absorption for solid chlorophyll b. The spectra of deuteriochlorophyll a and b are very similar to the ordinary hydrogen compounds in the carbonyl region in both polar and nonpolar solvents. This furnishes conclusive evidence that the four peaks in the 1750–1600 cm⁻¹ region do not involve hydrogen vibrations.

The carbonyl region spectra of pyrochlorophyll a in polar and nonpolar solvents are very similar to those of chlorophyll a. The three bands that appear in the spectra of pyrochlorophyll a in polar solvents (Fig. 9) can be assigned in the following way: 1737 cm⁻¹, to the propionic acid

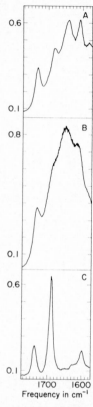

FIG. 9.   Infrared spectra of pyrochlorophyll a in the carbonyl region in various states of aggregation. (A) CCl_4 solution; (B) KBr disc; (C) tetrahydrofuran solution.

ester absorptions; 1688 cm⁻¹, to the ketone carbonyl absorption, and 1598 cm⁻¹ to the C=C absorption. The intensity of the ester carbonyl absorption for pyrochlorophyll a is approximately one-third that of the corresponding band in chlorophyll a. This is consistent with absence of the C-10 carbomethoxy group in the pyro compound. The presence of an aggregation band at 1643 cm⁻¹ in nonpolar solvents indicates that the C-10 carbomethoxy group is not essential for aggregate formation. Further, the spectra of pyropheophytin a in polar and nonpolar solvents are very similar to each other and to pheophytin *a*, and show no evidence for aggregation peaks.

The infrared spectra of chlorobium chlorophyll (660) in the carbonyl region (Fig. 10) can be interpreted in the same way as those of pyrochlorophyll *a*. The three major bands in the spectra of chlorobium chlorophyll (660) in polar solvents are readily assigned: 1735 cm^{-1} to the

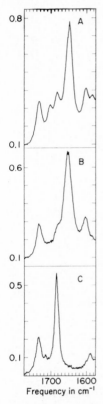

FIG. 10. Infrared spectra of chlorobium chlorophyll (660) in various states. (A) CCl_4 solution; (B) KBr disc; (C) tetrahydrofuran solution.

propionic acid ester carbonyl, 1685 cm^{-1} to the C-9 ketonic carbonyl, and 1594 cm^{-1} to the skeletal C=C absorption. In nonpolar solvents the spectra show an aggregation peak at 1655–1657 cm^{-1}.

In addition to the three strong absorptions the spectra of the aggregated chlorobium chlorophyll (660) show a weak band at 1704–1700 cm^{-1}, and the monomeric substances show a corresponding absorption at 1715 cm^{-1}. On the other hand, the spectra of the magnesium-free methyl chlorobium pheophorbide (660) and chlorobium pheophytin

(660) (17) in polar and nonpolar solvents do not show this absorption. It is possible that the expected absorption peak is masked by the much stronger free C-9 carbonyl band at ~1700 cm⁻¹. The position of the ~1710 cm⁻¹ band indicates that it could arise from an ester carbonyl absorption. The presence of two ester carbonyl absorptions in the spectra of chlorobium chlorophyll (660) is not consistent with the proposed

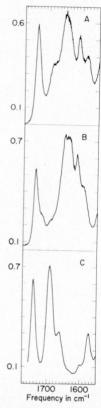

FIG. 11.   Infrared spectra of bacteriochlorophyll illustrating solvent effects on the carbonyl region spectra. (A) CCl₄ solution; (B) KBr disc; (C) tetrahydrofuran solution.

molecular structure, which allows only one propionic ester residue. The observation that both the aggregated and disaggregated materials show the 1710 cm⁻¹ band rules out the possibility that intermolecular hydrogen bonding with the hydroxyethyl group splits the propionic ester absorption into a free ester component and bonded ester component. Finally, the possibility exists that the 1710 cm⁻¹ band may be due to an impurity in the chlorobium chlorophyll (660) that possesses a C-10 ester or free carboxylic acid function. At this time the precise nature of the 1710 cm^{-1} absorption is unknown.

The spectra of bacteriochlorophyll (Fig. 11) in the 1750–1550 cm^{-1} region shows five bands in nonpolar solvents: 1743–1738, 1695–1692, 1670–1656, 1637, 1613 cm^{-1}, and three bands in polar solvents: 1743, 1693, and 1665 cm^{-1}. By analogy with the previous attributions the bands at 1743 and 1693 cm^{-1} are assigned to ester carbonyl and free ketone carbonyl absorptions. The band at 1670–1656 cm^{-1} is assigned to the acetyl carbonyl vibration (14). The interpretation of the spectra of bacteriochlorophyll in nonpolar solvents closely follows that of chlorophyll b. The presence of an acetyl substituent provides two ways to generate aggregation peaks. This is in fact what is observed. The band at 1637 cm^{-1} is due to the C-9 ketonic carbonyl aggregation peak, and the band at 1613 cm^{-1} is due to the acetyl carbonyl aggregation peak. As mentioned earlier, the spectra of monomeric bacteriochlorophyll lack the C=C skeletal absorptions at 1620 cm^{-1}. A band that can be assigned to the skeletal vibration of the tetrahydroporphyrin is seen at 1575 cm^{-1}. On the other hand, the spectra of bacteriopheophytin and methyl bacteriopheophorbide in polar and nonpolar solvents show skeletal absorptions at both 1620–1624 cm^{-1} and 1579–1582 cm^{-1}.

G. The 1600–1300 cm^{-1} Region

The positions of prominent absorption bands in solid-state spectra (taken in KBr discs or as films on salt plates) of the chlorophylls and their magnesium-free derivatives in the 1600–1300 cm^{-1} are given in Tables III and IV. The medium intensity absorption peaks that appear in this region most likely arise from (a) the skeletal vibrations of the tetrapyrrolic macrocycle, and (b) the carbon-hydrogen bending modes of the phytyl group and of the alkyl substituents of the chlorin ring. The spectra of the chlorophylls and the metal-free derivatives show bands 1560–1550, 1540–1520, 1500–1490, and 1355–1345 cm^{-1} that may be assigned to the C=C and C=N skeletal vibrations of the chlorin ring. Similar absorption bands in this general region of porphyrin (31) and pyrroles (32) have previously been assigned in the same way. However, the band at ~1550 cm^{-1} appears to be particularly characteristic of chlorins, since it is not observed in the spectra of porphyrins (although sometimes a weak band is observed), or in tetrahydroporphyrins (bacteriochlorophyll). The skeletal absorptions for bacteriochlorophyll are observed to occur at 1521, 1450, 1421, 1340 cm^{-1}, the first band being the strongest in intensity. The positions of the skeletal absorption for chlorobium chlorophyll (660) do not seem to be shifted by the presence

TABLE III

INFRARED ABSORPTION BANDS OF THE CHLOROPHYLLS IN THE 1600–1300 CM⁻¹ AND 1300–650 CM⁻¹ (FINGERPRINT) REGIONS

Chlorophyll *a*	Deuterio-chlorophyll *a*	Chlorophyll *b*	Deuterio-chlorophyll *b*	Bacterio-chlorophyll	Pyro-chlorophyll *a*	Chlorobium chlorophyll (660)
1553	1553	1553	1555	—	1550	1551
1536	1534	1524	1510	1521	1534	1538
1494	1496	1478	1499	—	1491	1492
1454	1438	1452	1437	1450	1451	1450
—	—	—	—	1421	—	—
1379	1316	1379	1311	1376	1380	1382
1347	1342	1348	1340	1340	1346	1349
1288	1287	1290	~1290	1285	1287	1284
—	—	—	—	1243	—	—
1191	1188	1191	1157	—	1185	1174
1163	1152	1166	1116	1145	1163	1150
1134	1133	1112	—	—	1130	1132
1070	1068	1065	1062	1058	1065	1068
—	—	—	—	—	—	1049
1043	1042	1042	1036	1028	1035	1023
990	907	988	907	—	988	—
—	981	—	989	—	—	986
915	814	923	820	—	918	—
890	893	—	889	895	884	893
850	855	839	—	845	852	848
800	758	798	770	795	802	794
745	690	760	716	682	745	749

of a *meso* substituent. This indicates that the location of the skeletal absorption bands is governed by the overall conformation of the macrocyclic ring, not by electronic effects.

The absorption bands that arise from the C-H bending modes are, as expected, more intense in the spectra of those materials that contain the phytyl (or farnesyl) group than is the case for the methyl pheophorbides.

TABLE IV

INFRARED ABSORPTION BANDS OF THE PHEOPHYTINS AND PHEOPHORBIDES IN THE 1600–1300 CM^{-1} AND 1300–650 CM^{-1} (FINGERPRINT) REGIONS

Pheophytin *a*	Pheophytin *b*	Methyl pheophorbide *a*	Methyl pheophorbide *b*
1584	1586	1582	1585
1555	1555	1553	1557
1539	—	1537	1533
1500	1498	1498	1497
1456	1455	1451	1452
1383	1385	—	—
—	—	1365	1367
1348	1354	1347	1353
1299	1303	1298	1302
1197	—	—	—
1162	1165	1166	1166
1125	1125	1123	1115
1067	1066	1062	1061
1038	1038	1035	1037
992	—	992	996
911	923	911	916
898	893	895	894
846	844	843	842
—	—	755	754
713	713	719	716
671	676	675	678

The 1456–1448 cm^{-1} peak is assigned to the CH_3 antisymmetric bending mode. Finally, the *gem*-dimethyl group present in phytyl (farnesyl) gives rise to a doublet at 1385–1375 cm^{-1}, which can be associated with the CH_3 symmetric bending mode. A single absorption peak is seen at 1368–1366 cm^{-1} in methyl pheophorbides. The expected isotope effect can be observed in the spectra of the fully deuterated chlorophylls *a* and *b*, which show the C-H bending absorptions at lower frequency than do ordinary hydrogen-containing chlorophyll; the —CH_3 symmetric bending absorption appears at 1320–1310 cm^{-1}, while the —CH_3 antisymmetric bending absorption appears at 1437–1438 in the deuteriochlorophylls.

H. The 1300–650 cm⁻¹ (Fingerprint) Region

The region below 1300 cm^{-1} is commonly known as the fingerprint region, even though it does include some stretching vibrations, e.g., carbon-oxygen modes. The absorptions in the $1300–650 \text{ cm}^{-1}$ region are generally associated with molecular motions involving many atoms; examples of these include in-and out-of-plane bending and breathing vibrations of ring structures. The spectra of the chlorophylls in the fingerprint region are complex, and only tentative assignments of the absorption bands can be made. The positions of the major bands in the spectra of the chlorophylls are given in Table III. Several medium intensity

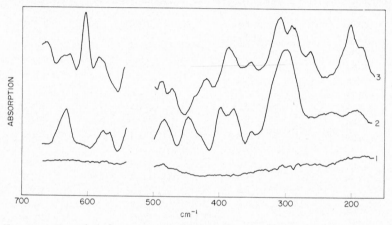

Fɪɢ. 12.   Far-infrared spectra of chlorophylls in cyclohexane. Curve *1*, cyclohexane solvent; curve *2*, chlorophyll *b*, suspension; curve *3*, saturated chlorophyll *a*.

bands can be observed throughout the series of compounds. The 1200–1160 cm⁻¹ peaks can be assigned to an ester antisymmetric carbon-oxygen stretching vibration, and the absorption peaks at 1170–1035 cm⁻¹ can be assigned to symmetric ester carbon-oxygen stretching modes. The 1049 cm⁻¹ peak in the spectrum of chlorobium chlorophyll (660) could be linked with the carbon-oxygen stretch of the hydroxy ethyl group. The vinyl C-H out-of-plane bending mode absorbs at 990–980 and at 923–910 cm⁻¹. In the fully deuterated chlorophylls these absorptions probably occur at 907 and 820–814 cm⁻¹. The peak at 1135–1115 cm⁻¹ has been previously assigned to a pyrrole ring-breathing mode (33). Other absorptions most likely associated with the tetrapyrrole ring system bending modes are found at 1305–1285, and 800–790 cm⁻¹. Finally, the N-H out-of-plane bending mode gives rise to an absorption at 720–710 cm⁻¹ for the metal-free derivatives.

I. The 650–200 cm⁻¹ (Far-Infrared) Region

The far-infrared region of the spectrum shows absorption peaks originating in metal-ligand fundamental stretching vibrations, as well as some that arise from bending or deformation modes of organic moieties. The spectra of the chlorophylls *a* and *b* (Fig. 12) in this region are complex and show a number of weak and medium intensity bands. Most of these bands are associated with the chlorin ring vibrations and are difficult to assign. However, a pair of medium intensity bands at 400–370 cm⁻¹ can be logically assigned to a pyrrole ring deformation, since it is common to chlorins, porphyrins, and pyrroles (17a).

The spectra of aggregated chlorophylls *a* and *b* in mineral oil mulls (Table V), cyclohexane, and benzene are significantly different from

TABLE V

INFRARED ABSORPTION BANDS OF CHLOROPHYLLS AND RELATED COMPOUNDS IN THE 667–160 CM⁻¹ REGION[a]

Chlorophylls				Pheophytins		Methyl pheophorbides	
Ordinary		Deuterio					
a	b	a	b	a	b	a	b
629m	624m	622m	624sh	613m	616m	615s	613s
604s	—	697m	—	—	—	—	—
579m	571w	572m	563m	586w	575w	587sh	568m
562sh	556w	552m	542w	568w	563w	554sh	557m
545m	537vw	—	—	540vw	—	—	—
522w	522vw	508w	519w	—	—	517w	—
504w	—	—	498w	—	—	—	506w
488w	480w	478w	480vw	482w	484w	478w	477w
469w	—	—	460m	464sh	464w	465sh	462sh
—	449w	—	—	—	433w	440vw	434w
418m	—	424m	409m	422w	—	—	—
393m	395m	369m	380m	393w	393m	389m	393m
—	378m	—	359m	372w	376m	372m	374m
350w	346w	334w	328w	—	—	—	—
—	—	—	—	343w	339w	—	—
—	—	—	—	—	—	330sh	332w
—	—	—	—	307w	305vw	308m	308m
308s	303sh	294s	289sh	—	—	—	—
290s	295s	275s	267s	—	—	—	—
—	—	—	—	283vw	280w	—	279vw
259m	—	259sh	—	—	—	—	—
234w	238w	—	—	—	—	—	241w
—	—	—	—	213w	—	213vw	—
197s	—	190s	183m	—	—	—	—
—	191w	—	—	189w	189w	—	189w

[a] For mineral oil mulls: s = strong; m = medium; w = weak; vw = very weak; sh = shoulder.

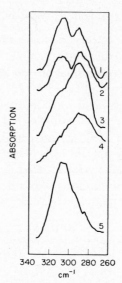

340 320 300 280 260
cm⁻¹

FIG. 13.   Far-infrared spectra of chlorophyll *a* in various media. Curve *1*, cyclohexane; curve *2*, Nujol mull; curve *3*, pyridine-cyclohexane (10% v/v); curve *4*, methanol-cyclohexane (1% v/v); curve *5*, benzene.

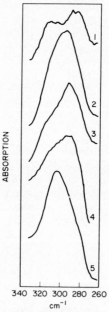

340 320 300 280 260
cm⁻¹

FIG. 14.   Far-infrared spectra of chlorophyll *b* in various media. Curve *1*, cyclohexane; curve *2*, Nujol mull; curve *3*, pyridine-cyclohexane (10% v/v); curve *4*, methanol-cyclohexane (1% v/v); curve *5*, benzene.

the spectra of the monomeric materials measured in methanol or pyri-dine-cyclohexane solution, particularly in the 320–260 cm^{-1} region (Figs. 13 and 14). The major bands at 306 cm^{-1} and 312 cm^{-1} for aggregated chlorophylls a and b are completely absent for monomeric solutions, as well as for the pheophytins and the methyl pheophorbides. We consider that the 306–312 cm^{-1} peak arises from the magnesium-carbonyl (or aldehyde) oxygen vibration in the aggregate, and that it is direct evidence for the existence of an Mg-O bond in chlorophyll aggregates. The spectra of chlorophylls a and b exhibit their most intense bands in the far-infrared at 292 and 195 cm^{-1}, and at 293 and 196 cm^{-1}, respectively. The spectra of the magnesium-free pheophytins and methyl pheophor-bides do not show strong absorption at these frequencies. These bands thus may be assigned to magnesium-nitrogen stretching and bending modes. The positions of both magnesium-oxygen and magnesium-nitrogen absorption peaks are consistent with data derived from model systems (17a). However, a shift to lower frequency of the metal-ligand vibrations in the deuterated chlorophylls a and b indicates that these vibrations are not totally independent of the deformation modes of the macrocyclic ring.

III. Nuclear Magnetic Resonance Spectra

A. Introduction

The availability of high-sensitivity, high-resolution magnetic reso-nance spectrometers has opened new avenues for the investigation of the chemical and biological properties of the chlorophylls. The sensitivity of modern spectrometers, particularly when coupled with time averag-ing techniques, has made it practical to detect less than 0.01 M hydrogen. High-resolution spectra of solids are always measured in solution. Be-cause only 0.2–0.3 ml of solution is required, a sample weight of only a few milligrams is sufficient, and as NMR spectroscopy is a nondestructive procedure, the sample may be recovered intact. NMR spectroscopy is a tool ideally suited to porphyrin and chlorin studies. The principles and practice of NMR spectroscopy are described in a number of excellent works (34–37).

Basically, a proton magnetic resonance (PMR, used interchangeably here with NMR) spectrum can be used to count the numbers of protons in different environments in an organic molecule. Three parameters usually suffice to characterize the NMR response of a proton or group of protons: the chemical shift of the proton; the spin coupling constant of the proton to other nuclei; and the relative area under the proton resonance peak.

Chemical shifts are a consequence of differences in nuclear environments. The nuclear resonance frequency of a particular nucleus occurs at different values of a given applied magnetic field, according to the nature of the chemical compound containing the nucleus. Protons poorly shielded by electrons (as a consequence, say, of binding to an electronegative group) come into resonance at lower magnetic fields than do protons well shielded by electrons (as are protons in a covalent bond).

As is generally the case in the application of spectroscopic methods to organic compounds, assignment of chemical shifts to particular proton(s) in a molecular formula involves empirical correlations between resonance behavior and structure. Chemical shifts are thus magnetic shielding factors, and in the case of protons are related not only to the immediate chemical environment but may also depend sensitively on the stereochemical arrangements of the molecule.

Chemical shifts are always measured relative to a standard. All chemical shifts in this chapter will be given in units of δ, parts per million (ppm), from tetramethylsilane (TMS) as an internal reference; a plus $(+)$ sign will identify resonances that occur at a higher magnetic field than required for the resonance of protons in TMS. For PMR spectra taken at 100 megacycles, as were most of the spectra used for illustration in this chapter, chemical shifts in ppm (δ) convert directly into cycles per second from TMS, when TMS is taken as zero.

Under high resolution, proton resonances are frequently observed as multiplets rather than single peaks. This type of multiplicity arises from magnetic interactions between the proton and neighboring nuclei that have magnetic moments. This interaction, called *spin-spin coupling*, is transmitted by valence electrons and is related to the degree of delocalization of electrons, hybridization and other factors. Since, in many cases, the effect of coupling on the resonance signal can be predicted from structural considerations, the multiplicity of observed resonance absorptions becomes a valuable aid in the conversion of NMR data to structural information. Spin-spin interactions are expressed as a coupling constant, J, in cycles per second (cps). Unlike chemical shifts, the magnitude of which are field dependent, J is invariant with applied magnetic field. Spin-spin interactions, so important in other areas of NMR, are relatively unimportant in the case of porphyrins and chlorins because of structural considerations.

The area under a resonance absorption band is directly proportional to the number of nuclei producing the resonance. Providing the signals are not saturated by application of excessive radiofrequency power levels, measurements of relative areas can be translated into relative number of protons. The area under a —CH_3 resonance will have three times the relative area of a single proton in the same compound. Such information is of obvious utility in the assignment of NMR spectra.

B. Special Features of the NMR Spectroscopy of Porphyrins and Chlorins

Most NMR studies are not greatly concerned with the question of sample purity. If the area under the resonance curve is directly proportional to the concentration of absorbing species, an impurity that is present even to the extent of 5% by weight will probably be below the limits of detection (assuming a signal-to-noise ratio of 20 to 1). In the case of porphyrins and chlorins, these considerations no longer are valid. The molecular weight of the substances we are concerned with is in the range 500–1000, and even 1% by weight of a compound of low molecular weight can produce an equimolar proton concentration of an impurity resonance. Thus, the proton resonance spectrum can be even more demanding of sample purity than are visible absorption extinction coefficients for the presence of colorless impurities in chlorins. High molecular weight impurities are also a serious problem in studies of chlorin NMR. A mixture of closely related, but not identical, chlorin compounds results in a spectrum of broad, poorly defined resonance peaks. Reaction with oxygen and methanol, for example, can contaminate chlorophyll with sufficient allomerized chlorophyll to make NMR spectra difficult to interpret. In practice, these considerations mean that the chlorins, even samples obtained in a state of purity by optical criteria, must be further purified. Samples must be freed from residual solvent by codistillation with carbon tetrachloride on the vacuum line. Samples for NMR must be dissolved in purified and nonreacting solvents, and the sample tubes sealed off in high vacuum. The manipulation of chlorophylls and related compounds in air and light should be as rapid as is consistent with an accurate definition of the concentration of the sample. Chlorophyll samples prepared for NMR and kept in air are altered so rapidly that they cannot be used after a few hours at best, but NMR samples prepared from pure components in sealed tubes show no changes for weeks or even months.

On an *a priori* basis the NMR spectra of porphyrins and chlorins derived from biological materials are expected to be simple from the point of view of the NMR spectroscopist. The hydrogen-containing groups in naturally occurring porphyrins should give simple spectra amenable to a first-order analysis. With the exception of the propionic acid side chain, which constitutes a five-spin system, and the 7 and 8 protons of ring IV, which is a seven-spin system, this is generally the case. The usual side chains encountered in chlorophyll and related compounds, e.g., methyl,

ethyl, vinyl, formyl, and hydroxyethyl, contain protons that do not experience spin-spin interactions with protons of any other functional group
and as a result produce resonances of low multiplicity and complexity.
The most complicated spin system in the side chains is that of the vinyl
residue, which is a three-spin system, but still amenable to "pencil-and-
paper" analysis.

The NMR spectra of naturally occurring porphyrins and chlorins exhibit extraordinarily large chemical shifts. The range of proton resonances
is roughly 14 ppm, or almost the entire range of the proton spectrum
(14 ppm wide). Indeed, the internal chemical shifts for protons in porphyrins are no doubt the largest that have been observed in any class
of compounds. These large chemical shifts may be correlated with the
structural features of porphyrins that give rise to magnetic nonequivalence of protons, and the unique nature of the shifts often makes it possible to identify directly single proton positions in the molecule.

The line width of the NMR response has an important bearing on the
nature of the data that may be obtained. The natural width of a spectral
line is inversely proportional to the average time the system spends in
the excited state. Sharp lines are observed only when the local magnetic
fields are averaged out. In solids and very viscous liquids, or for highly
anisotropic solutes, these criteria are not met, and the natural line widths
become large. As a general rule, the sensitivity and resolution are both
reciprocal functions of the natural line width at half height. As the following discussion will show, the line width (and chemical shift) for porphyrin and chlorin spectra depends upon both the solvent and the concentration. In addition to these factors, which may have major effects on
the line shapes, porphyrins and chlorins generally show relatively broad
lines (0.6–1 cps wide) because the molecule is large and molecular
tumbling is not as rapid as for a smaller molecule. Thus, there is intrinsic
line broadening because of the large size and anisotropy of porphyrins
and chlorins.

Rapid tumbling of the observed molecule relative to the magnetic
field is essential for high-resolution spectra. Observation in the liquid
phase is mandatory. Even viscous solutions pose problems. Consequently
there are strict natural limitations on the extension of high-resolution
NMR to chlorins in rigid structures such as chromatophores or chloroplasts.

C. Porphyrin NMR as a Basis for the Assignment of
Chlorin Resonances

The studies of porphyrin NMR that have been carried out by R. J.
Abraham, A. H. Jackson, G. W. Kenner, and their colleagues, and by
W. S. Caughey and W. S. Koski provide a very important key to the

NMR behavior of the chlorophylls. A brief review is justified as this work is important to an understanding of the assignment of the chlorophyll resonances.

The striking effects of "ring currents" on the NMR spectra of porphyrins were first reported by Becker and Bradley (38) and Ellis et al. (39). The fully conjugated macrocyclic ring, which gives porphyrins and chlorins their characteristic low-energy electronic absorption spectrum, also permits the formation of fairly large "ring currents" by the precession of the π-electrons about a closed conjugated path. The "ring current" approximation (40, 41) has successfully accounted for large paramagnetic anisotropy* at the periphery of the porphyrin ring and the correspond-

TABLE VI

CONCENTRATION DEPENDENCE OF CHEMICAL SHIFTS OF COPROPORPHYRIN
IV METHYL ESTER[a]

Proton	Concentration		Dilute[b]
	0.2 M		
Methine	9.82	triplet	10.20 singlet
	9.66		
	9.49		
—O—CH$_3$	3.64	doublet	3.72 singlet
	3.60		
β-CH$_3$	3.44	doublet	3.72 singlet
	3.20		
NH	+4.4		+3.72

[a] After Abraham et al. (42).
[b] Extrapolated to infinite dilution.

ingly large diamagnetic anisotropy* above the plane and in the center of the ring. Becker and Bradley's original observation showed that the methine protons in coproporphyrin I methyl ester (approx. 0.05 M in CDCl$_3$) absorbed at 9.96 ppm, or 2.7 ppm lower in field than the protons in benzene, and the imine protons in the same sample were in resonance at +3.89 ppm, roughly 13 ppm higher in field than the N-H resonance in pyrrole.

R. J. Abraham and his co-workers were the first to report the remarkable concentration effects on the NMR spectra of porphyrins in neutral

---

* The magnetic lines of force generated by the circulating electrons of the "ring current" augment the field experienced by protons on the periphery and in the plane of the ring, and oppose the magnetic field experienced by protons above, below, or in the center of the macrocyclic ring. The first class of protons comes into resonance at lower field, the second at higher field than would be expected in the absence of a ring current effect. A shift to lower fields is termed paramagnetic; to higher, diamagnetic.

TABLE VII

CHEMICAL SHIFTS IN NMR SPECTRA OF PORPHYRINS[a,b]

Compound	N-H	Methine	—CH₃	—CH₂CH₃	—CH₂CH₂COOCH₃	Others
Porphin	+4.40	11.22 α, β, γ, δ	—	—	—	9.92 (H)
Octamethylporphin	+4.80	10.98 α, β, γ, δ	3.78	—	—	—
Etioporphyrin I	+4.81	11.00 α, β, γ, δ	3.78 (1, 3, 5, 7)	4.30 1.84 (2, 4, 6, 8)	—	—
Etioporphyrin II	+4.81	11.02 α, β, γ, δ	3.81 (1, 4, 5, 8)	4.31 1.84 (2, 3, 6, 7)	—	—
Etioporphyrin III	+4.86	11.00 α, β, γ, δ	3.78 (1, 3, 5, 8)	4.29 1.82 (2, 4, 6, 7)	—	—
Tetramethyl tetrapropyl porphin	+4.76	10.98 α, β, γ, δ	3.77 (1, 3, 5, 8)	—	—	CH₂ CH₂ CH₃ 4.25 2.27 1.25 (2, 4, 6, 7)
Coproporphyrin II tetramethyl ester	+4.29	11.22 α, γ; 11.06 β, δ	3.83 (1, 4, 5, 8)	—	4.73, 3.30, 3.78 (2, 3, 6, 7)	—
Coproporphyrin III tetramethyl ester	+4.26	11.11 α, β; 11.21 γ; 11.02 δ	3.83 (1, 3, 5, 8)	—	4.67, 3.32, 3.78 (2, 4, 6, 7)	—
Uroporphyrin II	—	11.17 α, γ; 11.27 β, δ	—	—	4.71, 3.30, 3.79 (2, 3, 6, 7)	CH₂ CO₂ CH₃ 5.46 3.88 (1, 4, 5, 8)
Mesoporphyrin II dimethyl ester	+4.36 (1, 3)	11.07 α, γ; 11.00 β, δ	3.81 (1, 3, 5, 7)	4.29 1.85 (2, 6)	4.67, 3.34, 3.81 (4, 8)	—
Mesoporphyrin IX dimethyl ester	+4.48 (3, 4)	11.03 α, β, δ; 11.18 γ	3.81 (1, 3, 5, 8)	4.31 1.84 (2, 4)	4.72, 3.31, 3.76 (6, 7)	—

					CH=CH$_2$
Protoporphyrin IX dimethyl ester	+4.37	11.03 α, β, δ 11.21 γ	3.79 (1, 3, 5, 8)	—	4.68, 3.28, 3.75 (6, 7)	8.29 6.57 6.39 (2, 4)
Deuteroetioporphyrin IX	+4.64 (1, 2) +4.79 (3, 4)	11.09 α, β 11.02 γ, δ	3.88 (1, 3) 3.81 (5, 8)	4.33 1.85 (6, 7)	—	9.67 (H) —
Pyroetioporphyrin VII	— —	— —	3.91 (6) 3.83 (1, 4, 7)	4.35 1.87 (2, 3, 8) —	— —	9.65 (H) —

^{a} After Abraham et al. (45, 45a).
^{b} Chemical shift of protons (δ, ppm).

solution (42). These effects are of more than theoretical importance. The effects are so large that an assignment of the individual resonances in the spectrum demands a knowledge of the concentration. Any report of the chemical shift values for porphyrins or chlorins that does not include a precise statement of the concentration, or an extrapolation to infinite dilution, is of limited utility for studies of the structural dependence of the NMR (43, 43a). Table VI shows the concentration dependence of the NMR spectrum of coproporphyrin IV tetramethyl ester. Neglecting the imine protons, this molecule has an axis of symmetry. Thus, at best, one would expect to observe two magnetically nonequivalent sets of the three substituent groups (β-CH$_3$, β-CH$_2$CH$_2$COOCH$_3$) and three nonequivalent methine protons with relative areas 1:2:1; the magnetic difference between the two sets could well be too small for observation. At infinite

TABLE VIII

EFFECTS OF β SUBSTITUENTS ON CHEMICAL SHIFTS OF *Meso* (METHINE)-PROTONS IN ALKYL PORPHYRINS[a]

β substituent	Proton shift (ppm)
H(porphin)	0 (11.22)
Alkyl —CH$_2$—	0.11
CH$_3$OOCCH$_2$CH$_2$—	0.02
CH$_3$OOCCH$_2$—	−0.03
H$_2$C=CH—	0.11

[a] After Abraham et al. (45).

dilution the splitting between the peaks of the two groups is smaller than the line width, but in concentrated solution the anticipated fine structure was easily observed, and all the peaks were shifted to higher field. These effects of porphyrin aggregation appear to be quite general with the notable exception of N-alkylated porphyrins, the NMR spectra of which showed only trivial dependence on concentration (44).

Porphyrin dications would not be expected to form π-π aggregates, and in fact, the NMR spectra of porphyrins in solution in trifluoroacetic acid (TFA) vary only slightly with concentration. Formation of the dication species apparently increases the porphyrin ring current (41) and thus shifts the resonances arising from peripheral substituents to lower field than in the neutral molecule. This ring current effect should be uniform in a series of porphyrins, and thus porphyrin spectra in TFA have provided a means of estimating neighboring group anisotropies and substituent effects on the "ring current" in these compounds (45, 45a). Chemical shift data for a number of important porphyrins are given in Table VII.

From a study of the NMR of 12 different porphyrins in TFA, Abraham *et al.* (*45*) derived a set of additive constants for the effect of a β substituent on the chemical shift of an adjacent *meso* proton (Table VIII). Similar "neighboring group" parameters were also calculated for the N-H protons. These are additive corrections that are quite analogous to the well-known Shoolery rules (*46*).

Continuing investigation of the effect of *meso* substitution on the NMR spectra of porphyrins by Abraham *et al.* (*45a*) showed that substitution of an alkyl residue at a *meso* position invariably shifts the resonances due to peripheral substituents to higher fields, whereas the resonances due to the N-H protons are shifted to lower fields. These shifts are readily explained by the assumption that *meso* substitution, unlike β substitution, substantially reduces the "effective ring current" in the macrocycle.

The assignments of the main groups in the spectra of porphyrins to a variety of substituents depends only on first-order application of NMR techniques (*45*). The assignment of specific absorptions among these groups to specific positions in the molecule and estimation of neighboring group anisotropies is not, however, so straightforward. Once accomplished, however, this information helped considerably in the decipherment of the chlorophyll NMR.

D. The NMR Spectra of Metal-Free Chlorins (Pheophorbides)

The complete spectral assignment for the NMR of chlorophylls *a* and *b* by Closs and colleagues depended to a very large extent on the assignment of the resonances in the respective methyl pheophorbides (*47*). Prior to the full assignment of the pheophorbide spectra, incomplete assignments of the NMR spectra of chlorin e_6 trimethyl ester (*48*), rhodochlorin dimethyl ester (*48*), and chlorobium methyl pheophorbides 660 and 650 (*49*) had been made. Subsequently the NMR assignment for methyl pyropheophorbide *a* has appeared (*16*), and we have continued our investigations of the NMR behavior of methyl bacteriopheophorbide and chlorobium methyl pheophorbide 660 (*50*).

For ease of reference and ready comparison we will discuss the spectra of the five methyl pheophorbides that have been examined in detail as a group. The numbering of the protons is given in Fig. 15. The discussion will follow the regions of the spectra, proceeding from the low to the high field resonances. Table IX summarizes the chemical shift assignments for methyl pheophorbides *a*, *b*, methyl pyropheophorbide *a*, methyl bacteriopheophorbide, and chlorobium methyl pheophorbide (660) in deuteriochloroform solution at concentrations (moles per liter) indicated in the table. The spectra of these five pheophorbides are shown in Figs. 16–20.

TABLE IX
CHEMICAL SHIFTS (δ, PPM) IN METHYL PHEOPHORBIDES IN CDCl$_3$

Proton	Methyl pheophorbide a (47) (0.06 M)	Methyl pheophorbide b (47) (0.08 M)	Methyl pyropheophorbide a (16) (0.06 M)	Methyl bacteriopheophorbide (50) (0.04 M)	Methyl chlorobium pheophorbide (660) (50) (0.04 M)
3b	—	10.58	—	—	—
α	9.15	9.76	9.20	8.96	9.76
β	9.32	8.89	9.32	8.47	—
δ	8.50	8.47	8.50	8.40	9.84
2	7.85	7.75	—[a]	—	4.53
2″	6.12	6.16	—[a]	—	—
2′	6.04	6.08	—[a]	—	1.45[b]
10	6.22	6.22	5.13	6.08	5.19
8	4.40	4.45	4.42	4.28[c]	6.42
7	4.13	4.15	4.23	4.02[d]	~4.05
β	—	—	—	—	4.03[e]
11	3.88	3.95	—	3.84	—
5	3.62	3.46	3.58	3.48	3.45
12	3.57	3.62	3.58	3.57	3.61
4	3.48	3.37	—[a]	2.20	3.63
1	3.32	3.28	3.35	3.44	3.83
3a	3.05	—	3.13	1.72	3.24
2	—	—	—	3.15[f]	—
7′7″	~2.45	~2.50	—[a]	~2.35	~2.40
β′	—	—	—	—	1.95[g]
8′	1.82	1.88	1.72	1.79	2.09
4′	1.60	1.48	1.55	0.46	0.87
13	0.50	0.83	} ~+1.85	0.46	0.87
14	+1.75	+2.15		+0.96	+1.86

[a] Signal-to-noise ratio too low for accurate determination.

[b] CH$_3$ in hydroxyethyl group at position 2.

[c] Includes the 3° proton in ring II.

[d] Includes the 4° proton in ring II.

[e] The —CH$_2$— in the β position.

[f] The CH$_3$ resonance of the
CH$_3$—C— group at position 2.
$\overset{\|}{O}$

[g] The CH$_3$ of the ethyl group at β position.

The low field lines in all the spectra clearly belong to the methine bridge protons and the formyl proton of methyl pheophorbide b. The assignments for methyl pheophorbide a are based on the considerations that the δ proton is flanked by only one pyrrole ring [cf. Becker et al. (43) and Woodward and Škarić (48)], and should thus be the most shielded, and that the paramagnetic anisotropy of the ring V carbonyl group probably deshields the β proton more than the α, making the β proton the lowest peak in the spectrum (47). Methyl pyropheophorbide *a* was assigned by analogy to methyl pheophorbide *a* (16). In methyl pheophorbide *b*, the order of assignment of the α and β protons was reversed, following the presumption that the formyl group in the 3b position should exert a powerful deshielding effect on the α proton, thus shifting the α proton to lowest field (47).

In methyl bacteriopheophorbide, all three of the methine protons are adjacent to only one pyrrole ring, which accounts for the relatively high field location of these resonances. The α proton must certainly be the

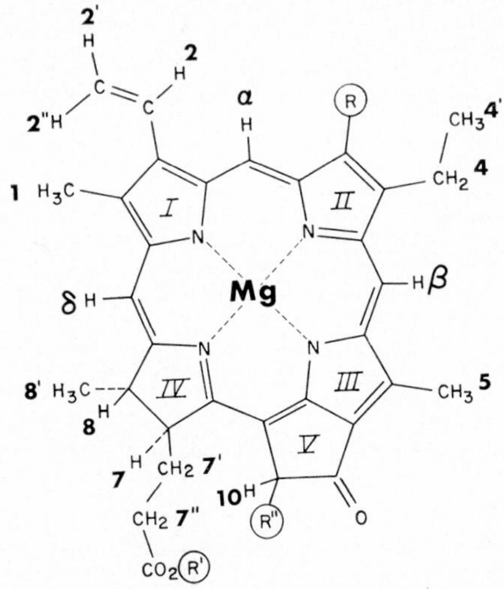

Compound[a]	R	Proton no. of R	R'	Proton no. of R'	R"	Proton no. of R"
Methyl pheo-phorbide *a*	CH_3	3a	CH_3	12	CO_2CH_3	11
Methyl pheo-phorbide *b*	CHO	3b	CH_3	12	CO_2CH_3	11
Methyl pyropheo-phorbide *a*	CH_3	3a	CH_3	12	H	10
Methyl bacterio-pheophorbide[b]	CH_3	3a	CH_3	12	CO_2CH_3	11
Methyl chlorobium pheophorbide[c] 660	CH_3	3a	CH_3	12	H	10

[a] In the magnesium-free pheophorbides and pheophytins, the >NH protons are designated 13, 14.

[b] In bacteriochlorophyll, the 2 position is occupied by an acetyl, CH_3CO-group; these protons are designated as 2; Ring II contains two extra hydrogen atoms, corresponding to protons 7 and 8, and these are designated as 3° and 4°.

[c] Position 2 in chlorobium chlorophyll 660 is presumed to contain an hydroxy ethyl group, CH_3CHOH-. These are designated as protons 2 and 2'.

FIG. 15.   Nomenclature and proton designations.

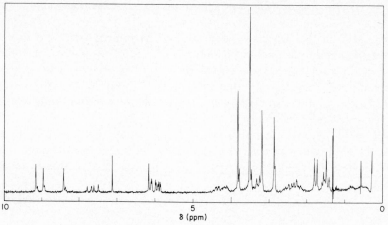

FIG. 16. NMR spectrum of methyl pheophorbide *a* in CDCl$_3$ (0.11 mole/liter).

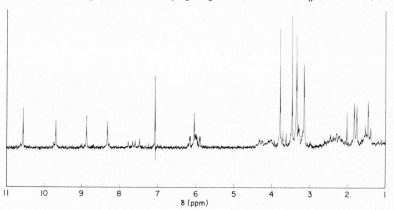

FIG. 17. NMR spectrum of methyl pheophorbide *b* in CDCl$_3$ (0.07 mole/liter).

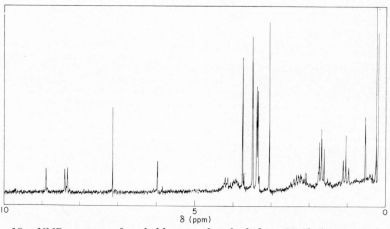

FIG. 18. NMR spectrum of methyl bacteriopheophorbide in CDCl$_3$ (0.04 mole/liter).

least shielded because of the acetyl group at position 2. The assignment of the β and δ resonances is less certain (50) but follows from the argument given above for methyl pheophorbide *a*. Chlorobium methyl pheophorbide (660) shows only two resonances in the low field region that may be assigned to methine protons. Holt and co-workers concluded (51) that the absence of a high field "δ-type" methine resonance indicates

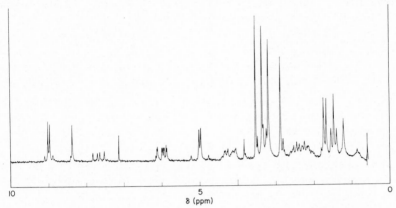

FIG. 19.   NMR spectrum of methyl pyropheophorbide *a* in CDCl₃ (0.16 mole/ liter).

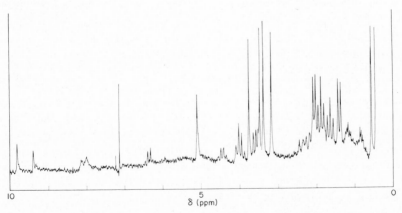

FIG. 20.   NMR spectrum of methyl chlorobium pheophorbide (660) in CDCl₃ (~0.08 mole/liter).

that the presumed extra alkyl group in this compound is located at the δ position. Mathewson *et al.* (52) have since shown that the lower field methine proton in this compound may be exchanged for deuterium by treatment with acidic exchange agents. This exchange behavior prompted them to the conclusion that the δ methine position was not blocked, since this is the proton that readily exchanges in other chlorins (cf. Section IV, B). This argument, based on the reactivity of the molecule, presumes that the structural differences between chlorobium methyl pheophorbide (660) and the other pheophorbides in the series could not account for the difference in reactivity. If this argument is accepted, the extremely low field value for the presumed δ methine resonance which is 1.3 ppm below the normal value, must be explained. We favor the assignment δ, α, in order of increasing shielding for these methine resonances (50); however, a final determination must await a satisfactory structural comparison with known material.

The resonances in the region 5–8 ppm have been assigned to vinyl or other strongly deshielded substituents. The vinyl groups in methyl pheophorbides a and b methyl pyropheophorbide a are easily recognized as ABX spin-splitting patterns (47). A standard analysis of the splitting pattern gave $|J_{2,2'}|$18.7,18.3; $|J_{2,2''}|$10.9,11.2; and $|J_{2',2''}|$1.6,1.6 cps, respectively, for methyl pheophorbides a and b (47).

The sharp one-proton singlet at 6.22 ppm in both methyl pheophorbides a and b, which coincides in both spectra with the AB part of the vinyl resonance, was assigned to the proton at position 10 on the basis of three deshielding effects that would be expected to account for the low field value. The strong "ring current" effect for a position close to the macrocyclic ring, the paramagnetic contribution of the two adjacent carbonyl groups, and the fact that the C-10 proton is tertiary suggest considerable deshielding for the C-10 proton. The exchange behavior of this proton confirmed the assignment (47).

Methyl pyropheophorbide a shows a doublet of doublets which center at ~4.89 ppm and must be assigned to the two C-10 protons. These protons appear to be magnetically nonequivalent (δ ~6 cps; J ~20 cps). The apparent coupling constant is quite close to that observed for 1,3 cyclopentane dione (53), ~22 cps, but the magnetic nonequivalence is somewhat larger than we would anticipate for a group so far removed from an asymmetric center. This peak appears as a broad singlet at lower concentrations, which suggests that π-π aggregation tends to increase the magnetic dissymmetry at position 10.

The resonance at 5.19 ppm in chlorobium methyl pheophorbide (660), which corresponds to two protons, was assigned to the 10 position on the basis of exchange behavior with base (16, 52) and of the structural considerations given above. The same arguments served to assign the one-proton singlet at 6.08 ppm in methyl bacteriopheophorbide to the hydrogen at position 10. The one-proton quartet ($|J| = 6.7$ cps) at 6.42 ppm in chlorobium methyl pheophorbide (660) has been assigned to the α hydrogen of the hydroxyethyl group at position 2 (49) on the basis of comparison with the spectrum of hematoporphyrin IX (43) and the fact that acetylation of the molecule shifted this resonance 1.15 ppm to lower field (49).

The low intensity multiplets that center near 4.35 ppm in all of the spectra have been assigned to the protons at positions 7 and 8, on the basis of the low-field values and complicated splitting patterns anticipated for these protons (16, 47, 50). In all cases, the assignments have been confirmed by double resonance techniques that leave no doubt that the resonance assigned to the 8 proton is coupled to the high-field methyl doublet. The resonance assigned to the 7 proton is coupled to the 7',7" multiplet at ∼2.50 ppm. These resonances appear in all the spectra. It has not been possible in any of these compounds to resolve either the 7 or 8 proton into the anticipated doublet by irradiation of either the 8' or 7' protons with a proper decoupling frequency. In the case of methyl pheophorbides a and b (47) and methyl pyropheophorbide a (50) the line width during decoupling indicated a maximum coupling constant of 2.8 cps for coupling between the 7 and 8 protons. In chlorobium methyl pheophorbide (660) the 8 proton appears as a distinct quartet, with a line width such that the maximum coupling between protons 7 and 8 must be less than 1 cps (50). The observed maximum for spin coupling of 2.8 cps between the 7 and 8 protons in all five methyl pheophorbides is consistent only with a *trans*-relationship for the 7 and 8 alkyl groups in all five samples (50). This result was anticipated by the degradative studies of Linstead et al. on chlorophyll (54).

The interesting, and as yet unresolved part of the 7 and 8 (and 3° and 4° in bacteriopheophorbide) resonance behavior, is the sizable difference in chemical shift for these two protons. If substituent effects are neglected and one assumes that the chlorin system has a plane of symmetry normal to the plane of the ring and bisecting rings II and IV, the conclusion follows that the "ring current" effect should make equal contributions to the magnetic environment of protons 7 and 8 (47). It is then necessary to attribute the observed chemical shift differences between these two protons to neighboring group anisotropies. Since the 8' methyl group and the propionic ester side chain at position 7 can be expected to provide similar shielding at both positions, ring IV and its substituents remain as the major contributing factor to the observed difference.* In order of decreasing chemical shift difference between the 7 and 8 protons, the chlorins in this series may be arranged as fol-

---

* For bacteriochlorophyll, it is the acetyl group at position 2 vs. the methyl group at position 5 that would contribute to deshielding of the 3° and 4° protons of ring II. Proton 3° is less shielded than 4°, which is the reverse of the assignment for the 7 and 8 protons.

lows: chlorobium methyl pheophorbide (660) > methyl pheophorbide
b > methyl pheophorbide *a* > methyl bacteriopheophorbide > methyl
pyropheophorbide *a* > chlorin e_6 trimethyl ester, with ~0.5 and 0.13
ppm as the extreme values. The order in the series makes it clear the
carbomethoxy group on ring V and its stereochemistry cannot account
for the observed chemical shift difference. The paramagnetic anisotropy
of the carbonyl groups on ring V also cannot account for the difference,
since the 7 position is expected to be much more deshielded by the ring
V carbonyl groups than is the 8 position. The anisotropy of the macrocyclic ring, a factor which we discounted earlier, then remains as the
only reasonable explanation for the chemical shift difference between
the 7 and 8 protons. It may be that the macrocyclic ring is not planar,
and that the differences in bulk of the two substituent groups on ring
IV forces the ring into a preferred bent conformation. Such a distortion
from planarity would bring one of the protons closer to the center of
the ring and push the other farther out than would be the case in a planar
conformation. The observation that the chemical shift difference for the
7 and 8 protons in chlorophyll *a* (*50*), which has a bulky phytyl group
at position 12, is very close to the same value as in methyl pheophorbide
a, which has only a methyl group at position 12, strongly discounts this
argument. Further, the substituents at 7 and 8 in the series above are
all identical, and a great deal more twist in one case than in the other
would not be expected. A possible key to this dilemma comes from the
studies carried out by Abraham and his colleagues (*45*) on the effect of
"meso" substitution on the chemical shift of other groups in porphyrins.
Their studies clearly indicate that the effect of *meso* (methine) substitution on porphyrin spectra (and chlorin spectra by analogy) are much
larger than can be accounted for by the introduced changes in bond
anisotropies. In the case in point it would appear that substitution on
the chlorin skeleton at the γ position skews the "ring current" anisotropy,
or the electron density, or both, in such a way that proton 8 is deshielded
with respect to 7. In chlorobium methyl pheophorbide (660), two of the
meso positions are substituted, and the chemical shift difference between
protons 7 and 8 is the largest in the series. This would suggest that the
"extra alkyl" group in chlorobium methyl pheophorbide (660) is located
at the β position. The relative order of the 8 and 7 and 3° and 4° resonances in methyl bacteriopheophorbide is also accommodated by this
assumption, since the 8 and 3° protons are both on the same side of the
ring.

Assignments of the intense envelope of lines located between 3 and
4 ppm have been made in all cases with a reasonable degree of certainty.
This so-called "low-field methyl" region includes most of the resonances

from groups that are attached directly to the macrocyclic ring, and also those of the carboxylic esters. The ester methyl group on the propionic acid side chain should exhibit the normal shielding value, since it is so distant from the ring. The assignments of the lines at 3.57 and 3.62 ppm in methyl pheophorbides a and b to the propionic ester methyl were made by comparison with the pheophytin spectra, from the spectrum of methyl (CH_3) deuteriopheophorbide b (47), and from a study of the dilution shifts for methyl pheophorbide a (Section IV, A, 2) (47). The assignments for the other three chlorins were by direct analogy ($16, 50$). The methyl ester group attached to ring V is deshielded by both the ring current and the adjacent ketone group, and thus the lines at 3.88, 3.95, and 3.84 ppm in methyl pheophorbides a and b and in methyl bacteriopheophorbide, respectively, have been assigned to the methyl group at position 11. The absence of the resonances at 3.88 ppm in methyl pyropheophorbide a gave sound confirmation to this assignment (16).

The sharp line at 3.15 ppm in methyl bacteriopheophorbide has been assigned to the acetyl group at position 2 from considerations of the probable effect of the ring current on a normal acetyl resonance. This assignment has been confirmed by studies with deuterio-(CH_3) bacteriopheophorbide that showed that the protons of this acetyl group are easily exchanged with hydrogen acids (50).

Once the ester and acetyl methyl resonances had been assigned, it remained to distribute the known groups that are directly attached to the macrocycle among the spectral lines in the region. In the cases of methyl pheophorbides a and b, this task was accomplished by consideration of neighboring group anisotropies, then checking the assignment by a careful study of the "aggregation maps" for the system. A wrong assignment would have resulted in an uninterpretable "aggregation map" (see Section IV, A), and thus this internal check gave a high degree of certainty to the otherwise speculative assignments (47). The assignment of these resonances for methyl pyropheophorbide a followed directly from the methyl pheophorbide a assignment. Methyl bacteriopheophorbide has only two ring methyl groups that must be assigned to lines in this region. Both (1 and 5) are flanked by ketone groups, but the deshielding effect of ring V ketone carbonyl group fixed in the plane of the ring should be somewhat larger than the freely rotating acetyl group at position 2. Thus, the lines at 3.44 and 3.48 ppm in methyl bacteriopheophorbide have been assigned to the methyl group at positions 1 and 5, respectively. The shift between these two methyl groups is larger in bacteriochlorophyll, and the assignment is again highly compatible with studies of aggregation (50).

The proper assignment of the ring methyl resonances in chlorobium methyl pheophorbide (660) is complicated by the difficulty of estimating the effects of the additional *meso* substituent in this compound on the ring anisotropy at different parts of the molecule. There can be little doubt that the effect can be large, as shown by the larger difference in the 7 and 8 proton resonances. The assignment given in Table VII is tentative, although it appears to be compatible with preliminary aggregation studies in this system. The chemical shift values and assignments for the ring methylene groups that absorb in the region 4 ppm have all been obtained by double resonance techniques. In methyl pheophorbides *a* and *b*, irradiation of the highest field triplet methyl resonance (assigned to position 4') gave the chemical shift values for the methylene group at position 4 as 3.48 and 3.37 ppm, respectively (47). It is interesting to note that the replacement of a methyl group at position 3 in methyl pheophorbide *a* by a formyl group decreased rather than increased the chemical shift value of the adjacent methylene. In methyl bacteriopheophorbide, the ethyl group at position 4 is attached to a dihydropyrrole ring; this shifted the 4 resonance toward a normal methylene value, i.e., 2.20 ppm. Chlorobium methyl pheophorbide (660) has two strongly deshielded ethyl resonances. One of these is only slightly lower than the value for methyl pheophorbide *a*, and has been assigned to the ethyl methylene at position 4. The remaining ethyl methylene resonance is shifted 0.4 ppm to lower field, and must be assigned to the presumed additional *meso* substituent. The chemical shift of this ethyl group furnishes no real help in the attempt to define the location of this *meso* substituent (α, β, or δ), since equally convincing arguments can be made for one or two adjacent pyrrole groups. On the basis of the arguments outlined above in discussing the 7 and 8 resonances, and the exchange behavior of this molecule, we favor an assignment to the β position for this group, but more detailed structural information is clearly needed.

The propionic ester methylene groups in all the chlorins should exhibit roughly the same shielding, and the difference between the two methylene groups will be compensated for by the fact that one group is adjacent to the carboxylic ester and the other is closer to the macrocyclic ring. The complicated pattern of lines that forms an envelope near 2.40 ppm has, therefore, been assigned to these methylene groups.

The assignment of the "high-field methyl" resonances was very straightforward, and indeed these groups were assigned prior to an interpretation of the rest of the spectrum. In methyl pheophorbides *a* and *b*, and methylpyropheophorbide *a*, the 8' and 4" methyl resonances are clearly distinguishable as a doublet and a triplet, respectively. The doublet of doublets due to the 8' and 3' methyl groups in methyl bacteriopheophorbide are not so easy to assign. Again, it was assumed that the paramagnetic effect of the groups clustered around ring V would predominate, and the assignment was consistent with subsequent aggregation studies (50). The 4″ methyl group of methyl bacteriopheophorbide was, however, very easily identified as the "normal" methyl resonance at 1.10 ppm. The assignments of the ethyl *methyl* resonances in chlorobium methyl pheophorbide (660) follows directly from the assignment given above for the methylene resonances that are coupled to them. The same holds true for the hydroxyethyl and 8-methyl resonances, giving the series of values shown in Table IX.

The imine hydrogens in all five of these chlorins appear as highly broadened lines at very high field values. The nonequivalence of the two signals indicates that the scrambling process, which exchanges the protons with each other, must be slow with respect to the frequency separating the two peaks.

E. Pheophytin NMR Spectra

The pheophytins differ from the methyl pheophorbides only in the structure of the esterifying alcohol group at position 12. The differences between the NMR spectra of the chlorin nucleus esterified with methyl or phytyl or farnesyl should be, and apparently is (47, 50), very small. The major complication in the spectra of the pheophytins is, of course, the presence of a large amount of hydrogen in the alcohol function,

TABLE X
CHEMICAL SHIFT (δ, PPM) ASSIGNMENTS IN PHEOPHYTINS[a]

Position	Pheophytin a	Pheophytin b	Pyropheophytin a
3b	—	10.58	—
α	9.14	9.76	9.10
β	9.32	8.92	9.20
δ	8.50	8.47	8.46
2	7.84	7.76	—
2′	6.05	6.12	—
2″	6.14	6.18	—
10	6.24	6.22	5.10
1	3.33	3.30	3.32
3a	3.07	—	3.02
5	3.63	3.48	3.53
11	3.88	3.95	—
7′, 7″	~2.47	~2.45	—
Phytyl (12)	5.14	5.17	—
Phytyl (C=CH)	4.45	4.48	—

[a] All samples at 0.06 M in $CDCl_3$.

which masks much of the high-field methyl region and portions of the low-field spectrum as well.

The NMR spectrum of phytol, the ester group in chlorophylls *a* and *b* and bacteriochlorophyll and *trans-trans* farnesol, the esterifying alcohol in chlorobium chlorophyll (660), have both been published. Phytol shows resonances at 0.78 ppm (terminal —CH₃); 0.87 ppm (side-chain —CH₃); 1.18 ppm (—CH₂—); 1.68 ppm (vinyl —CH₃); 4.13 ppm (primary alcohol —CH₂—); 5.42 ppm (vinyl H) (55). Farnesol shows a peak at 3.96 ppm (the methylene hydrogens of an allylic alcohol), 1.66 ppm, and 1.59 ppm (characteristic of olefinic methyl hydrogens) (56).

A complete assignment of the pheophytin NMR spectra thus involves only a combination of the pheophorbide and alcohol spectra. The spectra of pheophytins *a* and *b* (47), as well as pyropheophytin *a* (16), have been analyzed and are presented in Table X.

F. Chlorophylls. NMR Spectra of Magnesium-Containing Compounds

The presence of magnesium in pheophorbides or pheophytins makes remarkable changes in the physical properties of the compounds and even the reactivity with respect to acid-catalyzed proton-exchange at the δ position. These changes in physical properties can result in apparent drastic changes in the NMR spectrum. The NMR spectrum of chlorophyll *a* in concentrated solution in alcohol-free chloroform appears as a series of broad, poorly defined peaks that would seem to have no simple relation to the chemical shift parameters observed for methyl pheophorbide *a* and pheophytin *a*. As will be discussed in detail in Section IV, these apparent changes in the NMR are simply due to the reversible aggregation of the macrocyclic rings, and it is now quite easy to specify conditions under which the magnesium-containing compounds will be disaggregated in solution. Under disaggregating conditions, their NMR spectra correspond directly to the appropriate pheophorbide spectra. The NMR spectra of the methyl chlorophyllides, or chlorophylls, dissolved in electron-donating solvents such as acetone d₆, tetrahydrofuran d₈ or chloroform d₁–methanol d₄ mixtures clearly show that the presence of magnesium makes only small changes in the NMR spectrum of the chlorins. Indeed, the changes attendant on magnesium insertion are smaller than those found on dissolution of the pheophorbides in trifluoroacetic acid. Thus, the assignments of the separate resonances in the spectra of the magnesium-containing compounds are directly analogous to the assignments that were worked out for the methyl pheophorbides.

IV. Applications of Infrared and Nuclear Magnetic Resonance Spectroscopy

In the foregoing sections, the infrared and nuclear magnetic resonance spectra of the chlorophylls have been described and analyzed in some detail. With this as a point of departure, some of the problems to which such information is pertinent can now be briefly considered.

A. Chlorophyll Aggregation and Multiple Forms of Chlorophyll in Nature

Recent investigations have provided a basis for the supposition that chlorophyll exists *in vivo* in a variety of forms and that these forms may represent different states of aggregation of chlorophyll. Most of the evidence is based on absorption and fluorescence spectroscopy in the visible (*57–59*), and is discussed in detail elsewhere in this volume (Chapters 6 and 11). The aggregation behavior of chlorophyll has thus become a subject of keen interest, particularly in the context of the two-light-reaction hypothesis in photosynthesis (*60–62*). That aggregated chlorophyll functions as an auxiliary pigment in photosynthesis, and that both monomeric and polymeric (aggregated) chlorophyll are involved in photosynthesis places a premium on knowledge about the aggregation behavior of chlorophyll *in vitro*. Both IR and NMR spectra provide very pertinent information on this point. Although it cannot be alleged that information derived from chlorophyll studies in defined solution can easily be applied to situations involving the intact photosynthetic apparatus, nevertheless conclusions on the aggregation behavior of chlorophyll derived from *in vitro* studies must surely be pertinent to the larger situation of photosynthesis.

1. CHLOROPHYLL AGGREGATION FROM INFRARED SPECTROSCOPY

In Section II, F, it has been shown that the IR spectra of chlorophyll and its magnesium-containing derivatives in the carbonyl region (1750–1600 cm^{-1}) show an extraordinary solvent dependence. Whereas chlorophyll *a* dissolved in nonpolar solvents shows four absorption peaks in the carbonyl region, in polar (basic electron-donating) solvents the spectra show only three peaks. Magnesium-free derivatives have IR spectra in this region that are not solvent dependent. It is clear that the coordination properties of magnesium play a decisive role in the genesis of the chlorophyll IR absorptions in the carbonyl region.

Although the solvent dependence of the carbonyl region IR spectra has been interpreted in terms of keto-enol tautomerism (12), it appears that a more consistent explanation can be given in terms of the coordination properties of magnesium and its consequences on the IR absorption of the ketone carbonyl group in ring V.

According to this view, the coordination number of magnesium in chlorophylls is always greater than 4. In solutions of polar (or electron-donating) solvents, the coordination unsaturation of the magnesium is primarily satisfied by coordination with solvent molecules, and chlorophyll exists in these solutions in monomeric form. In nonpolar solvents the coordination unsaturation of the magnesium is satisfied by coordination to ketone carbonyl oxygen of ring V, or in the case of chlorophyll b, to aldehyde carbonyl oxygen as well. Such coordination obviously leads to aggregation of the chlorophyll. The carbonyl absorption frequency moves to lower frequencies, and in solutions in nonpolar solvents, where dimers exist, obviously half of the carbonyl oxygens of ring V are free and half coordinated to magnesium. The carbonyl absorption thus appears as two peaks: one is the normal ketone oxygen absorption, the other originating in C=O coordinated to magnesium, occurring at somewhat lower frequency. The latter may properly be termed an *aggregation* or *association* peak, and its detection in an infrared spectrum is a valid diagnosis for the presence of chlorophyll aggregates.

The effects of aggregation as the IR spectrum of chlorophyll in the carbonyl region is shown vividly in Fig. 21. Curve A of Fig. 21 is the spectrum of chlorophyll a dissolved in carbon tetrachloride. The spectrum is distinguished by an enormous aggregation peak at about 1640 cm^{-1}, and an almost vestigial (uncoordinated) C=O peak at 1695 cm^{-1}; evidently this chlorophyll solution is highly aggregated, with an average composition of trimer or greater. Curve B is the same solution warmed very briefly at 60°; the aggregation peak is markedly reduced in intensity, and the solution approximates a dimer. Curve C is the spectrum of the same chlorophyll sample in tetrahydrofuran; no aggregation peak is present, and the chlorophyll is present as monomer.

The correlation between the presence of an aggregation peak, the presence of chlorophyll aggregates in nonpolar solvents, and the disaggregating effects of polar solvents are in agreement with direct molecular weight determinations by Aronoff (63) and by Katz et al. (22).

The frequency and intensity of the aggregation peak make possible some interesting deductions about the aggregation behavior of various chlorophylls. The greater the frequency difference between the free carbonyl absorption peak and the aggregation peak, the greater the degree of aggregation. Thus, the aggregation peak of chlorophyll a dissolved in $CHCl_3$ and CCl_4 occurs at about 1650 cm^{-1}, whereas the aggregation peak in highly oriented solid chlorophyll *a* appears at 1640 cm^{-1}. The intensity of the aggregation band is also related to the extent of aggregation. The relative intensity of the $\sim$1650 cm^{-1} aggregation peak increases as the concentration of chlorophyll *a* in CCl$_4$ increases, presum-

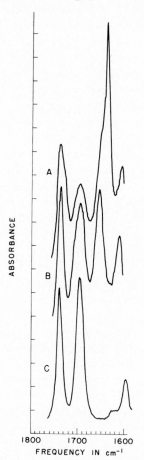

FIG. 21. Aggregation and disaggregation in chlorophyll *a*. Curve *A*, sample dissolved in CCl$_4$; curve *B*, solution warmed for 1 minute at 60°C; curve *C*, same material dissolved in tetrahydrofuran.

ably a result of increase in aggregates with increasing concentration. The intensity of the aggregation peak is also increased in less favorable solvents such as trichlorethylene or cyclohexane. For chlorophyll *b*, the relative intensity of the $\sim$1600 cm^{-1} band appears to parallel the degree of aggregation, and is greater for the oriented solid than for CCl$_4$ solution.

The spectra of pyrochlorophyll a in nonpolar solvents show an aggregation peak that is more intense and is shifted to lower frequencies than is the case for chlorophyll a. It is thus a fair presumption that pyrochlorophyll a is more highly aggregated in nonpolar solvents than is chlorophyll a. The absence of a carbomethoxy group at position C-10, then, facilitates aggregate formation, presumably on steric grounds. Methyl chlorophyllides likewise appear more highly aggregated than their phytyl-containing prototypes; here again, the steric hindrance to aggregate formation is diminished by replacement of the bulky phytyl by a methyl group. Chlorobium chlorophyll (660), a pyro compound lacking a C-10 carbomethoxy group, as does pyrochlorophyll a, similarly appears from its aggregation peak to be more highly aggregated than is chlorophyll a.

Of the two aggregation bands manifested by bacteriochlorophyll, only one overlaps another absorption. The $1610 \, cm^{-1}$ peak can be interpreted as an aggregation band, and is totally absent in polar solvents. The intensity of this band can be used to estimate the extent of bacteriochlorophyll aggregation. The relative intensity of this peak increases, as for the other chlorophylls, in the order: $CHCl_3 < C_2Cl_4 <$ solid. A comparison of the relative intensities of the bands in the $1750-1600 \, cm^{-1}$ region leaves the impression that bacteriochlorophyll is more highly aggregated than chlorophyll b in nonpolar solvents. Methyl bacteriochlorophyllide likewise appears more highly aggregated than bacteriochlorophyll (12).

The aggregation of chlorophylls appears to be a general phenomenon and is not dependent on the type of chlorophyll, but only on the presence of a magnesium atom and an intact carbonyl group in ring V. Further, aggregation is evident with metals other than magnesium. The spectra of the divalent metal-pheophytin a derivatives can be interpreted in the same way as for the true chlorophylls. In nonpolar solvents the zinc and copper pheophytin a, and zinc and copper pyropheophytin a spectra (19) show aggregation peaks at $1650-1640 \, cm^{-1}$. On the other hand, the mercury (II) and cadmium pheophytin a and silver (II) pyropheophytin a do not give evidence for aggregation bands in nonpolar solvents.

Copper (II) and zinc are known to coordinate strongly to both nitrogen-and oxygen-containing donors and to form penta- or hexa-coordinate complexes. Further, porphyrin complexes of these metals are reported to coordinate one or two pyridine (64) or piperidine (65) molecules in axial positions, thus demonstrating the coordination unsaturation of the metals in these systems. In nonpolar solvents this unsaturation, as in the case of the magnesium-containing chlorophylls, is overcome by coordination of the metal to the ketone carbonyl. From relative intensity considerations the order of the degree of aggregation of the pheophytin metal complexes is: $Mg > Cu > Zn >>> Cd \sim Hg \sim Ag$.

2. CHLOROPHYLL AGGREGATION FROM NMR SOLVENT DEPENDENCE.
   COORDINATION AGGREGATION

The NMR spectra of the chlorophylls exhibit an unusual solvent dependence, entirely parallel to the solvent dependence of the IR spectra. The addition of methanol d_4 to chloroform solutions of the chlorophylls and methyl chlorophyllides results in marked changes in the chemical shift parameters. In chloroform or carbon tetrachloride solution, the NMR spectra of the chlorophylls (or chlorophyllides) are poorly resolved. In tetrahydrofuran, or in $CDCl_3$-CD_3OD mixtures, the spectra become sharp, and a number of the proton resonance signals are observed to show different chemical shifts. These changes are particularly noticeable in the methyl resonance region and in the behavior of the C-10 proton signal. The C-10 proton signal, as a matter of fact, can scarcely be seen in the spectra of $CDCl_3$ solutions. In the case of chlorophyll b, the aldehyde proton completely loses its identity in $CDCl_3$. The addition of incremental amounts of CD_3OD to $CDCl_3$ solutions causes some of the proton resonances to shift, and the change in chemical shift parameters as a function of solvent composition permits some interesting conclusions to be drawn about the aggregation behavior of chlorophyll.

The effects of change in solvent composition on the NMR spectrum of chlorophyll a are shown in Fig. 22. The C-10 proton shows the largest paramagnetic shift with increasing methanol concentration. The methyl group resonances of protons 5 and 11 also show large paramagnetic shifts. Relatively minor changes occur at the methine proton resonances, and the resonances of the methyl groups at positions 1 and 3a are likewise insensitive to the solvent composition. The addition of methanol to $CDCl_3$ solutions of chlorophyll b (Fig. 23) causes a large paramagnetic shift in the position of the aldehyde proton; sizable shifts to lower field are observed as well for the α- and β-methine protons, the C-10 proton, and the methyl groups at positions 5 and 11.

The striking shifts in the spectra of chlorophylls a and b upon the addition of methanol d_4 to chloroform solutions must be attributed to specific interactions of the solute with the solvent. In $CDCl_3$ solutions, the chlorophyll molecules are aggregated by intermolecular coordination between carbonyl groups and the central magnesium atom. The addition of the first two molecules of methanol brings about most of the chemical shift change. The disaggregating effect of the methanol can be clearly linked to formation of methanol–chlorophyll complexes. However, so large is the magnitude of the chemical shifts changes on methanol addition that solute–solvent interactions alone cannot reasonably account for the observations. Solute–solute interactions must also be implicated.

The maximum chemical shift differences between aggregated chlorophyll in $CDCl_3$ solution and disaggregated chlorophyll in $CDCl_3$-CD_3OD solution are mapped onto the structural formulas of methyl chlorophyl-

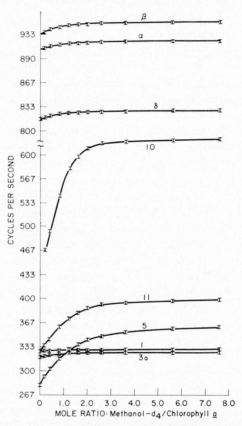

FIG. 22. Disaggregation of chlorophyll a (0.15 mole/liter) in $CDCl_3$ solution by CD_3OD from NMR data. Chemical shifts of designated protons are plotted as a function of methanol concentration (47). See Fig. 15 for proton numbering.

lide a (Fig. 24) and chlorophyll b (Fig. 25). These aggregation maps give some concrete indications of the nature of the aggregates. For both compounds, regions of mutual overlap occur in the vicinity of ring V. If the aggregates are considered to be, on the average, dimers, as can be inferred from the molecular weight measurements, then the half-circle represents the second monomer unit. The magnesium atom of one molecule then is in position to bond to the carbonyl group of the other.

In the case of chlorophyll b, two highly shielded regions exist in the aggregates, located in the vicinity of both carbonyl functions. In this situation, it is possible that two kinds of aggregate exist, of comparable stability, generated in one case by coordination of magnesium with ketone carbonyl, and in the other by coordination with aldehyde carbonyl.

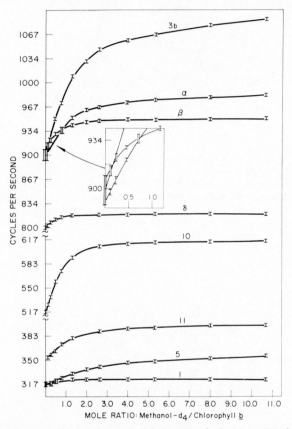

FIG. 23. Disaggregation of chlorophyll b (0.15 mole/liter) in $CDCl_3$ by CD_3OD. Chemical shifts of designated protons vs. CD_3OD concentration (47). See Fig. 15 for proton numbering.

At higher concentration trimers probably occur in which one molecule binds two other molecules by means of both its ketonic and aldehyde carbonyl function.

The geometry of the aggregates cannot be fully deduced from the NMR data now available. Steric considerations preclude a parallel geometry, for the phytyl and carbomethoxy groups will both project above and below the plane of the macrocycle. A labile equilibrium must exist

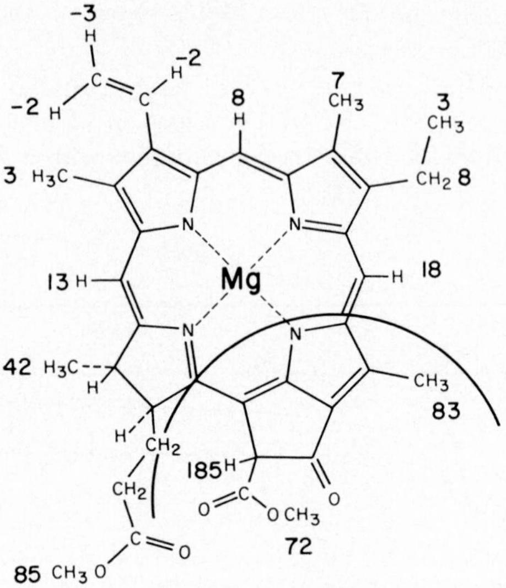

FIG. 24. Aggregation map of methyl chlorophyllide a constructed from methanol titration data. The numbers give maximum chemical shift differences for the indicated protons (in ppm $\times$ 100, or cps for 100 mc spectra) between highly aggregated ($CDCl_3$ solution) and fully disaggregated ($CDCl_3 + CD_3OD$) methyl chlorophyllide a.

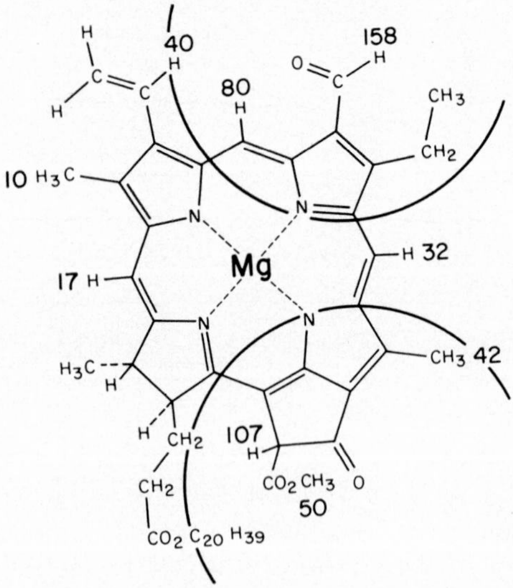

FIG. 25. Aggregation map of chlorophyll b from NMR chemical shift differences between $CDCl_3$ and $CDCl_3$-CD_3OD solutions for indicated protons. Chemical shifts given in cycles per second (cps), equivalent to parts per million (ppm) $\times$ 100.

between monomer and aggregate. If aggregates were long-lived, then partially aggregated solutions containing both monomer and aggregate would be expected to give spectra containing lines characteristic of both species. In fact, only one set of lines is observed, and the resonance positions are weighted averages between monomer and aggregate. The implication is that the aggregates form and dissociate with a rate that is high relative to the spin relaxation processes that are occurring.

The dilution behavior of both chloroform and chloroform–methanol solutions indicates that little change in the structure of the species in both solutions occurs over the concentration ranges up to about 0.3 M. No further aggregation to higher polymers occurs at higher concentrations, and little evidence is found for the dissociation of dimers within the concentration range studied in aggregating solvents. This is in sharp contrast to the concentration dependence behavior of magnesium-free derivatives described below, and emphasizes the essential differences between coordination aggregation and the forces responsible for the aggregation of metal-free chlorins. The absence of any significant concentration dependence for NMR spectra in chloroform-methanol solution strongly suggests the absence of association among methanol-chlorophyll complexes.

The Brody's (59) have been strong protagonists for the view that the substance responsible for the long wavelength absorption in chloroplasts is aggregated chlorophyll. The evidence for this opinion is mainly spectroscopic, and is derived from either visible absorption spectra in concentrated solution, or fluorescence spectra at low temperatures. The polar solvents ethanol, pyridine, and acetone have been used by them, and they have adduced evidence for the formation of chlorophyll aggregates at room temperature in these solvents. Infrared and NMR spectroscopy, however, emphatically establish, as expected, that chlorophyll is monomeric in these solvents at room temperature. We have examined the NMR spectrum of chlorophyll a in tetrahydrofuran down to $-90°$C, with no evidence for the formation of chlorophyll aggregates in tetrahydrofuran solution. There thus appears to be a serious inconsistency between the NMR results and the conclusions from optical spectroscopy that requires resolution.

3. CHLOROPHYLL AGGREGATION FROM NMR CONCENTRATION DEPENDENCE. π-π AGGREGATION

The NMR spectra of magnesium-free chlorophyll derivatives are highly concentration dependent. Concentration changes do not affect all the proton resonances in the same way, for the α- and β-methine protons are subject to particularly large paramagnetic shifts on dilution (47).

The methyl groups bonded to the conjugated ring also undergo paramagnetic shifts on dilution.

It is reasonable to assume that the observed concentration effects result primarily from intermolecular solute-solute interactions. These π-π

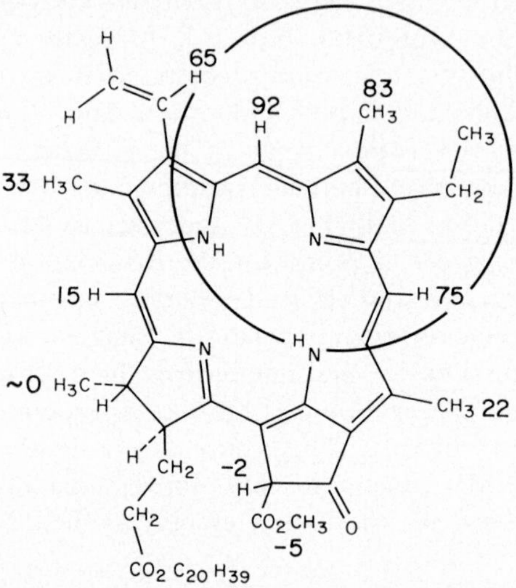

FIG. 26. Aggregation map of pheophytin a based on chemical shift changes as a function of concentration in $CDCl_3$ solution, in the concentration range 0.3 mole/liter to extrapolated values at zero concentration.

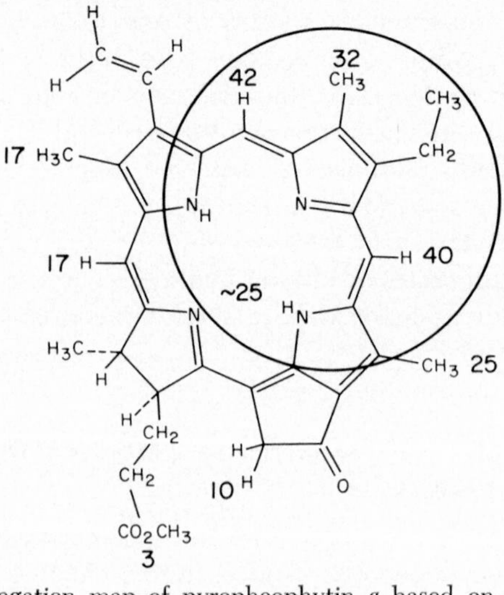

FIG. 27. Aggregation map of pyropheophytin a based on chemical shift differences between concentrated and dilute solutions in $CDCl_3$, in the range from 0.18 mole/liter to extrapolated values at zero concentration.

interactions are basically charge-transfer phenomena; stabilization occurs by formation of aggregates in which electron transfer occurs. An aggregation map of pheophytin *a* (*47*) (Fig. 26) and pyropheophytin *a* (*16*) in chloroform (Fig. 27) constructed from the concentration dependence of the chemical shifts provides grounds for the conclusion that in such solutions an appreciable fraction of the molecules is present as a dimer, in which the mutual diamagnetic shielding of the monomer units differs at the various substituent protons. A sandwich-type arrangement can be postulated in which the projections of the two centers of the macrocycles are somewhat displaced from each other. In the orientation shown in Figs. 26 and 27, in which the circle represents the π-electron system of a second pheophytin molecule, those protons located at positions of mutual overlap will experience an appreciable diamagnetic shielding. The largest chemical shifts are found in the vicinity of ring II, and this region should then correspond to the area of maximum overlap of the monomer units in the dimer.

As in the case of coordination aggregates, the aggregates resulting from the operation of π-π forces have NMR spectra with sharp lines, indicative of a very mobile equilibrium between monomer and dimer. Because of the sharpness of the lines, it can be inferred that the mean lifetime of the aggregated species must be very short. In contradistinction to coordination aggregation, however, π-π aggregation is highly sensitive to concentration. The π-π interactions are weaker than coordination forces, and thus the difference in concentration response.

B. Exchangeable Hydrogen in Chlorophyll

The question whether chlorophyll possesses exchangeable hydrogen has long aroused speculation. Efforts to detect labile hydrogen by tritium studies have been largely inconclusive (*66*), as were various infrared studies (*23b,c, 24*). The assignment of the proton magnetic resonance spectrum of chlorophyll has made a simple and direct answer possible (*67*). When chlorophyll is dissolved in a disaggregating solvent (acetone d_6, tetrahydrofuran d_8 or $CDCl_3$ to which methanol d_4 has been added), the NMR spectrum changes with time. The δ and the C-10 proton resonances diminish until they disappear altogether. The rate at which the resonance peak changes can readily be established. Evidently these two protons are labile, and exchange with the deuteron in the hydroxyl group of the alcohol. Since only protons are detected, the exchange results in the disappearance of the hydrogen resonance. The presence of labile hydrogen in chlorophyll has thus been firmly established.

Table XI summarizes the exchange situation for a variety of chloro-

TABLE XI

PSEUDO-FIRST-ORDER RATE CONSTANTS FOR HYDROGEN EXCHANGE IN METHANOL -d$_4$–ACETONE-d$_6$ (M/M) SOLUTION AT 40° BY NMR^a

Compound	Concn. (mole/l)	k^b for C-10 (sec^{-1})	k^b for δ (sec^{-1})	k^b for β (sec^{-1})	k^b for α (sec^{-1})
Chlorophyll a	0.050	$>3 \times 10^{-3}$	2×10^{-6}	$<<5 \times 10^{-8}$	$<<5 \times 10^{-8}$
Chlorophyll b	0.050	5×10^{-4}	4×10^{-6}	$<<5 \times 10^{-8}$	$<<5 \times 10^{-8}$
Bacteriochlorophyll	0.050	7×10^{-5}	2×10^{-7}	6×10^{-7}	1×10^{-6}
Pyrochlorophyll a	0.050	$<5 \times 10^{-8}$	6×10^{-6}	$<<5 \times 10^{-8}$	$<<5 \times 10^{-8}$
Pheophytin a	0.050	$>>3 \times 10^{-3}$	$<<5 \times 10^{-8}$	$<<5 \times 10^{-8}$	$<<5 \times 10^{-8}$

^{a} After Dougherty et al. (67).

^{b} k is calculated as a pseudo-first-order rate constant from k (sec^{-1}) $= \ln 2/t_{\frac{1}{2}}$.

phylls. The C-10 proton undergoes exchange at a rate at least two orders of magnitude greater than the exchange rate at the δ position. Exchange at the δ position is strongly influenced by the presence of magnesium. Removal of the magnesium reduces the exchange rate at the δ position to a very low value. Indeed, the presence of magnesium is essential for the δ proton to be labile under neutral conditions. The other two methine bridge positions undergo exchange also, but at a rate at least three orders of magnitude more slowly.

In bacteriochlorophyll, the 3 methine protons are all labile to about the same degree. We have also observed that the methyl hydrogens in the acetyl group at position 2 exchange rapidly in acid solution (50).

The role in photosynthesis of the labile hydrogens in chlorophyll, if any, still is obscure. Nevertheless the power and utility of NMR methods in chlorophyll studies are convincingly displayed in this straightforward resolution of an old and troublesome problem.

C. Keto-Enol Tautomerism in Chlorophyll

The possibility of keto-enol tautomerism in chlorophyll is another topic in chlorophyll chemistry that has stimulated much discussion. The existence of a β-keto ester grouping in ring V makes such speculations attractive, and Holt and Jacobs (12) made the enol form of chlorophyll the basis for an interpretation of the solvent dependence of the IR spectra in the carbonyl region. The IR and NMR results described here provide convincing evidence that chlorophylls a and b exist in solution predominately in the keto form, and that the equilibrium concentration of enol must be small.

Hydroxylic protons undergo instantaneous exchange when brought into contact with deuterated alcohols. If the proton resonance attributed to the C-10 proton really arose from an enol-OH group, it should disappear upon shaking the pigment solution with D_2O or CH_3OD. In all cases, even a large excess of CH_3OD causes only a slow decrease of the proton resonance in question. This behavior is quite different from what might be expected from an enol hydroxyl group.

Perhaps the most persuasive argument in favor of a keto configuration for chlorophyll in solution derives from the behavior of pyrochlorophyll a (16). In the pyro series, the carbomethoxy group of carbon 10 has been replaced by a proton; no β-keto ester group exists. Keto-enol tautomerism is clearly far less probable, and in agreement with this expectation, the two C-10 protons are found to be essentially nonexchangeable. The δ-proton in pyrochlorophyll, however, is exchangeable. The solvent dependence of the IR spectra in the carbonyl region is very similar for pyrochlorophyll *a* and chlorophyll *a*. As pyrochlorophyll *a* can exist only in the keto form, keto-enol tautomerism cannot play an important part in the genesis of its carbonyl region IR spectra. Chlorophyll *a* must also therefore exist with high probability predominantly in the keto form.

The C-10 proton in all chlorophyll compounds and derivatives containing a carbomethoxy group at position C-10 is readily exchangeable. This suggests that small amounts of enol in a relatively slowly established equilibrium with the keto form provides the mechanism for exchange.

D. Path of Hydrogen in Photosynthesis from PMR Studies on Deuterated Algae

The availability of fully deuterated organisms, including photosynthetic ones (*21*), in combination with NMR spectroscopy makes it possible to follow the path of hydrogen in photosynthesis. Hydrogen incorporated into the compounds of a fully deuterated organism, either by chemical exchange, or through metabolic activity, can be detected in the isolated compounds by PMR, and if the spectra have been assigned an integration makes it possible to ascertain where and how much hydrogen has been introduced into the molecule.

To determine whether the labile hydrogen atom in chlorophyll at the δ-methine position is involved in photosynthesis, and whether the hydrogen atoms at positions 7 and 8 in ring IV participate in a reversible hydrogenation-dehydrogenation cycle with water, fully deuterated *Scenedesmus obliquus* (grown in 99.7% D_2O) with CO_2 as the sole carbon source was harvested by centrifugation. The cells were then immediately resuspended in fresh H_2O medium, and allowed to continue photosynthesis. After 16 hours the cells were again harvested; the chlorophylls were extracted, purified by the usual methods, and examined by PMR spectroscopy. The only proton resonance that could be observed was at the C-10 position; since this proton was found to be exchanged even in the dark, the introduction of hydrogen here presumably is by direct exchange and does not necessarily involve photosynthesis. No evidence for the introduction of hydrogen into the δ, 7, or 8 positions could be detected. Under the conditions of the experiment, then, the bulk of the chlorophyll exists in a form that precludes hydrogen exchange at the δ, 7, or 8 positions, either by photosynthesis, or by subsequent hydrogen transport (*68*).

V. Concluding Remarks

Sufficient progress has now been made in the acquisition and interpretation of IR and NMR data to provide a sound basis for future application of these important spectroscopic techniques to problems of chlorophyll structure and function. Infrared spectra are most useful in the 1800–1600 cm^{-1} (carbonyl) and 600–160 cm^{-1} (low infrared) regions, but eventual interpretation of the fingerprint region now appears a possibility (69). The analysis of chlorophyll NMR spectra has now proceeded to the point where application to the *in vivo* situation is practical, and some such applications have been described in this chapter. The application of NMR techniques used in conjunction with isotopic hybrid organisms to problems of chlorophyll biogenesis has already been briefly described elsewhere by Katz (70). It would appear highly likely that other applications of IR and NMR to problems of biological interest involving porphyrins and chlorophylls will be soon forthcoming.

ACKNOWLEDGMENTS

Our IR and NMR studies on chlorophyll were carried out in close collaboration with Dr. Harold H. Strain, whose intimate familiarity with chlorophyll made a vital contribution to our work. Dr. Henry L. Crespi was responsible for the culturing of the organisms that were used as a source of chlorophyll, bacteriochlorophyll, and chlorobium chlorophyll. Our work also benefited greatly from the excellent preparative work of Mr. Walter Svec. Many of the NMR spectra were collected by Miss Gail Norman. Dr. Gerhard C. Closs of the Chemistry Department, University of Chicago, was largely responsible for the first assignment of the chlorophyll NMR spectrum.

REFERENCES

(1) C. E. Meloan, "Elementary Infrared Spectroscopy." Macmillan, New York, 1963.
(2) S. E. Wiberley, N. B. Colthup, and L. H. Daly, "Introduction to Infrared and Raman Spectroscopy." Academic Press, New York, 1964.
(3) C. N. R. Rao, "Chemical Applications of Infrared Spectroscopy." Academic Press, New York, 1964.
(4) L. J. Bellamy, "The Infrared Spectra of Complex Molecules," 2nd ed. Methuen, London, 1958.
(5) N. Nakanishi, "Infrared Absorption Spectroscopy-Practical." Holden-Day, San Francisco, California, 1962.
(6) R. G. White, "Handbook of Industrial Infrared Analyses." Plenum Press, New York, 1964.
(7) K. Nakamoto, "Infrared Spectra of Inorganic and Coordination Compounds." Wiley, New York, 1963.
(8) F. F. Bentley, E. F. Wolforth, N. E. Srp, and W. R. Powell, *Spectrochim. Acta* 13, 1 (1958); H. R. Wyss, R. D. Werder, and Hs. H. Günthard, *ibid.* 20, 573 (1964); W. T. Bolleter, *Appl. Spectry.* 18, 72 (1964).
(9) H. A. Willis, R. G. J. Miller, D. M. Adams, and H. A. Gebbie, *Spectrochim. Acta* 19, 1457 1963).
(10) R. Stair and W. W. Coblentz, *J. Res. Natl. Bur. Std.* 11, 703 (1933).
(11) J. W. Weigl and R. Livingston, *J. Am. Chem. Soc.* 75, 2173 (1953).
(12) A. S. Holt and E. E. Jacobs, *Plant. Physiol.* 30, 553 (1955).
(13) A. Stoll and E. Wiedemann, *Helv. Chim. Acta* 42, 679 (1959).
(14) J. E. Falk and J. B. Willis, *Australian J. Sci. Res.* A4, 579 (1951).
(15) H. H. Strain, M. R. Thomas, H. L. Crespi, M. I. Blake, and J. J. Katz, *Ann. N.Y. Acad. Sci.* 84, 617 (1960).

(16) F. C. Pennington, H. H. Strain, W. A. Svec, and J. J. Katz, *J. Am. Chem. Soc.* **86**, 1418 (1964).

(17) A. S. Holt and H. V. Morley, in "Comparative Biochemistry of Photoreactive System" (M. B. Allen, ed.), p. 169. Academic Press, New York, 1960.

(17a) L. J. Boucher, H. H. Strain, and J. J. Katz, *J. Am. Chem. Soc.* **88** (1966).

(18) H. R. Wetherell, M. J. Hendrickson, and A. R. McIntyre, *J. Am. Chem. Soc.* **81**, 4517 (1959).

(19) A. S. Holt, *Proc. 5th Intern. Congr. Biochem., Moscow, 1961* Vol. 6, p. 59. Pergamon Press, Oxford.

(20) A. N. Sidorov and A. N. Terenin, *Opt. Spectry.* (*USSR*) (*English Transl.*) **8**, 254 (1960).

(21) H. F. DaBoll, H. L. Crespi, and J. J. Katz, *Biotechnol. Bioeng.* **4**, 281 (1962).

(22) J. J. Katz, G. L. Closs, F. C. Pennington, M. R. Thomas, and H. H. Strain, *J. Am. Chem. Soc.* **85**, 3801 (1963).

(23) A. V. Karyakin, V. M. Kutyurin, and A. K. Chibisov, *Dokl. Akad. Nauk. SSSR* **140**, 1321 (1961).

(23a) A. V. Karyakin and A. K. Chibisov, *Biofizika* **8**, 441 (1963).

(23b) A. V. Karyakin and A. K. Chibisov, *Opt. Spectry.* (*USSR*) (*English Transl.*) **13**, 209 (1962).

(23c) A. N. Sidorov, *Opt. Spectry.* (*USSR*) (*English Transl.*) **13**, 206 (1962); **15**, 454 (1963).

(24) V. M. Kutyurin, A. V. Karyakin, A. K. Chibisov, and I. Yu. Artamkina, *Dokl. Akad. Nauk. SSSR* **141**, 744 (1961).

(25) S. Schwartz, M. H. Berg, I. Bossenmaier, and H. Dinsmore, *Methods Biochem. Anal.* **8**, 221 (1960).

(26) A. F. H. Anderson and M. Calvin, *Arch. Biochem. Biophys.* **107**, 251 (1964).

(27) V. M. Kutyurin and V. P. Knyazev, *Dokl. Akad. Nauk. SSSR* **149**, 456 (1963).

(28) J. J. Katz, M. R. Thomas, H. L. Crespi, and H. H. Strain, *J. Am. Chem. Soc.* **83**, 4180 (1961).

(29) S. E. Wiberley, N. B. Colthup, and L. H. Daly, "Introduction to Infrared and Raman Spectroscopy," p. 377. Academic Press, New York, 1964.

(30) N. Fuson and M. L. Josien, *J. Chem. Phys.* **20**, 1043 (1952).

(31) D. W. Thomas and A. E. Martell, *J. Am. Chem. Soc.* **78**, 1338 (1956); **81**, 5111 (1959); K. Ueno and A. E. Martell, *J. Phys. Chem.* **80**, 934 (1956).

(32) R. C. Lord and F. A. Miller, *J. Chem. Phys.* **10**, 328 (1942).

(33) S. F. Mason, *J. Chem. Soc.* p. 976 (1958).

(34) J. A. Pople, H. J. Bernstein, and W. J. Schneider, "High Resolution Nuclear Magnetic Resonance." McGraw-Hill, New York, 1959.

(35) L. M. Jackman, "Applications of Nuclear Magnetic Resonance Spectroscopy." Pergamon Press, Oxford, 1959.

(36) O. Jardetzky and C. D. Jardetzky, *Methods Biochem. Anal.* **9**, 235-410 (1962).

(37) R. H. Bible, "Interpretation of NMR Spectra." Plenum Press, New York, 1965.

(38) E. D. Becker and R. B. Bradley, *J. Chem. Phys.* **31**, 1413 (1959).

(39) J. Ellis, A. H. Jackson, G. W. Kenner, and J. Lee, *Tetrahedron Letters* **2**, 23 (1960).

(40) C. E. Johnson and F. A. Bovey, *J. Chem. Phys.* **29**, 1012 (1958).

(41) R. J. Abraham, *Mol. Phys.* **4**, 145 (1961).

(42) R. J. Abraham, P. A. Burbidge, A. H. Jackson, and G. W. Kenner, *Proc. Chem. Soc.* p. 134 (1963).

(43) E. D. Becker, R. B. Bradley, and C. J. Watson, *J. Am. Chem. Soc.* **83**, 3743 (1961).

(43a) W. S. Caughey and W. S. Koski, *Biochemistry* 1, 923 (1962).
(44) W. S. Caughey and P. K. Iber, *J. Org. Chem.* 28, 269 (1963).
(45) R. J. Abraham, A. H. Jackson, and G. W. Kenner, *J. Chem. Soc.* p. 3468 (1961).
(45a) R. J. Abraham, A. H. Jackson, G. W. Kenner, and D. Warburton, *J. Chem. Soc.* p. 853 (1963).
(46) J. N. Shoolery, in "NMR and EPR Spectroscopy" (Varian Associates Staff, eds.), pp. 100 *et seq.* Pergamon Press, Oxford, 1960.
(47) G. L. Closs, J. J. Katz, F. C. Pennington, M. R. Thomas, and H. H. Strain, *J. Am. Chem. Soc.* 85, 3809 (1963).
(48) R. B. Woodward and J. Škarić, *J. Am. Chem. Soc.* 83, 4676 (1961).
(49) J. W. Mathewson, W. R. Richards, and H. Rapoport, *J. Am. Chem. Soc.* 85, 364 (1963).
(50) J. J. Katz, R. C. Dougherty, H. L. Crespi, W. A. Svec, G. Norman, and H. H. Strain, to be published.
(51) A. S. Holt, D. W. Hughes, H. J. Kende, and J. W. Purdie, *J. Am. Chem. Soc.* 84, 2835 (1962).
(52) J. H. Mathewson, W. R. Richards, and H. Rapoport, *Biochem. Biophys. Res. Commun.* 13, 1 (1963).
(53) H. S. Gutowsky, M. Karplus, and D. Grant, *J. Chem. Phys.* 31, 1278 (1959).
(54) R. P. Linstead, U. Eisner, G. E. Ficker, and R. B. Johns, *Chem. Soc.* (*London*), *Spec. Publ.* 3, 83 (1955).
(55) N. S. Bhacca, L. F. Johnson, and J. N. Shoolery, "High Resolution NMR Spectra Catalog," Spectrum No. 346. Varian Associates, Palo Alto, California, 1962.
(56) H. Rapoport and H. P. Hamlow, *Biochem. Biophys. Res. Commun.* 6, 134 (1961).
(57) M. B. Allen, C. S. French, and J. S. Brown, in "Comparative Biochemistry of Photoreactive Systems" (M. B. Allen, ed.), Chapter 3, p. 33. Academic Press, New York, 1960.
(58) J. S. Brown, in "La Photosynthese," No. 119, p. 370. C.N.R.S., Paris, 1963.
(59) S. S. Brody and M. Brody, *Natl. Acad. Sci.—Natl. Res. Council, Publ.* 1145, 455 (1963); S. S. Brody, *J. Theoret. Biol.* 7, 352 (1964); S. S. Brody and M. Brody, *Arch. Biochem. Biophys.* 110, 583-585 (1965).
(60) G. E. Hoch, in "Biochemical Dimensions of Photosynthesis" (D. W. Krogman and W. H. Powers, eds.), p. 5. Wayne Univ. Press, Detroit, Michigan, 1965.
(61) E. I. Rabinowitch and Govindjee, *Sci. Am.* 213, 74 (1965).
(62) L. N. M. Duysens, *Progr. Biophys. Biophys. Chem.* 14, 1-104 (1964).
(63) S. Aronoff, *Arch. Biochem. Biophys.* 98, 344 (1962).
(64) J. R. Miller and G. D. Dorough, *J. Am. Chem. Soc.* 74, 3977 (1952).
(65) E. W. Baker, M. S. Brookhart, and A. H. Corwin, *J. Am. Chem. Soc.* 86, 4587 (1964).
(66) B. Coleman and W. Vishniac, *Natl. Acad. Sci.—Natl. Res. Council, Publ.* 1145, 213 (1963).
(67) R. C. Dougherty, H. H. Strain, and J. J. Katz, *J. Am. Chem. Soc.* 87, 104 (1965).
(68) J. J. Katz, R. C. Dougherty, W. A. Svec, and H. H. Strain, *J. Am. Chem. Soc.* 86, 4220 (1964).
(69) S. Kaufman, D. B. Hall, and J. J. Kaufman, *J. Mol. Spectry.* 16, 264 (1965).
(70) J. J. Katz, *Ann. Priestly Lectures* 39, 1-110 (1965); R. C. Dougherty, H. L. Crespi, H. H. Strain, and J. J. Katz, unpublished work (1965).

8

Some Properties of Chlorophyll Monolayers and Crystalline Chlorophyll*

BACON KE

Charles F. Kettering Research Laboratory, Yellow Springs, Ohio

I. Introduction

It is well known that photosynthesis takes place in the chloroplasts, where chlorophylls and other pigments are located. Considering the size of the granum ($0.1\,\mu$ in diameter), in which most of the pigments are actually concentrated, the chlorophyll concentration could be as high as $0.1\,M$. Such a high chlorophyll concentration in the photosynthetic apparatus naturally leads to the question of the manner of chlorophyll distribution. The chemical structure of chlorophyll suggests that these molecules could be located at some sort of interface formed between hydrophilic and hydrophobic phases. The existence of such interfaces in the photosynthetic apparatus is strongly suggested by the presence of lamellar structures in the chlorophyll-bearing grana in chloroplasts, as revealed by electron microscopy. It has been estimated from electron micrographs and chlorophyll content that the total available lamellar

* Contribution No. 207 from the Charles F. Kettering Research Laboratory, Yellow Springs, Ohio. The preparation of this review was supported in part by Public Health Service Research Grant GM-12275, from the National Institute of General Medical Sciences.

TABLE I
SUMMARY OF CHLOROPHYLL-MONOLAYER STUDIES

Authors	Year	Force-area isotherm	Absorption spectrum	Fluorescence	Surface potential	Other properties	Reference
Hughes	1936	X	—	—	X	—	(2)
Sjoerdsma	1936	X	—	—	—	—	(3)
Alexander	1937	—	—	—	X	Dark-field microscopy	(4)
Hanson	1937	X	—	—	—	—	(5)
Langmuir and Schaefer	1937	X	—	X	—	Film thickness; film viscosity	(6)
Rodrigo	1953	X	X	X	—	—	(7)
Jacobs et al.	1954	—	X	X	—	—	(8)
Trurnit and Colmano	1959	—	—	—	—	—	(9)
Van Winkle et al.	1963	—	X	X	—	—	(10)
Bellamy, Gaines, and Tweet	1963	—	X	X	X	—	(11)
Litvin and Gulyaev	1964	—	X	X	—	—	(12)
Rosoff and Aron	1965	—	—	—	—	Chemical reaction (pheophytinization)	(13)
McCree	1965	—	—	—	—	Photoconduction	(14)

area is just enough to accommodate all the chlorophyll molecules in a close-packed monomolecular layer (*1*).

These concepts of chlorophyll aggregation *in vivo* have led many investigators to simulate the functional structure of the photosynthetic apparatus with close-packed monomolecular layers of chlorophyll molecules, possibly oriented in some way, and to study their physical and chemical properties. Table I is a summary, listing the chlorophyll-monolayer studies that have appeared in the literature.

II. Chlorophyll Monolayers

A. Monolayer Preparation

In one earlier study, chlorophyll monolayers were prepared by evaporating the solvent from drops of a benzene solution of chlorophyll on a water surface (*6*). Other solvents such as acetone (*5*), ether, or petroleum ether containing small amounts of pyridine (*8*) were also used. Bellamy *et al.* consider benzene to be the best compromise, because chlorophyll is insufficiently soluble in some of the more preferred spreading solvents, while most good solvents (for example, ether) are slightly soluble in water (*11*).

The Langmuir balance is a basic and versatile apparatus for monolayer work: measurements of the pressure-area relationship (*5*), surface potential (*4*), film viscosity (*6*), and, more recently, absorption spectra and fluorescence (*15*), as well as the transfer of the monolayer onto another solid surface, can all be conveniently carried out in the balance. Many modifications of the balance allowing automatic recording of the pressure-area curves have been reported. The automatic film balance has been applied to chlorophyll by Trurnit and Lauer (*16*) and Gaines (*17*). Because chlorophylls are light-sensitive, it is generally recommended that monolayer preparations and measurements be carried out under a dim green light.

Compression and decompression of chlorophyll films spread from benzene indicate that benzene can remain trapped in the film and may be squeezed out after the first compression. It is therefore recommended that films spread from benzene should always be precompressed before the actual measurement is made.

B. Some Physical Properties

1. CHLOROPHYLL *a* MONOLAYERS

The state and properties of a monolayer are governed by the molecular structure of the material forming the layer, the temperature, the lateral pressure within the monolayer, and the composition of the sub-phase. Among the more important parameters for characterizing chlorophyll monolayers are the surface pressure vs. area (F-A) isotherms, surface potential, film viscosity, absorption spectrum, and fluorescence spectrum. The first three parameters for chlorophyll a (Chl a) monolayers will be described in this section; the optical properties will be treated in subsequent sections.

The more recent F-A isotherms are those reported by Trurnit and Colmano (9) for chlorophylls a and b and those reported by Bellamy et al. (11) for Chl a and pheophytin a. For Chl a, the compression and decompression curves are identical at compression rates from 0.5 to 20 $Å^2$/molecule/minute. However, marked hysteresis occurs if compression is carried close to the collapse point or higher. The plateau pressure for Chl a is in the neighborhood of 30 dyne/cm. Other reported values of the collapse pressure vary from 22–23 to slightly over 30 dyne/cm (at 20°C). The minor differences between the values reported by various workers can usually be attributed to differences in such experimental conditions as the pH and the spreading solvent and to the purity of the chlorophylls used.

Molecular areas of Chl a as derived from pressure-area measurements by different workers fall in the range of 75–125 $Å^2$. Since the area of the flat porphyrin ring is slightly over 225 $Å^2$, it is apparent that the porphyrin rings of the chlorophyll molecules are oriented either vertically or at a tilted angle instead of lying flat on the ring plane.

Hanson (5) reported earlier that chlorophyllide can form a monolayer consisting of close-packed porphyrin rings, and he assumed that the porphyrin planes are tilted at 55 degrees as in crystals. Chlorophylls contain, in addition, a hydrophobic phytol chain, the molecular area of which is 35–55 $Å^2$, as measured from the monolayer of free phytol under 25 dyne/cm pressure. Considering these facts, a probable configuration for chlorophyll would be a folded molecule, anchored at the water surface by the ester linkages of the molecule, allowing the porphyrin ring and phytol chain to be directed upward (11). The film thickness of 14 Å for a single monolayer, measured earlier by Langmuir and Schaefer (6), is compatible with this configuration. In this configuration, the presence of phytol would effectively prevent crystallization of the porphyrin rings. More recently, investigation on the angular distribution of polarized fluorescence indicates further that no preferred molecular orientation exists in the plane of the monolayer (18).

The surface potential, ΔV, is the change in potential difference between the water and the air caused by the presence of the monolayer. It is related to the structure of the molecule forming the monolayer.

The surface potential can be measured with a metal electrode coated with a radioactive substance such as polonium or with a vibrating condenser system (4, 11).

Alexander (4) first reported the surface potential for Chls a and b. The surface potential for Chl a was measured to be 305–386 mv. For a given type of chlorophyll, surface potential increases gradually but slightly with increasing compression. This slight change in dipole moment in the chlorophyll film upon compression indicates that no major change occurs in the configuration of the ester linkage with respect to the water surface as the films are compressed. On a more acidic (pH 6) subphase, the magnesium atom is extracted from chlorophyll and the surface potential increases markedly. More recently, Bellamy et al. reported a lower surface potential of 230–310 mv for Chl a (11). These authors could detect no change in surface potential of either Chl a or pheophytin a monolayers upon illumination.

2. MONOLAYERS OF CHLOROPHYLL a DERIVATIVES

Although Chl a monolayers have been studied most often, several other related compounds have also been used. Among the compounds investigated are Chl b, pheophytin a, methyl chlorophyllide a, and copper pheophytin a.

The F-A relationship and absorption spectra of Chl b monolayers were reported by Trurnit and Colmano (9). The F-A isotherms of Chl b monolayers are similar to those of Chl a, but occasionally the Chl b monolayer isotherms reach a much higher plateau pressure (45 dyne/cm). The small chemical differences between Chl a and b, and consequently, the intermolecular forces in their films, are manifested in the values of area/molecule and in the specific compressibilities. For instance, the specific compressibilities, defined as the decrease in area per unit pressure increase, reduced to unit area, are 18×10^{-3} and 23×10^{-3} cm/dyne respectively for Chl a and Chl b.

The surface potentials for Chl a and Chl b were reported by Alexander to be 305–385 and 250–300 mv, respectively (4). The difference was thought to be due to the extra polar group in Chl b. The lower moment in Chl b indicates that this polar group may act to oppose the keto and ester groups at the anchorage point of the molecule.

Pheophytin a monolayers are stable only at pressures below 11–12 dyne/cm, although the molecular areas of Chl a and pheophytin a differ by only a few per cent. Pheophytin a has a much higher surface potential (545–585 mv) than Chl a.

Monolayers of methyl chlorophyllide a are stable at pressures below 5 dyne/cm; slow collapse occurs upon further compression beyond this point. In contrast to Chl a monolayers, chlorophyllide monolayers are highly rigid. These results support the suggestion that, in the absence of the phytol chain, ordered packing or crystallization in the chlorophyllide monolayers is favored.

Copper pheophytin a was studied in conjunction with its use as a quenching agent for Chl a fluorescence (see below). Copper pheophytin a forms stable monolayers on phosphate buffer solution. It is nonfluorescent in solution or in the film state. Its F-A relationship is similar to that of Chl a, although it collapses at a lower pressure. Copper pheophytin a spread on phosphate buffer undergoes a slow chemical degradation, which is thought to be caused by the formation of allomerized products as a result of either opening or expansion of the cyclopentanone ring.

3. MULTICOMPONENT MONOLAYERS

Multicomponent monolayers containing chlorophylls a and b and β-carotene were first prepared by Colmano in 1962 (19) in an attempt to simulate the pigment assembly in the natural photosynthetic apparatus. The mixed monolayer was prepared by spreading a benzene solution containing chlorophylls a and b and β-carotene in the mole ratio of 6:3:1 on a phosphate buffer solution and compressing to 20 dyne/cm. The absorption spectrum of such multicomponent monolayers bears a striking resemblance to that of a unicellular alga containing similar pigments. The spectroscopic resemblance in both the peak positions and the peak height ratios has led the author to postulate that *in vivo* chlorophylls may also be present as close-packed monolayers at liquid interfaces.

Although pure β-carotene can be spread on a phosphate buffer solution, it does not spread on a pure water-air interface. Since carotene is a pure hydrocarbon, its molecules do not orient in a monolayer, but cluster into islands. With increasing film pressure, the area of the islands decreases and the thickness increases, as revealed by ellipsometry measurements.

More recently, multicomponent monolayers composed of Chl a plus inert diluents such as fatty acids, alcohols, and esters have been reported (20). These mixed monolayers represent a two-dimensional solution of the pigment. Depending on the pigment–diluent proportion, the intermolecular distance of the pigment in a two-dimensional layer can thus be effectively regulated.

Within experimental error, the area occupied by the Chl a-oleyl alcohol (or stearyl alcohol) mixed monolayers is almost identical to the area occupied by the same number of molecules of the components. This behavior indicates that either the components are immiscible or that the components can form ideal two-dimensional solutions. In other words, the excess free energy of mixing is zero.

Oleyl alcohol monolayer has a collapse pressure of 30 ± 1 dyne/cm; its collapse is rapid and reversible. Chl a monolayer has a collapse pressure of 23 dyne/cm. The collapse pressure of the Chl a-oleyl alcohol mixed layer varies with the composition, suggesting that the components are miscible. The collapse pressure data indicate that a two-dimensional solution exists at least up to a chlorophyll mole fraction of 0.1 (20). The increase in fluorescence yield of the pure Chl a film when diluted with oleyl alcohol also suggests that Chl a is effectively dissolved in oleyl alcohol (21) (see below).

Monolayers of saturated stearyl alcohol collapse slowly; they can be compressed beyond the equilibrium spreading pressure. Mixed layers containing stearyl alcohol and Chl a show no detectable collapse until the pressure exceeds 35 dyne/cm. On the other hand, when the mixed layers are held at constant area and at pressures above 22 dyne/cm, a gradual pressure decrease occurs indicating a slow collapse, while no such changes occur at pressures below 21 dyne/cm. These results indicate that the components are immiscible, because the mixed layer becomes unstable when the pressure is above the collapse pressure of the less stable component. The immiscibility of these two components is further indicated by the fact that no increase in fluorescence yield can be detected when the pure chlorophyll film is diluted with stearyl alcohol (21) (see below).

C. Absorption Spectra

The absorption spectrum is an important physical property of chlorophyll monolayers. The first extensive examination of the absorption spectra of chlorophyll microcrystals and monolayers was carried out by Jacobs and co-workers (8). The absorbance of a single monolayer is sufficiently strong that the monolayer can be visually detected and the absorption peak can be located with a spectrophotometer. To produce a complete absorption spectrum of significant magnitude, Jacobs et al. used a stack of 4–6 microscope cover glasses, on each of which one monolayer was deposited. Jacobs reported that apparently two kinds of monolayers are formed with Chl a: layers spread on a water subphase containing 10^{-6} M Ca^{++} had a high optical density and its red peak shifted to 735 mμ; the red peak of Chl a monolayers formed on Ca^{++}-free water remained at 670 mμ.

In these early experiments the amount of control over the surface pressure, and consequently the surface concentration, was not specified.

Furthermore, it was not stated whether the monolayers were at an air–water interface or dried during the measurements.

Subsequently two other groups independently re-examined the absorption spectra of chlorophyll monolayers (9, 11). Since the two groups used quite different approaches in obtaining the spectra, their techniques will be briefly described here. Trurnit and Colmano (9) used a stack of 15 microscope slides, mounted 3 mm apart on a Teflon holder. The

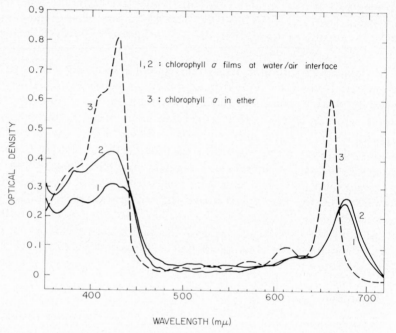

FIG. 1. Absorption spectra of 30 Chl *a* films in series and Chl *a* in solution (9). The surface concentration in the films are: 9.3×10^{13} molecules/cm² (curve 1); 12.1×10^{13} molecules/cm² (curve 2). Chl *a* concentration in ether, $9.82 \times 10^{-6} M$ (curve 3).

monolayer was spread on the aqueous surface in the film balance and compressed to the desired pressure. While still wet, the stack of glass slides was quickly transferred to the trough. As soon as the slides came in contact with the monolayer surface, the pressure dropped and the monolayer started to spread onto the new surface of the slides and eventually was stopped by the Teflon holder. The monolayer was then recompressed to the original pressure. The decrease in surface area in the film balance corresponded to the available surface area on the glass slides. Now each of the two sides of the slide was covered with a chlorophyll monolayer at a water–air interface and under a known surface pressure and surface concentration. The slides were quickly transferred to the sample compartment of a spectrophotometer and the spectrum taken while the slides were still wet. The reference compartment contained a stack of 15 dry glass slides.

To obtain the absorption spectra of the chlorophyll monolayers at a water–oil interface, the film-covered wet slides were inserted into a glass cell placed in the sample compartment and filled with Nujol oil.

Figure 1 shows the absorption spectra of 30 parallel monolayers of chlorophyll *a* under two different surface pressures. The absorption spectrum of chlorophyll *a* in solution is also shown for comparison. In the monolayer spectra, the absorption peaks are shifted toward longer wavelength and the half-band width becomes considerably broader. The band shift and band broadening both increase with the number of molecules per unit film area. At the same time, the ratio of the blue peak to blue-satellite peak of the Chl *a* monolayer decreases considerably. In Chl *b* monolayers the blue satellite peak decreases, and the increasing band width of the red absorption peak results in a comparatively high absorption extending to 700 mμ.

The absorption spectra of chlorophyll monolayers at the water–Nujol interface are similar to those at the water–air interface, except the change in the ratio of blue:blue satellite absorption is less pronounced.

The General-Electric group (*11, 15*) measured the absorption spectrum of a monolayer spread on a water surface in a film balance and under a given pressure. This approach appears to eliminate both the difficulty of film deterioration resulting from deposition on glass slides and the correction needed for the change in reflection of the substrate caused by the monolayers. The spectrometer used by these investigators, as shown in Fig. 2, is a multiple-pass instrument containing two horizontal silver or aluminum mirrors facing each other and parallel to within 0.001 radian. One mirror is immediately above and one immediately below the water surface, with a separation of 4.28 cm. Light from a tungsten lamp passes through a grating monochromator and is collimated and deflected by a tilted mirror to the horizontal mirror below the water surface. The light beam is then reflected back and forth between the two mirrors through the water–film–air interface 16–30 times. When the light beam finally emerges from the water, it is focused onto the 2-inch entrance hole of an integrating sphere 10 inches in diameter, the interior of which is coated with MgO. A photomultiplier is mounted at the exit hole of the integrating sphere. As a reference source, the collimated beam may be deflected directly into the integrating sphere through a pair of mirrors.

With the integrating sphere as the light collector, the noise problem created by the excursion of the light beam through the shimmering water surface is largely eliminated. The error caused by the reflectance change resulting from the introduction of the monolayer can be largely minimized by making the mirrors strictly parallel.

The effective optical density $(O.D.)_{eff}$ of the monolayer can be calculated from the relationship $(O.D.)_{eff} = (1/N) \log (fI_{00}/I_s)$, where N is the number of passes the beam travels through the interface and I_0

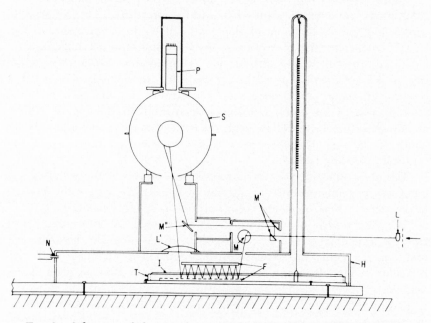

Fig. 2.   Side view of the monolayer spectrometer used by the General Electric group (15). The symbols are: L, L', lenses; M, M', M", mirrors; F, mirrors above and below the water surface; I, water-air interface; T, the Langmuir trough; H, light-tight housing; N, inlet for inert gas; S, integrating sphere; P, photomultiplier.

and I_{00} represent, respectively, the signals generated by the beams which either pass through the multiple-reflection mirror system or pass directly to the integrating sphere. The ratio I_0/I_{00}, designated as f, may be plotted as a function of wavelength, and should be stable to within $\pm 2\%$. With the monolayer under a given pressure, another set of readings of the multiply reflected beam signals is taken and designated as I_s.

The absorption spectra constructed from data points obtained for 26 passes through a single Chl a monolayer under 15 dyne/cm pressure on pH 8 phosphate buffer are shown in Fig. 3. The absorption spectrum of Chl a in benzene is also shown and arbitrarily matched at the blue peak

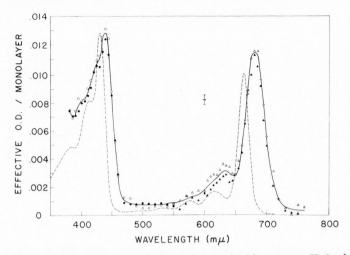

FIG. 3. Absorption spectrum of a monolayer of Chl *a* on a pH 8 phosphate buffer under nitrogen and at 15 dyne/cm surface pressure, measured by the spectrometer shown in Fig. 2 through 26 passes (data points are from different runs). Absorption spectrum of Chl *a* in benzene (dashed curve) is shown for comparison. The spectra are arbitrarily matched at the blue peak. Inset cross refers to error limits on the absorption coefficient and wavelength resolution (*11*).

of the monolayer spectrum. As seen from Table II, which compares the peak wavelength and optical density of several absorption peaks, the overall characteristics of the spectra reported by different workers are in good agreement.

TABLE II

COMPARISON OF THE ABSORPTION PEAK POSITIONS OF THE CHL *a*
MONOLAYER OBTAINED BY DIFFERENT INVESTIGATORS

	Red peak		Blue peak		Blue satellite peak	
Authors	λ, mμ	O.D.	λ, mμ	O.D.	λ, mμ	O.D.
Jacobs *et al.* (*8*)	675	0.011	435	0.0138	420	0.012
Trurnit and Colmano (*9*)	675	0.008	420	0.012	—	—
Bellamy *et al.* (*11*)	680	0.011	440	0.013	420	0.011

As found by the workers (*11*) at General Electric, the absorption spectra are stable over the pressure range of 1–8 dyne/cm for pheophytin *a* and 1–15 dyne/cm for Chl *a*. Dramatic changes occur when the surface pressure exceeds these limits. Figure 4 shows the absorption spectra of pheophytin *a* within (6.5 dyne/cm) and beyond (13.5 dyne/cm) the stable limit. The observed change is very similar to the large shift of the red peak for Chl *a* layers formed on the surface of water containing Ca^{++}.

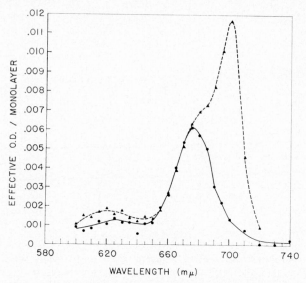

Fig. 4. Absorption spectra of a monolayer of pheophytin *a* at 6.5 dyne/cm (solid curve) and 13.5 dyne/cm (dashed curve). Other conditions the same as those for Fig. 3 (*11*).

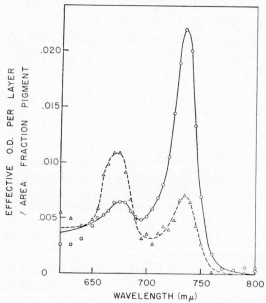

Fig. 5. Absorption spectra of monolayers of methyl chlorophyllide (solid curve) and a mixture of methyl chlorophyllide and triolein (mole ratio 1:2) (broken curve) (*21a*). Both monolayers on pH 8 phosphate buffer; surface pressure 3.5 dyne/cm. Area fraction, 1.0 for pure film, 0.36 for mixed film.

The absorption spectrum of methyl chlorophyllide a monolayer has a major peak at 735 mμ and a minor peak at 675 mμ (Fig. 5). The long-wavelength peak is similar to that of crystalline chlorophyllide a and is characteristic of a condensed film structure. When an inert diluent (triolein) is added to the film, the chromophore-chromophore interaction leading to the red shift is reduced as the chromophores are separated; the 735 mμ peak height decreases, while the 670 mμ peak height increases (21a).

The red shift is much less for Chl a monolayers than for crystalline Chl a. Jacobs et al. interpret the shift in terms of an interaction between the closely packed chromophores (22). The fact that the peak shift observed for chlorophyllide monolayers was as large as that observed for its crystals suggests that the interaction is predominantly confined to a two-dimensional plane. The much smaller peak shifts observed for the chlorophyll monolayers must have resulted from a smaller degree of order in the chlorophyll layer. The greater distance between the porphyrin rings and their random orientation resulting from the presence of phytol chains both influence the dipole-dipole interaction, which is believed to be responsible for the band shift in crystal spectra. The small increase in the red-band shift with increasing compression implies that the mutual transition moment interaction is not particularly important in determining the energy of the first singlet excited state of chlorophyll in the monolayer. It is also possible that the observed shifts may be solvent (water) effects rather than chromophore interaction involving the principal transition moments of the molecules.

The increase in optical density and the large shift of the red band for both Chl a on a Ca^{++}-containing subphase (8) and pheophytin a under a high surface pressure (11) have been explained by the formation of a bimolecular layer of more closely packed films, which is consistent with the large increase in optical density. However, Sperling recently measured the absorption spectra of Chl a multilayers prepared over a wide range of lateral pressures and found no peak shift. The major red and blue absorption bands are at 679 and 438 mμ, respectively. Within experimental error, the optical density per monolayer is constant for multilayers from 4 up to 128 layers; between 10 and 30 dyne/cm the optical density per layer ranges from 0.006 to 0.008 (23).

More recently, Litvin and Gulyaev (12) measured the absorption and fluorescence spectra of chlorophyll "monolayers" reportedly prepared in a manner similar to that used by Jacobs et al. (22). They used chlorophyll stock solutions of different concentrations (15×10^{-4} to 2×10^{-2} M) to control the packing density in the film produced. The packing density was expressed in terms of area occupied by chlorophyll molecules. At a packing density corresponding to an area of 70–130 Å² per chlorophyll molecule, the absorption spectrum appears similar to those reported in the literature (8, 9, 11). At a higher packing density corresponding to an area of 27 and 14 Å² per chlorophyll molecule, the major red peak at 672 mµ remained, but absorption in the 690–740 mµ region increased noticeably. When the packing density was increased to 7–4 Å² per molecule, a new absorption band occurred at 690 mµ. It was also reported that films produced on a subphase containing Ca^{++}, or films prepared in the regular manner but stored at 20°C for 10–12 days, show separate new absorption peaks at 700, 715, and 740 mµ. The various new peaks were attributed to different forms of chlorophyll aggregates and their resemblance to the various forms of chlorophyll *in vivo* was also suggested. These new peaks did not involve chemical changes in chlorophyll, since normal absorption spectra of Chl *a* were obtained for the chlorophyll film when rinsed off and examined in solution. Judging from the molecular areas reported, the films prepared by these authors could not be a truly monomolecular layer, but more likely a heterogeneous layer consisting of colloidal aggregates. It would be of interest to further investigate and define the actual structure of these films.

D. Fluorescence

Langmuir and Schaefer reported that while a benzene solution of chlorophyll illuminated by blue light shows red fluorescence, the fluorescence disappears as soon as the solution is spread on the water surface and the benzene has evaporated (6). Similar observations were reported later by Rodrigo (7) and Jacobs et al. (8). The absence of fluorescence in chlorophyll monolayers is obviously caused by concentration quenching. Assuming that energy migration among chromophores takes place by the inductive resonance transfer mechanism of Förster, the excitation energy in an aggregate may be trapped by a center which has a high probability for nonradiative transition to the ground state.

In very densely packed films, Litvin and Gulyaev observed fluorescence at −196°C (12). The fluorescence intensity increased with lowering of the temperature. As described in the preceding section, with increasing packing density the major red absorption band remained at 672 mµ, but the absorption level in the 690–740 mµ region increased noticeably. At the same time discrete fluorescence maxima appeared at longer wavelengths. At an intermediate packing density, fluorescence maxima appeared at 682–685, 700, and 722–728 mµ. At a packing density of 7–4 Å², a new fluorescence maximum at 738–740 mµ plus a shoulder at 750 mμ appeared. For films prepared on a subphase containing Ca^{++}, fluorescence maxima appeared at 752, 775, and 800–836 mμ.

Tweet et al. reasoned that if absence of fluorescence in a monolayer is caused by concentration quenching, then separating the chlorophyll molecules from each other in the monolayer should restore at least part of the fluorescence (24). They carried out the separation by spreading chlorophyll in a mixed monolayer containing an inert diluent. An apparatus similar to that shown in Fig. 3 was modified for fluorescence measurement. Light from an intense source was condensed, conducted through a light pipe immersed in the Langmuir trough, and reflected onto the monolayer at the air–water interface. The excitation beam was reflected internally and measured by a photodetector placed on the other end of the trough. The fluorescence emission produced by the monolayer was measured by a calibrated photomultiplier through appropriate filters. The Langmuir trough was enclosed in a light-tight housing provided with atmosphere control, and shutters were used to avoid over-illumination and decomposition by light.

Certain long-chain alcohols and esters serve as ideal two-dimensional diluents; they form stable monolayers, are transparent in the spectral region of interest, and are chemically inert toward chlorophyll. Although fatty acids also form stable monolayers, they are not considered as suitable diluents for chlorophyll because of their reaction with the magnesium of chlorophyll in the monolayer (21). Sperling (23) found that pheophytinization takes place at an appreciable rate only in solutions containing chlorophyll and a fatty acid. The rate of conversion to pheophytin also depends on the chlorophyll-to-fatty acid ratio, but the conversion takes place at a much lower rate for a mixed monolayer deposited on the glass slide.

Saturated alcohols such as stearyl alcohol (24) and heptadecanol are ineffective in increasing the fluorescence yield. The ineffectiveness is probably due to the immiscibility of these fatty alcohols with chlorophyll, and therefore the chromophores are not effectively separated. Diluents such as phytol, oleyl alcohol, and triolein increase the fluorescence of chlorophyll substantially if the distance between the chromophore molecules exceeds about 50 Å. For instance, the fluorescence yield of a pure chlorophyll monolayer can be increased 1000-fold when it is diluted with oleyl alcohol to reduce the chlorophyll area fraction to 0.01. For stearyl alcohol, the fluorescence yield is never greater than 1% of the maximum yield in oleyl alcohol. Experimental evidence from F-A measurements and absorption spectra indicate that those alcohols which increase the fluorescence yield actually form ideal two-dimensional solutions with chlorophyll (cf. Section II, B, 3).

Using a combination of interference and cutoff filters, the spectral distribution of fluorescence emitted by the diluted chlorophyll monolayer under a pressure of 5–7 dyne/cm (in which the chlorophyll molecules are spaced 75 Å apart) was obtained. The distribution of emission intensity at a given wavelength, E (λ), was calculated with the aid of the equation

$$S = C \int_0^\infty E(\lambda) \cdot R(\lambda) \cdot T(\lambda) \, d\lambda$$

where S is the photomultiplier signal with filter in place divided by the signal without filter, T (λ) the transmission of the filter, R (λ) the spectral response of the photomultiplier, and C a normalizing constant. An approximation obtained by fitting data to this equation by a trial-and-error procedure gave a fluorescence spectrum with a major peak at 681 mμ and minor peaks at 700, 730, and 750 mμ.

E. Chlorophyll-Quencher Interaction and Energy Transfer

Chlorophyll serves as the primary energy absorber of the photosynthetic apparatus. During photosynthesis the absorption of radiant energy is followed by a series of electron-transfer reactions. Certain quinones are known to be involved at one or more linkages in the electron-transfer chain (see Chapter 19). Vitamin K_1 is one of a number of naturally occurring quinone-type compounds found in photosynthetic organisms. Quenching of chlorophyll fluorescence in solution containing quinone (25), formation of photo-induced electron spin resonance (ESR) signals in chlorophyll solution containing quinones (26), and flash-induced transient absorption changes accompanying redox reactions in solutions containing chlorophyll and redox agents including quinones (27) have been reported. It appears that chlorophyll-quinone interaction in a monolayer would serve as a useful model system for energy transfer in photosynthetic reactions.

Tweet et al. recently reported that vitamin K_1 is an efficient quencher for chlorophyll fluorescence in a monolayer diluted with oleyl alcohol (28). In Fig. 6, the relative fluorescence (the ratio of the fluorescence yields in the presence and in the absence of vitamin K_1) in the diluted monolayer is plotted as a function of vitamin K_1 concentration expressed in numbers of molecules per unit area. Over a chlorophyll concentration range corresponding to a chlorophyll area fraction from 0.003 to 0.026 in the mixed layer, or a chlorophyll-chlorophyll separation from 175 to 75 Å, the agreement was within 10%. The solid line represents values calculated from the Stern-Volmer equation $\phi(Q)/\phi_0 = 1/(1 + k_Q C_Q)$, with the quenching constant $k_Q = 3.4 \times 10^{13}$ cm^2/molecule. The quenching of chlorophyll fluorescence by the nonfluorescent copper pheophytin a is shown in the same figure by the dashed curve. As with vitamin K_1, the chlorophyll fluorescence is progressively quenched with increasing copper pheophytin concentration.

Quenching of chlorophyll fluorescence by copper pheophytin may be interpreted by Förster's mechanism of inductive resonance transfer. The Förster mechanism is based on a near-field, dipole-dipole coupling be-

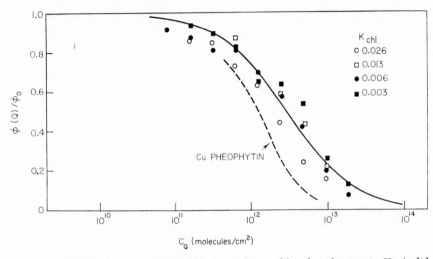

FIG. 6. Fluorescence yield for Chl a monolayers diluted with vitamin K_1 (solid curve and data points) or Cu-pheophytin (dashed curve) as quencher, ϕ (Q), divided by fluorescence yield without quencher, ϕ_0, as a function of molecular concentration of quencher in the film (28). Different data points represent measurements on films diluted to different chlorophyll area fractions (K_{chl}) with oleyl alcohol: $\bigcirc = 0.26$; $\square = 0.013$; $\bullet = 0.006$; $\blacksquare = 0.003$. The solid curve represents a fit of the Stern-Volmer equation to the data.

tween two oscillators with an inverse sixth-power dependence on their separation. The coupling is the greatest when the emission spectrum of the emitter coincides with the absorption spectrum of the quencher. An important parameter in the Förster theory is R_0, the range of the coupling interaction. R_0 is defined as the distance between the excited molecule and the acceptor molecule at which the probability per unit time of transfer is equal to the probability per unit time that the excited molecule will become de-excited by all other processes, including fluorescence. Tweet et al. have fitted the Förster equation, modified for the two-dimensional case, to the experimental data on fluorescence quenching by copper pheophytin and obtained 40 Å for R_0, which is in good agreement with values of 38–41 Å derived from absorption and emission spectra of diluted monolayers (24).

Although the quenching of chlorophyll fluorescence by copper pheophytin can be interpreted by the Förster mechanism, such a mechanism is ruled out for the quenching of chlorophyll fluorescence by vitamin K_1, because the latter lacks a suitably located absorption band. Two alternative quenching mechanisms are: (a) static quenching, in which the fluorescent molecule in the ground state and the quencher molecule form a nonfluorescent complex, and (b) dynamic quenching, which involves de-excitation by collision between the excited donor and the acceptor molecules. Within the chlorophyll concentration range used by Gaines et al., the quenching of fluorescence in the diluted monolayer by vitamin K_1 is largely independent of the chlorophyll concentration (28). The close fit of the Stern-Volmer equation to the quenching data supports the collisional-quenching mechanism, as static quenching would show a yield dependence on the initial concentration of the fluorescent species. At low quencher concentration, the quenching mechanism would require almost every quencher molecule in the monolayer to be a part of the nonfluorescent complex. Furthermore, there is no spectral evidence to suggest the formation of a ground-state complex.

In undiluted monolayers the lifetime of the excited state is very short, so that diffusion and molecular reorientation cannot effectively produce the configuration required for quenching within this lifetime. Quenching can occur only when the quencher molecule is already in close proximity to an excited chlorophyll molecule. In undiluted monolayers, the interaction among chlorophyll molecules becomes more important, and the sharing of energy among the pigment molecules leads to an efficient reduction of fluorescence yield.

F. Chemical Reactions

Colmano (29) first reported the rapid decomposition of a chlorophyll monolayer at a distilled water–air interface under zero surface pressure. The absorption spectrum of the conversion product bears some resemblance to that of pheophytin a. However, there were additional increases in absorption in the 470 and 700 mμ regions. In 30 minutes the blue maximum:blue satellite ratio decreased from 1.44 to 0.70 and the red maximum:blue minimum ratio decreased from 55 to 12. The decomposition was inhibited by a higher film pressure, a higher pH (2.5×10^{-4} M, phosphate buffer, pH 7–8), an inert atmosphere free of carbon dioxide, and darkness.

Rosoff and Aron (13) recently studied the conversion of Chl a to pheophytin *a* in a monolayer in the presence of H^+. Mixed films containing Chl *a* and pheophytin *a* over a wide range of compositions showed that the two components form a two-dimensional solution without mutual interference. The pressure as well as the collapse pressure for the mixed films is a linear function of the composition. Thus, the collapse pressure is sensitive enough to detect 2–3% pheophytin in the mixed film and is said to provide a more convenient technique for following the degree of conversion than the spectrophotometric technique.

Rosoff and Aron studied the change in rate of pheophytinization for a Chl *a* film spread on an aqueous subphase of pH 4 under two different initial pressures. At initial pressures of 6 and 16 dyne/cm, corresponding to molecular areas of 120 and 99 $Å^2$/molecule, respectively, the rate constants are 1.36×10^3 and 1.43×10^2 min^{-1}. These results agree with Colmano's earlier observation (29) that a higher pressure exerts a higher protective action on the decomposition of chlorophyll, probably resulting from a reduced accessibility of the reactive groups to the H^+.

One interesting observation is that the rate of pheophytinization in the monolayer is more than 10 times greater than that occurring in the bulk solution under comparable conditions (30). The difference was attributed to a high H^+ concentration at the monolayer-subphase interface, possibly the result of attraction by the negative end of the chlorophyll dipole oriented toward the subphase.

Evidence was obtained suggesting that oxidation of the cyclopentanone ring accelerates pheophytinization, since the rate constant was three times lower when the experiment was carried out in a nitrogen atmosphere. Divalent metal ions such as Ca^{++} and Mg^{++} present in the subphase also reduce the rate of pheophytinization.

III. Crystalline Chlorophyll

Solid chlorophyll is often described in the literature as amorphous or waxy, implying the noncrystalline nature of the material. Willstätter and Stoll earlier described the powdered chlorophyll precipitate obtained by evaporation from a mixture of ether and petroleum ether as "microcrystalline," but subsequent X-ray examination on such preparations by Hanson (5) yielded no sharp diffraction patterns characteristic of crystallinity. On the other hand, ethyl or methyl chlorophyllides crystallize readily. The lattice parameters of crystalline ethyl chlorophyllide were determined by X-ray measurements (5). These earlier results suggest that the presence of the long-chain phytol probably hampers the chlorophyll molecules from forming an ordered array.

A. Preparation and Absorption Spectra

Interest in crystalline chlorophyll was renewed more than ten years later when Jacobs *et al.* made spectroscopic measurements on colloidal suspensions of ethyl chlorophyllide and found that the red absorption band had shifted about 80 mμ toward longer wavelengths (*22*). The red-band shift was assumed to represent the formation of microcrystals and was used as an experimental criterion in further studies of chlorophyll crystallization.

Addition of water containing 100 ppm Ca^{++} to an acetone solution of pure Chl *a* leads to the formation of crystalline material (*22*). The red

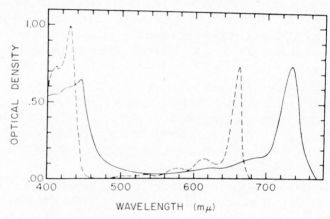

WAVELENGTH (mμ)

Fig. 7. Absorption spectra of Chl *a* crystals suspended in mineral oil (solid curve) and in solution (dashed curve) (*31*).

absorption band of crystalline Chl *a* lies near 735 mμ, as shown in Fig. 7. The Ca^{++} ions apparently act to neutralize the charges on the surface of the colloidal droplets, permitting their coalescence and crystallization. Several other methods were subsequently described for the preparation of crystalline chlorophylls. They include (a) evaporation of an ether solution of chlorophyll over a layer of water; (b) addition of an equal volume of hexane or pentane to an ether solution to induce the formation of microcrystals, followed by the addition of water to form a layer under the organic solvent and evaporation of the latter. The requirement for water in the formation of crystalline chlorophylls *a* and *b* and bacterio-chlorophyll was confirmed by the X-ray diffraction pattern of the product (*31*). Evaporating the solution in anhydrous solvents does not lead to an ordered structure. Although chlorophyllides crystallize much more easily, the presence of water was also found necessary for their crystallization.

On the other hand, pheophytins can be crystallized under strictly anhydrous conditions.

The change in absorption spectrum accompanying crystallization can
be followed spectrophotometrically. For ethyl chlorophyllide, the crystallization is carried out by first cooling a concentrated acetone solution
(O.D. at 660 mμ, approximately 500) to 0°C and diluting it rapidly with
water, also near 0°C. Absorption spectra of ethyl chlorophyllide at different stages of crystallization are shown in Fig. 8.

Two other groups also reported the preparation of crystalline chlorophylls (32, 33). In 1958, Zill et al. (32) prepared crystalline chlorophylls

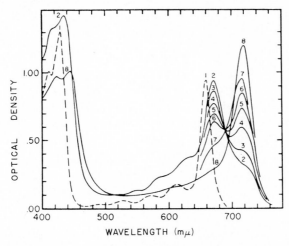

FIG. 8. Absorption spectra of ethyl chlorophyllide microcrystals at different
stages of growth (22). Curve 1: before addition of water to an acetone solution of
the pigment. Curves 2–8: 1, 5, 10, 15, 20, 30, and 50 minutes after the addition of
water.

a and b by a method similar to that described by Jacobs (22). They also
found that precipitation from a highly impure extract in organic solvents
by washing with water leads to a coprecipitation of crystalline Chl a and
amorphous Chl b. In 1959, Stoll and Wiedemann (33) reported the
crystallization of Chl a from 90% acetone and Chl b from 90% acetone or
a chloroform-methanol mixture. They also presented microphotographs of
well defined macrocrystals of both chlorophylls a and b.

Anderson and Calvin (34) described a modified method for the initial
separation of cholorophylls directly from an acetone extract of the plant
material by using a column containing powdered polyethylene of high
molecular weight. The separated chlorophylls were then transferred into
isooctane and chromatographed on a conventional sugar column. The

chlorophyll eluate was allowed to remain in the cold for the microcrystals to form. The larger crystals appear under the microscope as thin rectangular plates measuring 0.05 mm $\times$ 0.1 mm. The X-ray diffraction pattern of the larger crystals of chlorophyll a indicates a sheetlike structure with spacings at 10.6, 14, and 21 Å and relatively poor order within the sheets at spacing 4.23 and 7.4 Å (35). It was also shown by X-ray diffraction that the crystalline order can readily be broken down by drying under vacuum or by mechanical grinding. The absorption spectrum of Chl a crystals suspended in isooctane has maxima at 740 and 660 mμ. Absorption-spectrum measurements also show that heating breaks down the crystals, and that a partial recovery of the original spectrum takes place upon cooling, indicating a reversible aggregation (35).

B. Electric and Magnetic Properties of Crystalline Chlorophyll

In the transformation of light energy into chemical energy in photosynthesis, the primary act is light absorption and the formation of an oxidant and a reductant. In 1949, Katz first suggested that photoconduction and energy migration in the two-dimensional crystals of chlorophyll may play a central role in the primary act of photosynthesis (36). Later, Bradley and Calvin proposed that formation of oxidizing and reducing centers in the primary photosynthetic act may involve an initial creation of an electron and hole and their subsequent migration to different points, and that the chloroplast lamellar structure may provide separation for the oxidizing and reducing intermediates (37). As a result of these suggestions and implications, work on the photoconductivity of chlorophyll as well as chloroplast films has been undertaken by several investigators. Another method which gives direct information concerning the electronic configuration as well as the change in electronic configuration of chlorophyll is electron spin resonance (ESR). ESR depends on the fact that spinning electrons are magnetic, and when placed in a magnetic field, will interact with that field in a characteristic way. The applications of these techniques to chlorophyll in the crystalline state are described below.

1. Photoconductivity

The first evidence of photoconductivity in solid chlorophyll was reported by Nelson (38). Films of Chl a and methyl chlorophyllide a deposited on glass slides by evaporating solvent produced a steady-state current of the order of 10^{-12} amp under strong illumination and with 5 volts across the cell. The dark decay of the conductance was first order, with a half-life of 1–2 minutes, and the steady-state conductance was proportional to the first power of illumination intensity. The action spectrum of chlorophyll photoconductivity was similar to the absorption spectrum of the pigment. Methyl chlorophyllide *a* differed from Chl *a* in having two time constants for the rise and decay: one less than a second, and the other the order of several minutes.

Nelson's results were subsequently confirmed and extended by other workers (*39, 40*). Arnold and Maclay measured photoconductivity of dried films of both chloroplasts and chlorophyll (*39*). They also measured the photoconductivity of β-carotene, for which only blue light was effective. However, in a mixed film of β-carotene and chlorophyll at a ratio of 1:15, photoconductivity can also be sensitized by red light absorbed by chlorophyll. Since chlorophyll cannot transfer its singlet excitation energy to carotene, this sensitization was attributed to a transfer of an electron from chlorophyll to the carotene. Such a transfer was further supported by the junction potentials measured from overlapping chlorophyll and carotene layers under illumination.

Terenin *et al.* found similar results in studying the photovoltaic effect in microcrystalline powders of chlorophyll (*a + b*) and methyl chlorophyllide (*a + b*) by the condenser method (*40*). Again, the action spectrum of the photovoltaic effect resembled the absorption of the pigment. They also determined that the charge carriers were predominantly positive holes. A noticeable shift of the spectral sensitivity was observed for methyl chlorophyllide *a* during the formation of larger crystals induced by water-vapor treatment. This is consistent with the finding by Jacobs *et al.* (cf. Fig. 8) concerning the spectral shift for ethyl chlorophyllide *a* during crystallization (*22*).

Rosenberg and Camiscoli measured photoconductivity of crystalline chlorophyll *a* and *b* compressed between electrodes of Monel metal and conducting glass (*41*). Measurements were made at different temperatures and an activation energy of 0.36 ev was obtained for Chl *b* (similar measurements were not made for Chl *a* because of interference by a large dark current). This relatively large activation energy appears to rule out the involvement of photoconduction in photosynthesis, since the primary photophysical process in photosynthesis is known to be temperature independent.

More recently McCree (*42*) measured photoconductivity of a stack of chlorophyll monolayers deposited by the method of Blodgett (*43*) on a quartz plate with conducting grids. The magnitude of the photocurrent (10^{-12} to 10^{-13} amp) measured with such chlorophyll multilayers was essentially the same as that found for chlorophyll film prepared by evaporation from a solution. However, different grid materials produced quite different effects. With gold electrodes the photocurrent responded instantly to light, whereas with aluminum, silver, and colloidal graphite electrodes the photocurrent dropped from its initial value and reached an equilibrium slowly. The photocurrent increased with increasing number of monolayers and the applied voltage, and quantum-yield measurement showed that about 10^9 quanta were needed to produce one electron. This value appears to be too low for photoconduction to be of significance in photosynthesis.

In connection with the photoconductivity of chlorophyll, it is worthwhile mentioning briefly the novel experiments of Kearns *et al.*, who used thin layers of phthalocyanine and an electron acceptor (*o*-chloranil) to construct a model "photosynthetic cell" (*44*). In the presence of *o*-chloranil, both the dark conductivity and photoconductivity of phthalocyanine increased by several orders of magnitude. Electrostatic measurements showed that electrons were transferred from phthalocyanine to chloranil, and that the transfer was increased by illumination, leaving positive holes behind in the phthalocyanine film. Thus, by incorporating effective electron trapping centers into a pigment aggregate, the charge-carrier production becomes very efficient. For instance, the quantum efficiency for charge-carrier production with phthalocyanine alone is less than 10%, whereas it is almost 100% in the presence of *o*-chloranil. Another effect of introducing *o*-chloranil into phthalocyanine was to increase the lifetimes of the charge carriers.

The chlorophyll-carotene sandwich cell described at the beginning part of this section is similar to the phthalocyanine-chloranil system. In the former, however, both components are photoconductors, and no noticeable increase in photoconductivity results when the two components are placed together.

2. ELECTRON SPIN RESONANCE

Light-induced ESR signals have been observed in complete photosynthetic systems such as algae and in cell fragments such as chloroplasts or bacterial chromatophores (*45*). Photosynthetic materials from green plants exhibit two types of light-induced ESR signals. One is centered at $g = 2.0025$, is 7–9 gauss wide, decays rapidly (R signal), and shows no hyperfine structure; another is centered at $g = 2.0046$, is about 20 gauss wide, decays slowly (S signal), and shows partially resolved hyperfine structure. The R spin signal with narrow line width is produced maximally by light of wavelength absorbed by Chl *a* (in green plants) or BChl (in photosynthetic bacteria). Thus, some form or state of the respective chlorophylls must be involved in the formation of this radical species. The temperature independence of this signal suggests that the species could be active in the primary quantum conversion act. The assignment of chlorophyll to the R signal is further supported by the fact that the R signal is absent in etiolated leaves or chlorophyll-less algae mutants (45). One apparent exception was reported recently by Weaver and Bishop (46), who found that the *Scenedesmus* mutant which contains chlorophyll and other pigments but is incapable of reducing CO_2 also showed no R signal. They attributed the absence of the R signal in the mutant to a lack of P700, the terminal energy acceptor in photosynthesis. Beinert and Kok (47) have suggested the oxidized form of P700 to be responsible for the R signal and have recently shown a quantitative correlation between the spin concentration derived by double integration of the spin signal and the P700 concentration measured by difference spectroscopy. The origin of the S signal is less well defined.

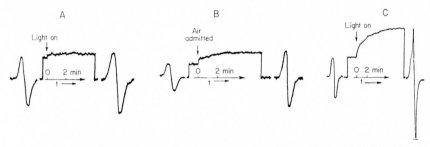

FIG. 9.   Electron spin resonance signals in crystalline Chl $(a + b)$ (49). (A) *In vacuo*. Dark signal and effect of illumination. (B) Effect of air on the dark signal. (C) Effect of illumination on the dark signal in air or in the presence of moisture.

Brody *et al.* first reported on light-induced ESR signals for Chl *a* crystals and in Chl *a* in pyridine (48). Chl *a* crystals show a weak dark signal, but a more intense signal appears upon illumination.

Holmogorov and Terenin also studied the light-induced ESR signals in crystalline Chl [a (80%) + b] (49). They reported a fairly intense dark signal for Chl $(a + b)$ *in vacuo* with $g = 2.0035$ and a line width of 11 gauss and a slightly more intense signal with the same g value and line width upon illumination (Fig. 9A). The magnitude of the dark signal was only slightly increased upon admission of air at a pressure of 10^{-4} mm Hg (Fig. 9B). On the other hand, a new, narrower signal ($g = 2.0030$, line width 7 gauss) appeared upon illumination and in the presence of air (Fig. 9C). It was subsequently found that moisture in the air was actually responsible for this new signal, since pure oxygen had only a slight effect. Crystalline ethyl chlorophyllide $(a + b)$ behaves similarly to Chl $(a + b)$, but crystalline pheophytin $(a + b)$ shows a dark signal only.

Although pure oxygen has little effect on the light-induced ESR signal of crystalline Chl $(a + b)$, it produces a strong enhancement in the light-induced narrow signal of Chl $(a + b)$ in the presence of water vapor. For instance, at a water vapor pressure of 18 mm Hg, oxygen at a pressure of 180 mm Hg increased the magnitude of the light-induced signal more than tenfold.

The dark ESR signal of chlorophyll is thought to be directly associated with its dark conductivity; the light-induced ESR signal is associated with the transfer of excitation in the crystal. Water may stabilize the light-induced narrow ESR band by accepting the electrons released from the pigment by light (49). Holmogorov and Terenin showed that another electron acceptor, p-benzoquinone (at 2×10^{-2} mm Hg) is also effective in stabilizing the light-induced ESR signal. Water molecules are thought capable of conferring to the molecular planes in crystalline chlorophylls the rigidity and orientation necessary for an undisturbed electron transport.

The enhancement of the light-induced ESR signal by water was confirmed by Anderson and Calvin for a dried film of pigments from an extract of *Chlorella* (50). However, the light-induced signal in the impure chlorophyll film was much narrower (3 gauss) than that produced in crystals. With pure Chl a crystals Anderson and Calvin reported that both the dark- and light-induced signals were at $g \cong 2$ and of the same line width at 15 gauss. The steady-state intensity of the light-induced signal was about four times that of the dark signal. The dependence of the ESR signal on water was not studied for pure Chl a crystals (50).

References

(1) J. B. Thomas, K. Minnaert, and P. F. Elbers, *Acta Botan. Neerl.* **5**, 315 (1956).
(2) A. Hughes, *Proc. Roy. Soc.* **A155**, 710 (1936).
(3) W. Sjoerdsma, *Nature* **138**, 405 (1936).
(4) A. E. Alexander, *J. Chem. Soc.* p. 1813 (1937).
(5) E. A. Hanson, *Rec. Trav. Botan. Neerl.* **36**, 183 (1939).
(6) I. Langmuir and V. J. Schaefer, *J. Am. Chem. Soc.* **59**, 2075 (1937).
(7) F. A. Rodrigo, *Biochim. Biophys. Acta* **10**, 342 (1953); Ph.D. Thesis, University of Utrecht (1955).
(8) E. E. Jacobs, A. S. Holt, and E. Rabinowitch, *J. Chem. Phys.* **22**, 142 (1954).
(9) H. J. Trurnit and G. Colmano, *Biochim. Biophys. Acta* **31**, 434 (1959).
(10) W. E. Ditmars, Jr., Q. Van Winkle, and H. F. Blank, Jr., *Ohio State Univ., Res. Found. Rept.* 1213-3 (1963).
(11) W. D. Bellamy, G. L. Gaines, Jr., and A. G. Tweet, *J. Chem. Phys.* **39**, 2528 (1963).
(12) F. F. Litvin and B. A. Gulyaev, *Dokl. Akad. Nauk SSSR* **158**, 460 (1964).
(13) M. Rosoff and C. Aron, *J. Phys. Chem.* **69**, 21 (1965).
(14) K. J. McCree, *Biochim. Biophys. Acta* **102**, 90 (1965).
(15) A. G. Tweet, *J. Chem. Phys.* **34**, 1412 (1963).
(16) H. J. Trurnit and W. E. Lauer, *Rev. Sci. Instru.* **30**, 975 (1959).

(17) G. L. Gaines, Jr., *Gen. Elec. Res. Lab. Rept.* 63-RL-3206c (1963).
(18) A. G. Tweet, G. L. Gaines, Jr., and W. D. Bellamy, *J. Chem. Phys.* **41**, 1008 (1964).
(19) G. Colmano, *Nature* **193**, 1287 (1962).
(20) G. L. Gaines, Jr., W. D. Bellamy, and A. G. Tweet, *J. Chem. Phys.* **41**, 538 (1964).
(21) A. G. Tweet, G. L. Gaines, Jr., and W. D. Bellamy, *J. Chem. Phys.* **40**, 2596 (1964).
(21a) G. L. Gaines, Jr., W. D. Bellamy, and A. G. Tweet, *J. Chem. Phys.* **41**, 2572 (1964).
(22) E. E. Jacobs, A. S. Holt, R. Kromhout, and E. Rabinowitch, *Arch. Biochem. Biophys.* **72**, 495 (1957).
(23) W. Sperling and B. Ke, *Photochem. Photobiol.* (in preparation) (1966).
(24) A. G. Tweet, W. D. Bellamy, and G. L. Gaines, Jr., *J. Chem. Phys.* **41**, 2068 (1964).
(25) R. Livingston and C.-L. Ke, *J. Am. Chem. Soc.* **72**, 909 (1950).
(26) G. Tollin and G. Green, *Biochim. Biophys. Acta* **60**, 624 (1962).
(27) B. Ke, L. P. Vernon, and E. R. Shaw, *Biochemistry* **4**, 137 (1965).
(28) G. L. Gaines, Jr., A. G. Tweet, and W. D. Bellamy, *J. Chem. Phys.* **42**, 2193 (1965).
(29) G. Colmano, *Biochim. Biophys. Acta* **47**, 454 (1961).
(30) S. H. Schanderl, C. O. Chichester, and G. L. Marsh, *J. Org. Chem.* **27**, 3771 (1962).
(31) E. E. Jacobs, A. E. Vatter, and A. S. Holt, *Arch. Biochem. Biophys.* **53**, 228 (1954).
(32) L. P. Zill, G. Colmano, and H. J. Trurnit, *Science* **128**, 478 (1958).
(33) A. Stoll and E. Wiedemann, *Helv. Chim. Acta* **42**, 676 (1959).
(34) A. F. H. Anderson and M. Calvin, *Nature* **194**, 285 (1962).
(35) A. F. H. Anderson, Ph.D. Thesis, University of California (1963).
(36) E. Katz, in "Photosynthesis in Plants" (J. Franck and W. E. Loomis, eds.), p. 291. Iowa State Coll. Press, Ames, Iowa, 1949.
(37) D. F. Bradley and M. Calvin, *Proc. Natl. Acad. Sci. U.S.* **41**, 563 (1955).
(38) R. C. Nelson, *J. Chem. Phys.* **27**, 864 (1957).
(39) W. Arnold and H. K. Maclay, *Brookhaven Symp. Biol.* **11**, 1 (1958).
(40) A. N. Terenin, A. K. Putzeiko, and I. Akimov, *Discussions Faraday Soc.* **27**, 83 (1959).
(41) B. Rosenberg and J. F. Camiscoli, *J. Chem. Phys.* **35**, 982 (1961).
(42) K. J. McCree, *Biochim. Biophys. Acta* **102**, 90 and 96 (1965).
(43) K. B. Blodgett, *J. Am. Chem. Soc.* **57**, 1007 (1935).
(44) D. R. Kearns, G. Tollin, and M. Calvin, *J. Chem. Phys.* **32**, 1020 (1960).
(45) M. S. Blois and E. C. Weaver, in "Photophysiology" (A. C. Giese, ed.), Vol. 1, p. 35. Academic Press, New York, 1964.
(46) E. C. Weaver and N. I. Bishop, *Science* **140**, 1095 (1963).
(47) H. Beinert and B. Kok, *Biochim. Biophys. Acta* **88**, 278 (1964).
(48) S. S. Brody, G. Newell, and T. Castner, *J. Phys. Chem.* **64**, 554 (1960).
(49) V. E. Holmogorov and A. N. Terenin, *Dokl. Akad. Nauk SSSR* **137**, 199 (1961); *Naturwissenschaften* **48**, 158 (1961); **50**, 299 (1963).
(50) A. F. H. Anderson and M. Calvin, *Nature* **199**, 241 (1963).

Section III
State of the Chlorophylls in the Cell

9

Chloroplast Structure

RODERIC B. PARK

Department of Botany and Lawrence Radiation Laboratory
University of California, Berkeley, California

I. Introduction

This chapter is concerned with a general description of the chloroplast and in particular the localization of chlorophyll within it. We shall briefly consider evidence which shows that the entire photosynthetic process from light absorption to CO_2 fixation is localized within the chloroplast. Then we shall review aspects of chloroplast structure as they have been revealed by light microscopy, electron microscopy, X-ray diffraction, and biochemical techniques. After assigning functions to major structures revealed by these methods, we shall concentrate on the morphology and biochemistry of the structures which contain chlorophyll—namely the internal membranes of chloroplasts. In bacteria and blue-green algae, bacteriochlorophyll and the plant chlorophylls, respectively, are present in the cell membrane itself and in membrane structure elaborated from the cell membrane. In other algae and in all other green plant cells, chlorophyll exists again in membranes, but in membranes contained within a specific cell organelle—the chloroplast. The chloroplast is easily resolved in living cells with the light microscope. It ranges in size from a few microns to tens of microns. Chloroplast shape is highly variable, especially in the algae, where it may exist in cup-shaped, spiral, or star-shaped forms. In bryophytes and higher plants the chloroplast is more generally saucer shaped, 5–10 μ in diameter, and several microns thick.

II. The Chloroplast as a Complete Photosynthetic System

There is abundant evidence that the materials within a chloroplast are fully competent to perform the entire photosynthetic process. The process begins with absorption of light by chlorophyll or an accessory pigment. Transfer of this energy to an energy trap where quantum conversion then takes place, and the subsequent use of chemical energy formed at the trap for the process of CO_2 fixation into carbohydrate. Evidence that all these reactions occur within the chloroplast comes both from *in vivo* and *in vitro* experiments which are reviewed below.

Among the first experiments directly demonstrating the association of photosynthetic functions with chloroplasts are those of Engelmann (*1*). Engelmann observed a number of fresh-water algae under the light microscope in the presence of bacteria which were chemotactic along a positive dissolved oxygen gradient. Some of his most elegant experiments were performed with the green alga *Spirogyra*. A figure from Engelmann's 1894 paper is reproduced in Fig. 1. When the entire filament of *Spirogyra* was illuminated, the bacteria congregated all along the spiral chloroplast. If only portions of the filament were illuminated, bacteria congregated only where the small beam of light impinged upon the chloroplast. This result had two implications. The first and most obvious was that a chloroplast must be illuminated in order to produce oxygen. The second implication was that oxygen was produced very close to the site of light absorption by chlorophyll. Prior to this time it might have been assumed that oxygen evolution was a reaction quite remote from the site of light absorption.

A second series of experiments relevant to the demonstration of complete photosynthetic capacity by chloroplasts were those of Robin Hill (*2, 3*) in 1937 and 1939. Hill showed that chloroplasts isolated from leaves in isotonic buffer and centrifugally concentrated evolved oxygen in the presence of light when supplied with an added electron acceptor such as ferric ion. This reaction, called the Hill reaction, is shown below.

$$2 H_2O + 4 Fe^{3+} \xrightarrow[\text{chloroplasts}]{\text{light}} O_2 + 4 H^+ + 4 Fe^{++}$$

Photosynthesis differs from the Hill reaction in that carbon dioxide rather than an artificial oxidant such as ferric ion is the electron acceptor. Hill was not able to demonstrate that CO_2 could serve as an electron acceptor in isolated chloroplasts, and such an experiment awaited the development of more elaborate biochemical techniques. Hill showed, however, that the light reactions of photosynthesis leading to the oxidation of water and the reduction of some added acceptor occurred within isolated chloroplasts.

Arnon *et al.* (*4*) in 1954 were the first workers to demonstrate conclusively that carbon dioxide could be reduced by isolated chloroplasts into intermediates of the carbon cycle of photosynthesis. In their work they demonstrated qualitatively, though not quantitively, that chloro-

FIG. 1. Localization of photosynthetic O_2 production in *Spirogyra*. After Engelmann (*1*).

plasts contained the entire photosynthetic apparatus leading from light absorption to fixation of carbon dioxide. However, one obvious deficiency existed, and continues to exist, in the acceptance of the chloroplast as the sole site of photosynthesis *in vivo*. This deficiency is that rates of photosynthesis in isolated chloroplasts on a unit chlorophyll basis seldom exceed 5% of the *in vivo* rate (*5*). *In vivo* rates of photosynthesis in both higher plants and algae can be as great as 200 μmoles CO_2 fixed per hour per milligram chlorophyll (*6*). The rate of CO_2 fixation in isolated chloroplasts seldom exceeds 10 µmoles/hr per milligram chlorophyll. This discrepancy may be in large part due to the damage which occurs to chloroplasts during the isolation process. Recent work by Spencer and Unt (7) shows that chloroplasts which are carefully isolated so as to retain their external membrane systems and stroma protein possess considerably higher CO_2 fixation capacities than chloroplasts isolated by ordinary procedures. Work such as that by Spencer and Wildman (8) and Spencer and Unt (7) may eventually lead to isolation procedures which yield chloroplasts capable of *in vivo* photosynthetic rates. Only then will we have conclusive proof that the chloroplast is fully competent to account for all aspects of photosynthesis as observed in intact systems.

Once the chloroplast was qualitatively accepted as a totally competent photosynthetic organelle, interest in distribution of photosynthetic function among chloroplast substructures increased. The remainder of this chapter is concerned with these studies of structure and distribution of function in mature chloroplasts. Aspects of chloroplast development are considered by Bogorad in Chapter 15.

III. Structure of Chloroplasts As Revealed by Light Microscopy

Chloroplasts are easily viewed in the *in vivo* state by light microscopy. As mentioned earlier, these organelles may assume bizarre shapes, especially in some algae. In many plants, however, the chloroplast appears as a green saucer-shaped body 5–10 µ in diameter. In green algae and some bryophytes the chloroplast contains an organized body called the pyrenoid, which is often surrounded by starch plates or lipid reserves. Chlorophyll as observed by light microscopy in the chloroplasts of algae and bryophytes appears uniformly distributed. In higher plants, however, the chloroplast from top view is seen to consist of a green field filled with small (0.2–1 µ) totally absorbing bodies called grana. The green field in which the grana lie is referred to as the stroma region of the chloroplast. Side views of the chloroplast show that the grana regions are interconnected by material indistinguishable from the grana themselves. These general observations were summarized by Heitz (9) in 1936. Two of his photographs illustrating these aspects of chloroplast morphology are reproduced in Fig. 2. Higher plant chloroplasts may be viewed by fluorescence microscopy using blue actinic light and observing the red fluorescence of chlorophyll. The chlorophyll fluorescence is seen to reside primarily in the grana stacks. Spencer and Wildman (8) have interpreted this to mean that the chlorophyll is localized in the grana regions of the chloroplast. However, we know from electron microscopy that not all membranes within the chloroplast are in the grana stacks, but that many run between grana stacks. Do these intergrana membranes contain chlorophyll? It is doubtful that fluorescence observations of whole chloroplasts will give us an answer since the electron micrographs

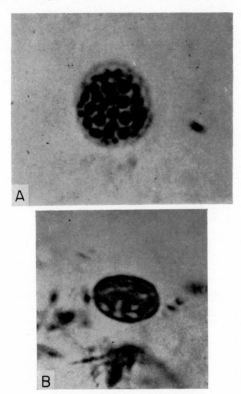

FIG. 2. Top (A) and side (B) views of a chloroplast as seen by light microscopy. The dark regions within the chloroplasts are grana. From Heitz (9).

show that the membrane concentration in the grana stack is much larger than the membrane concentration in the stroma. A similar ratio of fluorescence intensities might obscure fluorescence from the intergrana areas. Also, as mentioned in the next section, certain higher plant cells contain only large nongranal membranes and no grana membranes, though they appear to be photosynthetic. Obviously in these systems chlorophyll is distributed in the large membrane system of the chloroplast.

Some recent experiments by Lintilhac and Park (10) support the

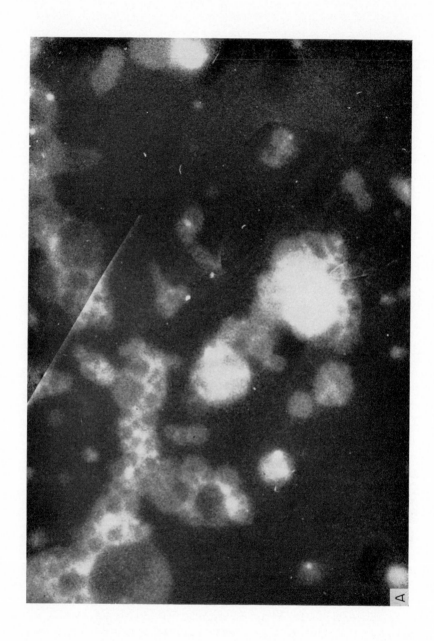

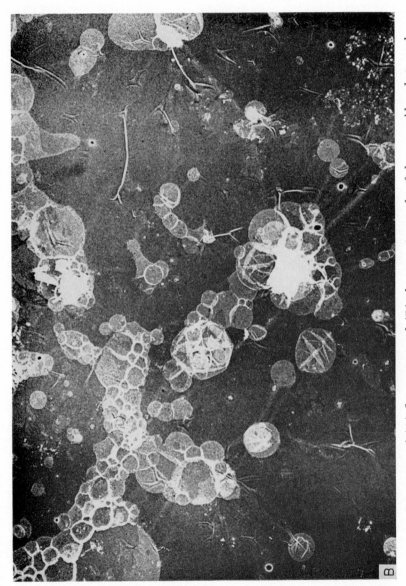

FIG. 3. Comparison of (A) fluorescence and (B) electron micrographs of the same chloroplast membrane specimen on Formvar film (10). Magnification × 2600.

arguments that chlorophyll is uniformly distributed throughout the internal membrane system. Chloroplast internal membranes placed on an electron microscope grid were observed by both fluorescence and electron microscopy (see Fig. 3). All the membranes, both small and large thylakoids (11), are seen to contain chlorophyll. Direct obervations of this sort are contrary to the conclusions of Spencer and Wildman.

The light microscope has been used to study both dichroism and birefringence in chloroplasts. Since dichroism is considered in Chapter 11 by Butler, we are only concerned with birefringence here. Menke (12) and Frey-Wyssling (13) both studied chloroplast birefringence in media of

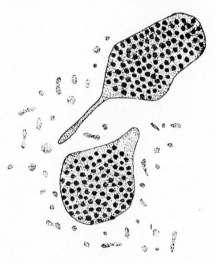

FIG. 4. Drawing of pseudopodia formation by chloroplasts. From Heitz (9).

varying refractive index. In this way they could differentiate between intrinsic and form birefringence. Form birefringence was interpreted as resulting from a layered system (12, 13) within the chloroplast. Frey-Wyssling proposed a model consisting of alternate layers of protein and lipid to account for the form birefringence. This model of layered structures was to a large extent realized with the application of electron microscopy to chloroplast structure.

Chloroplast structure and function are closely allied. For this reason it is important that the biochemist be aware of the morphological status of the chloroplasts with which he works. Initial studies by Kahn and von Wettstein (14), Spencer and Wildman (8), and Spencer and Unt (7) show that chloroplasts isolated in 0.4 M sucrose buffered with Tris or phosphate tend to be of two types. The first type retains its outer membrane and refractal jacket of stroma protein around the grana membrane and is called a class I chloroplast. The second type of chloroplast becomes ruptured during the isolation procedure and loses its outer membrane and stroma material. The latter type is referred to as a class II chloroplast. The biochemical assays by Spencer and Unt show that class I chloroplasts retain to the greatest extent the properties of *in vivo* chloroplasts—that is, comparatively high rates of CO_2 fixation (10 μmoles/hr per milligram chlorophyll), low rates of Hill reaction due to coupled phosphorylation, and ability to form pseudopodia when resuspended in appropriate media. Heitz had shown that chloroplast pseudopodia formation was a widespread and normal occurrence in plant cells. A drawing of this phenomenon taken from Heitz's paper appears in Fig. 4, in which the pseudopodia are shown extending into the cytoplasm. These observations of Heitz have been extended by Spencer and Wildman (8) and by Wildman *et al.* (15). Interestingly enough, the relatively high rates of CO_2 fixation in class I chloroplasts are attainable with no added cofactors. Thus it would seem that the integrity of the outer membrane has retained these cofactors in the chloroplasts, a situation that does not occur in other biochemical preparations. In conclusion, the light microscope is and continues to be a very valuable tool for studying the chloroplast in its *in vivo* environment. Light microscopy of chloroplast preparations is also a useful tool for the biochemist who wishes to better understand the photosynthetic capacities of his material.

IV. Structure of Chloroplasts As Revealed by Electron Microscopy

The electron microscope was first used to study chloroplast structure in 1940 (16). This early micrograph from Ruska's Laboratory showed that an isolated, dried chloroplast appeared to contain a number of internal membranes. The development of shadowing techniques by Williams and Wyckoff (17) in 1945 opened the way for the early ultrastructural investigations of chloroplast morophology by Granick and Porter (18), Frey-Wyssling and Mühlethaler (19), and Steinmann (20). These early studies were made on shadowed preparations of isolated chloroplasts. The shadow technique showed that the lamellar system of the chloroplast was a series of membranes that were piled upon one another much like a stack of coins. Steinmann (20) in 1952 showed the existence of these membranes in the first thin sections of chloroplasts, and confirmed the conclusions obtained from observation of shadowed preparations. Thus, the predictions from light microscopy (12, 13) were

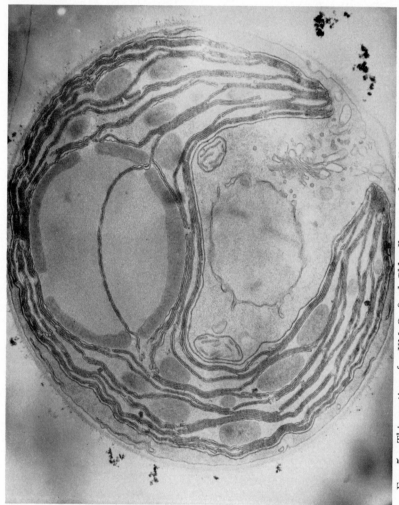

FIG. 5. Thin section of a KMnO$_4$-fixed *Chlorella pyrenoidosa* cell. The cup-shaped chloroplast contains a large pyrenoid. Magnification × 28,000.

to a large extent confirmed. The following fifteen years produced an enormous number of electron micrographs of chloroplast material, which has been subjected to various fixation and preparative procedures. In general, the results from sectioning are summarized in three micrographs,

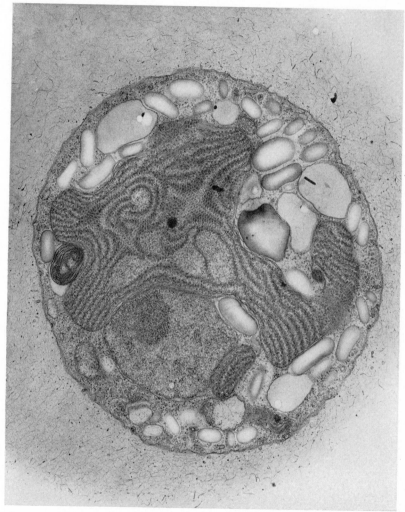

FIG. 6.   A thin section of glutaraldehyde osmium-fixed *Porphyridium cruentum*. Courtesy of Drs. E. Gantt and S. F. Conti. Magnification × 19,000.

shown in Figs. 5–7, in which a green alga, a red alga, and a higher plant chloroplast are compared. Each chloroplast is surrounded by a double membrane system. A high magnification picture of a *Chlorella* double

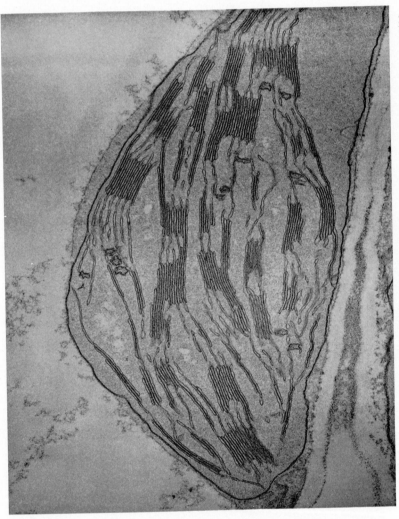

FIG. 7. Thin section of $KMnO_4$-fixed *Spinacea oleracea* chloroplast. Magnification × 32,500.

membrane shows that whereas the external membrane has the same morphological characteristics from $KMnO_4$ staining as the plasma membrane, the internal membrane of the double membrane appears identical in staining characteristics to the internal membranes of the chloroplast. In all the chloroplasts the internal membrane system is embedded in a matrix called the stroma. The internal membranes are actually closed, flattened sacs which have been termed thylakoids by Menke (11) and compartments by Weier et al. (21). The thylakoids of algae are much larger, in general, than those of higher plants. The thylakoids of red algae are separated by 325 Å particles which may contain the accessory (phycoerythrin) pigment system (22). The thylakoids of green algae are ap-

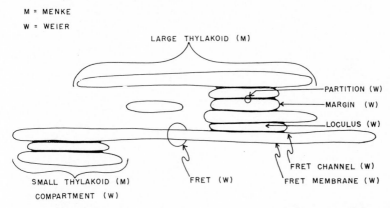

FIG. 8. Nomenclature of chloroplast internal membrane systems used by Weier (21) and Menke (11).

pressed in groups of two, three, or four, giving the structures seen in the *Chlorella* cross section. In any higher plant chloroplasts, on the other hand, small thylakoids are stacked to make grana structures, whereas the larger connecting membrane systems, termed the large thylakoids by Menke (11) and frets by Weier et al. (21), are much less frequent than they are in the algae. A summary of nomenclature given to these membranes by Weier and Menke is given in Fig. 8.

The degree of natural variation within a single plant is most graphically shown by the morphology of plastids in the neighboring mesophyll and bundle sheath cells of many monocots. Such a micrograph is shown in Fig. 9. The mesophyll cells are seen to contain the same kind of chloroplast as shown in Fig. 7 for a typical higher plant chloroplast. The adjacent bundle sheath cells, on the other hand, contain plastids that are indistinguishable from those present in certain algae. Thus, a higher plant appears to have a genetic capacity to produce a considerable variation

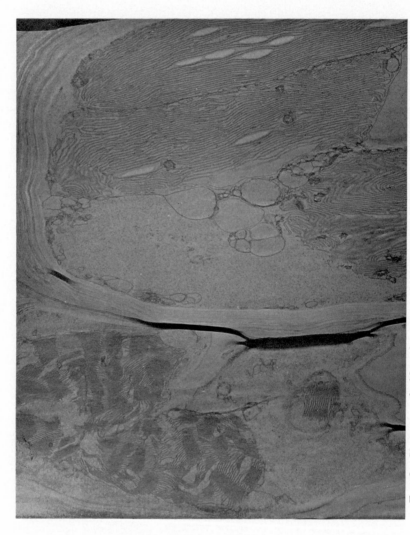

FIG. 9. Variation of chloroplast internal membrane structure in two adjacent cells (bundle sheath cell and mesophyll cell) in sugar cane leaf. Courtesy of Professor W. M. Laetsch. Magnification × 21,000.

in the detailed membrane system inside a chloroplast. Thomson and Weier (23) have shown that the nutritional status of bean plants can markedly affect the membrane arrangement within the chloroplast. Under conditions of low phosphate, the plastids of bean plants tend to assume the morphology of the bundle sheath cell plastids of monocots.

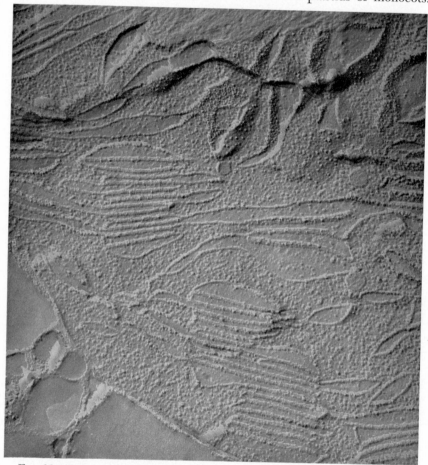

FIG. 10. Freeze-etch preparation of an isolated spinach chloroplast. Magnification × 51,000.

The gross morphology of a chloroplast may also be seen by the freeze-etch technique developed in Frey-Wyssling's laboratory (24). The freeze-etching of chloroplasts possibly gives the most accurate description of chloroplast ultrastructure yet obtained by the electron microscope, since the number of artifacts under conditions of rapid freezing is minimal (25). A micrograph of an isolated spinach chloroplast is presented in Fig. 10. It is seen in the isolated chloroplast that the thylakoids are somewhat swollen and that the protein and ribosomes in the stroma are quite evident. On the other hand, in *in vivo* material, the membranes and stroma material are so tightly appressed that the individual character of the stroma proteins is not so evident. This swelling which occurs during isolation may account for the loss of some of the photosynthetic activity of isolated chloroplasts. The swollen and shrunken states seen by freeze-etching may also explain volume and scattering changes observed in isolated chloroplasts as a consequence of osmotic changes, pH changes, and illumination. Tangential views of the internal membranes, evident in the upper portions of Fig. 10, show substructure within the chlorophyll-containing membrane. Evidence for the relationship of chlorophyll to these substructures is presented in the next section.

In summary, then, electron microscopy has shown that the chloroplast consists of two phases, a thylakoid phase which we shall see contains the chlorophyll, and a stroma phase which is the site of carbon cycle enzymes and other synthetic capacities of the chloroplast. Assignments of functions to these structures is considered next.

V. Distribution of Function within Chloroplasts

The work of Trebst *et al.* (26) and Park and Pon (27) demonstrates that the light reactions of photosynthesis and the associated electron transport reactions leading from the oxidation of water to the reduction of ferredoxin occur within the internal membrane system of chloroplasts, while the CO_2 fixation reactions of the carbon cycle occur within the stroma regions of the chloroplast. It should be added at this point that enzymatic systems other than CO_2 fixation systems are present in the stroma, the most notable being specific chloroplast ribosomes (28, 29) and an apparent ability to synthesize protein (30). Since these interesting capacities are not directly related to photosynthesis and chlorophyll, they are not considered here. If chloroplasts isolated in isotonic media are subjected to a hypotonic environment, the plastids are seen to swell and the stroma material leaks from the plastid. Centrifugation of this preparation yields a green precipitate and a soluble protein supernatant, and it is found that the protein is approximately equally distributed between the two phases (27). The green precipitate consists of membranes which are about 10% chlorophyll by weight, and the supernatant consists of the soluble stroma material. That both are needed for the photosynthetic process is shown in Fig. 11, in which two-dimensional chromatograms of the CO_2 fixation products of membranes alone, and the two mixed together, are shown. There is about a fiftyfold enhancement of

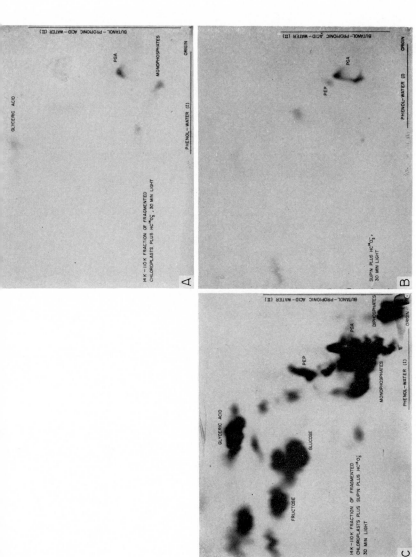

FIG. 11. Chromatograms of the products of $^{14}CO_2$ fixation by (A) chloroplast internal membranes (B) chloroplast stroma, and (C) internal membranes and stroma. From Park and Pon (27).

CO_2 fixation capacity when the two systems are mixed together. These results are diagrammatically presented in Fig. 12, in which the distribution of photosynthetic function between the membrane phase of the chloroplast and the stroma portion of the chloroplast is presented. Since it is the internal membrane system of the chloroplast which contains chlorophyll and performs the quantum conversion act of photosynthesis,

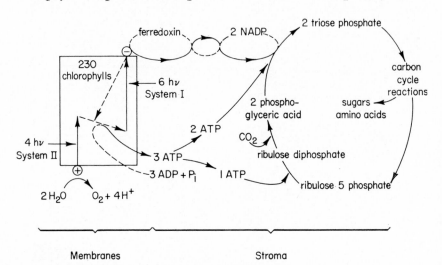

Membranes   Stroma

FIG. 12.   A schematic presentation of the distribution of photosynthetic function between chloroplast internal membranes and stroma.

it is to the detailed description and analysis of these internal membranes that the remainder of this chapter will be devoted.

Internal Membrane Systems of Chloroplasts

As already mentioned, chloroplast internal membranes are approximately 10% chlorophyll by weight. When illuminated, these membranes bring about the oxidation of water to produce oxygen gas and electron transport with accompanying phosphorylation to the level of a reducing agent which will reduce the soluble cofactor, spinach ferredoxin. The kinds of information available about these membranes are their chemical composition and the properties of membrane proteins, their enzymology and their morphology as seen in the light and electron microscopes and by X-ray diffraction. One of the most interesting and frustrating aspects of studying these membranes is the process of correlating these three factors, the composition, the enzymology, and the morphology, into a unified and consistent picture. We shall first discuss morphological features of these membranes, then the chemical composition and enzymology and finally we shall attempt to correlate these various kinds of information.

1. MORPHOLOGY

Our knowledge of chloroplast internal membrane substructure comes from several sorts of experiments. These involve electron and light microscopy and the process of X-ray diffraction. The electron microscope techniques are those of staining and section preparation, heavy metal shadowing, negative staining, and freeze-etching. These will be considered separately.

Typical examples of sectioned internal membrane systems of chloroplasts were shown in Figs. 5–7. If electron microscopy following potassium permanganate staining is done at high magnifications, a 75–90 Å periodicity is seen along the membranes (31–33). This periodicity has been described in both higher and lower plants. Such a periodicity is also beautifully evident in the preparations of Kahn and von Wettstein (14), although these authors do not comment upon it. These experiments, then, would tend to confirm models which have been advanced by Sjöstrand (34) and others, which suggest that membranes are built from micellar subunits. Such periodicity, however, could be an artifact caused by lipid micelle formation during fixation.

Heavy-metal shadowing of chloroplast internal membranes was first shown by Steinmann (20) to reveal a substructure on the membrane surface. This substructure consisted of a granularity with about a 200 Å periodicity. Following this, Steinmann and Frey-Wyssling (35) demonstrated similar structures in other plants and Park and Pon (27, 36) continued these studies with spinach chloroplasts. At times this substructure becomes very highly organized to give a paracrystalline array such as that shown in Fig. 13 (37). The fact remains that membranes occur in spinach chloroplasts which are apparently competent in quantum conversion and electron transport and yet contain no structure whatsoever, as seen by heavy-metal shadowing. On the other hand, there is evidence that the most efficient membranes, in terms of quantum conversion, are the highly structured ones which appear in spinach chloroplasts under short-day and perhaps some unknown additional conditions (38). The main subunits seen in Fig. 13 measure 185 by 155 by 100 Å. These units are termed quantasomes, and we have suggested that they may be the smallest units involved in photosynthetic conversion (37). It is also evident in the micrograph in Fig. 13 that a quantasome consists of subunits which are present on about 75–90 Å periodicities. These subunits may correspond to the subunits seen in the histological work

utilizing KMnO₄ fixation. Work with *Pharbitis* by Park (39) indicates
that the quantasome structure is evident in this plant on the external
portion of the thylakoid, whereas the internal portion of the thylakoid
as viewed after sonication consists of 90 Å particles distributed along the
surface. This is shown in Fig. 14, and would indicate that the thylakoid
membrane has two sides, a granular side with about a 90 Å periodicity

1000 Å

Fig. 13.   Paracrystalline quantasome array in spinach chloroplast internal membrane. Four subunits are seen per quantasome. From Park and Biggins (37).

which corresponds to the internal regions of the thylakoid, and a large
particle surface which corresponds to the external surface of the thylakoid and the quantasome structure. As we shall see, this small particle
surface and large particles are evident in both tissues in the freeze-etch
process. Again, it is not possible to exclude that these structures may in
part result from micelle formation or other artifact during preparation
of the specimen for microscopy.

FIG. 14. Shadowed preparation of ruptured *Pharbitis* thylakoids (39). Magnification × 32,300.

Negative staining of internal membranes of spinach chloroplasts has been performed by Park (40), and Oda and Huzisige (41), and Bronchart (42). These studies show that occasionally a 100 Å particle is attached to the internal membrane system in well preserved areas. How-

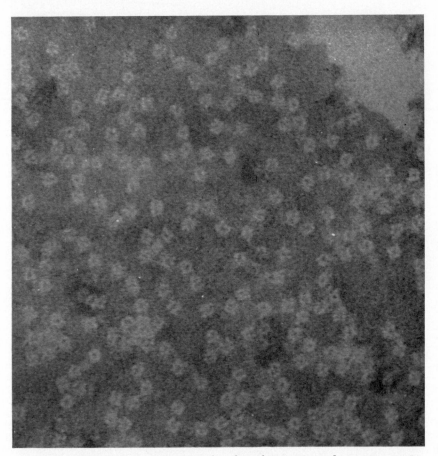

FIG. 15. Carboxydismutase as seen by photophotungstic acid negative staining. The proteins (550,000 mol. wt.) measure 80 × 110 Å and have an electron dense central core. Magnification × 480,000.

ever, the identification of such particles is in doubt since they appear to be indistinguishable from the size and structural morphology of the CO_2 fixation enzyme of photosynthesis which is located in the stroma. Both the particles described by Oda and Huzisige (41) and the colorless enzyme (carboxydismutase, see Fig. 15) are 80–110 Å in diameter and contain an electron dense central core. It is not known whether Oda's membranes were washed completely free from fraction I protein. The less well preserved areas of negative stained membranes do show substructure (42) that may be related to the subunits seen in freeze-etching.

The freeze-etch technique has been applied by Moor (25) and by Branton and Park (43) to spinach chloroplasts. One such picture of a chloroplast cross section was shown earlier in Fig. 10. A different kind of information is obtained if one observes the internal membranes of chloroplasts in tangential view rather than in cross section. A typical example from the work of Branton and Park is shown in Fig. 16. In this picture there are surfaces of small particles, surfaces with few or no particles, and surfaces with large particles. This view is typical whether one is looking at intact algal cells, intact higher plant cells, or isolated higher plant chloroplasts or chloroplast fragments. No other preparative technique gives such a detailed and consistent picture of membrane substructure. It is apparent from work with swollen, isolated plastids that the small particle surface may correspond to the internal, small particle surface seen in the *Pharbitis* thylakoid. Breakage then occurs stepwise down through a single membrane, yielding various layers within a 100 Å-thick membrane. The large particles (170 Å) are located within the membrane and are exposed when the surrounding material is removed by breakage.

Low-angle X-ray scattering experiments by Kreutz (44) have shown that there is a 37 Å periodicity along the thylakoid membrane. Menke has interpreted these experiments to mean that the membrane itself consists of a bimolecular leaflet of lipid covered on one side by protein in a way consistent with the Danielli-Davson model (45).

2. CHEMICAL COMPOSITION

The chemical composition of a single quantasome may be calculated from the size of the quantasome, its density, and a knowledge of membrane chemical composition. Such an analysis is given in Table I. It is seen from the data in Table I that the quantasome is sufficiently large to contain at least one of each of the components of the electron transport pathway of photosynthesis. The proteins in these membranes have been studied by Criddle and Park (57) and Biggins and Park (58). They are similar to the "structural protein" of mitochondria. Eighty per cent of the protein recovered in a detergent-solubilized preparation (58) gave a molecular weight of 20,000–40,000. The schlieren peak was heterogeneous and contained both cytochromes b_6 and f. Whether or not the quantasome corresponds to the photosynthetic unit of Emerson and Arnold (59) is less certain. Recent work by Izawa and Good (60) suggests on the basis of inhibitor evidence that the oxygen-evolving, photo-

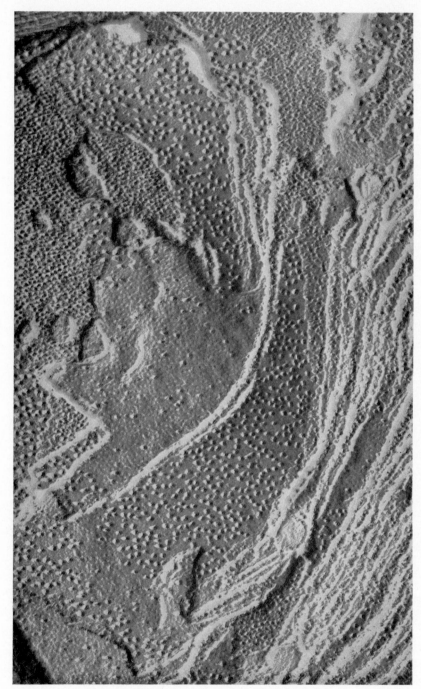

FIG. 16. A tangential view of chloroplast internal membranes by freeze-etching (*43*). Magnification × 90,000.

TABLE I

THE COMPOSITION OF THE QUANTASOME BASED ON ITS VOLUME AND DENSITY AND
ON THE CHEMICAL COMPOSITION OF CHLOROPLAST INTERNAL MEMBRANES [a]

Lipid[b] (Composition in moles per mole of quantasome)		
230 Chlorophylls		206,400
160 chlorophyll a	143,000	
70 chlorophyll b	63,400	
48 Carotenoids		27,400
14 β-carotene	7,600	
22 lutein	12,600	
6 violaxanthin	3,600	
6 neoxanthin	3,600	
46 Quinone compounds		31,800
16 plastoquinone A	12,000	
8 plastoquinone B	9,000	
6 plastoquinone C	3,000	
8–10 α-tocopherol	3,800	
4 α-tocopherylquinone	2,000	
4 vitamin K_1	2,000	
116 Phospholipids[c] (phosphatidylglycerols)		90,800
114 Digalactosyldiglyceride		134,000
346 Monogalactosyldiglyceride		268,000
48 Sulfolipid		41,000
? Sterols		15,000
Unidentified lipids		175,000
	Total lipid:	990,000
Protein		
9380 Nitrogen atoms as protein		928,000
2 Manganese		110
12 Iron including 1 as cytochrome b_6 and 1 as cytochrome f[d]		672
6 Copper		218
	Total protein:	930,000
Total lipid plus protein[e]		1,920,000

[a] See Park and Pon (36), Lichtenthaler and Park (46), and Park and Biggins (37) for original references.

[b] The fatty acid contribution to the molecular weight was determined from the analyses of Wolf et al. (47) and Debuch (48).

[c] The 116 phospholipids include 14 molecules of glycerophosphoryl inositol, 52 of glycerophosphoryl glycerol, 6 of glycerophosphoryl ethanolamine, 42 of glycerophosphoryl choline, and 2 of glycerophosphate.

[d] Lundegårdh (49, 50) reports the existence of cytochrome b_3 as well as cytochromes b_6 and f in chloroplasts. Other components of the electron transport chain which exist in a ratio of close to 1 per quantasome are plastocyanin [1 plastocyanin/300 chlorophylls, Katoh et al. (51, 52)], ferredoxin [1 ferredoxin/400 chlorophylls, Tagawa and Arnon (53)], and P-700 [1 P-700/400 chlorophylls, Kok and Hoch (54)].

[e] Amino acid analyses by Weber (55) show an enrichment in amino acids with nonpolar side chains similar to the amino acid analysis of structural protein of mitochondria (56).

synthetic unit may be considerably larger than the quantasome. It may be that the number of electron transport chains in photosynthesis considerably exceeds the number of oxygen evolution sites and that perhaps a number of quantasomes are attached to one oxygen-evolving site. These possibilities have been discussed by Park (39). It may be concluded, at any rate, that the photosynthetic unit is considerably larger than one of the 75–90 Å units as seen by histological techniques, heavy-metal shadowing in *Pharbitis*, and freeze-etching. On the other hand, a particle the size of the quantasome which is seen sometimes by heavy-metal shadowing and invariably by the freeze-etch procedure may correspond to a quantum conversion site in the membrane. One appealing thought that arises from knowledge of membrane breakage concerns experiments in which plastids have been broken into particles containing different chlorophyll *a* to *b* ratios by the process of freezing or thawing, or use of detergents. Knowledge that under freeze-etch conditions the membrane may break down the center rather than on either outer surface is an indication that similar breakages may occur during biochemical preparations. Thus, systems I and II of photosynthesis might be on opposite sides of the membrane, unable to transfer excitation energy, as shown by Sauer and Park (61).

The view of membrane structure presented in Fig. 16 must be correlated with a chemical composition and enzymological data presented in Table I. The question of how lipid and protein are localized within the membrane is only partially answered. A view of an acetone-extracted membrane is shown in Fig. 17, and it is seen that lipid removal from the membrane yields a series of particles corresponding to quantasome size, but in much greater relief (39). In many places it is possible to peer right down through the membrane to the plastic film background. Certain places in the membrane show a 90 Å periodicity. Thus, it appears that lipid and protein alternate with 90 Å periodicity along the membrane and that lipid may be regarded as wrapped around a protein matrix. Such a model would be consistent with the biochemical experiments of Sastry and Kates (62), who have shown that the lipid of chloroplast membranes is readily accessible for attack by lipases and galactosidases from *Phaseolus multiflorus* enzyme preparations. Bamberger and Park (63) approached this problem by partial enzymatic digestion of membranes and study of the freeze-etched residues. These studies suggested that chlorophyll is mainly associated with the large particles and their embedding matrix, as seen in freeze-etching. The smooth surface on which the large particles and embedding matrix lie appears to be composed of galactolipid. In general, however, conclusive localization of substances including chlorophyll within the membrane awaits future investigations.

In conclusion, we know a considerable amount about the chemical composition and morphology of the internal membrane system of the chloroplast. The greatest gap in our knowledge falls in the area which lies between solution chemistry, from which we know the chemical composition, and present electron microscope techniques, from which we know the morphology. It is the micromorphology of associations of discrete molecules within the membrane which will finally help us to explain not only the *in vivo* environment of chlorophyll molecules, but

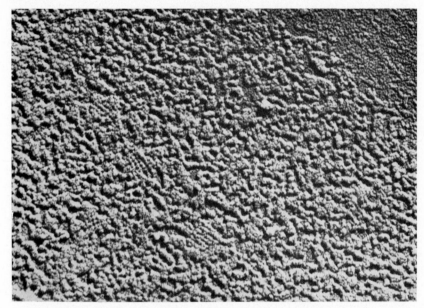

FIG. 17. A spinach chloroplast thylakoid membrane after acetone extraction (39). Magnification × 134,000.

the entire photosynthetic, quantum conversion, and electron transport process. We must count on the ingenuity of the investigators in the future to solve this problem.

REFERENCES

(1) T. W. Engelmann, *Arch. Ges. Physiol.* 57, 375 (1894).
(2) R. Hill, *Nature* 139, 881 (1937).
(3) R. Hill, *Proc. Roy. Soc.* B127, 192 (1939).
(4) D. I. Arnon, M. B. Allen, and F. R. Whatley, *Nature* 174, 394 (1954).
(5) R. Gee, G. Josh, R. F. Bils, and P. Saltman, *Plant Physiol.* 40, 89 (1965).
(6) J. A. Bassham, *Advan. Enzymol.* 25, 39 (1963).
(7) D. Spencer and H. Unt, *Australian J. Biol. Sci.* 18, 197 (1965).
(8) D. Spencer and S. G. Wildman, *Australian J. Biol. Sci.* 15, 599 (1962).
(9) E. Heitz, *Planta* 26, 134 (1936).

(*10*) P. Lintilhac and R. B. Park, *J. Cell. Biol.* (1966) (in press).

(*11*) W. Menke, *Experientia* 16, 537 (1960).

(*12*) W. Menke, *Protoplasma* 22, 56 (1934).

(*13*) A. Frey-Wyssling, *Protoplasma* 29, 279 (1937).

(*14*) A. Kahn and D. von Wettstein, *J. Ultrastruct. Res.* 5, 557 (1961).

(*15*) S. G. Wildman, T. Hongladarom, and S. I. Honda, *Science* 138, 434 (1962).

(*16*) G. A. Kausche and H. Ruska, *Naturwissenschaften* 28, 303 (1940).

(*17*) R. C. Williams and R. Wyckoff, *Proc. Soc. Exptl. Biol. Med.* 58, 265 (1945).

(*18*) S. Granick and K. R. Porter, *Am. J. Botany* 34, 545 (1947).

(*19*) A. Frey-Wyssling and K. Mühlethaler, *Vierteljahresschr. Naturforsch. Ges. Zuerich* 94, 179 (1949).

(*20*) E. Steinmann, *Experientia* 8, 300 (1952).

(*21*) T. E. Weier, C. R. Stocking, C. E. Bracker, and E. B. Risley, *Am. J. Botany* 52, 339 (1965).

(*22*) E. Gantt and S. F. Conti, *J. Cell Biol.* 26, 365 (1965).

(*23*) W. W. Thomson and T. E. Weier, *Am. J. Botany* 49, 1047 (1962).

(*24*) H. Moor, K. Mühlethaler, H. Waldner, and A. Frey-Wyssling, *J. Biophys. Biochem. Cytol.* 10, 1 (1961).

(*25*) H. Moor, Z. *Zellforsch Microskop. Anat.* 62, 546 (1964).

(*26*) A. V. Trebst, H. Y. Tsujimoto, and D. I. Arnon, *Nature* 182, 351 (1958).

(*27*) R. B. Park and N. G. Pon, *J. Mol. Biol.* 3, 1 (1961).

(*28*) J. W. Lyttleton, *Exptl. Cell Res.* 26, 312 (1962).

(*29*) A. B. Jacobson, H. Swift, and L. Bogorad, *J. Cell Biol.* 17, 557 (1963).

(*30*) R. I. B. Francki, N. K. Boardman, and S. G. Wildman, *Biochemistry* 4, 865 (1965).

(*31*) S. Murakami, Y. Morimura, and A. Takamiya, *Plant Cell Physiol.* (*Tokyo*) p. 65 (1963) (special issue).

(*32*) H. R. Hohl and A. Hepton, *J. Ultrastruct. Res.* 12, 542 (1965).

(*33*) T. E. Weier, A. H. P. Engelbrecht, A. Harrison, and E. B. Risley, *J. Ultrastruct. Res.* 13, 92 (1965).

(*34*) F. S. Sjöstrand, *J. Ultrastruct. Res.* 9, 340 (1963).

(*35*) A. Frey-Wyssling and E. Steinmann, *Vierteljahresschr. Naturforsch. Ges. Zuerich* 98, 20 (1953).

(*36*) R. B. Park and N. G. Pon, *J. Mol. Biol.* 6, 105 (1963).

(*37*) R. B. Park and J. Biggins, *Science* 144, 1009 (1964).

(*38*) R. B. Park and S. Drury, *Proc. Conf. Croissance Viellissement Chloroplastes, Gorsem, Belgium, 1965* (in press).

(*39*) R. B. Park, *J. Cell Biol.* 27, 151 (1965).

(*40*) R. B. Park, *in* "The General Physiology of Cell Specialization" (D. Mazia and A. Tyler, eds.), p. 219. McGraw-Hill, New York, 1963.

(*41*) T. Oda and H. Huzisige, *Exptl. Cell Res.* 37, 481 (1965).

(*42*) R. Bronchart, *Compt. Rend.* 260, 4564 (1965).

(*43*) D. Branton and R. B. Park, submitted to *J. Cell Biol.*

(*44*) W. Kreutz, *Z. Naturforsch.* 19b, 441 (1964).

(*45*) J. F. Danielli and H. Davson, *J. Cellular Comp. Physiol.* 5, 495 (1935).

(*46*) H. K. Lichtenthaler and R. B. Park, *Nature* 198, 1070 (1963).

(*47*) F. T. Wolf, J. G. Coniglio, and J. T. Davis, *Plant Physiol.* 37, 83 (1962).

(*48*) H. Debuch, *Experientia* 18, 61 (1962).

(*49*) H. Lundegårdh, *Physiol. Plantarum* 15, 390 (1962).

(*50*) H. Lundegårdh, *Physiol. Plantarum* 18, 269 (1965).

(51) S. Katoh, I. Suga, I. Shiratori, and A. Takamiya, *Arch. Biochem. Biophys.* **94**, 136 (1961).

(52) S. Katoh and A. Takamiya, *Plant Cell Physiol.* (*Tokyo*) **4**, 335 (1963).

(53) K. Tagawa and D. I. Arnon, *Nature* **195**, 537 (1962).

(54) B. Kok and G. Hoch, *in* "Light and Life" (W. D. McElroy and B. Glass, eds.), p. 397. Johns Hopkins Press, Baltimore, Maryland, 1961.

(55) P. Weber, *Z. Naturforsch.* **18b**, 1105 (1963).

(56) R. S. Criddle, R. M. Bock, D. E. Green, and H. Tisdale, *Biochemistry* **1**, 827 (1962).

(57) R. S. Criddle and L. Park, *Biochem. Biophys. Res. Commun.* **17**, 74 (1964).

(58) J. Biggins and R. B. Park, *Plant Physiol.* (1966) (in press).

(59) R. Emerson and W. Arnold, *J. Gen. Physiol.* **16**, 191 (1932).

(60) S. Izawa and N. E. Good, *Biochim. Biophys. Acta* **102**, 20 (1965).

(61) K. Sauer and R. B. Park, *Biochemistry* **4**, 2791 (1965).

(62) P. S. Sastry and M. Kates, *Biochemistry* **3**, 1280 (1964).

(63) E. S. Bamberger and R. B. Park, submitted to *Plant Physiol.*

—10—

The Procaryotic Photosynthetic Apparatus*

G. COHEN-BAZIRE AND W. R. SISTROM

Department of Bacteriology and Immunology,
University of California, Berkeley, California
and
Department of Biology,
University of Oregon, Eugene, Oregon

I. Introduction

There are three major groups of organisms in which photosynthesis is not associated with the presence of chloroplasts in the cells. These are: the blue-green algae, the purple bacteria, and the green bacteria. The absence of chloroplasts from the cells of these three groups is only one of many distinctive cellular properties which set apart bacteria and blue-green algae from all other groups of living organisms (1). The term *procaryotic* has been adopted to designate their cellular organization as opposed to the *eucaryotic* cellular organization characteristic of all the other organisms (1).

The differences in structural organization between eucaryotic and procaryotic cells raise a whole series of important questions concerning the relationships between structure and function. Here we will be concerned with the *nature of the structures responsible for the performance*

* Research in the authors' laboratories has been aided by grants from the National Science Foundation and the National Institutes of Health.

313

of photosynthesis in blue-green algae and bacteria, and their possible analogies to chloroplasts.

II. Structure

A. Blue-Green Algae

1. Cytology

Classical cytological investigations on the blue-green algae by light microscopy led some twenty years ago to the recognition of two main regions of the cell: a nonpigmented central region called the "centroplasm" and a cortical region, containing the photosynthetic pigments, called the "chromatoplasm" (2). Until this time it was generally assumed that the characteristic photosynthetic pigments of the blue-green algae were dispersed throughout the chromatoplasm.

This picture changed as a result of the development of thin sectioning in conjunction with electron microscopic examination. The first clear electron micrographs of thin sections were obtained by Niklowitz and Drews (3). A study of various representatives of three different families of blue-green algae showed that they characteristically contain an extensive system of lamellae (4), formed by closely apposed distinct unit membranes (5), sometimes separated by a very narrow gap which could enlarge to form intralamellar vesicles. The thickness of the lamellae was estimated to be between 140 and 160 Å. Serial sections (6) seemed to indicate that the lamellae are formed by large flattened sacs not dissimilar to the proposed structure of the discs forming the lamellar system of chloroplasts.

The lamellae of blue-green algae have a very striking arrangement within the cells. An examination of the published electron micrographs from several laboratories (5–7) shows that in all well preserved specimens prepared from healthy cells, the lamellae tend to be parallel to one another, not in closely packed stacks but with a relatively regular spacing between them. This can be seen with great clarity in Fig. 1.

In some cases, as in *Anacystis nidulans* (5), the lamellae are found only in the cortical region of the cytoplasm; more commonly, they are distributed throughout the cytoplasm, being absent only from the nuclear region of the cell. The interlamellar regions vary in width with the species examined and with the physiological state of the specimens. They often contain electron-opaque granules interpreted as ribosomes, as well as electron-transparent granules interpreted as storage products of photosynthetic activity (8) (see also Fig. 1).

Very little is known about the cytological origin of these lamellae. Connections between lamellae and the cytoplasmic membrane have never been definitely shown to exist. Pankratz and Bowen (9) have observed occasional short lamellar units continuous with the cytoplasmic membrane in the region of cross wall formation, which could be interpreted as newly forming lamellae. From examinations of dividing cells (9, 10) it is clear that during cell division, the developing cross wall and the adjacent cell membrane cut across some of the already existing lamellae and that further growth of these lamellae must occur.

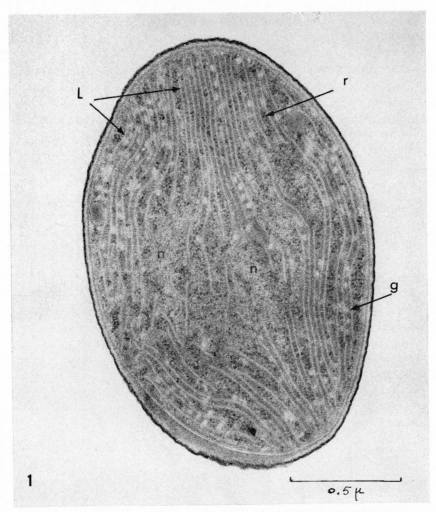

Fig. 1. Longitudinal thin section of a blue-green alga, *Gleocapsa*. Note the array of lamellae (*L*) with ribosomes (*r*) and electron-transparent granules (*g*) in the interlamellar spaces. The central region of the cell containing the nuclear material (*n*) is relatively free of lamellae. × 60,000. Courtesy of Mrs. M. Allen.

The photosynthetic pigments of blue-green algae include phyco-cyanins and/or phycoerythrins in addition to chlorophyll *a* and carot-enoids. The phycocyanins are loosely bound chromoproteins which are released along with the other soluble proteins of the cells if the integrity of the cytoplasmic membrane is destroyed (11); the other photosynthetic pigments are firmly attached to the cytoskeleton. In the case of violent cell breakage by sonic oscillation or alumina grinding, a particulate frac-tion can be isolated by differential centrifugation which contains the photosynthetic pigments free of phycocyanin (12, 13). This pigmented fraction is capable of performing a number of light-catalyzed reactions such as photophosphorylation (14) and the Hill reaction (15).

Examined in thin sections, this particulate fraction appears to consist of membrane-bounded vesicles of irregular size, which have been inter-preted as arising from the disruption of the intracellular lamellae visible in thin sections of intact cells (16). The chromoproteins must in some way be closely associated with the photosynthetic lamellae in the living cells.

2. ISOLATED PIGMENTED MATERIAL

Since the early report of Shatkin (16) there has been little work done on the isolation and chemical analysis of photosynthetically active cell fractions from blue-green algae. The recent work of Bloch and his co-workers suggests that the fatty acid composition of this material is similar to that of chloroplasts of higher green plants. Levin *et al.* (17) have shown that *Anabaena variabilis* contains a large amount of α-linolenic acid. This fatty acid is characteristic of oxygen evolving photosynthetic tissues (18). Levin *et al.* (17) further demonstrated that very nearly all the α-linolenic acid of *A. variabilis* is confined to the photosynthetically active particulate fraction of cell-free extracts and that all this fatty acid occurs in the form of galactolipids. In green plants, α-linolenic acid is a major constituent of galactolipids.

These chemical similarities of the photosynthetic apparatus of blue-green algae and of green plants suggest that, although the cellular or-ganization of these algae is very different from that of green plants, the molecular structure of their photosynthetic lamellae is probably very similar to that of chloroplast lamellae.

B. Purple Bacteria

1. CYTOLOGY

Even for the largest representatives of this group, light microscopy does not reveal the somewhat indefinite cortical localization of photosynthetic pigments observed in some blue-green algae; rather, the whole cell appears evenly pigmented. As early as 1925, it was recognized that the pigment system of purple bacteria was associated with cellular proteins extractable from ground cells (19). In 1938, French (20) obtained a pigmented cell-free extract of Rhodospirillum rubrum which showed the same absorption spectrum as the original cell suspension. The pigment complex could be precipitated by ammonium sulfate at half saturation, and attempts at fractionation did not separate bacteriochlorophyll from spirilloxanthin. French concluded that both pigments were attached to the same or to similar protein molecules. From these experiments, it was generally assumed that the photosynthetic pigments of bacteria were associated with soluble proteins dispersed in the cytoplasm. It was not until the early 1950's that new approaches were applied to this problem.

It had long been known (21) that some species of non-sulfur purple bacteria are facultative phototrophs. When such purple bacteria are grown aerobically in the dark, their photosynthetic pigment content becomes extremely low. Taking advantage of this fact, Pardee et al. (22) analyzed in the analytical ultracentrifuge extracts of R. rubrum grown photosynthetically and grown aerobically in the dark. From the deep red, cell-free extract of photosynthetically grown cells, a fraction having the typical in vivo absorption spectrum of the cells could be readily sedimented by high speed centrifugation. The sedimentation constant of this fraction was estimated to be 190–200 S. Electron micrographs of shadowed preparations of the isolated fraction showed disc-shaped structures having an apparent diameter of 1100 Å. Extracts of aerobically grown R. rubrum did not contain a fraction showing such a high sedimentation constant. These authors concluded that the photosynthetic pigments of R. rubrum were associated with particulate structures, which they termed "chromatophores," and inferred that these structures must be the site of the primary photochemical reactions in bacterial cells. Two years later, Frenkel (23) demonstrated that "chromatophores" could perform the light-catalyzed synthesis of adenosine triphosphate from adenosine diphosphate and inorganic phosphate, i.e., the reaction of photophosphorylation.

The first cytological study of purple bacteria by electron microscopic examination of thin sections was published in 1958 by Vatter and Wolfe (24). These authors observed that sections of photosynthetically grown cells of R. rubrum contain cytoplasmic structures very similar in size and shape to the isolated chromatophores described by Pardee, Schachman, and Stanier (see Figs. 2 and 3). In thin sections, these structures appeared more electron-transparent than the surrounding cytoplasm and

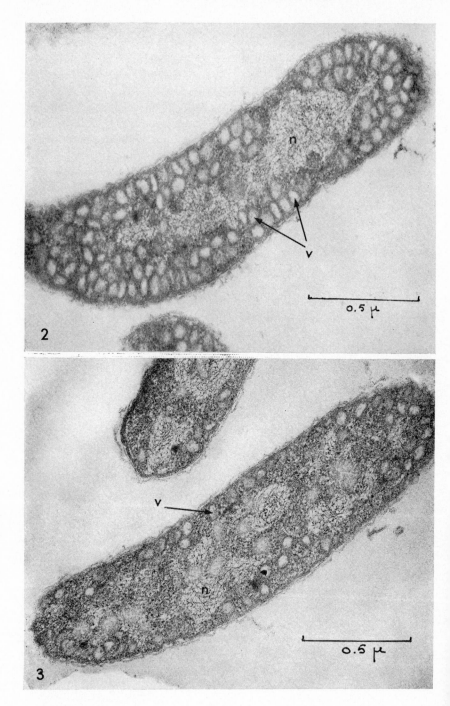

seemed to be bounded by a membrane. No such structures were visible in sections of the same organism grown aerobically, and therefore depleted of photosynthetic pigments. Thin sections of other purple bacteria (*Rhodopseudomonas spheroides, Chromatium* strain D) revealed numerous membrane-bounded vesicles smaller than those of *Rhodospirillum rubrum* but similar in appearance. The similarities of shape and dimension between the "chromatophores" isolated from cell-free extracts and the structures observed by Vatter and Wolfe in thin sections led to the general belief that the chromatophores were the structural units of photosynthetic activity in bacteria.

It soon became apparent, however, that the so-called "chromatophores" might be an artifact of preparation, resulting from the comminution of a larger structure by sonic waves or abrasion. In 1959, Tuttle and Gest (25) showed that mild osmotic lysis of *R. rubrum* cells did not release "chromatophores" in significant quantities. The photophosphorylating activity of lysed cells remained associated with structures sedimenting at low centrifugal forces. These authors suggested that "the photoactive pigment system of photosynthetic bacteria is associated with the (lipoprotein) cytoplasmic membrane and/or a reticulum of membranous extensions penetrating the cytoplasm." This view concerning the organization of the photosynthetic apparatus of purple bacteria was not, however, generally accepted at the time, and the idea of the chromatophore as an organized unit of structure and function continued to prevail (26).

In the meantime, fine structure studies by Drews (27) on *Rhodospirillum molischianum* and by Boatman and Douglas (28) on *Rhodomicrobium vannielii* showed that the electron-transparent, membrane-bounded vesicles so typical of *Rhodospirillum rubrum, Rhodopseudomonas spheroides*, and *Chromatium* are not a universal structural feature of purple bacteria. Thin sections of *Rhodospirillum molischianum* showed that the cytoplasm contains stacks of regularly disposed lamellae situated toward the periphery of the cell. *Rhodomicrobium vannielii* (Fig. 4) has a very extensive lamellar system reminiscent of the lamellar organization of some blue-green algae. The same type of organization encountered in *Rhodospirillum molischianum* has recently been observed

---

FIG. 2. Section of *Rhodospirillum rubrum* grown photosynthetically at low light intensity (50 ftc). Note the numerous membrane-bounded vesicles (v) except in the central region of the cell occupied by the nuclear material (n). × 60,000.

FIG. 3. Section of *Rhodospirillum rubrum* grown photosynthetically at moderate light intensity (1000 ftc). The membrane-bounded vesicles (v) are less abundant than in Fig. 2 and are mostly at the periphery of the cell. × 60,000.

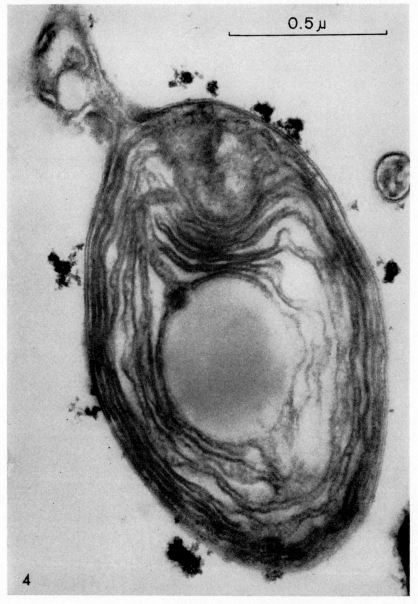

Fig. 4.   Longitudinal section of *Rhodomicrobium vannielii* showing the extensive system of paired membranes forming lamellae. × 86,000. Courtesy of Dr. E. S. Boatman (*28*).

in *Rhodospirillum fulvum* (*29*) (Fig. 5) and *Rhodospirillum photometricum* (*30*).

A very extensive system of closely packed lamellae similar to that of *Rhodomicrobium vannielii* is also found in two other non-sulfur purple bacteria: *Rhodopseudomonas palustris* (Fig. 6) and the *Rhodopseudomonas* sp. (*29*) which contains the newly discovered bacteriochlorophyll *b* (*31*) (Fig. 7). A mixture of vesicular and lamellar organization of intracytoplasmic membranes has been encountered in two purple bacteria: in *Chromatium* strain D grown at high light intensity (*32*) and in *Thiocapsa* (*33*) (Fig. 8).

If the lamellae observed in *Rhodospirillum molischianum*, *Rhodomicrobium vannielii*, and other purple bacteria are in fact the structures associated with the photosynthetic pigments in these organisms, the concept of "the chromatophore" as a unit of structure and function becomes difficult to accept.

Electron microscopic evidence to support the structural interpretation first proposed by Tuttle and Gest (*25*) has recently been obtained for *Rhodospirillum molischianum* (*34, 35*), *R. rubrum* (*36–38*), and *Rhodopseudomonas spheroides* (*37*) from examination of thin sections. In these three organisms, appropriate sections show clearly that the lamellar structures characteristic of *Rhodospirillum molischianum* and the vesicular structures characteristic of *R. rubrum* and *Rhodopseudomonas spheroides* are in physical continuity with the cytoplasmic membrane (see Figs. 9 and 10). The recent analysis (*38*) of the release of pigmented fractions by sonic disruption of *Rhodospirillum rubrum* and the electron microscopic examination of disrupted cells show that the photosynthetic pigments are associated with an extensive intracytoplasmic membranous system.

Although there is at present no direct cytochemical evidence that the structures, vesicular or lamellar, observed in thin sections of purple bacteria represent their photosynthetic apparatus, there exists a body of circumstantial evidence in support of this assumption. The photosynthetic pigment content of some purple bacteria can vary over a wide range in response to such external factors as light intensity and temperature (*35, 39*) and in the facultative aerobes, in response to oxygen tension (*40*). Thin sections of aerobically grown cells of *R. rubrum* contain very few membrane-limited vesicles. When pigment synthesis resumes under conditions of oxygen limitation, correlatively more vesicular invaginations of the cell membrane appear in the cytoplasm (*36*). Cells grown photosynthetically at high light intensities and containing a relatively small amount of photosynthetic pigments show in thin sections comparatively fewer membranous vesicles than cells grown at low light

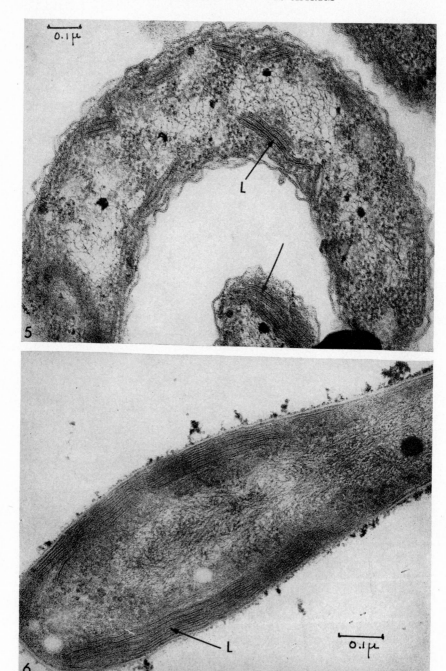

intensities containing large amounts of pigment (33, 36, 37). An example is given in Figs. 2 and 3. The cells shown in Fig. 2 were grown at a light intensity of 50 foot-candles (ftc) and their specific bacteriochlorophyll content was 14 μg per milligram of cellular proteins. The cells shown in Fig. 3 were grown at a light intensity of 1000 ftc and contained 8.5 μg of bacteriochlorophyll per milligram of cellular proteins. Several vesicles can arise from one invagination of the cytoplasmic membrane and remain connected with each other (37, 38). When the cells are disrupted by drastic means such as sonication, alumina grinding, or a French pressure cell, the tubular membranes are sheared preferentially at the sites of constrictions forming the vesicles, releasing a pigmented fraction relatively homogeneous in size.

In a detailed study on *Rhodospirillum molischianum*, Gibbs *et al.* (35) show that the shape, the size, the number, and arrangement of infoldings of the cell membrane vary with the photosynthetic pigment content of the cells. In cells low in bacteriochlorophyll, the infoldings are mostly tubular in shape, enlarging to flattened discs as the bacteriochlorophyll content increases. In cells containing high amounts of bacteriochlorophyll, several disc-shaped infoldings become appressed against one another, forming typical stacks of lamellae.

2. ISOLATED PIGMENTED FRACTIONS

a. Physicochemical Properties. The methods for isolation and purification of chromatophore material are similar to those developed for bacterial cell membranes. A variety of methods for rupturing the cells are available. These include passage through a French pressure cell, grinding with alumina, osmotic lysis, and sonication. Sonication leads to comminution of chromatophore material, which can be a serious disadvantage in some cases.

Differential centrifugation has been used almost universally as the sole method of purification. Simple centrifugation is not sufficient to free chromatophore material from ribosomes unless it is repeated 4 or 5 times.

FIG. 5. Longitudinal section of *Rhodospirillum fulvum*. Stacks of short lamellae (*L*) are present near the periphery of the cytoplasm. The lamellae are composed, as in *R. molischianum*, of flattened discs appressed against each other. A very electron-opaque dark line (arrow) is formed by the juxtaposition of two adjacent discs. × 90,000.

FIG. 6. Part of a longitudinal section of *Rhodopseudomonas palustris* passing through a pole of the cell. The extended system of closely packed lamellae (*L*) is parallel to the cell wall and appressed to the cell membrane on both sides of the cell. The pole of the cell does not contain lamellae. × 120,000.

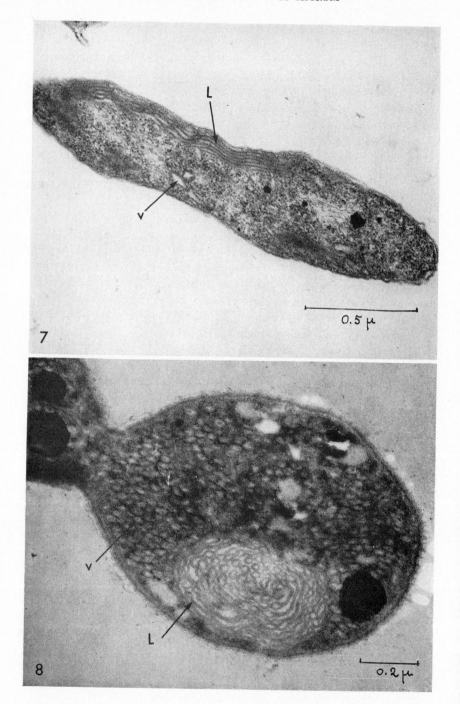

Removal of ribosomes may be effected by centrifugation through a sufficiently dense and nonviscous medium such as a RbCl solution (41) or by electrophoresis (42). The latter is probably the method of choice.

Several workers have reported that chromatophore material from *Rhodospirillum rubrum* and *Rhodopseudomonas spheroides* (40, 41) can be separated into two fractions by centrifugation through a sucrose density gradient. These fractions have been termed the "heavy" fraction and the "light" fraction according to their positions in the gradient. Worden and Sistrom (41) showed that the heavy and light fractions from *R. spheroides* differed not only in their rates of sedimentation, but also in such properties as the ratio of bacteriochlorophyll to carotenoid pigments and the absorption spectrum of bacteriochlorophyll. Thus, although the significance of the fractions is far from clear, they apparently are not merely artifacts of preparation.

Pardee *et al.* reported that the chromatophore material from *Rhodospirillum rubrum* had a sedimentation constant of about 190 S (22). Highly purified "light" fraction from *Rhodopseudomonas spheroides* has a sedimentation constant of 153 S at infinite dilution in water (41); Gibson has recently reported that chromatophore material from this same organism has a sedimentation constant of 165 S (43). Bergeron found that chromatophore particles isolated from *Chromatium* sp. had a sedimentation constant of about 120 S (44). Electron microscopic evidence (see below) suggests that at least a large fraction of the isolated chromatophore material is composed of nearly spherical particles approximately 50 mμ in diameter. This agrees with the X-ray diffraction data of Langridge *et al.* for the "light" fraction from *R. spheroides* (45). These data indicated that the particles were spheres with a diameter of 60 mμ; the X-ray scattering skin of the sphere was calculated to be about 80 Å thick.

Unfortunately, no thoroughgoing chemical analysis of chromatophore material is yet available. Newton's early analyses (46) of chromatophore material from *Chromatium* are still the most extensive. He found that there was about one-half as much lipid material as protein in the chromatophores and that the amount of lipid phosphorus was about 0.5% of the amount of protein. Sykes *et al.* (47) found a similar lipid content

---

Fig. 7. Longitudinal section of a *Rhodopseudomonas* sp. which contains bacteriochlorophyll *b*. Like *Rhodopseudomonas palustris*, it contains an extended system of lamellae (*L*) parallel to the long axis of the cell and adjacent to the cell membrane. A few irregular, membrane-bounded vesicles (*v*) are found in the cytoplasm. × 60,000.

Fig. 8. Section of the purple sulfur bacterium *Thiocapsa* showing a mixed vesicular (*v*) and lamellar (*L*) system of internal membranes. × 80,000.

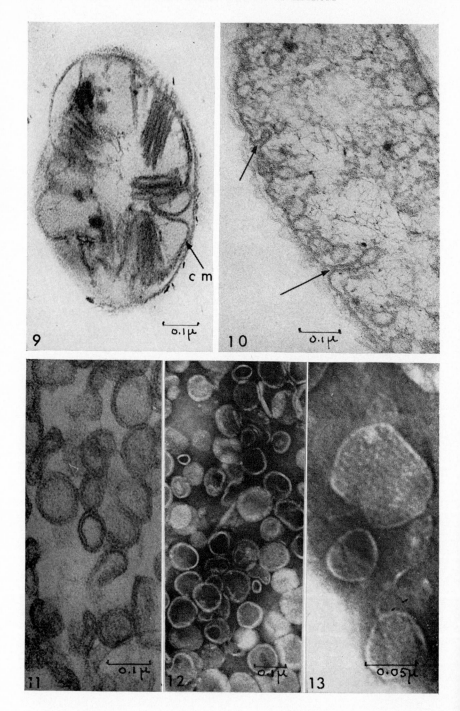

in chromatophore material from *Chlorobium thiosulfatophilum*. The lipid phosphorus content of chromatophore material from *Rhodopseudomonas spheroides* is similar to that of *Chromatium*, and protein represents about 50% of the dry weight of this fraction (*41*). Both lipid phosphorus and protein are constant proportions of the dry weight of chromatophore material regardless of the specific bacteriochlorophyll content of the cells. This is true even of the particulate fraction of cells grown under highly aerobic conditions and therefore devoid of bacteriochlorophyll (*41*).

The phospholipid composition of chromatophore material from *Chromatium* is remarkably simple. The only hydrolysis products detected by Newton were glycerol and ethanolamine (*46*). He did not analyze the fatty acids. The phospholipids of *R. spheroides* are composed largely of phosphatidylethanolamine (40%), phosphatidylcholine (20%), phosphatidic acid (25%), and phosphatidylglycine. The phospholipid composition of aerobically grown cells is the same as that of photosynthetically grown cells (*48*).

In common with other eubacteria and in marked contrast to all other photosynthetic organisms, photosynthetic bacteria do not contain polyunsaturated fatty acids and in particular do not contain α-linolenic acid (*49*).

Perhaps the most obvious and useful conclusion that may be drawn from this summary of almost nonexistent data is that those who are interested in understanding the photosynthetic apparatus of bacteria should devote less time to writing and reading reviews and more time to doing chemistry.

b. Structure. Comparatively little work has been done on the structure of isolated pigment fractions from purple bacteria since the first isolation of *Rhodospirillum rubrum* "chromatophores" (*12*). The first

---

Fig. 9. Cross section through an autolyzed cell of *Rhodospirillum molischianum* showing the continuity between the cytoplasmic membrane (*cm*) and the lamellar membranes. × 90,000. Courtesy of Drs. Giesbrecht and Drews (*34*).

Fig. 10. Section of *Rhodospirillum rubrum* treated with ribonuclease in order to show with greater clarity the continuity (arrows) between the cytoplasmic membrane and the membranes forming the vesicles. × 112,000.

Fig. 11. Thin section of the chromatophore fraction isolated from *Rhodospirillum rubrum* and fixed with osmium. The structure of the unit membrane bounding the vesicles is clearly distinguishable. × 120,000.

Fig. 12. Chromatophore fraction isolated from *Rhodospirillum rubrum* negatively stained with neutral potassium phosphotungstate. × 90,000.

Fig. 13. Rows of 50 Å subunits are visible on this negatively stained preparation of a purified chromatophore fraction isolated from *Rhodospirillum rubrum*. × 280,000. Courtesy of Dr. S. C. Holt.

electron micrographs of shadowed, air-dried preparations revealed disc-shaped structures relatively homogeneous in size and interpreted as flattened vesicles. Examined by the technique of negative staining, these vesicles do not appear closed, but very often show openings of varying diameters (33) (see Fig. 12). The diameter of these cup-shaped vesicles varies between 700 and 900 Å.

Thin sections show very clearly the unit membrane structure of their limiting membrane (33) (see Fig. 11).

Structural subunits 50 Å in diameter have been observed in negatively stained preparations (50). These subunits sometimes have the appearance and arrangement of those observed in negatively stained preparations of mitochondria (51) or of bacterial cytoplasmic membranes (52). Sometimes, the subunits appear as smooth knobs arranged with some regularity at the surface of the vesicles (42) (see Fig. 13) showing some similarity to the quantasomes observed in chloroplasts (53). It is still too early to know whether these substructures represent the ultimate photosynthetic units in purple bacteria.

Very little is known of the structure and composition of pigmented fractions isolated from purple bacteria containing internal lamellae. Hickman et al. (54) describe a fraction isolated from *Rhodospirillum molischianum* consisting of relatively large vesicles (100–200 mμ in diameter). Preliminary experiments performed with *Rhodopseudomonas palustris* show that even after an extensive sonification the major portion of the photosynthetic pigments remain associated with the cytoskeleton. A minor fraction of the pigments is released and appears to consist of large pieces of membranous material (55).

C. Green Bacteria

1. Cytology

The green bacteria are readily distinguishable from all other photosynthetic organisms by the chemical nature of their photosynthetic pigments. Each strain so far isolated in pure culture contains one of the two chlorobium chlorophylls (56), Cchl 650 or 660 (see Chapter 4), in addition to variable but comparatively small amounts of Bchl a. (57). Their carotenoids are also distinct from those encountered in all the other photosynthetic organisms (58).

Comparatively little work has been devoted to the photosynthetic apparatus of green bacteria. In the first attempt to study the fine structure of photosynthetic bacteria, Vatter and Wolfe (24) examined a strain of the green bacterium *Chlorobium limicola*. Thin sections did not reveal the vesicular membranous structures which are such characteristic cytoplasmic features of most purple bacteria. Electron-opaque granules 150–250 Å in diameter were seen scattered in the cytoplasm; they were tentatively interpreted by the authors as the "chromatophores" of these microorganisms.

A few years later, Bergeron and Fuller (59), working with a strain of *Chlorobium thiosulfatophilum*, were unable to observe in thin sections any special feature which would distinguish this organism from a nonphotosynthetic bacterium. The pigmented fraction isolated from *C. thiosulfatophilum* had a sedimentation constant of 50 S, corresponding to particles analogous in size to ribosomes, i.e., with maximal dimensions of 150 Å. This finding therefore agreed with the apparent fine structure of the cells.

The problem of the localization of the photosynthetic apparatus of green bacteria was investigated again very recently in a comparative study carried out on a variety of these organisms, including several strains of *C. limicola* and of *C. thiosulfatophilum* (60).

This work shows that green bacteria possess very distinctive features in their fine structure which differentiate them from all the other photosynthetic organisms so far examined (see Figs. 14 and 15).

The most striking elements are large, oblong vesicles (300–400 Å wide and 1000–1500 Å long) disposed immediately under the cytoplasmic membrane throughout the cortical region of the cells. These vesicles, which have been termed "chlorobium vesicles" (33) are surrounded by a thin (30 Å) electron-opaque membrane. In the outer region of each vesicle, this membrane is closely appressed to the cell membrane proper but distinct from it (see Figs. 14 and 15). In specimens embedded in Vestopal or methacrylate for electron microscopic examination, the vesicles appear electron-transparent and empty (as in Fig. 14); however, when epoxy resins are used as embedding media, the contents of the vesicles appear electron-opaque and show indications of a fibrillar fine structure (Fig. 16).

Apart from the chlorobium vesicles, the other peculiar elements observed in thin sections of green bacteria are intracytoplasmic membranous structures (Fig. 14) similar in appearance and localization to the mesosomes (61) which have been observed in many nonphotosynthetic bacteria; they appear to be formed by invagination of the cytoplasmic membrane, a unit membrane 80 Å thick, and are often associated, like typical mesosomes, with transverse wall formation (60).

2. ISOLATED PIGMENTED FRACTIONS FROM GREEN BACTERIA

a. Physicochemical Properties. Sykes *et al.* (47) have analyzed in the ultracentrifuge cell-free extracts of *C. thiosulfatophilum,* and find the

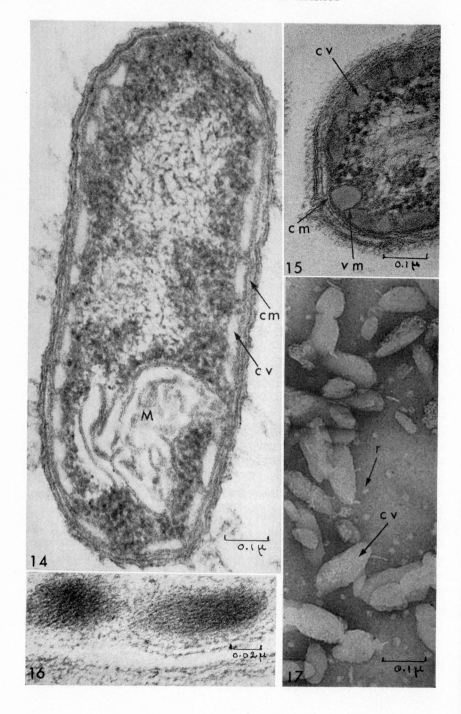

bulk of the photosynthetic pigments associated with material sedimenting with a S_{20} of 116 S. The sedimentation of this fraction is strongly dependent upon concentration, suggesting a dissymmetry in the shape of the individual particles. The study of release of pigment following osmotic lysis or sonication of the cells indicates that the photosynthetic apparatus of *C. thiosulfatophilum* is not associated with an integrated cellular structure, but rather with isolated cytoplasmic elements.

The results of chemical analyses by Sykes *et al.* have already been alluded to.

b. Structure. As a rule, most strains of green bacteria grow best at relatively low light intensities (less than 100 ftc) and their specific chlorophyll content is 3–5 times that of purple bacteria grown in the same conditions (*60*).

From cell-free extracts of green bacteria, a pigmented particulate fraction can be isolated which contains two to three times the specific pigment content of the original cells (*60*). This fraction, examined in the electron microscope by the technique of negative staining, is mostly composed of large oblong vesicles, 1000–1500 Å long and 500–750 Å wide (Fig. 17). The vesicles appear filled with fine fibrillar elements and correspond in shape and dimensions to the chlorobium vesicles observed in thin sections. Their appearance differs markedly from that of the pigmented fractions isolated from purple bacteria (compare Fig. 17 with Figs. 12 and 13).

Although the mesosomes observed in green bacteria have the same origin and fine structure as the membranous elements (vesicular or lamellar) of purple bacteria, there is no evidence to suggest that they are the structural elements of photosynthetic activity in these organisms.

Fig. 14.   Longitudinal section of *Chlorobium thiosulfatophilum* embedded in Vestopal. The electron-transparent chlorobium vesicles (*cv*) disposed at the periphery of the cell are contiguous to the cytoplasmic membrane (*cm*). A large mesosomal element (*M*) is present in the lower part of the cell. × 120,000.

Fig. 15.   Section through the pole of the cell of a green bacterium embedded in Maraglas. The chlorobium vesicles (*cv*) are electron opaque and surrounded by a very electron-dense membrane (*vm*) distinct from the cytoplasmic membrane (*cm*). × 120,000.

Fig. 16.   Section of part of a cell of *Chlorobium thiosulfatophilum* embedded in Epon showing the fine fibrillar elements present in the chlorobium vesicles. × 400,000.

Fig. 17.   Pigmented fraction isolated from an extract of *Chlorobium thiosulfatophilum* negatively stained with phosphotungstate, showing large oblong vesicles (*cv*) similar in shape and dimensions to the chlorobium vesicles seen in thin sections. Ribosomes (*r*) are also visible in this preparation. × 120,000.

III. Control of Photopigment Synthesis

A. Blue-Green Algae

In blue-green algae both the intensity and color of the incident light can influence the pigmentation of the cells. The early work of Engelmann and the ensuing controversy have been well reviewed by Rabinowitch (62). Engelmann suggested that blue-green algae tend to assume a color complementary to that of the light to which they are exposed. Recent work has substantiated this notion in part and has revealed that among the blue-green algae there are three distinct kinds of response to different colors and intensities of light.

In the first place, the amount of chlorophyll per unit dry weight is inversely related to light intensity (63). In *Anacystis nidulans*, phycocyanin varied more or less in parallel with chlorophyll: a 12-fold increase in light-intensity caused a 2-fold decrease in chlorophyll content and a 2.2-fold decrease in phycocyanin. It should be noted that these changes in pigmentation occurred at light intensities less than that necessary to saturate the growth rate. Furthermore, a culture grown at a low temperature had less chlorophyll and phycocyanin than one grown at a higher temperature in the same light intensity. As will be seen below, precisely analogous effects are observed in photosynthetic bacteria.

The second way in which incident light influences pigmentation in the blue-green algae was demonstrated by Hattori and Fujita (64, 65). They showed that in *Tolypothrix tenuis*, which contains both phycocyanin and phycoerythrin, the relative amounts of these two pigments depend on the color of the incident light. In green or blue light there is a preferential synthesis of phycoerythrin, whereas in red or purple light phycocyanin predominates. The mechanism apparently involves an interconversion of the precursors of the two pigments. The action spectrum for formation of each precursor is similar to the absorption spectrum of the final pigment. This form of control is direct: the pigment with the greater absorption for the wavelengths of highest intensity is formed preferentially.

An inverse chromatic control has recently been uncovered by Jones and Myers in *Anacystis nidulans* (66). Here, exposure to red light leads to a decrease in the amount of chlorophyll with little change in that of phycocyanin. Jones and Myers suggested that this control operates so as to keep the rates of the two photochemical reactions of green plant photosynthesis more or less equal. The elegance and precision of the control are well illustrated by the data shown in Table I. Cells grown in red light when exposed to red light have the same distribution of absorbed quanta between chlorophyll and phycocyanin as do cells grown in white light when exposed to white light.

TABLE I

DISTRIBUTION OF ABSORBED QUANTA BETWEEN CHLOROPHYLL AND PHYCOCYANIN
IN *Anacystis nidulans* GROWN IN WHITE AND IN RED LIGHT[a]

Incident illumination	Cell type[b]	Per cent of absorbed quanta absorbed by	
		Chlorophyll	Phycocyanin
Daylight	W	37	43
	R	7	74
Red	W	86	14
	R	53	47

[a] Adapted from data of Jones and Myers (66).
[b] W cells were grown in low intensity white light, R cells in red light.

B. Purple Bacteria

1. EFFECT OF ENVIRONMENT ON BACTERIOCHLOROPHYLL CONTENT

The pigmentation of photosynthetic bacteria is not known to be influenced by the color of the incident light. These bacteria do, however, show marked variations in bacteriochlorophyll content in response to changes in such factors as light intensity, oxygen tension, and growth rate.

Bacteriochlorophyll has most often been estimated from the optical density at 775 mμ of methanol or acetone-methanol extracts using the extinction coefficients previously published by us (39). These extinction coefficients are in error. The correct values are: for methanol, 60 mM^{-1}cm^{-1} and for acetone-methanol, 75 mM^{-1}cm^{-1} (67). In this chapter we have recalculated published bacteriochlorophyll estimates when it was clear from the context that the erroneous extinction coefficients had been used in the original calculations.

Figure 18 shows how the specific growth rates and bacteriochlorophyll contents of *Rhodospirillum rubrum* (68) and *Rhodopseudomonas spheroides* (69) vary with light intensity. In both organisms the pigment content is markedly dependent on light intensity even at intensities which are not saturating for the growth rate. In other words, any given light intensity determines both a unique growth rate and a unique pigment content. It will be recalled that Myers and Kratz observed that *Anacystis nidulans* responds in the same way (63). This anomalous relation can be emphasized by plotting specific bacteriochlorophyll content against growth rate as is done in Fig. 19. As can be seen, a nearly linear relationship is found with either organism. There is no readily apparent explanation for this effect.

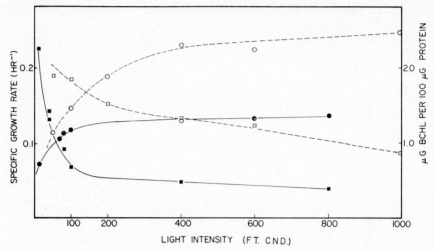

Fig. 18. Growth rate and specific bacteriochlorophyll content of *Rhodopseudomonas spheroides* and *Rhodospirillum rubrum* as functions of light intensity. Open symbols, strain Ga of *R. spheroides*; closed symbols, *R. rubrum*; (○ and ●), growth rate (hr^{-1}); (□ and ■), specific bacteriochlorophyll content (micrograms bacteriochlorophyll per 100 μg protein). Adapted from Holt and Marr (*68*) and Sistrom (*69*).

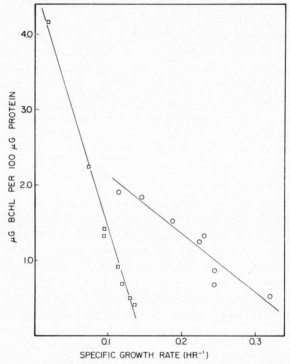

Fig. 19. Specific bacteriochlorophyll content and growth rate of *Rhodopseudomonas spheroides* and *Rhodospirillum rubrum*. The data of Fig. 18 have been replotted to show relationship between specific growth rate constant and specific bacteriochlorophyll content. (○-○), *R. spheroides*; (□-□), *R. rubrum*.

The situation becomes clearer if one studies the relationship of light intensity and bacteriochlorophyll content at constant growth rate. This can be done by using continuous cultures maintained in a Monod-Novick chemostat. Preliminary results in the laboratory of one of us (W.R.S.) using cultures of R. *spheroides* limited by succinate show that the specific bacteriochlorophyll content is given by the expression: $K + A/I$, where I is light intensity and K and A are empirical constants. Since K is not zero there is a finite bacteriochlorophyll content at infinite light intensity.

It has been known since the early observations of van Niel that oxygen reduces pigmentation in facultative aerobic species such a *Rhodospirillum rubrum* and *Rhodopseudomonas spheroides*. The effect of oxygen has been studied in detail by Cohen-Bazire et al. (39) and by Lascelles (70). Introduction of air into a culture growing anaerobically in the light causes an immediate inhibition of bacteriochlorophyll synthesis without any destruction of the pigment. Continued growth under highly aerobic conditions, in either light or dark, leads to a bleached culture. At lower oxygen tensions, the bacteriochlorophyll content is, at least roughly, inversely related to the oxygen tension. For technical reasons, the precise relationship has not been worked out. Bacteriochlorophyll synthesis under semi-aerobic conditions is completely independent of light intensity (40).

At a given light intensity, the bacteriochlorophyll content of R. *spheroides* is directly related to the specific growth rate. This effect can be seen most clearly in the chemostat. Thus at a specific growth rate of $0.024 \, hr^{-1}$ and a light intensity of 400 ftc, the specific bacteriochlorophyll content of this organism is 1.1 µg bacteriochlorophyll per 100 µg protein; while at a growth rate of $0.036 \, hr^{-1}$ and the same light intensity, it is 0.8 µg per 100 µg protein. The obligatory coupling of bacteriochlorophyll formation with protein synthesis (71, 72) is possibly related to the effect of growth rate on bacteriochlorophyll content.

It seems likely that the responses to changes in light intensity, oxygen tension, and growth rate all reflect the operation of a single control mechanism. The kinetics of bacteriochlorophyll synthesis after a sudden change in light intensity or oxygen tension indicate that the response involves the inhibition (or the release from inhibition) of enzymes concerned in the synthesis of bacteriochlorophyll rather than repression of the formation of these enzymes (39). Although the specific activities of some of the enzymes of porphyrin biosynthesis vary with growth conditions, the variations are not sufficient to account for the observed changes in bacteriochlorophyll content (73).

It was suggested some time ago (39) that the cell detects sudden changes in light intensity or oxygen tension because these changes alter the oxidation-reduction level of a component of the electron-transport chain. Recent experiments on the inhibition of bacteriochlorophyll synthesis in *Rhodospirillum molischianum* (74) which cannot grow aerobically in the dark, have corroborated this notion and have ruled out the possibility that changes in the amount of ATP in the cell control bacteriochlorophyll synthesis (75).

2. CONTROL OF REACTION CENTER BACTERIOCHLOROPHYLL

The earlier work on the control of bacteriochlorophyll synthesis by light intensity or oxygen tension was concerned only with the total amount of bacteriochlorophyll, and did not distinguish between bulk bacteriochlorophyll and reaction center bacteriochlorophyll (see Chapter 19). Obviously, in any consideration of the control of bacteriochlorophyll synthesis, it is necessary to take into account the relative amounts of these two species of bacteriochlorophyll; in other words, it is necessary to ask whether the size of the photosynthetic unit in bacteria can vary.

Unpublished experiments by Aagaard (77) in the laboratory of one of us (W.R.S.) have revealed several interesting facts about the control of the amount of reaction center bacteriochlorophyll in *Rhodopseudomonas spheroides*. Aagaard used as a measure of the amount of reaction center bacteriochlorophyll (in *R. spheroides* this is designated P870) the extent of the light-induced bleaching at 870 mμ. Some typical results are shown in Table II.

The amount of P870 varied from 0.4% to 1.5% of the total bacteriochlorophyll, while the specific total bacteriochlorophyll content varied from about 5 to 0.25 μg bacteriochlorophyll per 100 mg protein. In other words, the photosynthetic unit varied from about 250 molecules of bacteriochlorophyll per molecule of P870 to about 70. Since this is considerably less than the variation in specific bacteriochlorophyll content, it follows that the number of photosynthetic units changed by a factor of about 5.

The absorption spectrum of *R. spheroides* has three maxima due to bacteriochlorophyll at 800 mμ, 850 mμ, and 875 mμ. The absorbancy at 875 mμ relative to that at 800 and 850 mμ increases with decreasing cellular bacteriochlorophyll content; the absorbancy at 800 mμ relative to that at 850 mμ is constant. We can ascribe the absorption at 875 mμ to a bacteriochlorophyll component designated B875 and that at 800 and 850 mμ to a second component, B800–850. Aagaard found that the ratio of P870 (reaction center bacteriochlorophyll) to B875 does not vary with changes in the total cellular bacteriochlorophyll content. The spectrum of carotenoid-less mutants of *R. spheroides* indicates that they possess only B875. In these mutants, the ratio of P870 to total bacteriochlorophyll is independent of the specific bacteriochlorophyll content of the cells. Furthermore, the value of the ratio of P870 to B875 in the wild type is about equal to the ratio of P870 to total bacteriochlorophyll in carotenoid-less mutants.

TABLE II

REACTION CENTER AND TOTAL BACTERIOCHLOROPHYLL IN CHROMATOPHORE
MATERIAL FROM *Rhodopseudomonas spheroides*
GROWN UNDER VARIOUS CONDITIONS

Strain	Growth condition	Specific Bchl content[a]	ΔOD at 870 mμ per 10 μg Bchl	P870/Bchl[b] (%)
Ga	60 ftc	4.65	−0.005	0.4
	Semi-aerobic	0.25	−0.014	1.2
R-22	250 ftc	1.93	−0.018	1.5
(carotene-less)	6000 ftc	0.25	−0.022	1.8

[a] Micrograms Bchl per 100 μg cell protein.
[b] Molecules P870 as percentage of total number of Bchl molecules. The amount of P870 was calculated on the assumption that the molar extinction coefficient of P870 is the same as that of Bchl at 870 mμ (12×10^4).

These results suggest that the control of bacteriochlorophyll synthesis in *R. spheroides* is compounded of two distinct parts: the first having to do with the control of the number of photosynthetic units, and the second with the control of the size of the units.

3. STRUCTURAL BASIS OF CONTROL OF BACTERIOCHLOROPHYLL CONTENT

A priori, one might imagine that variation of cellular bacteriochlorophyll content could reflect either a change in the amount of chromatophore material with a fixed bacteriochlorophyll content or a change in the bacteriochlorophyll content of a fixed amount of chromatophore material or a combination of these two extremes.

The first alternative leads to the expectation that the specific bacteriochlorophyll content of chromatophore material isolated from cultures with different specific bacteriochlorophyll contents will be constant and that the amount of the chromatophore material will vary according to the cellular bacteriochlorophyll content. The second alternative leads to precisely the opposite expectations.

It would be possible to test these expectations without isolating chromatophore material if a specific chemical marker for this material were available. It is likely that lipid phosphorus can be used as such a marker because of the membranous nature of chromatophore material. Indeed, Worden and Sistrom (*41*) have indicated that chromatophore material from *R. spheroides* has a constant lipid phosphorus content regardless of the bacteriochlorophyll content of the cells.

With these ideas in mind, analyses of cells of *Rhodopseudomonas palustris* and *Rhodospirillum rubrum* grown in various light intensities for lipid phosphorus, DNA, and bacteriochlorophyll (Table III) have been

TABLE III

AMOUNTS OF BACTERIOCHLOROPHYLL AND LIPID PHOSPHORUS RELATIVE TO THE DNA CONTENT OF CELLS OF *Rhodopseudomonas palustris* AND *Rhodospirillum rubrum* GROWN AT LOW AND HIGH LIGHT INTENSITIES

	R. palustris			*R. rubrum*		
Component	100 ftc	6000 ftc	Ratio[a]	100 ftc	6000 ftc	Ratio[a]
Bchl:DNA (µg/µg)	4.06	0.51	(8.0)	5.6	1.1	(5.1)
Lipid P:DNA (mµmole/µg)	13.2	2.0	(6.7)	27.8	14.4	(1.92)
Bchl:lipid P (µg/µmole)	310.0	260.0	(1.2)	201.0	77.5	(2.6)

[a] Figures in parentheses are values for 100 ftc cells relative to those for 6000 ftc cells.

performed (76). The ratio of bacteriochlorophyll to DNA may be taken as an index of the specific cellular bacteriochlorophyll content; both species show a wide variation in this ratio. The ratio of lipid phosphorus to DNA may be taken as an index of the amount of membranous material per cell. In the case of *R. palustris* the change in amount of membranous material is nearly equal to the change in the cellular bacteriochlorophyll content. It follows that the bacteriochlorophyll content of the chromatophore material should be constant; this seems to be the case since the ratio of bacteriochlorophyll to lipid phosphorus is roughly constant. The variation in cellular bacteriochlorophyll content in *R. palustris* is, therefore, entirely accounted for by the change in the amount of chromatophore material per cell. On the other hand, in *R. rubrum* the change in the amount of membrane material per cell (lipid phosphorus per DNA) accounts for only about one third of the change in the amount of bacteriochlorophyll per cell. The larger part of the latter change is due to the variation in the amount of bacteriochlorophyll per unit chromatophore material (bacteriochlorophyll per lipid phosphorus).

Analyses of isolated chromatophore material from *Rhodospirillum molischianum* (35) and *Rhodopseudomonas spheroides* (37) indicate that these organisms follow the same pattern as *Rhodospirillum rubrum*.

That is, the major part of the change in cellular bacteriochlorophyll is due to a change in the specific bacteriochlorophyll content of the chromatophore material, accompanied by a relatively minor change in the amount of chromatophore material per cell.

Holt and Marr (68) have performed similar analyses on isolated chromatophore fractions of R. rubrum and reached an opposite conclusion. They found that the bacteriochlorophyll content per unit chromatophore material is constant when extracted from cultures grown at low and moderate light intensities. However, there is a marked decrease in the bacteriochlorophyll content per unit chromatophore material from cells grown at high light intensities. Hence, their conclusion that R. rubrum behaves according to the first alternative mentioned above is true only over a relatively narrow range of specific bacteriochlorophyll contents.

We conclude that there are at least two ways in which photosynthetic bacteria can vary the amount of bacteriochlorophyll per cell. The first, exemplified by the behavior of Rhodopseudomonas palustris, involves an increase in the amount of chromatophore material with no change in the amount of bacteriochlorophyll per unit chromatophore material. The second, exemplified by Rhodospirillum rubrum, involves an increase in both the amount and specific bacteriochlorophyll content of the chromatophore material.

So far, all the comparative electron microscopic examinations have been performed on organisms for which analytical data suggest only minor variations in the amount of chromatophore material, e.g., R. rubrum and R. molischianum. As has been observed repeatedly (35–37), the electron micrographs show a much greater change in the amount of membranous material than would be expected from the analytical data. No explanation for this discrepancy is possible at the present time.

REFERENCES

(1) R. Y. Stanier and C. B. van Niel, Arch. Mikrobiol. **42**, 17 (1962).
(2) F. E. Fritsch, in "Structure and Reproduction of the Algae," Vol. II, p. 770. Cambridge Univ. Press, London and New York, 1945.
(3) W. Niklowitz and G. Drews, Arch. Mikrobiol. **24**, 134 (1956).
(4) W. Niklowitz and G. Drews, Arch. Mikrobiol. **27**, 150 (1958).
(5) H. Ris and N. Singh, J. Biophys. Biochem. Cytol. **9**, 63 (1961).
(6) D. C. Wildon and F. V. Mercer, Australian J. Biol. Sci. **16**, 585 (1963).
(7) J. A. Chapman and M. R. J. Salton, Arch. Mikrobiol. **44**, 311 (1962).
(8) P. Eshlin, Protoplasma **58**, 439 (1964).
(9) H. S. Pankratz and C. C. Bowen, Am. J. Botany **50**, 387 (1963).
(10) W. T. Hall and G. Claus, J. Cell Biol. **19**, 551 (1963).
(11) H. L. Crespi, S. F. Mandeville, and J. J. Katz, Biochem. Biophys. Res. Commun. **9**, 569 (1962).

(12) H. K. Schachman, A. B. Pardee, and R. Y. Stanier, *Arch. Biochem. Biophys.* **38**, 245 (1952).
(13) J. B. Thomas, *Koninkl. Ned. Akad. Wetenschap., Proc.* **C55**, 207 (1952).
(14) B. Petrack and F. Lipmann, in "Light and Life," (W. D. McElroy and B. Glass, eds.), p. 621. Johns Hopkins Press, Baltimore, Maryland, 1961.
(15) W. A. Susor and D. W. Krogmann, *Biochim. Biophys. Acta* **88**, 11 (1964).
(16) A. J. Shatkin, *J. Biophys. Biochem. Cytol.* **7**, 583 (1960).
(17) E. Levin, W. J. Lennarz, and K. Bloch, *Biochim. Biophys. Acta* **84**, 471 (1964).
(18) J. Erwin and K. Bloch, *Biochem. Z.* **338**, 496 (1963).
(19) R. Wurmser, R. Levy, and G. Tessier, *Ann. Physiol. Physicochim. Biol.* **1**, 298 (1925).
(20) C. S. French, *Science* **88**, 60 (1938).
(21) C. B. van Niel, *Bacteriol. Rev.* **8**, 1 (1944).
(22) A. B. Pardee, H. K. Schachman, and R. Y. Stanier, *Nature* **169**, 282 (1952).
(23) A. W. Frenkel, *J. Am. Chem. Soc.* **76**, 5568 (1954).
(24) A. E. Vatter and R. S. Wolfe, *J. Bacteriol.* **75**, 480 (1958).
(25) A. L. Tuttle and H. Gest, *Proc. Natl. Acad. Sci. U.S.* **45**, 1261 (1959).
(26) J. A. Bergeron and R. C. Fuller, in "Macromolecular Complexes" (M. V. Edds, Jr., ed.), p. 179. Ronald Press, New York, 1961.
(27) G. Drews, *Arch. Mikrobiol.* **36**, 99 (1960).
(28) E. S. Boatman and H. C. Douglas, *J. Biophys. Biochem. Cytol.* **11**, 469 (1961).
(29) G. Cohen-Bazire, unpublished experiments (1965).
(30) W. De Boer, unpublished experiments (1966).
(31) K. E. Eimhjellen, O. Aasmundrund, and A. Jensen, *Biochem. Biophys. Res. Commun.* **10**, 232 (1963).
(32) R. C. Fuller, S. F. Conti, and D. M. Mellin, in "Bacterial Photosynthesis" (H. Gest, A. San Pietro, and L. P. Vernon, eds.), p. 71. Antioch Press, Yellow Springs, Colorado, 1963.
(33) G. Cohen-Bazire, in "Bacterial Photosynthesis" (H. Gest, A. San Pietro, and L. P. Vernon, eds.), p. 89. Antioch Press, Yellow Springs, Colorado, 1963.
(34) P. Giesbrecht and G. Drews, *Arch. Mikrobiol.* **43**, 152 (1962).
(35) S. P. Gibbs, W. R. Sistrom, and P. B. Worden, *J. Cell Biol.* **26**, 395 (1965).
(36) G. Cohen-Bazire and R. Kunisawa, *J. Cell Biol.* **16**, 401 (1963).
(37) G. Drews and P. Giesbrecht, *Zentr. Bakteriol., Parasitenk., Abt. I. Orig.* **190**, 508 (1963).
(38) S. C. Holt and A. G. Marr, *J. Bacteriol.* **89**, 1402 (1965).
(39) G. Cohen-Bazire, W. R. Sistrom, and R. Y. Stanier, *J. Cellular Comp. Physiol.* **49**, 25 (1957).
(40) G. Cohen-Bazire and R. Kunisawa, *Proc. Natl. Acad. Sci. U.S.* **46**, 1543 (1960).
(41) P. B. Worden and W. R. Sistrom, *J. Cell Biol.* **23**, 135 (1964).
(42) S. C. Holt and A. G. Marr, *J. Bacteriol.* **89**, 1413 (1965).
(43) K. D. Gibson, *Biochemistry* **4**, 2027 (1965).
(44) J. A. Bergeron, in "The Photochemical Apparatus, its Structure and Function," p. 118. Brookhaven Natl. Lab., Upton, New York, 1959.
(45) R. Langridge, P. D. Barron, and W. R. Sistrom, *Nature* **204**, 97 (1964).
(46) J. W. Newton and G. A. Newton, *Arch. Biochem. Biophys* **71**, 250 (1957).
(47) J. Sykes, J. A. Gibson, and D. S. Hoare, *Biochim. Biophys. Acta* **109**, 409 (1965).
(48) J. Lascelles and J. F. Szilágyi, *J. Gen. Microbiol.* **38**, 55 (1965).
(49) G. Constantopoulos, personal communication (1965).
(50) H. Löw and B. A. Afzelius, *Exptl. Cell Res.* **35**, 431 (1964).

(51) W. Stoeckenius, *J. Cell Biol.* **17**, 443 (1963).
(52) D. Abram, *J. Bacteriol.* **89**, 855 (1965).
(53) R. B. Park and N. G. Pon, *J. Mol. Biol.* **6**, 105 (1963).
(54) D. D. Hickman, A. W. Frenkel, and K. Cost, *in* "Bacterial Photosynthesis" (H. Gest, A. San Pietro, and L. P. Vernon, eds.), p. 111. Antioch Press, Yellow Springs, Colorado, 1963.
(55) G. Cohen-Bazire, unpublished observations (1965).
(56) R. Y. Stanier and J. H. C. Smith, *Biochim. Biophys. Acta* **41**, 478 (1960).
(57) J. M. Olson and C. A. Romano, *Biochim. Biophys. Acta* **59**, 726 (1962).
(58) S. Liaaen Jensen, *Ann. Rev. Microbiol.* **19**, 163 (1965).
(59) J. A. Bergeron and R. C. Fuller, *in* "Biological Structure and Function" (T. W. Goodwin and O. Lindberg, eds.), Vol. 2, p. 307. Academic Press, New York, 1961.
(60) G. Cohen-Bazire, N. Pfennig, and R. Kunisawa, *J. Cell Biol.* **22**, 207 (1964).
(61) P. C. Fitz-James, *J. Biophys. Biochem. Cytol.* **8**, 507 (1960).
(62) E. I. Rabinowitch, *in* "Photosynthesis and Related Processes," Vol. 1, p. 424. Wiley (Interscience), New York, 1945.
(63) J. Myers and W. Kratz, *J. Gen. Physiol.* **39**, 11 (1955).
(64) A. Hattori and Y. Fujita, *J. Biochem.* **46**, 521 (1959).
(65) Y. Fujita and A. Hattori, *Plant Cell Physiol.* (*Tokyo*) **1**, 293 (1960).
(66) L. W. Jones and J. Myers, *J. Phycol.* **1**, 6 (1965).
(67) R. K. Clayton, *Biochim. Biophys. Acta* **75**, 312 (1963).
(68) S. C. Holt and A. G. Marr, *J. Bacteriol.* **89**, 1421 (1965).
(69) W. R. Sistrom, *J. Gen. Microbiol.* **28**, 607 (1962).
(70) J. Lascelles, *Biochem. J.* **72**, 508 (1959).
(71) W. R. Sistrom, *J. Gen. Microbiol.* **28**, 599 (1962).
(72) M. J. Bull and J. Lascelles, *Biochem. J.* **87**, 15 (1963).
(73) J. Lascelles, *J. Gen. Microbiol.* **23**, 487 (1960).
(74) W. R. Sistrom, *J. Bacteriol.* **89**, 403 (1965).
(75) W. R. Sistrom, *in* "Bacterial Photosynthesis" (H. Gest, A. San Pietro, and L. P. Vernon, eds.), p. 53. Antioch Press, Yellow Springs, Colorado, 1963.
(76) R. Kunisawa and G. Cohen-Bazire, unpublished experiments (1965).
(77) J. Aagaard, unpublished experiments (1965).

—11—

Spectral Characteristics of Chlorophyll in Green Plants

WARREN L. BUTLER

Department of Biology
University of California, San Diego–La Jolla, California

I. Introduction

In recent years a number of different forms of chlorophyll *a* have been observed spectroscopically *in vivo*. These results are in marked contrast to the earlier literature, in which it was assumed that only one form of chlorophyll *a* existed. In the early literature, extraction, purification, and chemical identification established the existence of several chemically distinct forms of chlorophyll, labeled *a*, *b*, *c*, and *d*, but there was little reason to expect or look for more than one modification of chlorophyll *a*. The photochemical function of the pigments up to the actual photochemical event was generally agreed upon. Chlorophyll *a*, which was found in all photosynthetic tissue and organisms (except bacteria), was assumed to be the primary pigment of photosynthesis. The other major pigments, which were characteristic of the species, such as carotenes, chlorophylls *b* and *c*, and phycobilin pigments, were classified as accessory pigments.

Englemann (*1*) showed very early that light absorbed by the phycobilin accessory pigments of marine algae was used for photosynthesis. Later Dutton and Manning (*2*) measured quantum yields of photosynthesis of diatoms in weak monochromatic light and found that the carotenoid accessory pigments, mainly fucoxanthol, were as efficient as chlorophyll in sensitizing photosynthesis. Similar results on the photosynthetic efficiency of accessory pigments in algae were reported by Emerson and Lewis (*3*). Dutton *et al.* (*4*) then found that light absorbed by fucoxanthol gave rise to chlorophyll fluorescence, thus showing that the accessory pigments functioned in photosynthesis by transferring energy to chlorophyll *a*. French and Young (*5*) and Duysens (*6*) demonstrated accessory-pigment sensitization of chlorophyll fluorescence in red and blue-green algae, and Duysens (*6*) made a comprehensive quantitative study of energy transfer in photosynthetic organisms. The efficiency of energy transfer to chlorophyll *a* ranged from about 50% for carotenoid pigments to 100% for chlorophyll *b*. The action of accessory pigments supported the concept of a chlorophyll *a*-dominated photochemical system. On teleological grounds the *raison d'être* for accessory pigments was to make a greater portion of the visible spectrum useful for photosynthesis. The teleological argument was particularly convincing in the case of phycobilin-containing marine algae which lived at depths in the ocean which could be penetrated only by green light.

The mechanism of energy transfer was generally assumed to be inductive resonance. The efficiency of transfer depended on the overlap between the fluorescence emission spectrum of the donor molecule and the absorption spectrum of the acceptor molecule and the proximity of the molecules (*7*). The idea of resonance transfer of energy from pigment to pigment, always toward the longer wavelength-absorbing form, became the dominating concept in photochemical mechanisms of photosynthesis. Emerson and Arnold's (*8*) experiments on photosynthesis in flashing light could be explained by assuming that several hundred chlorophyll molecules acted together as a unit. It was suggested (*9*) that a few special chlorophyll molecules which had a longer wavelength maximum than chlorophyll *a* acted as energy traps by virtue of their strategically placed absorption maximum. The energy gathered by the chlorophyll *a* molecules, either by direct photon absorption or by resonance transfer from accessory pigment molecules, was transferred by inductive resonance to the long-wavelength chlorophyll molecules, where the actual photochemistry took place.

These were the prevailing views up to the time of Emerson's classical experiments on the synergistic effects of simultaneous irradiation with two wavelengths of monochromatic light (10). In retrospect, it is apparent how a powerful principle, such as energy transfer, can form a conceptual straitjacket that is difficult to break out of. The development of a dual-pigment system concept of photosynthesis as opposed to energy transfer within a single, chlorophyll a-dominated pigment system was concomitant with the accumulation of evidence that more than one form of chlorophyll a existed in vivo. The present chapter will review the literature on multiple forms, or physical states, of chlorophyll a in vivo and attempt to relate these different forms to the current two-pigment system concept. The concepts of the relationship between chlorophyll fluorescence and photosynthesis were also undergoing a major revision during the period that the two-pigment system theory was being developed. The fluorescence yield of chlorophyll in vivo has been related to a specific point in the photosynthetic electron transport chain and, as a result, fluorescence measurements have become an important tool in the study of photosynthetic electron transport. The recent literature on chlorophyll fluorescence will be discussed from the viewpoint of the two-pigment theory of photosynthesis.

II. Early Evidence for an "Inactive" Form of Chlorophyll a

Not all the experimental findings could be readily explained by the generally accepted scheme of a single chlorophyll a-mediated photochemical system. Emerson and Lewis (11) measured the quantum yield of photosynthesis of Chlorella in monochromatic light and found that the yield declined at wavelengths longer than 680 mμ even though chlorophyll a absorbed wavelengths in this region. On simple theory, any quantum of red light that was absorbed should excite a chlorophyll molecule to its lowest excited singlet state and should be as photochemically active as any other absorbed quantum.

A few years later, Haxo and Blinks (12), working with the red alga Schrizymenia, reported another anomaly that was difficult to explain with the accepted concept that chlorophyll a was the primary pigment of photosynthesis. Based on equal incident quantum flux, green light absorbed by the accessory pigment, phycoerythrin, was three times as effective in photosynthesis as red light absorbed by chlorophyll a. The

anomaly was shown to obtain in a large number of red algae, and preliminary measurements on blue-green algae indicated the same phenomenon (13).

Duysens (6) confirmed that the phycobilin accessory pigments in
red and blue-green algae were more active photosynthetically than the
chlorophyll and further showed that the action spectrum for the excitation of chlorophyll fluorescence was essentially the same as the action
spectrum for photosynthesis. That light absorbed by the phycobilin
should excite chlorophyll fluorescence more efficiently than light absorbed directly by the chlorophyll was even more of an anomaly. Duysens suggested that the chlorophyll in red and blue-green algae consisted
approximately of equal parts of a fluorescent, photochemically active
form and a nonfluorescent, inactive form and that the accessory pigment
preferentially transferred excitation energy to the active form. Duysens
(6, 14) also found evidence in the fluorescence emission spectrum of
Porphyra lacineata for the presence of small amounts of a long-wavelength absorbing form of chlorophyll which was sensitized to fluorescence by energy transfer from chlorophyll *a*. He postulated that half of
the chlorophyll *a* was inactive photochemically because it transferred
excitation to an energy sink which was photochemically inactive. The
hypothesis of an inactive form of chlorophyll absorbing in the long wavelength region could also account for the decline in the quantum yield of
photosynthesis found by Emerson and Lewis in far-red light. Lavorel
(15) suggested, on the basis of a study of the fluorescence of dyes as a
function of concentration, that the "red drop" was due to the presence
of nonfluorescent, photochemically inactive chlorophyll dimers absorbing on the long-wavelength side of the chlorophyll absorption band.

Krasnovsky and Kosobutskaya (16) studied the formation of chlorophyll in cell-free extracts of dark-grown bean leaves. They found that
protochlorophyll was transformed to a 670 mμ-absorbing form of chlorophyll which was subject to a destruction by photooxidation. As the
chlorophyll accumulated in the leaf, the absorption maximum gradually
shifted to 678 mμ and the chlorophyll became more resistant to photooxidation. Early work by a number of investigators [see Rabinowitch
(17)] showed that the absorption maximum of chlorophyll shifted to
longer wavelengths when the chlorophyll was precipitated as a colloidal
suspension or when it was absorbed onto a solid phase. Krasnovsky and
Kosobutskaya suggested that (a) chlorophyll existed *in vivo* in different
states of aggregation, (b) the shift of the absorption maximum of chlorophyll in greening leaves toward longer wavelengths was due to aggregation, and (c) the monomeric 670 mμ-absorbing chlorophyll was a
fluorescent, photochemically active form, whereas the colloidally aggre-

gated longer-wavelength forms were relatively nonfluorescent and in-active. The absorption maximum at 678 mμ in mature leaves would represent an equilibrium between the monomeric and aggregated forms. The relationship between the absorption maximum and the degree of aggregation was supported by the later work of Trurnit and Colmano (18). They showed that the absorption maximum of chlorophyll in mono-molecular layers shifted toward longer wavelengths as the layer was compressed on a film balance.

For some time the theoretical arguments indicating more than one form of chlorophyll a in vivo were not supported by direct spectroscopic measurements on photosynthetic tissue. Some of the early measurements of the absorption spectra of leaves and algae [see Rabinowitch (17)] showed an asymmetry in the long-wavelength chlorophyll absorption band which, in retrospect, can be attributed to the presence of two forms of chlorophyll a. The measurements lacked a high degree of resolution, and uncertainties as to the effects of light scattering gave rise to caution in the interpretation of the spectra. Spectroscopic techniques capable of resolving different forms of chlorophyll a in vivo began to be applied in the latter half of the 1950's.

III. Absorption Spectroscopy

The measurement and interpretation of the absorption spectra of plant tissue is complicated by the light-scattering properties of the ma-terial. For our purposes the most serious of the artifacts are associated with the wavelength-selective scattering arising from the anomalous dispersion at absorption bands (19). Fortunately, these artifacts can be eliminated by using a spectrophotometer which collects a reasonably large solid angle from the sample. The Shibata opal-glass technique (20) is a convenient method by which commercial split-beam spectrophotom-eters can be used to measure absorption spectra of relatively trans-parent light-scattering material. A more efficient method is to place a large-area phototube directly behind the sample (21, 22). Spectro-photometers employing integrating spheres to collect all the scattered light were used much earlier. Rabinowitch (17) has reviewed the early literature on the absorption spectra of intact leaves and algae suspensions obtained with integrating-sphere spectrophotometers. The integrating sphere is the most reliable method of measuring the total amount of light absorbed by a light-scattering sample, but it is not the best method of measuring absorption spectra for the study of pigments. The absorption bands in a spectrum obtained with an integrating sphere, which measures the sum of the transmitted and reflected light, will be flatter and show less detail than the absorption bands in a spectrum measured by transmitted light only. The reader is referred to a recent review (23) on absorption spectroscopy of intact tissue for a more thorough discussion of the techniques of measurement and the artifacts which may arise.

Given adequate instrumentation, the problem of measuring different forms of chlorophyll *in vivo* is mainly one of resolution and spectrophotometric detectability.

A. Derivative Spectra

French and Harper (24) developed a derivative spectrophotometer to look for different forms of chlorophyll *in vivo*. The derivative spectrum does not contain any information that is not present in the absorption spectrum, but the derivative presentation accentuates and calls attention to small shoulders and asymmetries that might be overlooked in the absorption spectrum. With this technique, French and Huang (25) confirmed Krasnovsky's proposal that the chlorophyll *a* absorption band *in vivo* was comprised of two absorption bands with maxima near 670 and 680 mμ. Figure 1 shows the derivative spectrum of *Chlorella* along with the derivative spectrum of pure chlorophyll *a* in ether. The derivative spectrum of the algae shows chlorophyll *b* with an absorption maximum at about 650 mμ and two components absorbing in the chlorophyll *a* region. The absorption maxima of the latter two components were estimated to be 673 and 683 mμ. French and Elliot (26) found an additional absorption band with a maximum at 695 mμ in the derivative spectrum of *Euglena* and *Ochromonas*. The amount of 695-mμ-absorbing pigment varied considerably in *Euglena*, depending on growth conditions and the age of the culture, and could account for an appreciable part of the chlorophyll absorption band. Extracts of *Euglena* in 80% acetone, however, showed only one form of chlorophyll *a* and the small amount of chlorophyll *b* typical of *Euglena*. The three pigments, denoted in the present chapter as Chl *a* 673, Chl *a* 683, and Chl *a* 695, were assumed to be different physical states of chlorophyll *a*.

The derivative spectra of *Euglena* and *Ochromonas* have a definite shoulder which indicates the presence of Chl *a* 695. The derivative spectra for other green algae are essentially the same as that shown in Fig. 1 for *Chlorella*, where the presence of Chl *a* 695 is not apparent. Brown and French (27) used a computer to analyze the derivative spectra of *Chlorella* for the individual components. The absorption bands were assumed to have Gaussian distributions, and the parameters of wavelength maximum, band width, and absorbancy were varied to give a best fit to the experimental curve. The analysis for *Chlorella* indicated 15% Chl *b* 653 with a half-width of 15 mμ, 40% Chl *a* 673 with a half-width of 18 mμ, 40% Chl *a* 683 with a half-width of 14 mμ, and 5% Chl *a* 695 with a half-width of 15 mμ. It was recognized that the presence of Chl *a* 695 in *Chlorella* was not proved, but depended upon the shape assumed for the absorption bands.

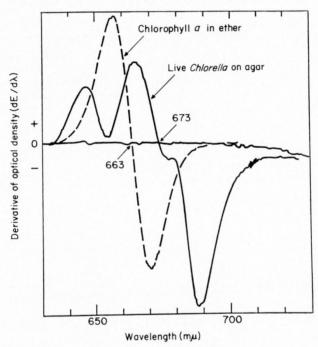

FIG. 1.   Derivative absorption spectra of *Chlorella* on agar and chlorophyll *a* in ether (25).

Brown and French (28) found an additional absorption band at 710 mμ in some old cultures of *Euglena* that had started to turn brown. The 695 mμ-absorbing band appeared to decrease as the 710 mμ band accumulated. Extracts of cultures showing the 710 mμ pigment showed relatively large amounts of pheophytin *a*. It was proposed that the pigment was a complex of pheophytin *a*, P *a* 710, which was formed from Chl *a* 695 in old cultures of *Euglena*.

B. Low-Temperature Spectra

Butler (29, 30) showed by low-temperature spectroscopy that several pigments contribute to the red-absorption band of chlorophyll. At

—196°C the absorption bands are sufficiently sharpened that the individual maxima can be resolved in the absorption spectra. The absorption
and derivative spectra of two strains of *Scenedesmus* at —196°C are
shown in Fig. 2. The absorption bands of chlorophyll *b* and two types of
chlorophyll *a* are clearly resolved in the spectra of both the wild-type and
the mutant strains. In addition, a small amount of a pigment absorbing
near 700 mμ is apparent as a small shoulder on the absorption spectrum

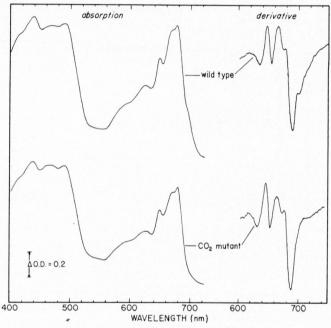

FIG. 2. Absorption spectra and derivative absorption spectra at —196°C of
wild-type strain and CO_2 mutant no. 8 of *Scenedesmus*.

and as a peak in the derivative spectrum of the wild-type cells. This pigment was first noted in the absorption, derivative-absorption and fluorescence-excitation spectra of green leaves at —196°C and was denoted
C-705 (29, 30). The mutant, which lacks the ability to fix CO_2 photosynthetically, does not have the C-705.

The relationship of C-705 to Chl *a* 695 may be a question of definition.
The low-temperature absorption spectra of *Euglena* cells showed small
shoulders which indicated the presence of both C-705 and Chl *a* 695
(31). (In old brownish cultures of *Euglena* an additional band at 710 mμ
corresponding to P *a* 710 was observed.) Thus, in *Euglena*, C-705 and
Chl *a* 695 are different pigments. However, in other green algae where
the presence of Chl *a* 695 has only been inferred from a computer analy-

sis, it is likely that C-705 and Chl *a* 695 are the same pigment. It was suggested that the Chl *a* 695 that is observed in the spectra of *Euglena* and *Ochromonas* is not a ubiquitous pigment and that it is different from the C-705 (*31*). C-705 has been detected as an absorption band or a fluorescence-excitation band in all green plants examined except for the mutant of *Scenedesmus* which fails to fix CO_2. The appearance of C-705 has been correlated with the onset of photochemical activity in a greening leaf (*30*). It should be kept in mind that the absorption band of C-705 has been resolved only at low temperatures. At room temperature the absorption band might be somewhat shifted.

C. Reversible Absorbancy Changes of Chlorophyll *in Vivo*

Several laboratories looked for forms of chlorophyll that were reversibly bleached by light. It was reasoned that those chlorophyll molecules that were involved in the photochemical conversion of energy should be altered during the conversion and restored in the dark. Coleman *et al.* (*32*) measured the difference of absorbancy between an illuminated and a dark sample of *Chlorella*. They found small negative bands in the light minus dark difference spectrum at 650, 680, and 700 mμ which they attributed to the bleaching of chlorophyll.

Kok (*33*) looked for reversible absorbancy changes by repetitively irradiating the sample with brief high-intensity flashes. The absorbancy was measured just before and just after each flash and the difference was integrated over a large number of flashes. This type of measurement detects reversible light-induced absorbancy changes that have a characteristic dark-decay time. Kok and Hoch (*34*) also found small negative bands at 650, 680, and 705 mμ with green algae.

The reversible bleaching at 650 and 680 mμ has been ascribed to artifacts arising from light-induced fluorescence-yield changes (*35*). Butler (*35*) showed that irradiation with red light increased the fluorescence yield of chlorophyll *in vivo* and irradiation with far-red light decreased the yield. In the experiments of Coleman *et al.* (*32*) and of Kok and Hoch (*34*), a small part of the measured signal was undoubtedly chlorophyll fluorescence excited by the measuring beam. Actinic irradiation of the sample with red or white light would increase the fluorescence yield of the chlorophyll so that the measurement would appear to show a slight bleaching. The magnitude of the bleaching at 650 and 680 mμ was well within a conservative estimate of the apparent optical-density changes resulting from fluorescence-yield changes (*35*). The fluorescence-yield changes would not account for the reversible bleaching at 705 mμ however.

On the assumption that the 705 mμ band represented the photochemically active pigment at the reaction center of a photosynthetic unit, Kok concentrated his efforts on defining the properties and function of this pigment, which he called "P700." The reversibly bleached absorption band was found in the region of 700–705 mμ (34), in all photosynthetic organisms tested and was particularly pronounced in blue-green and red algae. The ratio of P700 to chlorophyll a, assuming equal extinction coefficients, was about 1:400 (36). The bleaching of P700 was sensitized by light absorbed by chlorophyll a and carotenoids. The quantum yield for the bleaching was close to unity (36). P700 could be bleached chemically by ferricyanide, and the bleaching was reversed by ferrocyanide. The E'_0 of P700 was titrated against the ferricyanide-ferrocyanide redox couple and found to be 0.43 volt (37). Hexane-soluble components and 85% of the chlorophyll could be removed from chloroplast fragments without destroying the P700 (37). Attempts to solubilize P700 or to further enrich the particles resulted in a loss of P700. Acetone extracts of the P700-enriched fragments showed chlorophyll a, but no special long-wavelength-absorbing pigments.

The relationship between P700 and C-705 is questionable. Butler and Baker estimated the amount of C-705 in spinach chloroplasts from the low-temperature absorption spectrum to be 1–2% of the chlorophyll (31). Kok estimated that C-705 was 2–5% of the chlorophyll in *Scenedesmus* (38). The estimate of P700 based on the reversible bleaching of 705 mμ is 0.2–0.3% of the chlorophyll. The most striking difference, however, is that P700 is bleached by light even at −196°C while C-705 is not. It is possible, however, that C-705 is similar to P700, but that only a part of it is connected to the photosynthetic electron transport chain. It is also possible that at low temperature, where the C-705 absorption band is measured, more of the pigment is present than at room temperature. The main evidence for relating the two pigments comes from a mutant of *Scenedesmus* which lacks the low temperature absorption band of C-705 (39) and the reversible absorption band of P700 as well (38).

D. Nonspectroscopic Evidence for Heterogeneity of Chlorophyll *in Vivo*

Shlyk and Nikolayeva (40) showed by differential extraction techniques and by differential labeling with $^{14}CO_2$ that both chlorophyll a and chlorophyll b might exist in more than one state *in vivo*. They found that 20–25% of both chlorophyll a and b of green leaves was readily extracted with a weak polar solvent (less than 1% ethanol in petroleum ether) while extraction of the rest of the chlorophyll required a more polar solvent. They also incubated the leaves in the light with $^{14}CO_2$ for

10–30 minutes and compared the specific activity of the two fractions obtained by differential extraction. With both chlorophyll *a* and *b*, the specific activity of the fraction that was readily extracted with the weak polar solvent was significantly higher than the non-readily extractable fraction. They concluded that both chlorophylls *a* and *b* existed in at least two different states representing different ages of molecules. The newer chlorophyll molecules were more readily extracted and more readily labeled than the older molecules. This is one of the few pieces of evidence that indicates chlorophyll *b* may exist in two forms. French and Elliot (*26*) noted a double character to be chlorophyll *b* band on the derivative spectrum of *Ulva* and cautiously suggested the possibility of two types of chlorophyll *b*.

E. Spectroscopic Evidence for Many Forms of Chlorophyll

There have been a few reports that the chlorophyll absorption bands *in vivo* are even more complicated than has been indicated thus far. Thomas (*41*) measured the absorption spectrum of *Aspidistra elatior* chloroplasts and noted a number of very small shoulders on the main chlorophyll absorption band in the red region. He concluded that at least six forms of chlorophyll contributed to the absorption band. These results, however, have not been confirmed by other investigators even though techniques capable of greater resolution, e.g., low-temperature absorption and derivative spectroscopy, have been employed.

Metzner (*42*) measured the reflectance spectrum of a homogeneous mixture of dried *Chlorella* cells and MgO powder which had been pressed into a tablet. The instrument used an integrating sphere to collect diffuse reflectance and a xenon arc for the measuring light. Metzner noted nine different shoulders in the chlorophyll absorption band between 665 and 685 mμ. It is possible, however, that the red emission lines in the xenon arc contributed to the complexity of the spectra. Menke *et al.* (*43*) mixed dried algae and chloroplast preparations with $CaCO_3$ powder in order to dilute the very strong absorbancy of the preparations with a dry non-absorbing medium. Absorption and derivative spectra of these highly scattering samples, measured at $-196°C$ with transmitted rather than reflected light, showed the same bands that are observed *in vivo*. Dried chloroplasts of *Antirrhinum magus* showed Chl *a* 673, Chl *a* 683, and C-705 in the low-temperature derivative spectrum, and dried *Euglena* showed Chl *a* 673, Chl *a* 683, Chl *a* 695, and C-705. Some difference in the relative amounts of the pigments was noted between freeze-dried and air-dried preparations.

Michel-Wolwertz and Sironval (*44*) purified chlorophylls *a* and *b* from mature green leaves. Both chlorophylls were separated into five isomers each by repeated chromatography. In both cases four of the isomers were present in very small amounts relative to the predominant form. Difference spectra between the various isomers revealed small spectral differences. These authors concluded that the various isomers represented different chemical forms of chlorophyll *a* and *b* that were present *in vivo*. The possibility that isomerization occurred during the purification procedures could not be ruled out, however (see Chapter 2).

Although the work of Michel-Wolwertz and Sironval may be taken as partial support for the contention by Thomas (*41*) and by Metzner (*42*) that a large number of forms of chlorophyll exist *in vivo*, the main body of the spectral measurements on intact plant material indicates that the bulk of the chlorophyll *a* exists in two forms with absorption maxima near 673 and 683 mμ. A small fraction of the chlorophyll has an absorption maximum at about 700 mμ. In addition, *Euglena* and *Ochromonas* have a form of chlorophyll *a* which absorbs maximally at 695 mμ.

IV. Two Photosynthetic Pigment Systems

Emerson's investigations of the "red drop" in the quantum efficiency of photosynthesis in *Chlorella* led him to the discovery that far-red light could be used efficiently if it was accompanied by light of shorter wavelength. The action spectrum for the ability of short wavelength light to enhance the efficiency of far-red light indicated that chlorophyll *b* was involved. Emerson *et al.* (*10*) suggested that photosynthesis required two different photochemical steps: one energized by chlorophyll *a*, and the other by chlorophyll *b*. Emerson and Rabinowitch (*45*) went on to investigate the "red drop" in other algae which contained accessory pigments other than chlorophyll *a*. In each case the short-wavelength light enhanced the action of wavelengths in the red-drop region and the action spectrum for enhancement corresponded to the absorption spectrum of the accessory pigment (phycocyanin in *Anacystis*, phycoerythrin in *Porphyridium*, and fucoxanthol in *Navicula*). Govindjee and Rabinowitch (*46, 47*) showed that the enhancement spectra for the various algae also had a peak at 670 mμ in addition to the maxima of the accessory pigment. It was suggested (*45*) that the two photochemical reactions were carried out by two types of chlorophyll *a*. The accessory pigments transferred excitation energy to Chl *a* 673 in one photochemical system while Chl *a* 683 and Chl *a* 695 (or C-705) were a part of the other photochemical system.

Evidence relating the two photochemical systems accumulated from several laboratories. Kok and Hoch (34) showed that P700 was affected differently by the two pigment systems. The long-wavelength system bleached (oxidized) P700 while the short-wavelength system restored (reduced) P700. Duysens (48) showed that in *Porphyridium* the action spectrum for the light-induced oxidation of cytochromes corresponded to the absorption spectrum of chlorophyll whereas the action spectra for O_2 evolution and for chlorophyll fluorescence correspond to phycoerythrin absorption. French and Myers (49) measured the action spectrum for O_2 evolution from *Chlorella* obtained by adding monochromatic light on top of a constant background of either 650 or 710 mμ light. The action spectrum obtained with the 650 mμ background was shifted appreciably to longer wavelengths than the action spectrum obtained with the 710 mμ background.

Duysens called the two pigment systems I and II (48). His action spectra on blue-green and red algae indicated that the accessory pigments and chlorophyll were found in both pigment systems, but that system II had most of the accessory pigment while system I had most of the chlorophyll. The chlorophyll in system II, chlorophyll a_2, had a higher fluorescence yield than the chlorophyll in system I, chlorophyll a_1. Duysens and Amesz (50) suggested an electron transport scheme which was similar to that presented by Hill and Bendall (51).

$$H_2O \rightarrow (\text{system II}) \rightarrow Q \rightarrow Cyt \rightarrow P \rightarrow (\text{system I}) \rightarrow XH \rightarrow PNH$$

where P is the P700 of Kok.

A number of action spectra indicate that Chl *a* 673 is associated with system II and that Chl *a* 683 is associated with system I. P700 is closely associated with system I and has been postulated to be the energy sink and reaction center site for system I. Spectroscopic evidence for a similar sink for system II has not been found. The Chl *a* 695 which is found in *Euglena* is difficult to place within the pigment systems because enhancement has not been found in *Euglena* (28).

Separation of the Two Pigment Systems by Fractionation

The concept that several hundred chlorophyll molecules acted in concert suggested that the photosynthetic unit might be a morphological entity. Thomas *et al.* (52) measured the Hill reaction of small chloroplast fragments as a function of particle size and concluded that the photochemical activity was maintained down to a particle size of 1000 mμ³. These particles were estimated to contain approximately 100 chlorophyll molecules. Park and Pon (53) obtained similar results, but found that somewhat larger particles were required for full activity. Direct morphological evidence for a unit was also found by Park *et al.* (53, 54). They noted in electron micrographs that the inner structure of chloroplast lamellae had a cobblestone appearance with units of about 100 Å to 200 Å. These units, called quantasomes, appear to contain the photochemical apparatus.

Theoretical photochemical mechanisms involving two separate pigment systems suggest the possibility of two photosynthetic units of different composition and function. Various attempts have been made to fractionate chloroplast fragments into particles with different pigment composition and different photochemical activities. Butler and Baker (31) sonicated *Chlorella* and *Euglena* cells and spinach chloroplasts and separated the chloroplast fragments by differential centrifugation. The various fractions were examined by low-temperature absorption and fluorescence-excitation spectroscopy for the presence of Chl *b*, Chl *a* 673, Chl *a* 683, C-705, and, in the case of *Euglena*, Chl *a* 695. Even the smallest particles (the supernatant from a 173,000 g centrifugation) had essentially the same pigment composition as the intact chloroplasts. The ratio of Chl *a* 683 to Chl *a* 673 was slightly lower in the smaller particles, but no major change of pigments was noted.

Allen et al. (55, 56) devised a technique of repeated freezing and grinding followed by sonication and density-gradient centrifugation of the supernatant from a 144,000 g centrifugation to obtain particles which had a higher chlorophyll *b* to chlorophyll *a* ratio and a higher Chl 673 to Chl 683 ratio than the larger particles. The particles exhibited a light-induced electron spin resonance (ESR) signal that remained in the dark, but could be discharged by reducing reagents. On concentration the particles tended to form membranes, and as the membranes formed the absorption maximum shifted from 672–673 mμ to 677–678 mμ.

Boardman and Anderson (57) fractionated chloroplast fragments by treatment with 0.5% digitonin and differential centrifugation. The small particles obtained by this procedure were enriched in chlorophyll *a* in contrast to the small particles obtained by Allen et al., which were enriched in chlorophyll *b*. The larger particles which sedimented at 10,000 g could reduce trichlorophenolindophenol (TCIP) and nicotinamide adenine dinucleotide phosphate (NADP) in a normal Hill reaction with water as the electron donor. The smaller particles lacked the capacity to evolve O_2. The smaller particles, however, were more active in the reduction of NADP when the electron donor was TCIP and ascorbic acid. Boardman and Anderson suggested that the digitonin treatment separated the two pigment systems and that the small particles contained only pigment system I. Further work is needed on fractionation of the chloroplast fragments, but the approach appears to be promising.

V. Chlorophyll Transformation during Chloroplast Development

Several spectroscopically distinguishable forms of chlorophyll have been noted in dark-grown bean leaves at different stages of greening. Shibata (58) showed that protochlorophyll with an absorption maximum at 650 mμ in dark-grown leaves, Pchl 650, was transformed by light to a form of chlorophyll absorbing maximally at 684 mμ, Chl 684. Chl 684 was converted to a 673 mμ-absorbing form of chlorophyll, Chl 673, in a dark reaction that required about 20 minutes to go to completion. Shibata also reported that Chl 673 was converted to a longer wavelength absorbing form, Chl 678, as a dark reaction during the following period of about 1 hour. Krasnovsky and Kosobutskaya (16) first noted that Chl 673 was converted to Chl 678 as the chlorophyll accumulated in the light, and they ascribed the spectral shift to an aggregation. The time course for the dark reactions associated with greening depends on the age of the etiolated leaves. The transformation of Chl 684 to Chl 673 and the re-formation of protochlorophyll were considerably slower in 9-day-old dark-grown bean leaves than in 6-day-old leaves (59).

The different forms of chlorophyll which appeared during greening were generally assumed to represent different chlorophyll-protein interactions. An alternative or more specific suggestion, presented below, is that the different absorbing forms of chlorophyll and protochlorophyll represent different states of chromophore aggregation. It should be emphasized that the experimental evidence does not distinguish uniquely between these two explanations.

Wolff and Price (60) showed that the photochemically active protochlorophyll and the newly formed chlorophyll were actually protochlorophyllide and chlorophyllide, respectively, i.e., the molecules were not esterified with phytol. They determined the time course of phytylization of the newly formed chlorophyll in 6-day-old dark-grown bean leaves by extracting the leaves with 80% acetone at various times after transformation of the protochlorophyllide. The degree of phytylization was inferred from the fraction of the acetone-soluble chlorophyll that could be transferred to isooctane. Virgin (61) determined the time course of phytylization in dark-grown wheat leaves by chromatographing extracts taken at various times after transformation. Both Wolff and Price and Virgin found that the phytylization reaction required about 1 hour to go to completion. The reaction would probably require more time in older etiolated tissue.

A. Influence of Structure

Butler and Briggs (62) examined the influence of proplastid structure on the absorption bands of protochlorophyllide and the newly formed chlorophyllide. If the different pigment forms were due to physical differences related to proplastid structure, disrupting the structure might alter the pigments. The leaves were either frozen to −196°C and thawed or ground with sand and buffer before irradiation ((Pchl 650), immediately after irradiation (Chl 684) and 20 minutes after irradiation (Chl 673). The absorption maximum of Pchl 650 shifted to 635 mμ, the maximum of Chl 684 to 673 mμ and that of Chl 673 was unchanged. The 635 mμ-absorbing protochlorophyllide was converted by light to a 673 mμ-absorbing chlorophyllide. Figure 3 shows the low-temperature absorption spectrum of pieces of dark-grown barley leaves before irradiation, Pchl 650; after thawing and refreezing to −196°C, Pchl 635; and after thawing, irradiation, and refreezing, Chl 673. In summary,

$$
\begin{array}{ccc}
 & h\nu & \\
\text{Pchl 650} \xrightarrow{\quad} & \text{Chl 684} \rightarrow & \text{Chl 673} \\
\downarrow & \downarrow & \downarrow \\
 & h\nu & \\
\text{Pchl 635} \xrightarrow{\quad} & \text{Chl 673} & \text{Chl 673}
\end{array}
$$

where the vertical arrow represents changes brought about by freezing to −196°C and thawing or by grinding.

It was suggested (62) that Pchl 650 was aggregated protochlorophyllide in the prolamellar body and that Chl 684 was similarly aggregated chlorophyllide in the prolamellar body. The physical disruption caused by freezing and thawing or by grinding led to the disaggregation and a shift of the absorption maxima to shorter wavelength. It was also suggested that the normal dark conversion of Chl 684 to Chl 673 represented a similar change of structure and disaggregation. Earlier work (59) showed that the fluorescence yield of the chlorophyllide more than doubled during the Chl 684 to Chl 673 conversion. Both the wavelength shift and the increased fluorescence yield are consistent with a disaggregation.

The proplastids of etiolated Euglena cells do not contain prolamellar bodies and show very little internal structure (63). Low-temperature absorption spectra showed that the protochlorophyll (the question of protochlorophyllide versus protochlorophyll has not been resolved in Euglena) absorption maximum was near 630 mμ whereas that of the newly formed chlorophyll was near 670 mμ (62). Freezing and thawing did not alter the absorption band of the newly formed chlorophyll, and no dark reaction analogous to conversion of Chl 684 to Chl 673 in etiolated leaves was observed. The correlation between the absence of the longer wavelength-absorbing forms of protochlorophyll and the newly formed chlorophyll and the absence of prolamellar bodies in the proplastid of *Euglena* is consistent with the conjectures relating pigment form to proplastid structure in dark-grown leaves.

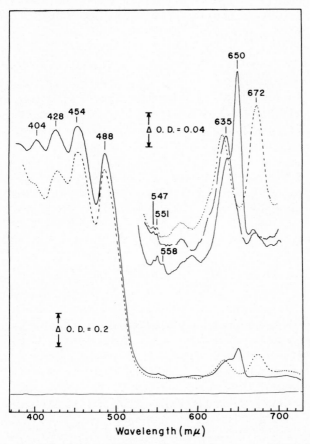

FIG. 3.   Absorption spectra of dark-grown barley leaves at −196°C. Solid curve (main figure and inset): spectrum of frozen leaves before irradiation. Broken curve (inset): spectrum of same leaves after they were thawed and refrozen to −196°C in darkness. Dotted curve (main figure and inset): spectrum of same leaves after they were thawed, irradiated with red light, and refrozen.

Protochlorophyllide is transformed to chlorophyllide in less than 10^{-3} second (*64*). Structural rearrangements could not occur in so short a time, so that the newly formed chlorophyllide must reside in the same structure that contains the protochlorophyllide. Electron microscopy has shown that the tubules which make up the prolamellar body are broken as a result of the transformation of the protochlorophyllide and that the prolamellar body disperses into individual vesicles (65, 66). The breaks in the prolamellar body appear in the micrographs even when the leaves are fixed immediately after transformation of the protochlorophyllide. The conversion of Chl 684 to Chl 673 probably coincides with the dispersal of the vesicles into unaggregated monomer units.

The greening process was also correlated with the development of energy transfer from carotene to chlorophyll (59). Energy transfer was noted by the presence of fluorescence-excitation bands corresponding to carotene absorption bands in the excitation spectrum of chlorophyll fluorescence. In a 6-day-old dark-grown bean leaf, no energy transfer was detected for about 20 minutes after a brief irradiation which converted the protochlorophyllide. By 20 minutes, the Chl 684 to Chl 673 conversion had gone to completion. During the next 40 minutes in the dark, fluorescence excitation bands attributable to carotene absorption bands developed. The absorption maximum of the chlorophyll did not change during this period. With further irradiation and accumulation of chlorophyll the efficiency of the energy transfer from carotene to chlorophyll increased. The approximate correlation between the time course of phytylization and the appearance of energy transfer from carotene to chlorophyll suggested that the two processes might be related. The lipophilic phytyl tail on the chlorophyll molecule would allow the lipid-dissolved carotene molecule to come into close proximity with the chromophoric porphyrin head of the molecule. The efficiency of energy transfer increases as the chlorophyll and carotene are brought together in the developing lamellar structures. It was suggested on the basis of these experiments (59) that the Chl 673, as it was formed from Chl 684, was still chlorophyllide and that phytylization did not alter the visible absorption spectrum of the molecule. It might be appropriate to represent the phytylization reaction as Chl 673 → Chl a 673. Phytylization may play a key role in the development of structure.

Sironval et al. (67) suggested a different interpretation for the transformation of Chl 684 to Chl 673. They measured the time course for phytylization of the newly formed chlorophyllide and found that it was similar to the time course of the spectral shift. Difference spectra between extracted Chl 684 and Chl 673 also showed small differences. Sironval et al. postulated that the shift of the absorption maximum was due to the phytylization of Chl 684. They also postulated that phytylization converted protochlorophyllide, Pchl 650, to protochlorophyll with an absorption maximum at 635 mμ, Pchl 635. Pchl 635 was transformed on irradiation to Chl 673. The observation that Chl 684 and Pchl 650 can be converted to Chl 673 and Pchl 635, respectively, by disrupting the structure argues against the direct influence of phytylization on the spectral shift. On the other hand, the alternative suggestion that the Chl 684 to Chl 673 reaction represents a disaggregation of the newly formed chlorophyllide, and that the phytol is put on chlorophyllide 673, would not be inconsistent with the data of Sironval *et al.* provided the phytylization reaction was fairly rapid.

Sironval *et al.* also suggested that in mature green cells, Chl *a* 673 was chlorophyll and Chl *a* 683 was chlorophyllide. Clearly, there is much too little chlorophyllide in mature green cells to account for the rather large amount of Chl *a* 683. Apparently, they did not distinguish between the newly formed Chl 684 in etiolated leaves and the Chl *a* 683 of mature green cells.

B. Correlation between Pigment Synthesis and Activity

The onset of photochemical activity in a greening leaf has been correlated with changes of pigment composition (68). Activity was assayed by light-induced fluorescence-yield changes of the chlorophyll, and low-temperature absorption and derivative spectra showed the pigment composition. The derivative spectra were particularly useful in detecting the onset of pigment formation. Low-temperature fluorescence-excitation spectra gave a qualitative measure of formation of C-705. The low-temperature excitation spectra of long-wavelength fluorescence ($\lambda > 730$ mμ) is a much more sensitive method of detecting C-705 than absorption spectroscopy. Figure 4 shows the low-temperature absorption and derivative spectra of four 8-day-old dark-grown bean leaves. The leaves, denoted 1, 2, 3, and 4 had been irradiated for 1.5, 2, 3, and 5 hours respectively, with white light. Up to 1.5 hours of illumination the absorption spectra showed a single chlorophyll which did not change in amount. The spectrum taken after 2 hours, however, shows the first indication of chlorophyll *b* and Chl *a* 683. These are best seen in the low-temperature derivative spectrum. With further greening, both Chl *a* and Chl *b* 683 increase. The low-temperature fluorescence-excitation spectra showed a similar time course for the formation of C-705. Thus, the particular pigments associated with systems I and II begin to form after about 2 hours of illumination in an 8-day-old dark-grown bean leaf.

The photochemical activity of the leaves was assayed by light-induced fluorescence-yield changes. The relation of fluorescence to photochemical activity will be discussed in the Section VI, B. In brief, the fluorescence yield acts as if it were determined by the oxidation state of a quenching substance, Q, which is a member of the electron transport chain. Q quenches fluorescence but QH does not. Q is reduced by system II and oxidized by system I, so that activity of systems I and II can be demonstrated by light-induced fluorescence-yield changes. The same approach was used earlier to examine mutants of *Scenedesmus* for the activity of systems I and II (39).

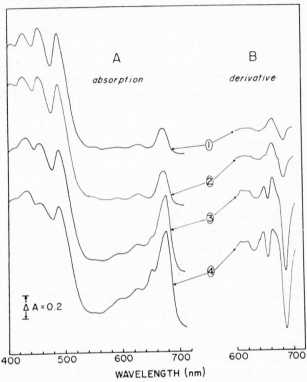

FIG. 4. Absorption spectra and derivative absorption spectra at −196°C of 8-day-old dark-grown bean leaves at different stages of greening. The leaves, denoted *1, 2, 3,* and *4,* received 1.5, 2, 3, and 5 hours illumination just prior to measurement.

Figure 5 shows the intensity of fluorescence excited by a weak measuring light (8 ergs/sec/cm^2) following 10-second irradiations with high-intensity 650- and 710-mµ light. The leaves were examined in N$_2$ at room temperature just prior to freezing to −196°C for the spectral measurements. The same leaves were used for the measurements shown in Figs. 4 and 5. The fluorescence of the leaves was maximal in N$_2$ (Q was reduced) so that actinic light could only decrease the fluorescence. Leaf No. 1 showed no change of fluorescence after irradiation, and, therefore, no indication of light-driven electron transport. The first sign of activity appeared with leaf No. 2. This same leaf showed the first indication of

Chl *b* and Chl *a* 683. With further greening the fluorescence-yield increased markedly. In leaves 3 and 4, the higher yield after 650 mμ light than after 710 mμ indicated that system II was functional as well as system I. System II was probably functional in leaf No. 2 also, but the fluorescence-yield changes were too small to differentiate between the actions of 650 and 710 mμ light. Oxygen evolution, which signified system II action, begins in a greening leaf after about 2 hours of illumination

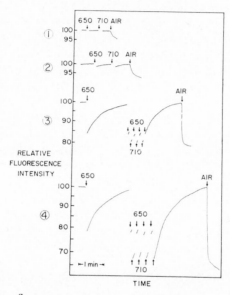

Fig. 5. Relative fluorescence intensity from dark-grown bean leaves at room temperature in N₂ (the same leaves were used in Figs. 4 and 5) excited by weak (8 ergs/sec/cm²) red light. Arrows marked "650" and "710" denote 10-sec irradiations with 650- and 710-mμ actinic light (800 ergs/sec/cm²). The measurements were not made during the actinic irradiation.

(69). The onset of activity is closely correlated with the synthesis of the pigments of systems I and II.

The formation of Chl *a* 683 is thought to represent an aggregation of the chlorophyll that may be related to structural development. A dark-grown bean leaf was illuminated for 2.5 hours to form small but definite amounts of Chl *a* 683 and Chl *b*. The leaf was then frozen (to −196°C) and thawed in an attempt to disrupt the early states of structure. This treatment resulted in the loss of Chl *a* 683; Chl *a* 673 and Chl *b* remained. At later stages of greening when the lamellar system is more developed Chl *a* 683 is resistant to freezing and thawing.

The Chl 684 which is formed directly from the protochlorophyll is different from the Chl a 683 of mature green leaves. The former is probably aggregated chlorophyllide in the prolamellar body while the latter is probably the aggregated chlorophyll in the lamellae. The function of these two pigments in the photosynthetic apparatus is quite different.

VI. Fluorescence

It was generally assumed, before the adoption of the concept of two pigment systems, that the chlorophyll fluorescence from green plants emanated from a single form of chlorophyll a. The emission spectra of chlorophyll a both *in vitro* and *in vivo* show a relatively sharp peak at 685 mμ with a broad, less intense band in the 730-mμ region. The fact that the ratio of the 685-mμ emission to the 730-mμ emission varies considerably from sample to sample has been ascribed to optical artifacts such as variations in the depth of penetration of the exciting beam and reabsorption of the fluorescence. For instance, the 685-mμ emission band of green-light-excited fluorescence from leaves is subject to more reabsorption that that of blue-light-excited fluorescence because the green light penetrates the sample farther (70). Virgin (71) demonstrated that the optical artifacts were affected by light scatter and that the emission spectrum of a leaf could be altered by changing the light-scattering properties of the leaf. In general, difference in emission spectra were attributed to such artifacts rather than to pigment differences. Recently, however, differences in the emission spectra of red and blue-green algae between accessory-pigment excitation of fluorescence and chlorophyll excitation have been ascribed to differences in the emission spectra of chlorophyll associated with pigment systems I and II (38, 72, 73).

A. Emission Spectra

Duysens (6, 14) showed that the fluorescence emission spectra of *Oscillatoria* and *Porphyra lacineata* were different when fluorescence was excited by light absorbed by the phycobilin accessory pigment rather than by light absorbed by chlorophyll a. Even after a correction for the effects of self-absorption the ratio of the 730-mμ emission to the 685-mμ emission was greater when fluorescence was excited by chlorophyll a-absorbed light. Duysens suggested (6) that a part of the emission of the 730-mμ region was due to small amounts of an unknown pigment (perhaps chlorophyll d) which drained excitation energy out of chlorophyll a. This was the first suggestion that the fluorescence emission band at 730 mμ *in vivo* might be due in part to a pigment other than chlorophyll a.

Brody (74) discovered that, on cooling to $-196°C$, the emission spectrum of *Chlorella* cells developed a strong band at about 720 mμ which was much greater than the 685 mμ-emission band at room temperature. He found the same low-temperature emission in concentrated, but not dilute, solutions of chlorophyll and concluded that the emission was from chlorophyll dimers. Butler (30) showed that a similar emission band appeared when green leaves were cooled to $-196°C$. The excitation spectrum of the long-wavelength emission from leaves at $-196°C$ showed a primary excitation band at 705 mμ as well as excitation by the bulk of the chlorophyll. The low-temperature absorption spectrum revealed a small shoulder at 705 mμ. It was concluded that the intense, low-temperature emission was due to small amounts of a 705 mμ-absorbing pigment, C-705, which received excitation energy from chlorophyll *a*. The fluorescence yield of C-705 was quite small at room temperature, but the excitation spectrum for long-wavelength fluorescence (λ 730 mμ) showed the 705 excitation band even at room temperature. Thus, in green leaves the excitation spectra also indicated that a part of the 730 emission could be attributed to a pigment other than chlorophyll *a*. C-705 can be detected much more readily in the fluorescence-excitation spectrum than in the absorption spectrum because the measurement can be made to be preferential for C-705. If a 730-mμ cutoff filter is placed in front of the phototube, most of the C-705 emission will be measured while about 98% of the chlorophyll *a* emission will be blocked.

Litvin *et al.* (75) also used low-temperature fluorescence spectroscopy to detect different forms of chlorophyll *in vivo*. The low-temperature emission spectra of green leaves by these investigators showed peaks at about 685, 695, and 720 mμ, the 720-mμ peak being the most intense. Thus, they confirmed Brody's observation and showed the presence of still another band at 695 mμ. In their study of the low-temperature fluorescence emission spectra of etiolated leaves during the greening process, Litvin and Krasnovsky (76) found that the first form of chlorophyll, corresponding to Shibata's Chl 684, had an emission band at 695 mμ. They attributed the 695 mμ-emission band from mature green leaves at $-196°C$ to a similar chlorophyll intermediate.

The presence of the 695 mμ-emission band at low temperature has been confirmed by several laboratories using green, blue-green, and red algae (38, 72, 73, 77). This emission band is under intensive investigation currently by several laboratories on the assumption that it is due to emission from the hypothetical energy trap for system II (72, 73, 77). Brody and Brody (77) noted a similar emission band in the low-temperature fluorescence spectra of purified chlorophyll *a*. The band appeared only if crystalline chlorophyll was washed exhaustively with petroleum ether.

Apparently, a substance which quenched the 695 mμ emission was present in the crystalline preparation. Michel-Wolwertz *et al.* (78) have also found small amounts of a 695 mμ-fluorescing chlorophyll after repeated chromatography of chlorophyll *a*.

Goedheer (79) also measured the fluorescence spectra of green, blue-green, and red algae at −196°C and found emission bands at 686, 695, and 720 mμ. He suggested that the three emission bands were due to Chl *a* 673, Chl *a* 683, and Chl *a* 695, respectively.

B. Relation of Fluorescence to Photosynthesis

Chlorophyll fluorescence from green plants has been used for many years as a tool in the study of the photochemistry of photosynthesis; in part, because of the possibility that the efficiency of fluorescence might be related to the efficiency of photosynthesis; in part, because it was a readily measured parameter. Kautsky and Hirsch (80) observed transient changes of chlorophyll fluorescence when plants were illuminated after a period of darkness. In general, the fluorescence transient (Kautsky effect) showed a rapid rise to a maximum followed by a slower decay to a lower steady-state value. The shape of the fluorescence intensity curve and the level of the steady-state fluorescence were sensitive to external conditions such as O_2 and CO_2 concentration and to photosynthetic poisons. During the same period (induction period of photosynthesis), photosynthesis increased and reached a steady-state rate at about the same time that the fluorescence reached a steady-state level. In general, the yield of fluorescence during the induction period was inversely related to the rate of photosynthesis. In detail, however, the empirical relationships between fluorescence and photosynthesis were not sufficiently simple and invariant to permit a generalized theoretical treatment that was not complicated and hypothetical. McAlister and Myers (81) made simultaneous recordings of fluorescence and rates of CO_2 uptake during the induction period of photosynthesis under a wide variety of experimental conditions. They concluded that at least two processes were involved during the induction period: one showed an inverse relationship between fluorescence yield and rate of CO_2 uptake; the other, a direct relationship. The early theories of the relationship between fluorescence and photosynthesis will not be reviewed here because they have not stood the test of time.

1. FLUORESCENCE YIELD AND ELECTRON TRANSPORT

Fresh concepts of the relationship of fluorescence to photosynthesis began to be introduced about the time that the two-pigment system hypothesis was being developed. Lavorel (*82*) suggested that the fluorescence from green plants could be divided into two parts: a constant fluorescence originating from the chlorophyll molecules which absorbed excitation energy and a fluorescence of variable yield from the photochemically active molecules. Lavorel calculated from the rate of the initial increase of fluorescence yield and certain assumptions that excitation energy from 100 to 200 chlorophyll molecules was transferred to each photochemically active molecule.

Kautsky *et al.* (*83*) re-evaluated the time course of the initial fluorescence transients on the assumption that the fluorescence yield was determined by the redox state of a compound, A, which was a member of the photosynthetic electron transport system. The oxidized state, A_0, quenched chlorophyll fluorescence, but the reduced state, A_1, did not. The initial rapid increase of fluorescence yield which occurs when a plant is first illuminated was ascribed to a light-driven reduction of A_0 to A_1. The decrease of the fluorescence yield following the initial maximum was ascribed to the oxidation of A_1 by the next member of the electron transport chain, B. B was oxidized by another light reaction from B_0 to B_1. At the steady-state the two light reactions maintained a flow of electrons from water, through the A and B couples to the CO_2 fixation reaction. Thus, Kautsky *et al.* proposed two photoreactions, the photoreduction of A and photooxidation of B, but not two pigment systems. This scheme was remarkably successful in accounting for the time course of the fluorescence changes.

Kautsky *et al.* (*83*) assumed that fluorescence quenching occurred because excitation energy was transferred from chlorophyll to the A_0 molecules. They calculated from the rate of the initial fluorescence increase (after poisoning of photosynthesis with phenylurethane) that the ratio of chlorophyll to A molecules was at least 400:1.

2. FLUORESCENCE AND THE TWO-PIGMENT MECHANISM

The first indication that two pigment systems of photosynthesis affected the fluorescence yield of chlorophyll differently was reported by Govindjee *et al.* (*84*). They measured the intensity of fluorescence at 685 mμ excited by red light, 670 mμ, by far-red light, 700 mμ, and by simultaneous excitation with both red and far-red. The intensity of fluorescence excited by both beams given simultaneously was less than the sum of the intensities from the individual beams.

Butler (*35*) measured the relative fluorescence yield of a green leaf excited with low-intensity red light before and after irradiation with high-intensity monochromatic light. The fluorescence yield was shown to be increased two- to threefold by a brief irradiation with a red actinic source. The fluorescence-yield changes persisted for a longer time in N_2 than in air so that most of these experiments were carried out in N_2. The high fluorescence yield induced by red light could be reversed by a subsequent brief irradiation with far-red light (λ 730 mμ). The increased yield could also be decreased by introducing O_2. Fluorescence-excitation spectra (for fluorescence of wavelengths longer than 730 mμ) measured under the conditions of high and low fluorescence yield are shown in Fig. 6. The excitation bands at 650 and 680 mμ due to light absorption by chlorophyll b and a, respectively, result in emission from chlorophyll a while

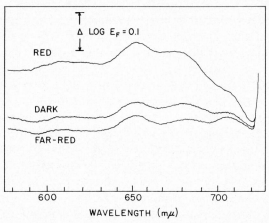

FIG. 6. Fluorescence excitation spectra of a mature green bean leaf in N_2 before irradiation (dark), after 5-sec irradiation with red light (650 mμ, 10^4 ergs/sec/cm²), and after 5-sec irradiation with far-red light ($\lambda > 720$ mμ, 10^5 ergs/sec/cm²). Fluorescence was excited by weak red light (650 mμ, 10 ergs/cm²) and was measured at wavelengths longer than 730 mμ.

the excitation band at 705 mμ indicates emission from C-705. The difference between spectra shows that the increased fluorescence is due solely to an increase in the yield of chlorophyll a fluorescence. The yield of C-705 fluorescence does not appear to be altered by irradiation. Some years earlier, Virgin (71) measured the fluorescence emission spectra of algae shortly after the onset of illumination, when the fluorescence yield was high, and at the steady-state conditions and found that the ratio of the emission at 685 mμ to that at 730 mμ was appreciably greater during the period of the high fluorescence yield. He concluded that the difference in the ratio was due to an unspecified light-scattering change that caused less of the 685-mμ fluorescence to be reabsorbed when the yield was high. The difference in the excitation spectra, however, shows that the difference in emission spectra was real and not an artifact. Lavorel (85, 86) and Rosenberg et al. (87) also found that the ratio of the fluorescence emission at 680 mμ to that at 730 mμ was higher at the peak of the Kautsky effect than during the steady state. The variations in the shape of the emission spectra are consistent with the observations of excitation spectra (35), which indicated that the fluorescence of variable yield was chlorophyll *a* fluorescence.

The action spectrum for the ability of far-red light to decrease the fluorescence yield was shown to have a maximum at 705 mμ provided the intensity of the actinic light was not too great (35). It was concluded that the decrease of yield was due to action by the long-wavelength pigment system of the Emerson effect. Teale (88), Duysens (89), Duysens and Sweers (90), and Butler and Bishop (39) showed that the action spectrum for the light-induced fluorescence increase corresponded to system II.

Duysens and Amesz (50) had proposed an electron transport scheme in which the two pigment systems moved electrons from water to pyridine nucleotide (PN).

$$H_2O \rightarrow (\text{system II}) \rightarrow Q \rightarrow Cyt \rightarrow P700 \rightarrow (\text{system I}) \rightarrow XH \rightarrow PN$$

The components in the chain between the two systems were oxidized by system I and reduced by system II. They had also suggested that the chlorophyll *a* in system II (chlorophyll a_2) was fluorescent while that in system I (chlorophyll a_1) was not. The scheme of Kautsky et al. (83) was modified to bring the fluorescence-yield changes into the two-pigment mechanism. The function of compound A was ascribed to Q and the two light reactions proposed by Kautsky were ascribed to system I and system II. Q quenches the fluorescence of chlorophyll a_2, but QH does not. Light absorbed by system I or by P700 oxidizes QH to Q causing the fluorescence to be quenched. It was proposed that DCMU blocked electron transport just after Q. After the addition of DCMU, system II could reduce Q so that the fluorescence yield would increase in response to light, but system I could not oxidize QH. QH was oxidized to a small extent by dark metabolism even after the addition of 3-(3,4-dichlorophenyl-1,1-dimethylurea (DCMU). As a result the fluorescence yield relates to a specific point in the photosynthetic electron transport chain, and the measurement of fluorescence has become an exceedingly useful tool in probing the pathway of electron transport.

The relatively constant background level of fluorescence which has been ascribed tentatively to the chlorophyll a_1 presents more of a problem. P700 has been proposed as an energy sink for system I (34) and, as such, it should quench the fluorescence of chlorophyll a_1. In the bleached state, however, the quenching should be relieved. Experimental attempts to demonstrate the quenching effect of P700 on chlorophyll a_1 have failed. Lavorel (85, 86) showed that the emission spectrum of the constant part of the fluorescence had relatively more emission at 720 mμ compared with 685 mμ than the emission spectrum of the variable fluorescence. Vredenberg and Duysens (91) confirmed these spectral differences and further showed that the long-wavelength emission did not increase as P700 was bleached. They suggested that, if the 720 mμ emission was due in part to chlorophyll a_1, this chlorophyll did not transfer excitation energy to P700. Rather it reacted first to reduce X. The quenching of chlorophyll a_1 by X would be analogous to the quenching of chlorophyll a_2 by Q. The resolution of anomalous fluorescence behavior of chlorophyll a_1 could lead to a significant advance in the formulation of the two-pigment mechanism.

3. The Franck-Rosenberg Theory

Franck and Rosenberg (92) developed a different theory to account for enhancement phenomena in photosynthesis. They proposed two photoreactions, but only one type of reaction center. In one photoreaction, triplet-excitation energy of the chlorophyll at the reaction center is used to oxidize cytochrome and reduce a diffusible oxidant X. The long-lived triplet state was invoked for this reaction to allow time for X to diffuse to the reaction center during the lifetime of the excited state. In the other photoreaction, singlet-excitation energy is used to reduce cytochrome and oxidize an enzyme, Y, connected with O_2 evolution. Cytochrome, Y and the reaction-center chlorophyll are complexed so that this reaction can occur within the lifetime of the excited singlet state.

$$X + (Y - Chl^*_{met} - cyt_{Red} - H_2O) \rightarrow (Y - Chl - cyt_{Ox}) + XH + OH^-$$
$$2\,XH \rightarrow X + XH_2 \tag{1}$$

$$(Y_{Red} - Chl^*_{sing} - cyt_{Ox}) \rightarrow (Y_{Ox} - Chl - cyt_{Red})$$
$$4\,Y_{Ox} \rightarrow 4\,Y_{red} + O_2 \tag{2}$$

Franck and Rosenberg postulated two pigment systems, but one system is capable of sensitizing both reactions. The short-wavelength system, consisting of monomeric Chl a 673, transfers all its excitation energy as singlet energy to the reaction-center chlorophyll. This transfer of energy is so rapid that the Chl a 673 cannot fluoresce. If the chlorophyll reaction center is complexed with reduced Y and oxidized cytochrome, the energy is used immediately with no chance for fluorescence from the reaction-center chlorophyll. If Y or cyt are not in the proper oxidation state, the singlet energy cannot be used and the excited reaction-center chlorophyll molecules will fluoresce or go over to the triplet state, where the energy can be utilized to oxidize cytochrome and reduce X. Thus, the short-wavelength system can sensitize both photoreactions. Fluorescence emanates from the reaction-center chlorophyll molecules.

The long-wavelength pigment system consisting of amorphous Chl a 683 and semicrystalline P700 has its own energy traps. These traps can transfer energy to the reaction-center chlorophyll molecules, but only triplet energy. Thus, the long-wavelength pigment system can sensitize reaction 1, but not reaction 2. The participation of both pigment systems accounts for the Emerson enhancement phenomenon. If triplet energy supplied by the long-wavelength system is used to oxidize cytochrome, singlet energy supplied by the short-wavelength system can be utilized more efficiently. The participation of the long-wavelength system will decrease the fluorescence yield because of the more efficient utilization of singlet energy.

Franck and Rosenberg proposed that resonance transfer of singlet energy could connect a number of reaction centers, n, into a super unit. Thus, singlet energy, arriving at a reaction center which was not in the proper state for reaction 2, could roam over n reaction centers within its lifetime. The super unit increases the efficiency of singlet energy. If there were no super unit and each reaction center had to alternate between oxidation and reduction of cytochrome, one-half of the cytochrome would be oxidized at the steady state. Also, the fluorescence yield of the reaction-center chlorophyll would increase by a factor of 2 in going from normal steady-state photosynthesis to a transient condition or to a poisoned state where none of the singlet energy could be used. (For example, poisons such as DCMU which block O_2 evolution prevent the regeneration of reduced Y so that singlet energy cannot be used.) With the super unit, the steady-state fraction of oxidized cytochrome will be $1/2n$ and the fluorescence yield of the reaction-center chlorophyll can increase by a factor of $2n$.

The long-wavelength pigment system is assumed to contribute a constant-yield fluorescence which is relatively rich in the far-red wavelengths. The fluorescence of variable yield emanating from the reaction-center chlorophyll molecules has an emission maximum at 685 mμ. Thus, the total emission spectrum of green tissue will be relatively enriched in the 685-mμ fluorescence when the fluorescence yield is high.

C. Delayed Light

Strehler and Arnold (93) discovered that plants continue to emit light at very low intensity for several minutes after the cessation of illumination. The emission which is known as chemiluminescence or delayed light was shown to have the spectrum of chlorophyll fluorescence (94).

The delayed light thus arises from the lowest excited-singlet state of chlorophyll which has been excited chemically. The chemical excitation is generally ascribed to a recombination of metastable states, e.g., triplet states or holes and electrons.

The studies of delayed light led to analogies between the photosynthetic apparatus and semiconductors. These analogies were also supported by studies of thermoluminescence (95), photoconductivity (96), electron spin resonance (97), and the temperature dependence of delayed light (98). Arnold suggested that the photoionization of chlorophyll gave rise to electrons and holes which could recombine to give excited chlorophyll molecules. Thermoluminescence and the complex decay kinetics of the delayed light indicated that the holes and electrons were trapped at various energy levels below the first excited singlet state.

Various investigations have shown relationships between photosynthesis and chemiluminescence. Goedheer (99) related delayed light to the two-pigment systems of photosynthesis. Delayed light excited by 659 mμ irradiation could be partially quenched by simultaneous irradiation with longer-wavelength light. The action spectrum of the quenching of delayed light in *Chlorella* had a maximum at 700 mμ (99). Butler showed essentially the same action spectrum for the quenching of chlorophyll fluorescence (35). Sweetser *et al.* (100) showed that the delayed light decayed much more slowly after the addition of low concentrations of CMU. Bertsch (101) confirmed the effects of poisons on the delayed light and the antagonistic effects of red and far-red light and suggested that the delayed light emanated from the chlorophyll molecules which were photochemically active.

The relationship of delayed light to the photochemistry of photosynthesis is not clear. The most direct evidence relating the two processes in purple photosynthetic bacteria was the finding that a nonphotosynthetic mutant of *Rhodopseudomonas spheroides* which lacked the P870 reaction-center pigment did not emit delayed light (102) whereas the wild-type strain did. However, the relationship between delayed light and photosynthesis in purple photosynthetic bacteria may be different from that in green plants. At least, the relationship between delayed light and fluorescence in purple bacteria is different from that in green plants (see Clayton, Chapter 19). The various correlations that exist between delayed light and photosynthesis in green plants are suggestive of a meaningful relationship but, as will be discussed later in this section, these correlations may only reflect a dependence of delayed light on fluorescence yield.

Arnold and Davidson (103) raised an interesting question as to the relationship between chlorophyll fluorescence and chemiluminescence.

They plotted the intensity of emission D, versus the delayed time, t, at which the measurement was made in a log-log plot and found essentially a straight-line relationship; log $D = -$log $t +$ constant, over the time range of 5×10^{-5} to 5×10^3 sec. It was suggested, on the basis of an extrapolation of the delayed-light decay curve back to the time interval of fluorescence, 1 to 2×10^{-9} sec, that an appreciable part of what was ordinarily called fluorescence was, in fact, a fast component of the delayed light. However, in more recent work Arnold reported (103a) that when light intensities were used such that delayed light was proportional to the intensity of exciting light, the curve of log D versus log t leveled off at short times so that extrapolation of these measurements did not indicate appreciable intensities of fast-decaying delayed light.

Clayton (104) has made a strong case for equating fluorescence and delayed light. Fluorescence and the 3-msec component of delayed light from *Chloropseudomonas ethylicum* (a green bacterium), *Chlorella pyrenoidosa*, and *Anacystis montana* were measured following the onset of illumination consisting of repetitive flashes at a rate of 160/sec. The time course of the fluorescence-yield changes was essentially the same as that shown by Kautsky for continuous light. In the three species tested the time course for the fluorescence intensity was very similar to the time course for delayed light. In *C. ethylicum* the intensity dependence for the initial fluorescence increase was similar on a relative scale to the intensity dependence for the excitation of delayed light. In addition, the spectrum of delayed light in *C. ethylicum* was identical to the spectrum of the variable fluorescence but different from the overall fluorescence-emission spectrum. All these similarities between delayed light and the variable-yield fluorescence led Clayton to propose that the fluorescence of variable yield was in fact a fast-decaying component of the delayed light. Clayton has reviewed this work and presented some of the data in Chapter 19 of this volume.

It is possible, however, to interpret the relationship between fluorescence and delayed light in a different way. The yield of fluorescence of the chemically excited chlorophyll molecules should be subject to the same constraints as the yield of the photon-excited molecules. Those conditions and reagents which alter the fluorescence yield should similarly alter the delayed light. The yield of the variable component of the fluorescence can be related to the oxidation state of Duysens' quencher Q. Thus, irradiation with far-red light which decreases the fluorescence yield by oxidizing QH to Q should also decrease the intensity of delayed light. The slow decay of delayed light after addition of CMU is also consistent with action of CMU on fluorescence. The changes of fluorescence and delayed light should follow a similar time course after the onset of illumination in Clayton's experiments because the intensities of both emissions are determined by the oxidation state of Q. The intensity dependence for the excitation of delayed light and fluorescence-yield changes will be similar for the same reason that both are determined by Q. The comparison between the various emission spectra from *C. ethylicum* (see Fig. 5, Chapter 19 in this volume) does not necessarily indicate the same mechanism for the variable-yield fluorescence and the delayed light. The similarity between the spectra of these two emissions shows only that both emanate from bacterial chlorophyll. The overall fluorescence spectrum is different because it also includes emission from chlorobium chlorophyll which acts as an accessory pigment. Chlorobium chlorophyll may transfer only 30–40% of its excitation energy to bacterial chlorophyll (*105*) so that it can contribute to the overall fluorescence. Analogous relationships obtain in red and blue-green algae where the overall fluorescence spectrum is due to emission from both the phycobilin accessary pigment and chlorophyll *a* whereas the fluorescence of variable yield (and undoubtedly the delayed light as well) shows only the chlorophyll fluorescence spectrum.

VII. Orientation of Chlorophyll *in Vivo*

Attempts to use polarized light to deduce structural characteristics of the photosynthetic apparatus were made quite early. Menke (*106*) detected birefringence and a weak dichroism with chloroplasts of *Closterium* and concluded that a lamellar structure consisting of a series of parallel planes gave rise to the optical anisotropy. Later Goedheer (*107*) made an extensive series of measurements with polarized light and found a dichroic ratio of $1.13 + 0.05$ at 680 mμ with chloroplasts of *Mougeotia*. This magnitude of dichroism could be attributed to a dichroism of shape. Goedheer concluded that the chlorophyll molecules were complexed to the proteins of the lamellar structure in such a way that very little net orientation resulted.

Olson et al. (*108*), in an attempt to observe the fluorescence of C-705 in chloroplasts microscopically, found that the fluorescence in the long wavelength region was highly polarized with the electric vector parallel to the planes of the lamellae. The polarized emission was scarcely detectable when the microscopic observations were made by eye, but was very striking when observations were made with an infrared image-converter tube. Observation of the fluorescence through cutoff filters showed little polarization at the shorter wavelengths of chlorophyll fluorescence where the eye is more sensitive, but a strong polarization at longer wavelengths. They suggested that the oriented chlorophyll molecules were C-705 molecules. Olson *et al.* (*109*) also showed a strong polarized absorption (dichroism) at wavelengths beyond the chlorophyll *a* absorption maximum. The strongest absorption of polarized light occurred when the plane of polarization was parallel to the planes of the lamellae. The plane of polarized absorption was in agreement with the plane of polarized emission. Dichroic ratios as high as 4.0 were observed. The wavelength dependence of the dichroic ratio had a maximum at about 705 mμ in measurements made on chloroplasts *in situ* in *Euglena*, *Mougeotia*, *Spirogyra*, *Syragonium*, and *Mesotaenium* (*110*). The maximum in the spectrum for the dichroic ratio does not necessarily indicate the absorption maximum of the oriented pigment. The qualitative discrepancy between the high degree of dichroism and polarized fluorescence and the small amounts of C-705 with its low fluorescence yield at room temperature gave rise to some doubt to ascribing the oriented oscillators to C-705.

Butler *et al.* (*111*) observed polarized fluorescence and dichroism in a mutant of *Scenedesmus* which lacked C-705 (*39*), by criteria of low-temperature absorption and derivative spectroscopy (see Fig. 2), and lacked P700 (*38*), by the criterion of reversible absorbancy changes at 700 mμ. C-705 and P700 were demonstrated in the wild-type cells. The small size of the *Scenedesmus* chloroplasts precluded quantitative measurements of polarized fluorescence and dichroism, but qualitative observations with a microscope indicated that the degree of pigment orientation was as great in the mutant as in the wild type. The absorption spectroscopy showed that the mutant contained normal amounts of Chl *a* 673 and Chl *a* 683. The oriented chlorophyll was ascribed to Chl *a* 683. Chl *a* 673 was assumed to be unoriented. The previous assignment of Chl *a* 673 and Chl *a* 683 to monomeric and aggregated forms, respectively, of chlorophyll *a* suggests that aggregation leads to orientation of the molecules in the lamellae.

Sauer and Calvin (*112*) measured the dichroism of chloroplast fragments that were partially oriented in an electric field. They found appreciable dichroism in the long wavelength region with a maximum of 1.27 at 695–700 mμ. Sauer (*113*) also attempted to orient chloroplast fragments in a flow system. The flow dichroism showed a maximum dichroic ratio of 1.02 at 695 mμ. The relatively low dichroic ratios compared to measurements of chloroplasts *in situ* are probably due to an inability to orient the fragments. Disruption of the chloroplasts could also result in less orientation.

Sauer (*114*) also measured optical rotatory dispersion (ORD) spectra of chloroplast fragments and solutions of chlorophyll. Rather strong

Cotton effects were found with the chloroplast fragments which appeared to be related to the aggregation of the pigments in the lamellar fragments. This technique may prove valuable in studying the structural relationships of the chloroplast fragments. Ke has also studied these phenomena, as reported in Chapter 13, part III.

VIII. Conclusion

The data on different forms of chlorophyll *in vivo* are consistent with an organization of pigments into two photochemical systems. The data also suggest that chlorophyll is in different physical states in the two systems. System I contains an aggregated, low-fluorescent, oriented form of chlorophyll *a* with an absorption maximum at 683 mμ. System II contains a monomeric, fluorescent, unoriented form of chlorophyll *a* with an absorption maximum at 673 mμ.

Various stratagems have been employed in attempts to separate the two systems as morphological entities. A study of the development of the two systems in greening leaves (68) indicated, within the timing precision of the experiment, that the systems were formed at the same time. This one attempt, however, does not exhaust the possibilities of obtaining a relatively pure system I or system II at a particular stage of development. In another study (39), mutants of *Scenedesmus* which exhibited a deficiency of either system I or system II were examined spectroscopically to determine whether any of the mutations were reflected in the pigment composition. The mutations appeared to be due to deficiencies in the electron transport chain. All the mutants showed the same relative amounts of Chl *b* 650, Chl *a* 673, and Chl *a* 683. One mutant (see Fig. 2) lacked C-705 (and P700). Other mutants of photosynthetic organisms should be screened for their pigment composition *in vivo*. The most promising approach to date to the separation of systems I and II was an attempt at direct physical separation by digitonin treatment and differential centrifugation (57). The successful separation and recombination of the two photochemical systems and their related photochemical activities would represent a major advance and an important tool in the study of the mechanism of photosynthesis.

References

(1) T. W. Engelmann, *Botan. Z.* **42**, 81 (1884).

(2) H. J. Dutton and W. M. Manning, *Am. J. Botany* **28**, 516 (1941).

(3) R. Emerson and C. M. Lewis, *J. Gen. Physiol.* **25**, 579 (1942).

(4) H. J. Dutton, W. M. Manning, and B. M. Duggar, *J. Phys. Chem.* **47**, 308 (1943).

(5) C. S. French and V. K. Young, *J. Gen. Physiol.* **35**, 873 (1952).
(6) L. N. M. Duysens, Thesis, University of Utrecht (1952).
(7) T. Förster, Z. *Naturforsch.* **4a**, 321 (1949).
(8) R. Emerson and W. J. Arnold, *J. Gen. Physiol.* **15**, 39 (1931).
(9) H. Gaffron and K. Wohl, *Naturwissenschaften* **24**, 81 (1936).
(10) R. Emerson, R. Chalmers, and C. Cedarstrand, *Proc. Natl. Acad. Sci. U.S.* **43**, 113 (1957).
(11) R. Emerson and C. M. Lewis, *Am. J. Botany* **30**, 165 (1943).
(12) F. Haxo and L. R. Blinks, *Am. J. Botany* **33**, 836 (1946).
(13) F. Haxo and L. R. Blinks, *J. Gen. Physiol.* **33**, 389 (1950).
(14) L. N. M. Duysens, *Nature* **168**, 548 (1951).
(15) J. Lavorel, *J. Chem. Phys.* **55**, 911 (1958).
(16) A. A. Krasnovsky and L. M. Kosobutskaya, *Dokl. Akad. Nauk SSSR* **85**, 177 (1952).
(17) E. Rabinowitch, "Photosynthesis and Related Processes," Vol. 2, Part 1. Wiley (Interscience), New York, 1951.
(18) H. J. Trurnit and G. Colmano, *Biochim. Biophys. Acta* **30**, 434 (1958).
(19) P. Latimer, *Plant Physiol.* **34**, 193 (1959).
(20) K. Shibata, A. A. Benson, and M. Calvin, *Biochim. Biophys. Acta* **15**, 461 (1954).
(21) W. L. Butler and K. H. Norris, *Arch. Biochem. Biophys.* **87**, 31 (1960).
(22) K. H. Norris and W. L. Butler, *IRE, Trans. Bio-Med. Electron.* **8**, 153 (1961).
(23) W. L. Butler, *Ann. Rev. Plant Physiol.* **15**, 451 (1964).
(24) C. S. French and G. E. Harper, *Carnegie Inst. Wash. Year Book* **56**, 281 (1957).
(25) C. S. French and H. S. Huang, *Carnegie Inst. Wash. Year Book* **56**, 267 (1957).
(26) C. S. French and R. F. Elliot, *Carnegie Inst. Wash. Year Book* **57**, 278 (1958).
(27) J. S. Brown and C. S. French, *Plant Physiol.* **34**, 305 (1959).
(28) J. S. Brown and C. S. French, *Biophys. J.* **1**, 539 (1961).
(29) W. L. Butler, *Biochem. Biophys. Res. Commun.* **3**, 685 (1960).
(30) W. L. Butler, *Arch. Biochem. Biophys.* **93**, 413 (1961).
(31) W. L. Butler and J. E. Baker, *Biochim. Biophys. Acta* **66**, 206 (1963).
(32) J. W. Coleman, A. S. Holt, and E. Rabinowitch, *Science* **123**, 795 (1956).
(33) B. Kok, *Plant Physiol.* **34**, 184 (1959).
(34) B. Kok and G. Hoch, *in* "Light and Life" (W. D. McElroy and B. Glass, eds.), p. 397. Johns Hopkins Press, Baltimore, Maryland, 1961.
(35) W. L. Butler, *Biochim. Biophys. Acta* **64**, 309 (1962).
(36) B. Kok and W. Gott, *Plant Physiol.* **35**, 802 (1960).
(37) B. Kok, *Biochim. Biophys. Acta* **48**, 527 (1961).
(38) B. Kok, *Natl. Acad. Sci.—Natl. Res. Council, Publ.* **1145**, 45 (1963).
(39) W. L. Butler and N. I. Bishop, *Natl. Acad. Sci.—Natl. Res. Council, Publ.* **1145**, 91 (1963).
(40) A. A. Shlyk and G. M. Nikolayeva, *Biofizika (USSR) (English Transl.)* **8**, 261 (1963).
(41) J. B. Thomas, *Biochim. Biophys. Acta* **59**, 202 (1962).
(42) H. Metzner, *in* "Studies on Microalgae and Photosynthetic Bacteria" (Japan. Soc. Plant Physiol., eds.), p. 227. Univ. of Tokyo Press, Tokyo, 1963.
(43) W. Menke, C. S. French, and W. L. Butler, Z. *Naturforsch.* **20b**, 482 (1965).

(44) M. R. Michel-Wolwertz and C. Sironval, *Biochim. Biophys. Acta* **94**, 330 (1965).

(45) R. Emerson and E. Rabinowitch, *Plant Physiol.* **35**, 477 (1960).

(46) R. Govindjee and E. Rabinowitch, *Science* **132**, 355 (1960).

(47) R. Govindjee and E. Rabinowitch, *Biophys. J.* **1**, 377 (1961).

(48) L. N. M. Duysens, *Proc. 3rd Intern. Cong. Photobiol.*, Copenhagen, 1960, p. 135. Elsevier, Amsterdam, 1961.

(49) C. S. French and J. Myers, *Carnegie Inst. Wash. Year Book* **58**, 323 (1959).

(50) L. N. M. Duysens and J. Amesz, *Biochim. Biophys. Acta* **64**, 243 (1962).

(51) R. Hill and F. Bendall, *Nature* **186**, 136 (1960).

(52) J. B. Thomas, O. H. Blaauw, and L. N. M. Duysens, *Biochim. Biophys. Acta* **10**, 230 (1953).

(53) R. B. Park and N. G. Pon, *J. Mol. Biol.* **3**, 1 (1961).

(54) R. B. Park and J. Biggins, *Science* **144**, 1009 (1964).

(55) M. B. Allen, J. C. Murchio, S. W. Jeffrey, and S. A. Bendix, *in* "Studies on Microalgae and Photosynthetic Bacteria" (Japan. Soc. Plant Physiol., eds.), p. 407. Univ. of Tokyo Press, Tokyo, 1963.

(56) M. B. Allen and J. C. Murchio, *Natl. Acad. Sci.—Natl. Res. Council, Publ.* **1145**, 486 (1963).

(57) N. K. Boardman and J. M. Anderson, *Nature* **203**, 166 (1964).

(58) K. Shibata, *J. Biochem.* (*Toyko*) **44**, 147 (1957).

(59) W. L. Butler, *Arch. Biochem. Biophys.* **92**, 287 (1961).

(60) J. B. Wolff and L. Price, *Arch. Biochem. Biophys.* **72**, 293 (1957).

(61) H. I. Virgin, *Physiol. Plantarum* **13**, 155 (1960).

(62) W. L. Butler and W. R. Briggs, *Biochim. Biophys. Acta* **112**, 45 (1966).

(63) H. T. Epstein and J. A. Schiff, *J. Protozool.* **8**, 427 (1961).

(64) A. Madsen, *Physiol. Plantarum* **16**, 470 (1963).

(65) H. I. Virgin, A. Kahn, and D. von Wettstein, *Photochem. Photobiol.* **2**, 83 (1963).

(66) S. Klein, G. Byron, and L. Bogorad, *J. Cell Biol.* **22**, 433 (1964).

(67) C. Sironval, M. R. Michel-Wolwertz, and A. Madsen, *Biochim. Biophys. Acta* **94**, 344 (1965).

(68) W. L. Butler, *Biochim. Biophys. Acta* **102**, 1 (1965).

(69) J. H. C. Smith and A. Benitez, *Plant Physiol.* **29**, 135 (1954).

(70) C. S. French, *in* "Luminescence of Biological Systems," Publ. No. 41, p. 51. Am. Assoc. Advance. Sci., Washington, D.C., 1955.

(71) H. I. Virgin, *Physiol. Plantarum* **7**, 560 (1954).

(72) J. A. Bergeron, *Natl. Acad. Sci.—Natl. Res. Council, Publ.* **1145**, 527 (1963).

(73) R. Govindjee, *Natl. Acad. Sci.—Natl. Res. Council, Publ.* **1145**, 318 (1963).

(74) S. S. Brody, *Science* **128**, 838 (1958).

(75) F. F. Litvin, A. A. Krasnovsky, and G. T. Rikhireva, *Dokl. Akad. Nauk SSSR* **135**, 287 (1960).

(76) F. F. Litvin and A. A. Krasnovsky, *Dokl. Akad. Nauk SSSR* **120**, 764 (1958).

(77) S. S. Brody and M. Brody, *Natl. Acad. Sci.—Natl. Res. Council, Publ.* **1145**, 455 (1963).

(78) M. R. Michel-Wolwertz, C. Sironval, and J. C. Goedheer, *Biochim. Biophys. Acta* **94**, 584 (1965).

(79) J. C. Goedheer, *Biochim. Biophys. Acta* **88**, 304 (1964).

(80) H. Kautsky and A. Hirsch, *Naturwissenschaften* **19**, 964 (1931).

(*81*) E. D. McAlister and J. Myers, *Smithsonian Inst. Misc. Collections* **99**, No. 6 (1940).

(*82*) J. Lavorel, *Plant Physiol.* **34**, 204 (1959).

(*83*) H. Kautsky, W. Appel, and H. Amann, *Biochem. Z.* **332**, 277 (1960).

(*84*) Govindjee, S. Ichimura, C. Cedarstrand, and E. Rabinowitch, *Arch. Biochem. Biophys.* **89**, 322 (1960).

(*85*) J. Lavorel, *Biochim. Biophys. Acta* **60**, 510 (1962).

(*86*) J. Lavorel, *in* "La Photosynthèse," p. 161. C.N.R.S., Paris, 1963.

(*87*) J. L. Rosenberg, T. Bigat, and S. Dejaegere, *Biochim. Biophys. Acta* **79**, 9 (1964).

(*88*) F. W. J. Teale, *Biochem. J.* **85**, 14P (1962).

(*89*) L. N. M. Duysens, *Proc. Roy. Soc.* **B157**, 301 (1963).

(*90*) L. N. M. Duysens and H. E. Sweers, *in* "Studies on Microalgae and Photosynthetic Bacteria" (Japan. Soc. Plant Physiol., eds.), p. 353. Univ. of Tokyo Press, Tokyo, 1963.

(*91*) W. J. Vredenberg and L. N. M. Duysens, *Biochim. Biophys. Acta* **94**, 355 (1965).

(*92*) J. Franck and R. L. Rosenberg, *J. Theoret. Biol.* **7**, 276 (1964).

(*93*) B. Strehler and W. Arnold, *J. Gen. Physiol.* **34**, 809 (1951).

(*94*) W. Arnold and J. Davidson, *J. Gen. Physiol.* **37**, 677 (1954).

(*95*) W. Arnold and H. Sherwood, *J. Phys. Chem.* **63**, 1314 (1959).

(*96*) W. Arnold and H. K. Maclay, *Brookhaven Symp. Biol.* **11**, 1 (1959).

(*97*) B. Commoner, J. J. Heise, and J. Townsend, *Proc. Natl. Acad. Sci. U.S.* **42**, 710 (1956).

(*98*) G. Tollin and M. Calvin, *Proc. Natl. Acad. Sci. U.S.* **43**, 895 (1957).

(*99*) J. C. Goedheer, *Biochim. Biophys. Acta* **64**, 294 (1962).

(*100*) P. B. Sweetser, C. W. Todd, and R. T. Hersh, *Biochim. Biophys. Acta* **51**, 509 (1961).

(*101*) W. F. Bertsch, *Proc. Natl. Acad. Sci. U.S.* **48**, 2000 (1962).

(*102*) R. K. Clayton and W. F. Bertsch, *Biochem. Biophys. Res. Commun.* **18**, 415 (1965).

(*103*) W. Arnold and J. Davidson, *Natl. Acad. Sci.—Natl. Res. Council, Publ.* **1145**, 698 (1963).

(*103a*) W. Arnold, private communication (1965).

(*104*) R. K. Clayton, *J. Gen. Physiol.* **48**, 633 (1965).

(*105*) J. M. Olson and C. Sybesma, *in* "Bacterial Photosynthesis" (H. Gest, A. San Pietro, and L. P. Vernon, eds.), p. 413. Antioch Press, Yellow Springs, Ohio, 1963.

(*106*) W. Menke, *Kolloid-Z.* **85**, 256 (1938).

(*107*) J. C. Goedheer, Ph.D. Thesis, University of Utrecht (1957).

(*108*) R. A. Olson, W. L. Butler, and W. H. Jennings, *Biochim. Biophys. Acta* **54**, 615 (1961).

(*109*) R. A. Olson, W. L. Butler, and W. H. Jennings, *Biochim. Biophys. Acta* **58**, 144 (1962).

(*110*) R. A. Olson, *Natl. Acad. Sci.—Natl. Res. Council, Publ.* **1145**, 545 (1963).

(*111*) W. L. Butler, R. A. Olson, and W. H. Jennings, *Biochim. Biophys. Acta* **88**, 651 (1964).

(*112*) K. Sauer and M. Calvin, *J. Mol. Biol.* **4**, 451 (1962).

(*113*) K. Sauer, *Biophys. J.* **5**, 337 (1965).

(*114*) K. Sauer, *Proc. Natl. Acad. Sci. U.S.* **53**, 716 (1965).

—12—

Absorption and Fluorescence Spectra of Bacterial Chlorophylls in Situ*

JOHN M. OLSON AND ELIZABETH K. STANTON

Biology Department, Brookhaven National Laboratory, Upton, New York

I. Introduction

Pioneer studies of absorption spectra of bacterial chlorophylls *in situ* were carried out in 1939 by Katz *et al.* (*1, 2*) in Utrecht. After World War II, Duysens (*3*) undertook the study of energy transfer in photosynthesis using fluorescence of the chlorophylls *in vivo* as the key to success. The absorption and fluorescence spectra obtained by Duysens for purple bacteria have appeared in earlier reviews of this subject by French and Young (*4*) and by Rabinowitch (*5*). Absorption spectra for both purple and green bacteria have been recently published by Schlegel and Pfennig (*6*) and by Clayton (*7*).

A. Environment *in Situ*

Spectroscopically the bacterial chlorophylls in living cells and in aqueous suspensions of chromatophores differ significantly from solutions of these chlorophylls in various organic solvents. In general the

---

* Research carried out at Brookhaven National Laboratory under the auspices of the U. S. Atomic Energy Commission.

position of the red absorption band for each chlorophyll is shifted toward the blue when the chlorophyll is removed from its environment *in situ* and dissolved in alcohol, acetone, or ether. This blue shift is even more noticeable for the bacterial chlorophylls than for the chlorophylls of algae and higher plants.

Three types of interaction with the local environment have been postulated to explain the absorption spectra of the bacterial chlorophylls *in situ:* (a) chlorophyll-protein interaction, (b) chlorophyll-chlorophyll interaction and (c) chlorophyll-carotenoid interaction.

Interaction of chlorophyll with protein and lipid, as is observed in distinct complexes such as the Bchl-protein complexes obtained from green bacteria (see Chapter 13), is undoubtedly an important factor in determining the spectra *in situ.* The environment provided by such complexes is probably responsible for wavelength differences of 15–30 mμ between the location of the red band *in vivo* and in solution.

Chlorophyll-chlorophyll interaction resulting from the formation of large chlorophyll aggregates *in vivo* has been suggested to explain the very large wavelength differences (80–100 mμ) often existing between several of the red bands *in vivo* and the single band observed in solution (*8–10*). The presence in most purple bacteria of two or three Bchl absorption bands between 800 and 900 mμ is consistent with the theory that Bchl exists in two or three states of aggregation (corresponding to dimensionality) *in vivo.*

Chlorophyll-carotenoid interaction has been proposed to explain the great variability in the fine structure of the near-infrared absorption spectrum of Bchl *a* in purple bacteria (*11, 12*). However, the evidence against a specific carotenoid effect being a general phenomenon is increasing (*13–15*). It seems more reasonable to consider the carotenoids along with lipids as nonpolar elements of the environment of the chlorophylls *in vivo.*

B. Chromatophores versus Whole Cells

For absorption characteristics of bacterial chlorophylls *in situ,* washed chromatophore fractions offer certain advantages over whole cell suspensions in most cases: (a) There is little or no light-scattering problem. (b) The resolution is better; there is less optical flattening of absorption peaks. (c) UV-absorbing material not associated with the photosynthetic apparatus is absent [*Rhodopseudomonas* chromatophore preparations require special methods to avoid contamination with ribosomes (*16*)]. Comparison of spectra obtained with whole cells and with chromatophores shows that in most cases the same absorption bands appear at the same wavelength positions (within $\pm 1\,\text{m}\mu$) in each spectrum. This indicates that the entire pigment complex of the organism remains more or less intact in the chromatophore fraction.

The twice washed chromatophore fractions used for absorption spectra described in this chapter were prepared as follows unless otherwise noted:

Step 1. Whole bacterial cells were suspended in 0.05 M tris(hydroxymethyl)aminomethane(Tris) (pH 7.8) and broken in a French pressure cell.

Step 2. The crude extract from the French pressure cell was centrifuged twice at 2.5 $\times$ 10^4 g to sediment unbroken cells and large cell fragments.

Step 3. The supernatant was then centrifuged for 60 minutes at 1.4 $\times$ 10^5 g to sediment the chromatophore fraction.

Step 4. The chromatophore fraction was suspended in 0.05 M Tris, centrifuged for 60 minutes at 1.4 $\times$ 10^5 g, and resuspended in 0.05 M Tris.

Absorption spectra were obtained with a Cary 14R recording spectrophotometer. Locations of absorption maxima were obtained at a scan rate of 10 Å/sec.

For fluorescence emission spectra of chlorophylls *in situ*, whole-cell suspensions are often better than cell extracts, because the fluorescence yield of the bacterial chlorophylls is very sensitive to the chemical environment. Unless chromatophores are prepared very carefully and maintained under reducing conditions, the fluorescence yield of the chlorophyll is likely to drop drastically from the level observed *in vivo* (17). In addition, accessory chlorophylls which do not fluoresce *in vivo* may fluoresce in extracts because energy transfer to the "acceptor chlorophyll" is disrupted. Corrected fluorescence spectra in terms of relative quantum emission per unit wavenumber were obtained with a type of fluorometer described previously (18).

II. Bacteriochlorophylls

Spectroscopically the bacteriochlorophylls are characterized by a prominent absorption band in the orange in addition to the main absorption band in the near-infrared region. Two types of Bchl have been found in photosynthetic bacteria. Most purple bacteria and several strains of green bacteria contain Bchl *a*, which is closely related to Chl *a* chemically. Bacteriochlorophyll *b* is the name given to a newly discovered chlorophyll found in two strains of purple bacteria which do not contain Bchl a: *Rhodopseudomonas* sp. NHTC 133 and a green photosynthetic rod of Lascelles (19).

A. Bacteriochlorophyll a

Dissolved in organic solvents Bchl a has a single near-infrared absorption band located at about 770 mμ and a fluorescence band at about 800 mμ (20). *In vivo* near-infrared absorption bands of Bchl a are found from 800 to 890 mμ, and the corresponding fluorescence bands at slightly higher wavelengths (see Table I).

1. PURPLE BACTERIA

The spectra of Bchl a in purple bacteria (Figs. 1–4) all show a Soret band (370–376 mμ), an orange band (~590 mμ), and multiple near-infrared bands (800–890 mμ). In the spectra of chromatophore fractions the absorption bands located between the Soret band and the orange band of Bchl a belong to carotenoids. These carotenoid bands are absent in bacteria grown in the presence of diphenylamine (14, 15, 21, 22) and in a number of mutants of purple bacteria in which carotenoid synthesis is blocked (7, 11, 14). These mutants show the spectra of Bchl a *in vivo* with almost no interference from other pigments. The absorption band at ~265 mμ in chromatophore spectra is probably due in part to Bchl a, and the small peaks and shoulders in the 310–320 mμ region also appear to be associated with Bchl a.

The fluorescence band of Bchl a in purple bacteria is shifted about 0.03 μ⁻¹ (20 mμ) to the red side of the longest wavelength absorption band of Bchl a (see Figs. 5 and 6 and Table III). The fluorescence of only the longest wavelength form of Bchl a appears *in vivo* because of the high efficiency of energy transfer from the shorter-wavelength forms to the longest-wavelength form (3).

Krasnovskii and co-workers (8, 9) have attempted to explain the multiple absorption bands *in vivo* and the large wavelength differences between Bchl a *in vivo* and in solution in terms of various states of Bchl a aggregation. The spectral characteristics of thin films of Bchl a and other model systems mimic those of Bchl a *in vivo* in several important respects, but often the absorption and fluorescence bands in the model systems are much broader than the bands observed *in vivo*.

The Bchl a spectrum characteristic for *Rhodospirillum rubrum* has only two near-infrared absorption bands, a lesser band at about 800 mμ and a greater band at about 880 mμ (see Fig. 1). Figure 5 shows the mirror symmetry of absorption and fluorescence bands separated by 0.032 μ⁻¹ in *R. rubrum* (see also Table III). The fluorescence emission from *R. rubrum* at 77°K has been studied by Brody and Linschitz (23)

TABLE I

SPECTRAL CHARACTERISTICS OF BACTERIOCHLOROPHYLLS *in Situ*

Organism	Wavelengths (mμ) of absorption maxima			Wavelength (mμ) of fluorescence maximum[a]	Source
	UV band	Orange band	Infrared bands		
Rhodospirillum rubrum[b]	376	589	804 · 880	906 (~920)	Fig. 1; (23)
Rhodomicrobium vannielii[b]	376	590	800 · 870	900 ± 4	(7)
Rhodopseudomonas palustris[b]	372	587	804, 858, ~880	—	Fig. 4
Rhodopseudomonas spheroides[b]	376	589	800, 850, 875	—	Fig. 3; (14)
Chromatium D[b]	371	590	806, 850, 890	912	Fig. 2; (2)
Chlorobium thiosulfatophilum,[b] strain L	370[c]	602[c]	809[c]	816 ± 1 (825)	(23a)
Chloropseudomonas ethylicum,[b] strain 2K	371[c]	603[c]	809[c]	818[c] (827)[c]	Chapter 13
Rhodopseudomonas sp. NHTC 133[d]	399	604	~830[e] · 1014	1040	Fig. 11

[a] Wavelengths within parentheses are for 77°K.
[b] Contains Bchl *a*.
[c] Determined from purified Bchl-protein complex.
[d] Contains Bchl *b*.
[e] Identification with Bchl *b* not yet certain.

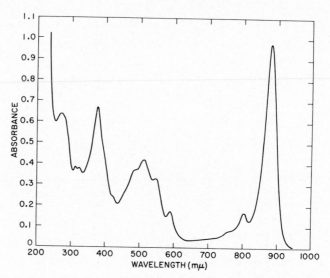

Fig. 1. Absorption spectrum of washed chromatophore fraction from *Rhodospirillum rubrum*, American Type Culture Collection 11170. Peaks (and shoulders) are located at 271, 312, 323, 376.4, (488), 511, 546, 589, 804, and 880.4 mμ.

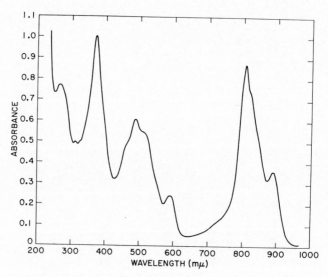

Fig. 2. Absorption spectrum of washed chromatophore fraction from *Chromatium* sp., strain D. Peaks (and shoulders) are located at 265, 312, (324), 371, (~460), 488, (520), 590, 806, and 889 mμ.

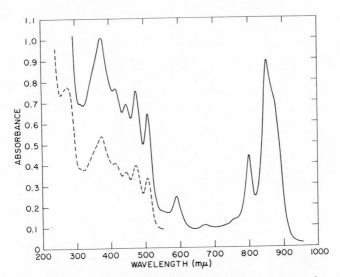

FIG. 3. Absorption spectrum of washed chromatophore fraction from *Rhodopseudomonas spheroides*, strain 2.4.1. Peaks (and shoulders) are located at (~265), ~314, 376, (~415), 447, 475, 508.5, 589, 800, 852, and (~880). The very high absorbance in the 260-mμ region (dashed curve) is due to contamination by ribosomal RNA (*16*).

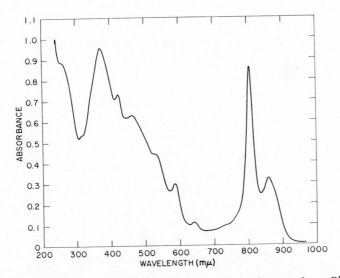

FIG. 4. Absorption spectrum of washed chromatophore fraction from *Rhodopseudomonas palustris*, van Niel strain 2137. Peaks (and shoulders) are located at 258, 314, 372, 423, 457, (~535), 587, 641, 804, 858, and (~880) mμ.

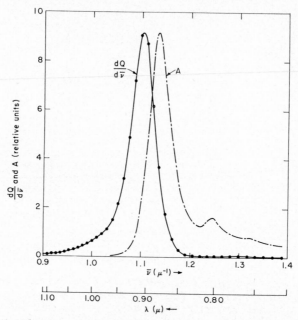

FIG. 5.   Fluorescence (dQ/dv̄) and absorption (A) bands of R. *rubrum*.

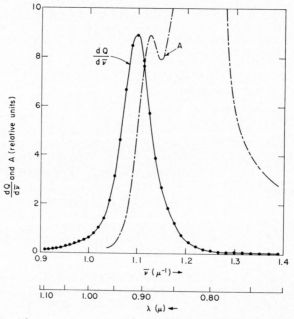

FIG. 6.   Fluorescence (dQ/dv̄) and absorption (A) bands of *Chromatium* sp.,
strain D.

and is listed in Table I. Figure 6 shows the absorption and fluorescence bands of Bchl 890 in *Chromatium* (see also Table III).

Most purple bacteria show at least three absorption bands in the near-infrared region. *Chromatium*, strain D (Fig. 2) and *Rhodopseudomonas spheroides* (Fig. 3) are well studied examples of this phenomenon. The locations of the bands are listed in Table I. The relative peak heights of the near-infrared absorption bands in both *Chromatium* and *R. spheroides* are extremely variable depending on the history of the culture examined. The original interpretation of the complex near-infrared spectra of these bacteria in terms of three basic forms of Bchl *a* was given by Wassink *et al.* (2). Considerable work has been done to demonstrate a relationship between the carotenoid content and the ratios of the near-infrared peak heights of the Bchl *a in vivo* (11, 12). The most recent work of Crounse *et al.* (14), however, indicates that the light intensity during growth and the total Bchl content of the cells are the most important factors in determining the relative contributions of the three forms of Bchl *a* in *R. spheroides*. Similar effects were noted in both wild-type bacteria and a carotenoidless mutant strain.

The studies of Bril (15, 17) on the absorption and fluorescence characteristics of chromatophores from *Rhodospirillum rubrum*, *Chromatium*, and *Rhodopseudomonas spheroides* have shown that detergents can change these characteristics by blocking energy transfer to the longest wavelength form of Bchl *a* and by actual transformation of the longest wavelength form. This latter observation seems to support the idea that the longest-wavelength form of Bchl *a in situ* is in an aggregated state which is disrupted by detergent action. This interpretation seems preferable to the concept of a specific protein bearer which would somehow be modified by the detergent to account for the disappearance of the longest wavelength absorption and fluorescence bands.

The absorption spectra of *Rhodopseudomonas palustris* (Fig. 4) and *Rhodomicrobium vannielii* (7) appear superficially to follow the pattern of *Rhodospirillum rubrum* in having only two bands in the near-infrared region. This is especially true for *Rhodomicrobium vannielii*, since the 870-mμ band is about twice as high as the 800-mμ band. However, for both *Rhodopseudomonas palustris* and *Rhodomicrobium vannielii* the near-infrared spectra show a shoulder on the long-wavelength side of the higher-wavelength peak. This indicates the existence of a third peak which remains unresolved because of the relatively low concentration of the longest-wavelength form of Bchl *a*.

2. GREEN BACTERIA

Bchl *a* is present in green bacteria to the extent of about 5–10% of the total chlorophyll content (19, 24, 25). Its presence is indicated by the

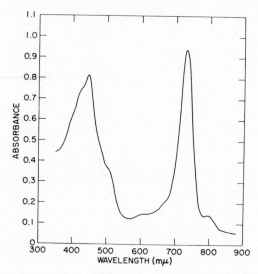

FIG. 7. Absorption spectrum of *Chlorobium thiosulfatophilum*, strain L (courtesy of J. A. Bergeron). Peaks (and shoulders) are located at (~420), 446, (~510), 730, and ~805 mμ.

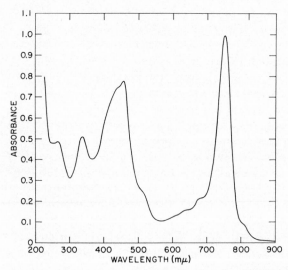

FIG. 8. Absorption spectrum of crude extract of *Chloropseudomonas ethylicum*, strain 2K. Peaks (and shoulders) are located at 265, 336, (~430), 455, (~515), (675), 750, and (~810) mμ.

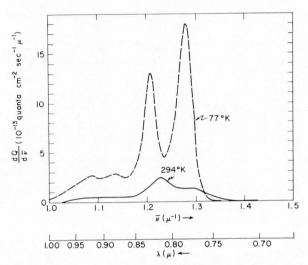

FIG. 9. Fluorescence spectra of *Chloropseudomonas ethylicum,* strain 2K, at 294°K and 77°K (*26*).

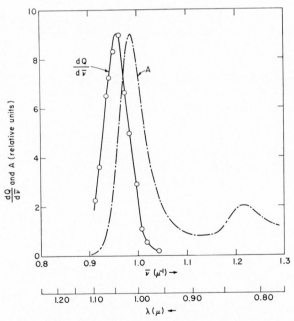

FIG. 10. Fluorescence (dQ/dv) and absorption (A) bands of *Rhodopseudomonas* sp., NHTC 133. Fluorescence data courtesy of R. K. Clayton.

minor absorption band (often seen as a shoulder only) at 809 mμ in the spectra of whole cells (Fig. 7) and cell extracts (Fig. 8). The fluorescence at 816–818 mμ (room temperature) or 827 mμ (77°K) is unmistakable evidence of its presence and function as an energy acceptor (see Fig. 9) (9, 26).

The absorption characteristics of Bchl *a* in green bacteria are almost completely masked *in vivo* by the high concentration of chlorobium chlorophyll. Since the Bchl *a* in the form of a chlorophyll-protein complex can be separated from the Cchl, it is possible to infer the spectral characteristics of Bchl *a in vivo* from the characteristics of the isolated Bchl *a*-protein complex in aqueous solution (see Chapter 13). This is done in Tables I and III. The main differences between the spectral characteristics of Bchl *a* in green bacteria and in purple bacteria are these: In green bacteria the orange band is at about 603 mμ instead of 590 mμ and a single near-infrared band appears at 809 mμ instead of 2–3 bands from 800 to 890 mμ; also the energy separation between absorption and fluorescence bands is about half the separation in purple bacteria (see Table III).

B. Bacteriochlorophyll *b*

In acetone solution Bchl *b* has a Soret band at 368 mμ, an orange band at 582 mμ, and a single near-infrared absorption band at 795 mμ

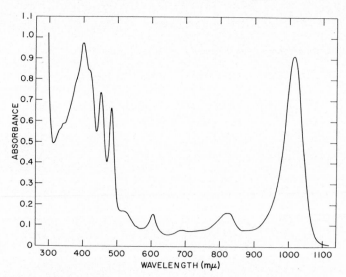

Fig. 11. Absorption spectrum of washed chromatophore fraction from *Rhodopseudomonas* sp., NHTC 133. Peaks (and shoulders) are located at (326), 340, (∼375), 399, (∼420), 451, 482, (∼530), 604, ∼820, and 1014 mμ.

(25). *In vivo* the near-infrared absorption band is shifted 222 mμ to 1017 mμ (27), the longest wavelength yet observed for an electronic transition in a chlorophyll. The absorption characteristics of Bchl *b* *in situ* are shown in Fig. 11 and listed in Table I. The fluorescence emission band is compared to the near-infrared absorption band in Fig. 10 and Table III. The energy separation between absorption and fluorescence bands is about the same as for Bchl *a* in purple bacteria.

III. Chlorobium Chlorophylls

The chlorobium chlorophylls are the characteristic chlorophylls of the green bacteria and function as the main light-collecting pigment *in vivo*. In healthy cells the red absorption band of Cchl is shifted 80–90 mμ into the near-infrared region from the position in organic solvents. In old cells and in extracts, a second form of Cchl appears with its red band shifted only 10–20 mμ from the location in organic solvents. Krasnovskii and co-workers (9, 28) have introduced the terms "aggregate" and "monomer" to describe the longer-wavelength form and the shorter-wavelength form, respectively, since the spectral characteristics of thin films of Cchl are similar to those of the longer-wavelength form and the spectral characteristics of dilute solutions of Cchl in organic solvents are similar to those of the shorter-wavelength form.

The chlorobium chlorophylls are divided into two groups based on spectral properties (29). The Cchls 650 show an absorption maximum at 650 mμ in ether solution and an absorption maximum at 730 ± 5 mμ *in vivo*. The Cchls 660 have corresponding maxima at 660 mμ and 750 ± 5 mμ. Any given green bacterium contains one or the other chlorophyll type, but not both.

A. Chlorophylls 650

Dissolved in diethyl ether, Cchl 650 has its major absorption peaks at 425 and 650 mμ and its fluorescence maximum at 653 mμ (29). *In vivo* the aggregated form absorbing at 730 ±5 mμ fluoresces at about 770 mμ. The absorption spectrum of *Chlorobium thiosulfatophilum*, strain L, which contains Cchl 650, is shown in Fig. 7. Spectral properties *in situ* are listed in Tables II and III. Cchl 650 appears to have been present in a strain of *Chlorobium limicola* used by Krasnovskii and Pakshina (30) as well as in the better known strain L of *C. thiosulfatophilum*.

B. Chlorophylls 660

Chlorobium chlorophyll 660 dissolved in ether has its main absorption peaks at 432 mμ and 660 mμ and its fluorescence maximum at 663 mμ

JOHN M. OLSON AND ELIZABETH K. STANTON

TABLE II

SPECTRAL CHARACTERISTICS OF CHLOROBIUM CHLOROPHYLLS *in Situ*

Organism	Wavelengths (mμ) of absorption maxima			Wavelength (mμ) of fluorescence maximum[a]	Sources
	UV band	Blue band	Red band		
Chlorobium thiosulfatophilum,[b] strain L	~320	446 ± 1	730 ± 5	771 ± 1 (~775)	Fig. 7; (31, 23a)
C. thiosulfatophilum,[c] strain PM	338	457	747	—	(29, 32)
Chloropseudomonas ethylicum,[c] strain 2K	336	457 ± 1	750 ± 5	770 ± 1 (780 ± 1)	Fig. 8; (26, 33, 34)

[a] Wavelengths within parentheses are for 77°K.
[b] Contains Cchl 650.
[c] Contains Cchl 660.

TABLE III

COMPARISON OF ABSORPTION AND FLUORESCENCE BANDS[a]

Chlorophyll	Organism	Absorption band				Fluorescence band			Separation
		λ_a (μ)	$\bar{v}_a$ (μ^{-1})	$\Delta\bar{v}_a$ (μ^{-1})	$\bar{v}_a' - \bar{v}_a$ (μ^{-1})	λ_f (μ)	$\bar{v}_f$ (μ^{-1})	$\Delta\bar{v}_f$ (μ^{-1})	$\bar{v}_a - \bar{v}_f$ (μ^{-1})
Bchl a	*Rhodospirillum rubrum*	0.880	1.136	0.053	0.16	0.906	1.104	0.057	0.032
	Chromatium	0.889	1.125	—	0.17	0.912	1.097	0.070	0.028
	Chloropseudomonas ethylicum[b]	0.809	1.236	0.040	0.06	0.818	1.222	0.045	0.014
	Chlorobium thiosulfatophilum,[b] strain L	0.809	1.236	0.046	0.06	—	—	—	—
Bchl b	*Rhodopseudomonas* NHTC 133	1.014	0.986	0.061	0.27	1.044	0.958	0.062	0.028
Cchl 660 Aggregate	*Chloropseudomonas ethylicum*	0.750	1.333	0.084	0.18	0.770	1.299	0.08	0.034
Monomer	*Chloropseudomonas ethylicum*	0.674	1.484	0.07	0.03	0.679	1.473	0.07	0.011
Cchl 650 Aggregate	*Chlorobium thiosulfatophilum* strain L	0.730	1.370	0.084	0.17	0.771	1.297	0.09	0.073

[a] Symbols have the following meanings: λ_a, wavelength of absorption maximum; $\bar{v}_a$, wavenumber of absorption maximum; $\Delta\bar{v}_a$, half-width of absorption band; $\bar{v}_a' - \bar{v}_a$, difference between band position in organic solvents and band position *in situ*; λ_f, wavelength of fluorescence maximum; $\bar{v}_f$, wavenumber of fluorescence maximum; $\Delta\bar{v}_f$, half-width of fluorescence band; $\bar{v}_a - \bar{v}_f$, difference between absorption band position and fluorescence band position.

[b] Characteristics are those of the purified Bchl-protein complex described in Chapter 13.

(29). The aggregated form *in vivo* has its maximum absorption at 750 ± 5 mμ and fluoresces at about 770 mμ. The absorption spectrum of a crude extract of *Chloropseudomonas ethylicum*, containing Cchl 660, is shown in Fig. 8. The slight shoulder at approximately 675 mμ is due to the monomer of Cchl 660. The fluorescence emission spectra of *C. ethylicum* at 294°K and 77°K are shown in Fig. 9. The band appearing at shorter wavelengths (higher wavenumbers) belongs to Cchl 660; the other bands belong to Bchl *a*. The spectral characteristics of Cchl 660 *in situ* are summarized in Tables II and III.

IV. Discussion

The generalization of Wassink *et al.* (2) was that the near-infrared absorption spectra of Bchl *a* in purple bacteria could be analyzed in terms of three symmetrical bands of approximately $0.04\,\mu^{-1}$ half-width. The first part of the generalization seems valid, but the values of $\Delta\bar{\nu}_a$ and $\Delta\bar{\nu}_f$ for Bchl *a* in purple bacteria (Table III) are substantially larger than $0.04\,\mu^{-1}$. The absorption and fluorescence bands of Bchl *a* in green bacteria are sharper than in purple bacteria, and the separation $\bar{\nu}_a - \bar{\nu}_f$ is much less than in purple bacteria. The values of $\Delta\bar{\nu}_a$, $\Delta\bar{\nu}_f$, and $\bar{\nu}_a - \bar{\nu}_f$ for Bchl *b in situ* are about the same as for Bchl *a*. The long wavelength absorption and fluorescence bands of the chlorobium chlorophylls *in situ* are noticeably broader than those for the bacteriochlorophylls.

The proposal that chlorophyll aggregation is largely responsible for the spectral characteristics *in vivo* is most satisfactory for explaining the characteristics of Cchl in green bacteria. For the explanation of the two- and three-banded near-infrared spectra of Bchl in purple bacteria, a theory is needed which is more specific than those already proposed. Such a theory should explain, for example, the empirical observation that the 800-mμ band of Bchl *a* is associated with the 850-mμ band in *Rhodopseudomonas spheroides* (35, 36), but is independent of the 850-mμ band in *Chromatium*, strain D (15).

Acknowledgments

The authors wish to thank Dr. Robert G. Bartsch for a culture of *Rhodopseudomonas palustris*, and Dr. Roderick K. Clayton for a culture of *R. spheroides*. Dr. Clayton also supplied the fluorescence data shown in Fig. 10. The absorption spectrum in Fig. 7 was obtained by Dr. John A. Bergeron.

References

(1) E. Katz and E. C. Wassink, *Enzymologia* 7, 97 (1939).
(2) E. C. Wassink, E. Katz, and R. Dorrestein, *Enzymologia* 7, 113 (1939).

(3) L. N. M. Duysens, Thesis, University of Utrecht (1952).
(4) C. S. French and V. M. K. Young, in "Radiation Biology" (A. Hollaender, ed.), Vol. III, p. 343. McGraw-Hill, New York, 1956.
(5) E. I. Rabinowitch, "Photosynthesis and Related Processes," Vol. II, Part 2, Chapter 37C, p. 1793. Wiley (Interscience), New York, 1956.
(6) H. G. Schlegel and N. Pfennig, Arch. Mikrobiol. 38, 1 (1961).
(7) R. K. Clayton, in "Bacterial Photosynthesis" (H. Gest, A. San Pietro, and L. P. Vernon, eds.), p. 495. Antioch Press, Yellow Springs, Ohio, 1963.
(8) A. A. Krasnovskii, K. K. Voinovskaya, and L. M. Kosobutskaya, Dokl. Akad. Nauk SSSR 85, 389 (1952); "Fluorescence and Photochemistry of Chlorophyll" (English translation by E. I. Rabinowitch), AEC Translation 2156, p.105. Oak Ridge, Tennessee, 1956.
(9) A. A. Krasnovskii, Yu. E. Erokhin, and Kh. Yui-Tsun, Dokl. Akad. Nauk SSSR 143, 456 (1962).
(10) A. A. Krasnovskii, Yu. E. Erokhin, and B. A. Gulyaev, Dokl. Akad. Nauk SSSR 152, 1231 (1963).
(11) M. Griffiths, W. R. Sistrom, G. Cohen-Bazire, and R. Y. Stanier, Nature 176, 1211 (1955).
(12) J. A. Bergeron and R. C. Fuller, Nature 184, 1340 (1959).
(13) E. C. Wassink and G. H. M. Kronenberg, Nature 194, 553 (1962).
(14) J. Crounse, W. R. Sistrom, and S. Nemser, Phototchem. Photobiol. 2, 361 (1963).
(15) C. Bril, Biochim. Biophys. Acta 66, 50 (1963).
(16) P. B. Worden and W. R. Sistrom, J. Cell Biol. 23, 135 (1964).
(17) C. Bril, Thesis, University of Utrecht (1964).
(18) L. N. M. Duysens and J. Amesz, Biochim. Biophys. Acta 24, 19 (1957).
(19) A. Jensen, O. Aasmundrud, and K. Eimhjellen, Biochim. Biophys. Acta 88, 466 (1964).
(20) J. H. C. Smith and A. Benitez, in "Moderne Methoden der Pflanzenanalyse" (K. Paech and M. V. Tracey, eds.), Vol. IV, p. 142. Springer, Berlin, 1955.
(21) G. Cohen-Bazire and R. Y. Stanier, Nature 181, 250 (1958).
(22) R. C. Fuller and I. C. Anderson, Nature 181, 252 (1958).
(23) M. Brody and H. Linschitz, Science 133, 705 (1961).
(23a) J. M. Olson and C. Sybesma, unpublished data, 1962.
(24) J. M. Olson and C. A. Romano, Biochim. Biophys. Acta 59, 726 (1962).
(25) A. S. Holt, D. W. Hughes, H. J. Kende, and J. W. Purdie, Plant Cell Physiol. (Tokyo) 4, 49 (1963).
(26) C. Sybesma and J. M. Olson, Proc. Natl. Acad. Sci. U. S. 49, 248 (1963).
(27) J. M. Olson and K. D. Nadler, Photochem. Photobiol. 4, 783 (1965).
(28) A. A. Krasnovskii, Yu. E. Erokhin, and I. B. Fedorovich, Dokl. Akad. Nauk SSSR 134, 1232 (1960).
(29) R. Y. Stanier and J. H. C. Smith, Biochim. Biophys. Acta 41, 478 (1960).
(30) A. A. Krasnovskii and E. V. Pakshina, Dokl. Akad. Nauk SSSR 127, 913 (1959).
(31) J. A. Bergeron and R. C. Fuller, in "Biological Structure and Function" (T. W. Goodwin and O. Lindberg, eds.), Vol. II, p. 307. Academic Press, New York, 1961.
(32) H. Larsen, Kgl. Norske Videnskab. Selskabs, Skrifter No. 1 (1953), as quoted in E. I. Rabinowitch, "Photosynthesis and Related Processes," Vol. II, Part 2, p. 1854. Wiley (Interscience), New York, 1956.

(33) V. V. Shaposhnikov, E. N. Kondratieva, and V. D. Fedorov, *Nature* **187**, 167 (1960).
(34) E. N. Kondrat'eva and L. V. Moshentseva, *Dokl. Akad. Nauk SSSR* **135**, 460 (1960).
(35) R. K. Clayton, *Photochem. Photobiol.* **1**, 305 (1962).
(36) W. R. Sistrom, *Biochim. Biophys. Acta* **79**, 419 (1964).

—13—

Chlorophyll-Protein Complexes
Part I
Complexes Derived from Green Plants

J. C. GOEDHEER

Biophysical Research Group, Physics Institute,
University of Utrecht, the Netherlands

I. Natural Chlorophyll-Protein Complexes

A. Smallest Photosynthetic Unit

1. ABSORPTION PROPERTIES

Thomas *et al.* (*1*) found that sonication of spinach chloroplasts yielded particles which were still active in performing the Hill reaction, as long as the particle size exceeded 10^6 Å^3. The smallest active fragments obtained by sonication ["quantasomes," cf. Park and Pon (*2*)] still have, according to Sauer and Calvin (*3*), an absorption spectrum similar to that of intact chloroplasts (Fig. 1). These particles are assumed to contain the whole photochemical apparatus of the chloroplast. In the presence of soluble, colorless stroma material they are capable of supporting CO_2 fixation as well. The molecular weight of such a "quantasome" was found to be of the order of 2×10^6. Protein accounts for about 50% of this value, while chlorophyll and carotenoids take up another 17% (*4*). About 230 Chl *a* molecules are present on a single fragment (*5*).

According to French *et al.* (*6*), the major Chl *a* forms *in vivo* are Chl *a* 673 and Chl *a* 683, named after the probable position of their absorption maxima. These designations do not give the exact positions for these chlorophyll forms in all plants, and by some authors the values are given as 670 and 680. In most, if not all, species a long wavelength form Chl *a* 695 is also present, but in a much lower concentration (7). From measurements of electric dichroism Sauer and Calvin (8) concluded that an oriented form of chlorophyll which absorbs around 700 mμ is present in spinach quantasomes. They estimated that about 5% of chlorophyll is present in this form.

The percentage of chlorophyll molecules present as Chl *a* 673, Chl *a* 683 and Chl *a* 695 varies with growing conditions and from species to species. It seems probable that more Chl *a* forms exist *in vivo* [cf. Thomas (9)]. Whether such forms are functional in photosynthesis or

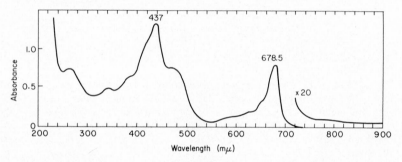

Fig. 1. Absorption spectrum of spinach "quantasomes" (3). The spectrum is similar to that of intact chloroplasts, except in the uv region. The ratio of pigment: protein absorption is high compared to that of the pigment-protein complexes of Figs. 3, 4, and 6.

other processes, or are produced by enzymatic degradation in a certain fraction of chloroplasts cannot be concluded. Rumberg and Witt (10) concluded from measurements on chloroplasts and digitonin-treated chloroplasts that bands in the difference spectrum at 703 and 430 mμ obtained by illumination are caused by oxidation of a special Chl *a* form, P700 (Fig. 2.). These changes occurred also at −150°C, though irreversibly. It was estimated that about 0.1% of the total chlorophyll was present in this form. Kok and Hoch (11) estimated about the same percentage. This would mean that only one P700 molecule could be present for each four quantasomes, and further that P700 is a form different from Chl *a* 695 in the elementary photosynthetic particles.

2. Fluorescence Properties

Fluorescence at room temperature, which has its maximum at about 684 mμ in dilute suspensions of chloroplast fragments, can be assumed to be emitted mainly by Chl *a* 673 for the following reasons. The "Stokes shift" (distance between fluorescence and absorption maximum) for Chl *a* 673 is 12 mμ, while for Chl *a* 683 it would be only 1 mμ. The latter value for a Stokes shift is highly unlikely in view of values measured with chlorophyll in organic solvents. The shift of 12 mμ is of the same order of magnitude as that of artificially prepared chlorophyll-protein complexes (*12*). Furthermore, the absorption maximum of a chloroplast suspension can be shifted from 680 to 675 mμ as a result of disappearance of the Chl *a* 683 form by means of photobleaching, while the fluorescence maximum remains at 684 mμ.

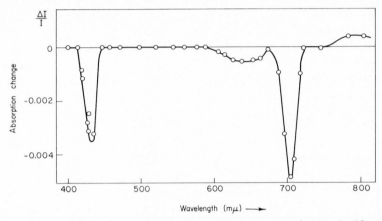

FIG. 2.   Absorption difference spectrum of illuminated aged spinach chloroplasts vs. unilluminated ones [phenazine methosulfate and ascorbic acid added, after Rumberg and Witt (*10*)]. Only the changes ascribed to oxidation of a chlorophyll *a* form, P700, are seen.

Some fluorescence emitted around 715 mμ at room temperature, but not visible as a separate band, is due to a chlorophyll form absorbing around 700 mμ, most probably Chl *a* 695. Cooling to 77°K results in a strong increase in fluorescence (40–100 times) to a high band at around 730 mμ (*13–15*). A fluorescence band at 695 mμ is also present at this temperature, which is ascribed to Chl *a* 683 (*16, 17*). Such a complex spectrum is also observed in suspensions of small fragments of spinach chloroplasts. The form P700 is assumed to be nonfluorescent even at 77°K (*18*).

When chloroplasts are irradiated with linearly polarized red light at room temperature, the degree of fluorescence polarization is very low. This is also the case if very small fragments of chloroplasts are measured. Even with quantasomes, the particle size would be sufficient to prevent rotational depolarization. Hence depolarization most likely is due to

energy transfer between differently oriented molecules. The possibility
that the Chl a 673 molecules, which are responsible for room temperature
emission, are grouped in systems with differently fixed orientation cannot
be excluded by these measurements.

The quantum yield of total chlorophyll fluorescence is about 1–2%
for chloroplasts (*19*). Since mainly Chl a 673 fluoresces at room temperature, the quantum yield of this form is about 4–8%. This value
agrees reasonably well with the quantum yield derived from measurements of fluorescence lifetime *in vivo* (*20, 21*). The relatively high yield
indicates that energy transfer from Chl a 673 to Chl a 683 is rather inefficient. However, the dimensions of a particle with molecular weight
of 2×10^6 are such that some energy transfer between those forms is
expected. If these pigment forms participate in systems II and I of photosynthesis, some separation should occur. Several attempts have been
made to separate the two functional systems of photosynthesis by splitting up the small chloroplast fragments.

B. Chlorophyll-Protein Complexes Derived from *Chlorella*

Allen *et al.* (*22*) isolated from *Chlorella* a photochemically active
chlorophyll-protein complex with a red absorption maximum at 672 mμ.
This complex had a chlorophyll $a:b$ ratio of about 1. The particles did
not sediment after 1 hour of centrifugation at 144,000 g. The electron
microscope showed that particles with a diameter of about 7–30 mμ were
in the supernatant. It was suggested that they might be combinations
of units of about 7 mμ or less. As not all these units contained chlorophyll,
it was not certain that all were derived from chloroplasts. The yield of
the 672 mμ complex was only 1–3% of total chlorophyll. Since the particles could be obtained only under special conditions of grinding, breaking, and centrifugation, the low yield was likely caused by the isolation
procedure. The heavier material from which the material was separated
had its absorption maximum at 678–680 mμ.

Allen and Murchio (*23*) also noted the tendency of the particle to
aggregate irreversibly into membranes. As a result of this aggregation, the
absorption maximum shifted from 672 to 677 mμ. The single particles
were able to perform the Hill reaction with DPIP,* but not with quinone.
No fluorescence measurements were reported.

The fraction with the low chlorophyll $a:b$ ratio might be a functional
complex active in system II. According to action spectra of the Hill reaction (*24*) and fluorescence (*25*), the bulk of Chl b transfers light

---

* The following abbreviations are used in this chapter: DCMU, 3-(3,4-dichloro-
phenyl)-1,1-dimethylurea; DPIP, 2,6-dichlorophenolindophenol; TPIP, 2-3'-6 trichlo-
rophenol indophenol, SDS, sodium dodecylsulfate.

energy to a short-wave Chl *a* form, probably Chl *a* 673, which is active in system II (see Chapter 18).

C. Chlorophyll-Protein Complexes Derived from Spinach Chloroplasts

Various investigations have been directed toward the splitting of spinach chloroplasts with the aid of detergents. The older literature concerning detergent fragmentation, which in general leads to nonfluorescent, clear, chlorophyll-containing suspensions, is reviewed by Kupke and French (12). Ke and Clendenning (26) found that chloroplast preparations lost their capacity for the Hill reaction after addition of the detergents sodium dodecyl sulfate (SDS), Tween 20, or α-picoline. Wolken and Schwerz (27) determined molecular weights for the chlorophyll-protein complexes extracted with 2% digitonin. They reported oxygen production by illumination of the detergent-treated preparation.

Boardman and Anderson (28) recently isolated fractions of different size from spinach chloroplasts after treatment with 0.5% digitonin. They found a heavy-weight fraction with a lower Chl *a*:Chl *b* ratio than occurs in the intact chloroplasts, and a light-weight fraction with a much higher Chl *a*:Chl *b* ratio. The heavy particles reduced TCIP or NADP in a Hill reaction. The light particles reduced NADP$^+$ effectively only when ascorbate plus TCIP was added as an electron donor, whereas the heavy-weight particle was relatively inactive in this reaction. The light-weight particle, however, was unable to reduce either NADP$^+$ or TCIP in a Hill reaction. The position of the absorption bands and the low percentage of Chl *b* might suggest that the light-weight particle corresponds to the Chl *a* form Chl *a* 683 *in vivo*. However, it should be emphasized that extraction of chlorophyll-protein complexes from the chloroplast can change both the position and shape of absorption bands, as well as the fluorescence yield.

Sironval *et al.* (29) used centrifugation and agar gel electrophoresis to separate two fractions from spinach chloroplasts treated with digitonin and SDS. The "heavy" fraction performed the Hill reaction with DPIP; this reaction could be stopped by addition of DCMU or SDS. The Chl *a*: Chl *b* ratio was about 2. The supernatant (Chl *a*:Chl *b* ratio = 7) reduced NADP with DPIP-ascorbate as electron donor in the presence of crude PPNR. This reaction was not inhibited by DCMU.

Wessels (30) used digitonin fragmentation to isolate functionally the two photochemical systems of photosynthesis. The water photolysis system (system II) was strongly inactivated by addition of digitonin or DCMU. Cyclic photophosphorylation and reduction of NADP$^+$ (system I) can proceed independently of system II by addition of a suitable redox couple, such as DPIP-ascorbate. Wessels (*31*) mentions that a definite weight ratio of digitonin:chlorophyll seems to be essential. When the digitonin exceeds 0.5%, the particles are no longer able to perform photophosphorylation, though reduction of $NADP^+$ remains possible if the necessary constituents, including plastocyanin, are available. Electron micrographs of these fractions do not show the 80 Å knobs which are visible in the preparations still able to perform photophosphorylation.

The absorption spectrum of fragments obtained from 0.27% digitonin treatment closely corresponds to that of whole chloroplasts and "quantasomes" (absorption maxima of chlorophyll at 678.5 and 437 mμ). The absorption spectrum of the 80,000 g supernatant of the 1–1.3% digitonin preparation was different (Fig. 3), with a red chlorophyll maximum at

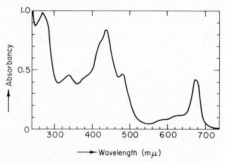

FIG. 3. Absorption spectrum of the 80,000 g supernatant obtained after treatment of spinach chloroplasts with 1–1.3% digitonin (*31*).

675 mμ. The Chl *a*:Chl *b* ratio of the latter fraction was about 6, while the ratio of the band at 280 mμ indicates a high ratio of protein/chlorophyll compared to quantasomes (*31*). By rate zonal centrifugation this fraction could be separated into three subfractions, the lightest of which showed a red absorption maximum at 670 mμ.

Kahn and Bannister (*32*) used 1% Triton X-100 to extract a soluble chlorophyll-protein complex from spinach chloroplasts. This complex could not be sedimented by a 1-hour centrifugation at 190,000 g. It contained about 3–8% of the total chlorophyll in the chloroplast and was able to reduce potassium ferricyanide, but not DPIP or $NADP^+$ in the light. The ability to reduce ferricyanide was not lost after 5 minutes' boiling, a result indicating that it was not an enzyme-mediated reaction.

Kok (*33*) attempted to isolate a long-wavelength form of chlorophyll (P700) by acetone treatment. He found that a mixture of 85% acetone and 15% water extracted 85% of the chlorophyll, while the light-induced changes at 700 mμ (by which this pigment form is characterized) were decreased less than twofold. Other solvents completely abolished the absorption changes. The particles could be fractionated, but they still were not soluble proteins.

Rumberg and Witt (10) and Ke (34) observed the difference spectrum of P700 without accompanying cytochrome changes, upon illumination of chloroplasts treated with digitonin or Triton X-100. It remains to be investigated in which of the fractions measured by Wessels, Boardman, and Anderson or Sironval et al. the 700 mμ absorption changes are incorporated.

Chloroplasts may be broken up without the use of detergents by overnight storage of a crude suspension of chloroplast fragments in the dark at 25°C. This results in the formation of chlorophyll-protein or pheophytin-protein complexes with red absorption maxima at about 669 mμ, which appear to be formed from Chl a 683. These complexes, which do not sediment after 1 hour at 44,000 g, are weakly fluorescent and have an emission maximum at 680 mμ. Cooling to −196°C sharpens the fluorescence bands, but no new bands appear.

D. Chlorophyll-Protein Complexes Derived from *Chenopodium album*

Yakushiji et al. (35) isolated a special chlorophyll-protein complex from Chenopodium album. This complex does not sediment at any speed in a Spinco ultracentrifuge, Model E. The molecular weight of the particle thus cannot exceed 2×10^5. The absorption spectrum shows main chlorophyll maxima at 667 and 428 mμ, and a protein band at 277 mμ (Fig. 4). Secondary maxima at 615 and 575 mμ are minor bands of Chl a, while the secondary band at 460 mμ probably represents the Soret band of Chl b.

The most remarkable property of this chlorophyll-protein complex is its change in absorption spectrum upon illumination. The Chl a bands show a strong decrease, while new bands arise at 743, 565, 399, and 364 mμ. The 277-mμ protein band remains unaltered. The sequence of the new bands strongly resembles that of bacteriochlorophyll. This pigment has bands of similar intensity ratios at 769, 574, 392, and 357 mμ in ether solution. The fluorescence spectrum of the illuminated protein-chlorophyll complex shows two bands at 677 and 746 mμ. Terpstra (36) showed that the Chenopodium chlorophyll-protein complex could be separated into two fractions: one which showed the transformation 677 → 743 mμ, and one in which the 677-mμ chlorophyll band bleached upon illumination; no 743-mμ form occurred.

The fluorescence action spectrum of the 746-mμ emission shows the presence of all bands measured in the absorption spectrum. This indi-

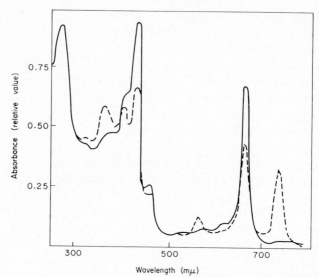

Wavelength (mμ)

Fig. 4. Absorption spectrum of a *Chenopodium* chlorophyll-protein complex before (solid line) and after (broken line) illumination (35). The sequence of absorption bands which appear as a result of illumination resembles that of bacteriochlorophyll (743 vs. 770, 566 vs. 580, 365 vs. 365). Bands which are presumably due to chlorophyll *b* (645 and 470 mμ) are not changed by illumination.

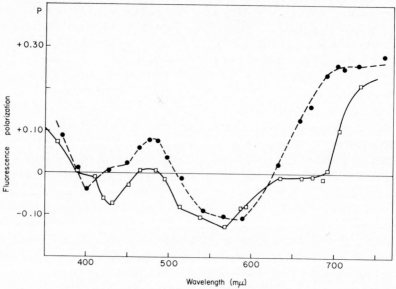

Wavelength (mμ)

Fig. 5. Fluorescence polarization spectrum of illuminated *Chenopodium* chlorophyll-protein (solid line) and bacteriochlorophyll (broken line) in ricinus oil. The spectra show similarities in polarization of the 743–770, 566–580, and 365–365 mμ bands.

cates that the fluorescence at 746 mμ is also excited by that fraction of chlorophyll which is not transformed into the 743-mμ form.

The fluorescence polarization spectrum of the 746-mμ emission shows positive values beyond 700 mμ and below 400 mμ, and a negative polarization in the region of the yellow band. In Fig. 5 the polarization spectrum in ricinus oil is given, together with that of bacteriochlorophyll. It seems likely that only one molecule of chlorophyll is transformed into the 743-mμ form in each protein-chlorophyll complex. The low polarization values at 668 and 435 mμ possibly indicate that one or a few 668 pigment molecules remain unchanged. Takamiya et al. (37) state that the transformation is related to oxidizing conditions. The photoconversion of Chenopodium chlorophyll was found to be reversible in the absence of oxygen, if sodium hydrosulfite or ascorbate was added. With the first reductant, 97% of the 668-mμ band could be regenerated, though the weak band at 458 and the shoulder at 645 (probably due to Chl b) disappeared. Cooling of this chlorophyll-protein complex to liquid nitrogen temperature resulted in an increase and sharpening of the fluorescence bands, but no new bands were formed. No light-induced changes were observed at temperatures below −80°C.

Lippincott et al (38) and Aghion (39) describe another chlorophyll complex with absorption maximum at about 740 mμ. They obtained chloroplast fragments from leaves treated with hot 85% methanol, followed by homogenization at 2–4°C and centrifugation by differential centrifugation. No bands at 566 mμ and 360 mμ were present, and the preparations were nonfluorescent. This procedure of isolating chloroplast fragments may produce some crystalline type of chlorophyll, detached from its place on the protein moiety. Chlorophyll crystals and oriented chlorophyll monolayers exhibit a similar shift of their absorption bands toward the red and are also nonfluorescent (see Chapter 8). Nishizaki (40) followed a similar treatment and could obtain band positions as far as 760 mμ. He observed an abolishment of Hill reaction capacity during the formation of the preparation.

E. Chlorophyll-Protein Complexes Derived from Etiolated Bean Leaves

Another chlorophyll-protein complex which has been studied extensively is the protochlorophyll holochrome. Smith et al. (41) extracted this protochlorophyll-protein complex from etiolated bean leaves. The absorption spectrum, given in Fig. 6, shows protochlorophyll bands at 640 and 440 mμ, and a protein band is present at 275 mμ. The properties of these complexes are considered by Boardman in Chapter 14, but a few remarks are appropriate here. The isolated holochrome still possesses the property of transforming protochlorophyll into chlorophyll upon illumination. During the conversion a porphin-type spectrum is transformed into a dihydroporphin type of spectrum. Thus a hydrogenation occurs on two carbon atoms in the conjugated chain of the pigment.

In the *Chenopodium* chlorophyll-protein complex the spectroscopic changes indicate a change from a dihydroporphin type into a tetrahydroporphin type of spectrum. However, the tetrahydroporphin-type of pigment cannot be extracted easily with organic solvents, unlike the

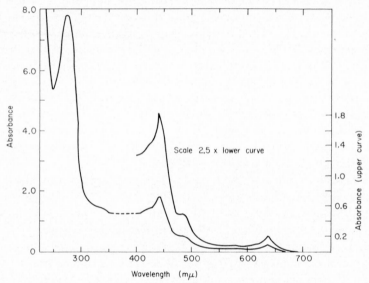

Fig. 6. Absorption spectrum of the protochlorophyll holochrome prepared from etiolated bean leaves (*41*).

chlorophyll formed by transformation of protochlorophyll. This difference might indicate that in the *Chenopodium* chlorophyll-protein complex the spectral changes are due to changes in the strongly complexed protein moiety, which changes might influence the spectral properties of the pigment. Such a hypothesis could explain also their occurrence under oxidizing conditions.

II. Artificial Chlorophyll-Protein Complexes

Chlorophyll can be easily attached to various substances, such as proteins, silica gel, or filter paper. Usually such adsorbed chlorophyll is non-fluorescent. If, however, some lipid is present, fluorescence of an adsorbed chlorophyll can be detected [cf. Seybold and Egle (*42*)].

Various attempts have been made to prepare artificial chlorophyll-protein complexes which give rise to oxygen evolution upon illumination. Eisler and Portheim (43) reported vigorous oxygen evolution with a horse serum-chlorophyll complex. This phenomenon could not be repeated by other workers. The oxygen evolution might have been caused by a peroxide decomposition catalyzed by chlorophyll [cf. Oster and Lu (44)]. Rodrigo (45) prepared an artificial chlorophyll-protein complex from chlorophyll-deficient *pelargonium* leaves, to which acetonic chlorophyll was added. He measured a slight oxygen production with quinone as an electron donor.

In photosynthesis, oxygen production results from decomposition of water, and it is a phenomenon that occurs with the help of specific enzymes. It needs the cooperation of at least several light quanta, which seems difficult to obtain in artificially prepared systems. Vishniac (46) described experiments in which he added an alcoholic chlorophyll extract to buffer solutions which contained chloroplast material from which chlorophyll had been removed by acetone treatment. For a Hill reaction he measured reduction of oxidized glutathione in the presence of NADP+ and glutathione reductase (in a helium atmosphere). Photochemical activity was measured as a decrease in redox potential. Boiled or frozen extracts proved to be inefficient. It does not seem very likely that in these experiments water actually functioned as an electron donor to produce oxygen. More likely, electrons were provided by other constituents in the extract. Experiments similar to those of Vishniac were done by Lynch and French (47). They intended to measure dye reduction photochemically instead of glutathione reduction. No dye reduction, however, could be detected with the reconstituted system.

Sapozhnikov and Maslova (48) made [according to Kupke and French (12)] absorption and fluorescence measurements of various artificial chlorophyll-protein complexes. The absorption maxima were at about 672 mμ, the fluorescence maxima at about 685 mμ. This indicates a "Stokes shift" of about 13 mμ. They also reported some Hill reaction capacity (reduction of potassium ferricyanide) in these artificially prepared protein-chlorophyll complexes.

The chlorophyll-protein crystals obtained by Takashima (49) from clover extracts probably are not naturally occurring chloroplast constituents. The carotenoid-free crystals were produced by treatment with a high concentration of α-picoline. In contrast to the usual behavior of chlorophyll-containing clear suspensions obtained by a treatment with detergents, these crystals are fluorescent. The absorption maximum is located at 670 mμ.

Chiba (50) prepared from such crystals fractions which contained

Chl *a* and *b* separately. Anderson *et al* (*51*) noted the variability of the chlorophyll-protein ratio, and even reported that chlorophyll-free crystals could be obtained. Also chlorophyll and protein moved separately in paper electrophoresis and paper chromatography. Krasnovskii and Brin (*52*) could remove protein from the crystals by washing them with water, and they concluded that the Takashima crystals are merely aggregated forms of chlorophyll. As the crystals are fluorescent, however, this does not seem very likely. It seems more plausible to assume that they represent a form of chlorophyll attached to oriented detergent micelles.

REFERENCES

(*1*) J. B. Thomas, O. H. Blaauw, and L. N. M. Duysens, *Biochim. Biophys. Acta* **10**, 230 (1953).
(*2*) R. B. Park and N. G. Pon, *J. Mol. Biol.* **3**, 1 (1961).
(*3*) K. Sauer and M. Calvin, *Biochim. Biophys. Acta* **64**, 324 (1962).
(*4*) R. B. Park and J. Biggins, *Science* **144**, 1009 (1964).
(*5*) K. Sauer and R. B. Park, *Biochim. Biophys. Acta* **79**, 467 (1964).
(*6*) C. S. French, J. S. Brown, M. B. Allen, and R. F. Elliot, *Carnegie Inst. Wash. Year Book* **58**, 327 (1959).
(*7*) Y. F. Frei, *Biochim. Biophys. Acta* **57**, 82 (1963).
(*8*) K. Sauer and M. Calvin, *J. Mol. Biol.* **4**, 451 (1962).
(*9*) J. B. Thomas, *Biochim. Biophys. Acta* **59**, 202 (1962).
(*10*) B. Rumberg and H. T. Witt, *Z. Naturforsch.* **19b**, 693 (1964).
(*11*) B. Kok and G. Hoch, *in* "Light and Life" (W. D. McElroy and B. Glass, eds.), p. 397. Johns Hopkins Press, Baltimore, Maryland, 1960.
(*12*) W. Kupke and C. S. French, *in* "Handbuch der Pflanzenphysiologie" (W. Ruhland, ed.), Vol. V, p. 298. Springer, Berlin, 1961.
(*13*) S. S. Brody, *Science* **128**, 838 (1958).
(*14*) J. C. Goedheer, *Biochim. Biophys. Acta* **53**, 420 (1961).
(*15*) W. L. Butler, *Arch. Biochem. Biophys.* **93**, 413 (1961).
(*16*) J. C. Goedheer, *Biochim. Biophys. Acta* **88**, 304 (1964).
(*17*) B. Kok, *Natl. Acad. Sci.—Natl. Res. Council, Publ.* **1145**, 45 (1963).
(*18*) J. C. Goedheer, *in* "Proceedings of Biochemistry of Chloroplasts," N. A. T. O. Conference Aberystwyth. Academic Press, New York, 1966, p. 75.
(*19*) P. Latimer, T. T. Bannister, and E. Rabinowitch, *Science* **124**, 585 (1956).
(*20*) S. S. Brody and E. Rabinowitch, *Science* **125**, 555 (1957).
(*21*) W. L. Butler and K. H. Norris, *Biochim. Biophys. Acta* **66**, 73 (1963).
(*22*) M. B. Allen, J. C. Murchio, S. W. Jeffrey, and S. A. Bendix, *in* "Studies on Microalgae and Photosynthetic Bacteria" (Japan. Soc. Plant Physiol., eds.), p. 407. Univ. of Tokyo Press, Tokyo, 1963.
(*23*) M. B. Allen and J. C. Murchio, *Natl. Acad. Sci.—Natl. Res. Council, Publ.* **1145**, (1963).
(*24*) J. Biggins and K. Sauer, *Biochim. Biophys. Acta* **88**, 655 (1964).
(*25*) J. C. Goedheer, *Biochim. Biophys. Acta* **102**, 73 (1965).
(*26*) B. Ke and K. A. Clendenning, *Biochim. Biophys. Acta* **18**, 74 (1958).
(*27*) J. J. Wolken and F. A. Schwerz, *J. Gen. Physiol.* **37**, 111 (1953).
(*28*) N. K. Boardman and J. M. Anderson, *Nature* **203**, 166 (1964).

(29) C. Sironval, H. Clysters, J. M. Michel, R. Bronchart, and M. R. Michel-Wolwertz, *in* "Currents in Photosynthesis." Donker Publ., Rotterdam, 1966, p. 111.

(30) J. S. C. Wessels, *Biochim. Biophys. Acta* **65**, 561 (1962).

(31) J. S. C. Wessels, *in* "Currents in Photosynthesis." Donker Publ., Rotterdam, 1966, p. 129.

(32) J. S. Kahn and T. T. Bannister, *Photochem. Photobiol.* **4**, 27 (1965).

(33) B. Kok, *Biochim. Biophys. Acta* **48**, 527 (1961).

(34) B. Ke, *Biochim. Biophys. Acta* **88**, 297 (1964).

(35) E. Yakushiji, K. Uchino, Y. Sigimura, I. Shiratori, and A. Takamiya, *Biochim. Biophys. Acta* **75**, 293 (1963).

(36) W. Terpstra, *in* "Currents in Photosynthesis." Donker Publ., Rotterdam, 1966, p. 157.

(37) A. Takamiya, H. Obata, and E. Yakushiji, *Natl. Acad. Sci.—Natl. Res. Council, Publ.* **1145**, 479 (1963).

(38) J. A. Lippincott, J. Aghion, E. Porcile, and W. Bertsch, *Arch. Biochem. Biophys.* **98**, 17 (1962).

(39) J. Aghion, *Biochim. Biophys. Acta* **66**, 212 (1963).

(40) Y. Nishizaki, *Plant Cell Physiol.* (*Tokyo*) **5**, 373 (1964).

(41) J. H. C. Smith, D. Kupke, J. E. Loeffler, A. Benitez, J. Ahrne, and A. T. Giese, *in* "Research in Photosynthesis" (H. Gaffron *et al.*, eds.), p. 464. Wiley (Interscience), New York, 1957.

(42) A. Seybold and K. Egle, *Botan. Arch.* **41**, 578 (1940).

(43) M. Eisler and L. Portheim, *Biochem. Z.* **135**, 293 (1923).

(44) G. Oster and R. Lu, *in* "Currents in Photosynthesis." Donker Publ., Rotterdam, 1966, p. 1.

(45) F. A. Rodrigo, Thesis, University of Utrecht (1955).

(46) W. Vishniac, *in* "Research in Photosynthesis" (H. Gaffron *et al.*, eds.), p. 285. Wiley (Interscience), New York, 1959.

(47) V. H. Lynch and C. S. French, *Arch. Biochem. Biophys.* **70**, 382 (1957).

(48) D. I. Sapozhnikov and T. C. Maslova, *Trans. Botan. Inst., Akad. Nauk SSSR* [4] **11**, 97 (1956).

(49) S. Takashima, *Nature* **169**, 182 (1952).

(50) Y. Chiba, *Arch. Biochem. Biophys.* **54**, 83 (1955).

(51) D. R. Anderson, J. S. Spikes, and R. Lumry, *Biochim. Biophys. Acta* **15**, 298 (1954).

(52) A. A. Krasnovski, and G. P. Brin, *Dokl. Akad. Nauk SSSR* **95**, 611 (1954).

—13—

Chlorophyll-Protein Complexes
Part II
Complexes Derived from Green Photosynthetic Bacteria*

JOHN M. OLSON

Biology Department, Brookhaven National Laboratory, Upton, New York

I. Introduction

A. Occurrence

Large macromolecular complexes of chlorophyll, carotenoid, polysaccharide, lipid, and protein can be obtained from photosynthetic bacteria simply by breaking the cells near neutral pH and centrifuging the crude extract between 1 and 2×10^5 g. Particles containing the photosynthetic pigments sediment readily under these conditions and are called the chromatophore fraction. The chromatophores range in weight from 10^6 to 10^8 depending on the species of bacterium, growth conditions, and method of cell breakage.

Soluble chlorophyll-protein complexes (defined arbitrarily as those of molecular weight less than 10^6) have been isolated from alkaline

---

* Research carried out at Brookhaven National Laboratory under the auspices of the U.S. Atomic Energy Commission.

extracts of green bacteria, but not from purple bacteria. Bacteriochloro-phyll-protein complexes can be obtained from *Chloropseudomonas ethyli-cum*, strain 2K, and *Chlorobium thiosulfatophilum*, strain L (*1*). At-tempts to release soluble complexes from *Rhodopseudomonas* sp. NHTC 133, *Chromatium*, and *Rhodospirillum rubrum* by alkaline extraction have met with little success.

B. Extraction and Purification (*1, 2*)

Steps 1–3 apply to both *Chloropseudomonas ethylicum* 2K and *Chlo-robium thiosulfatophilum* L.

Step 1. Cells are suspended in 0.2 M $Na_2B_2O_4$ (pH 9.2) or in 0.4 M Na_2CO_3 (pH ~11) and then broken by sonication.

Step 2. A green fraction (chlorobium chlorophyll plus Bchl) is pre-cipitated by the addition of 30 gm $(NH_4)_2SO_4$ per 100 ml of crude extract.

Step 3. This precipitate is dissolved in 0.01 M buffer (pH 7.8–8.5) and chromatographed on diethylaminoethyl cellulose (DEAE). A blue fraction containing the Bchl-protein complex is obtained by elution of the

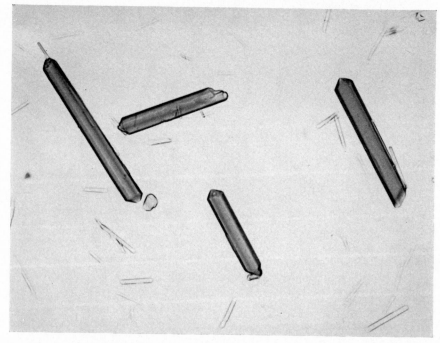

Fig. 1.　Crystals of Bchl-protein (*Chloropseudomonas ethylicum*) from a solution in 0.2 M $(NH_4)_2CO_3$.

DEAE with a gradient of NaCl in 0.01 M buffer. The chlorobium chlorophyll remains bound to the DEAE.

Steps 4–6 apply only to *Chloropseudomonas ethylicum* 2K.

Step 4. The blue fraction is rechromatographed on DEAE to obtain an estimated purity of 99% based on disc electrophoresis (3, 4) of the final Bchl-protein fraction.

Step 5. The blue fraction is then passed through a column of Sephadex G-100 in 1 M NaCl in 0.01 M phosphate buffer (pH 7.8) to remove any traces of photooxidized material.

Step 6. The Bchl-protein solution is concentrated to approximately 30 gm/1 and crystallized by slow dialysis versus 5–10 gm ($NH_4)_2SO_4$ per 100 ml of 1 M NaCl in 0.01 M buffer at 4°C (see Fig. 1).

II. Bacteriochlorophyll-Protein Complex from *Chloropseudomonas ethylicum*

A. Spectral Properties

1. ABSORPTION

a. Evidence for Bacteriochlorophyll a. The spectra of the chlorophyll-protein complexes from green bacteria indicate the presence of Bchl by the characteristic orange band at 603 mμ in addition to the far-red band at 809 mμ (see Fig. 2). There is no chlorobium chlorophyll associated with the complex. When the Bchl is removed from the complex, the absorption spectrum in ether is virtually identical to that of Bchl *a* from *Chromatium* (2). Furthermore, Jensen *et al.* (5) have shown that bacteriochlorophylls from green bacteria have the same R_f values as Bchl *a* from purple bacteria in paper chromatography. The chemical and physical studies of Holt and co-workers (6) on the Bchl of green bacteria have established its identity as Bchl *a* beyond any reasonable doubt.

b. Absorptivity Values. Figure 2 shows the absorption spectrum of Bchl-protein from *Chloropseudomonas ethylicum* in aqueous solution (7). This spectrum is maintained from pH 3 to 11 and in the presence of 9 M urea. Table I gives the locations of absorption maxima and the corresponding absorptivity values. These values were calculated from the average absorptivity ratios listed in Table I and the absorptivity at 415.0 mμ. This wavelength was chosen as a useful reference because it is an isosbestic wavelength for Bchl and its photooxidation product in the protein complex. The value of ε_{415} based on Bchl was found to be $(1.81 \pm 0.03) \times 10^4\,M^{-1}\,cm^{-1}$. The Bchl determination was made by quantitative extraction into methanol at 40°C, transfer to ether, and conversion to bacteriopheophytin at 4°C. The spectrophotometric determination of bacteriopheophytin a was based on the absorptivity value of 128 (gm/liter)$^{-1}$ cm^{-1} at 357.5 mµ found by Smith and Benitez (8).

c. *Oscillator Strength*. The values of transition moment (m) and oscillator strength (f) were calculated for the 603-mµ band and the 809-

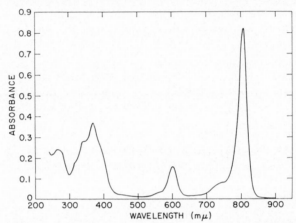

FIG. 2. Absorption spectrum of Bchl-protein dissolved in 0.25 M NaCl and 0.02 M phosphate buffer, pH 7.8 (7).

mµ band from the data in Fig. 2 and Table I. The area under the 603-mµ band plotted versus wavenumber was measured from 510 to 642 mµ and gave $m_{603} = 4.3 \times 10^{-18}$ esu cm and $f_{603} = 0.14$. Likewise the area under the 809-mµ band from 650 to 870 mµ gave $m_{809} = 8.3 \times 10^{-18}$ esu cm and $f_{809} = 0.40$.

TABLE I

MOLAR ABSORPTIVITY VALUES BASED ON BACTERIOCHLOROPHYLL

λ (mµ)	267	343	371	603	745	809
ε (mM^{-1} cm^{-1})[a]	37	49	67	28.4	13.4	154
$\varepsilon/\varepsilon_{415}$[b]	2.03	2.70	3.69	1.57	0.74	8.5

[a] Limit of error estimated to be ±4%.

[b] These are average values. The standard deviation in each case is about 2%.

d. *Effect of Acidification*. In 0.1 N HCl the Bchl-protein is immediately converted to a blue form with absorption bands at 267, 360, 593, and 790 mµ. The blue form gradually disappears and a pink bacteriopheophytin-protein appears with maxima at 363, 534, and 760 mµ.

e. *Spectrum at 77°K*. At the temperature of liquid nitrogen the 809-mµ band splits into 3 peaks at 805, 814, and 824 mµ, and the 603-mµ band splits into two peaks at 601 and 607 mµ. No such splittings at 77°K occur in the absorption spectrum of Bchl dissolved in methanol–ethanol–water mixtures.

2. FLUORESCENCE (7)

Emission spectra for the complex are shown in Fig. 3. At room temperature the fluorescence maximum occurs at 818 mμ, and the fluorescence yield is 0.19. Upon cooling to 77°K, the maximum shifts to 827 mμ, and the yield increases to 0.29. In addition, two minor bands at 880 and 917 mμ are resolved. The absence of fine splitting in the main band at 827 mμ is ascribed to the wide bandpass ($\Delta\lambda = 10$ mμ) of the fluorometer used to obtain these spectra.

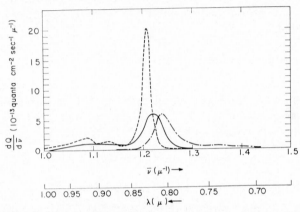

FIG. 3. Fluorescence emission spectra at 294°K (————) and 77°K (- - -) caused by 366-mμ excitation, and the 809-mμ absorption band (—·—·—) normalized to the 818-mμ fluorescence band (7).

B. Composition

1. AMINO ACID ANALYSIS (9)

The composition of the protein moiety of the complex is given in Table II. The six analyses upon which Table II is based were done on

TABLE II
AMINO ACID ANALYSIS[a]

Glycine	21.7	Lysine	10.1
Aspartic acid	20.7	Phenylalanine	9.2
Valine	18.4	Proline	8.8
Glutamic acid	18.2	Threonine	7.7
Serine	14.4	Tyrosine	5.0
Isoleucine	12.3	Histidine	4.1
Alanine	11.7	Tryptophan[b]	2.3
Arginine	11.3	Methionine[c]	2.2
Leucine	10.9	Half-cystine[c]	1.0

[a] The number of residues for each amino acid is given.
[b] Determined from alkaline hydrolysis.
[c] Determined from acid hydrolysis after oxidation with performic acid.

lyophilized Bchl-protein carried through the first 4 steps of purification described in Section I, B. The residue numbers were obtained by multiplying the relative amounts of each amino acid by a factor chosen to give approximately one half-residue of cystine (equivalent to one residue of cysteine) and nearly integral values for all other amino acids. On this basis the minimum molecular weight of the protein amounts to 2.1×10^4 per half-residue of cystine. The specific volume of the protein moiety is calculated to be 0.731 cm^3/gm from the specific volumes of the individual amino acid residues (10). Nitrogen content of the protein is calculated to be 0.173 ± 0.002 on the estimate of 11–18 amide groups per half-residue of cystine.

2. Bacteriochlorophyll-Nitrogen Stoichiometry

The nitrogen content of the complex was determined by micro-Kjeldahl analysis (11), and the absorptivity at 415 mμ based on nitrogen (a^N_{415} was found to be 14.8 ± 0.2 (gm/liter)$^{-1}$ cm^{-1}. The chlorophyll: nitrogen ratio was then calculated.

$$\text{Bchl:N} = a^N_{415}/\varepsilon_{415} = (8.2 \pm 0.2) \times 10^{-4} \text{ mole/gm}$$

3. Bacteriochlorophyll-Protein Stoichiometry

On the assumption that only chlorophyll and protein contribute to the nitrogen content of the complex, the ratio of Bchl to protein was calculated from the chlorophyll:nitrogen ratio and the nitrogen contents of Bchl a and the protein.

$$\text{Bchl:P} = (1.49 \pm 0.05) \times 10^{-4} \text{ mole/gm}$$

This is equivalent to 3.1 mole Bchl per 2.1×10^4 gm of protein and gives a minimum molecular weight of 2.4×10^4 per half-residue of cystine for protein and chlorophyll.

C. Molecular Weight

1. Ultracentrifuge Studies (2)

Solutions of Bchl-protein in $0.2 M$ NaCl in $0.01 M$ tris (hydroxymethyl)-aminomethane (Tris) (pH 8.0) were studied in the Spinco Model E Ultracentrifuge. The sedimentation coefficient S_{20} was approximately 7 Svedberg units from measurements of sedimentation velocity. The effective molecular weight $[M_{eff} = M(1 - \overline{V}d_o)]$ was $(3.5 \pm 0.1) \times 10^4$ gm/mole from studies of the equilibrium between sedimentation and diffusion.

2. PARTIAL SPECIFIC VOLUME

Measurements of $\overline{V}$ have been carried out on three samples of Bchl-protein purified through step 5 of Section I, B. Solutions were prepared in 0.2 M $(NH_4)_2CO_3$ or NH_4OAc, and the densities of solution (d) and solvent (d_o) were measured by precision pycnometry. Dry weights for solution (x) and solvent (x_o) were obtained by evaporation of samples at 60°C and heating *in vacuo* over P_2O_5 at 110°C until constant weights were obtained. Values for x_o were zero because of the disappearance of the ammonium salts during the evaporation at 60°C. Values of $\overline{V}$ were calculated from the equation

$$d_o\overline{V} = 1 - (d - d_o)/(x - x_o)$$

and gave an average value of 0.79 ± 0.02 cm³/gm. The molecular weight of the complex as it exists in solution was calculated to be $(1.67 \pm 0.17) \times 10^5$ gm/mole on the basis of M_{eff} and $\overline{V}$. This value is approximately seven times the minimum molecular weight calculated from the amino acid composition and chlorophyll content.

3. NUMBER OF CHLOROPHYLL GROUPS IN THE COMPLEX

The number (n) of Bchl a molecules in the complex as it exists in solution can be estimated from M_{eff} in two different ways.

a. From the chlorophyll-to-protein stoichiometry it is known that there are 3 moles of Bchl per 2.4×10^4 gm of protein plus chlorophyll. If the complex is composed of only these two components, then

$$n = \frac{M}{2.4 \times 10^4} \times 3 = 21 \pm 2$$

b. From the density measurements it is known that $[Bchl]/(d\text{-}d_o) = (6.2 \pm 0.2) \times 10^{-4}$ mole/gm. Since the concentration (x) of the complex in solution is related to the density values by the equation

$$x = (d - d_o)/(1 - d_o\overline{V})$$

and is related to the chlorophyll concentration by the equation

$$x = [Bchl] M/n$$

it follows that

$$n = \frac{M}{1 - d_o\overline{V}} \times \frac{[Bchl]}{d - d_o} = \frac{M_{eff}[Bchl]}{d - d_o} = 22 \pm 1$$

The value of 21 required for seven hypothetical subunits each containing

3 chlorophylls falls nicely within the range determined by density measurements.

D. Hypothetical Structure

1. BACTERIOCHLOROPHYLL CORE

The chlorophyll in the complex is thought to be buried inside the protein on the basis of the high solubility of the complex in aqueous solution and the low solubility of Bchl alone in such media. The specific volume of such a chlorophyll core is calculated to be $1.3 \text{ cm}^3/\text{gm}$. The density of the core (0.8 gm/cm^3) is thus much lower than the density of crystalline chlorophyll (1.08 gm/cm^3 for Chl a) (12) and is consistent with a loose packing arrangement in which the chlorophyll-chlorophyll interactions are weak in the sense defined by Kasha (13). The absence of strong interactions between pigment molecules in the complex is consistent with the sharpness of the absorption bands in the absorption spectrum (Fig. 1).

2. NONPOLAR AMINO ACID LAYER

The chlorophyll core of the complex is thought to be surrounded by a layer of nonpolar amino acid residues. A nonpolar environment for the chlorophyll molecules is consistent with the observed fine splittings in the 603-mμ and 809-mμ bands at 77°K. The binding of the chlorophyll to the protein is thought to be the result of noncovalent interaction between the phytol tails of the chlorophylls and the nonpolar residues of the protein. This view is consistent with the ability of methanol to remove Bchl completely from the complex at 0°C.

3. POLAR AMINO ACID COAT

The high solubility of the complex in aqueous media is attributed to an outer layer of polar amino acid residues. Of the amino acid residues in the protein moiety, 48% fall into the polar category by the criteria of Nemethy and Scheraga (14).

4. LIPIDS OR OTHER COMPONENTS

Within the experimental uncertainty of the molecular weight determination and the composition data, the chlorophyll and protein alone can account for the total weight of the complex. This means that the content of lipids or other components cannot exceed 9% of the total weight. No attempts have yet been made to detect the possible presence of minor constituents in addition to Bchl a and protein.

5. Spherical Model

The hypothetical structure of the Bchl-protein complex is visualized in the spherical model shown in Fig. 4. In this model the core of Bchl has a radius of 22 Å and is surrounded by a 10-Å layer of nonpolar amino acid residues. The outer shell of polar amino acid residues is 6 Å thick. A sphere was chosen to simplify the calculation of dimensions from the

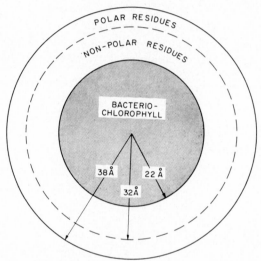

FIG. 4.   Spherical model of Bchl-protein complex.

composition and molecular weight (*15*). *The shape of the model should not be taken literally;* the model is only an attempt at a crude first approximation.

III. Modified Complex

A. Procedure

A conformational change in the Bchl-protein complex apparently is effected by a slow precipitation of concentrated material at room temperature at low ionic strength (*16*). The fine precipitate formed under these conditions is completely noncrystalline.

B. Evidence of New Conformation

1. Decreased Solubility

Although the unmodified complex is easily concentrated to a 3% solution in 1 M NaCl in 0.01 M buffer (pH 7.8), the modified material is only sparingly soluble in salt solution at pH 8.0. It appears that the degree of modification is correlated with the decreased solubility. That fraction of precipitated material which does not dissolve in 1 M NaCl at pH 8.0, but does dissolve in 0.4 M Na$_2$CO$_3$ at pH 11, shows spectral evidence of greater modification than that fraction of precipitated material which does dissolve readily in 1 M NaCl at pH 8.0.

2. Change in Absorption Spectrum

In the absorption spectrum of the modified complex shown in Fig. 5, both the 603-mμ band and the 809-mμ band are skewed toward the blue. Also a distinct shoulder appears at about 840 mμ, and the ratio of

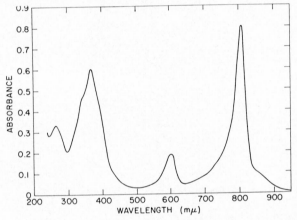

Fig. 5. Absorption spectrum of modified complex dissolved in 0.4 M Na$_2$CO$_3$ (16).

peak heights at 809 mμ and 371 mμ drops considerably from the ratio observable in Fig. 2. The changes in absorption spectrum cannot be ascribed merely to a general lowering and broadening of absorption bands upon modification of the complex. The unsymmetrical skewing of the 603- and 809-mμ bands in addition to the appearance of a distinct shoulder at 840 mμ suggests the presence of new peaks in the absorption spectrum.

In order to show the precise change in absorption spectrum upon modification of the complex, the difference spectrum shown in Fig. 6 was obtained by comparing matched solutions of modified and unmodified complex directly in the recording spectrophotometer. This difference spectrum shows the appearance of new bands at 366, ~440, 585, 618, 785, and 840 mμ. (The dashed portion of the curve in Fig. 6 is considered to be an artifact.) Furthermore it seems clear that the new bands at 585 and 618 mμ have arisen from a splitting of the 603-mμ band, and that the new bands at 785 and 840 mμ have arisen from a splitting of the 809-mμ band. These apparent band splittings are interpreted to indicate a strong interaction (*13*) between some of the 21 Bchl groups within one complex or between exposed Bchl groups in adjacent complexes.

C. Chlorophyll Aggregation

The specific mechanism proposed to explain the apparent band splittings is the formation of chlorophyll aggregates such as dimers (*17*). Application of the molecular exciton theory of McRae and Kasha (*18*) to the data in Figs. 2 and 6 enables one to calculate a value for θ, the angle between the 809-mμ transition dipoles in an aggregate, and a value for θ', the angle between the 603-mμ transition dipoles. The values for

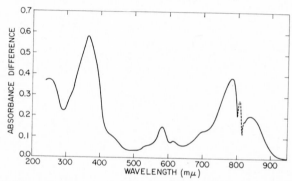

Fɪɢ. 6.   Difference spectrum: modified complex versus unmodified complex.

both θ and θ' are approximately 75°. According to Rabinowitch (*19*) the 603-mμ transition dipole is parallel to the long axis of the Bchl molecule whereas the 809-mμ transition dipole is perpendicular to the long axis. Thus the moments for the two transitions are perpendicular to each other in the plane of the Bchl molecule (see Fig. 7). This condition and the requirement that θ equal θ' limit the mode of aggregation to two basic arrangements.

In the first arrangement the chlorophyll molecules are cocked slightly with respect to one another [cf. Brody and Brody (*17*)]. The planes of adjacent chlorophylls in the aggregate must intersect so that each transition dipole (see Fig. 7) of each molecule makes a 45-degree angle with the line of intersection and a 75-degree angle with the corresponding dipole in the adjacent molecule. The angle between the planes of adjacent chlorophylls would then be 117 degrees. Such an aggregate can extend in one dimension only.

In the second arrangement the chlorophyll molecules are stacked with their planes parallel. Alternate molecules in a stack are rotated 75 degrees out of phase with respect to their nearest neighbors. Although the molecular exciton theory has not yet been applied quantitatively to this second arrangement, a stacking arrangement for aggregates of large planar molecules appears to require less of a conformation change in a chlorophyll-protein complex than does the first arrangement of cocked planes.

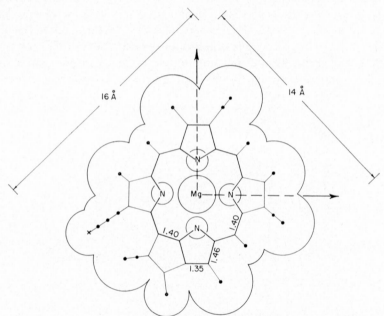

Fig. 7.   Bacteriochlorophyll *a* molecule in two dimensions. The perimeter of the molecule conforms to the van der Waals' radii of the ring substituents. Ring conjugation is indicated by heavy straight lines. Bond lengths (in Ångstroms) were obtained from crystalline tetraphenylphorphin (20). Circles around the N and Mg atoms indicate covalent radii. Arrows denote the long and short axes of the ring. The x indicates the oxygen atom to which the phytyl tail (not shown) is attached.

D. Biological Significance

Sybesma and Vredenberg (21, 22) have shown that illumination of whole cells of *Chloropseudomonas ethylicum* or *Chlorobium limicola* causes an absorbance decrease centered at 840 mμ. The relationship between the bleaching at 840 mμ and the fluorescence yield of Bchl *a in vivo* led to the inference that the absorbance changes in the 840-mμ region were due to a reaction center pigment P840, present in very low concentration. Further studies showed that the light-induced bleaching of P840 occurs at $-170°C$ and suggested that a cytochrome may be an electron donor for P840 (22).

The appearance of a new absorption band at 840 mμ in the modified Bchl-protein complex suggests the possibility that this form of the complex may contain the photochemical reaction center observed *in vivo*. If this be true, then P840 is a specialized form of Bchl *a*, and the modified form of the complex may be expected to sensitize photochemical electron transfer reactions *in vitro* as well as *in vivo*.

ACKNOWLEDGMENT

The procedure for purifying and crystallizing the Bchl-protein was worked out by Dr. Janet Scott and Miss Frances Roskosky. Many of the data presented in this chapter were obtained by Mr. George Latham and Miss Peggy Frandsen.

REFERENCES

(1) J. M. Olson and C. A. Romano, *Biochim. Biophys. Acta* **59**, 726 (1962).

(2) J. M. Olson, D. Filmer, R. Radloff, C. Romano, and C. Sybesma, in "Bacterial Photosynthesis" (H. Gest, A. San Pietro, and L. P. Vernon, eds.), p. 423. Antioch Press, Yellow Springs, Ohio, 1963.

(3) L. Ornstein, *Ann. N.Y. Acad. Sci.* **121**, 321 (1964).

(4) B. J. Davis, *Ann. N.Y. Acad. Sci.* **121**, 404 (1964).

(5) A. Jensen, O. Aasmundrud, and K. Eimhjellen, *Biochim. Biophys. Acta* **88**, 466 (1964).

(6) A. S. Holt, D. W. Hughes, H. J. Kende, and J. W. Purdie, *Plant Cell Physiol. (Tokyo)* **4**, 49 (1963).

(7) C. Sybesma and J. M. Olson, *Proc. Natl. Acad. Sci. U.S.* **49**, 248 (1963).

(8) J. H. C. Smith and A. Benitez, in "Moderne Methoden der Pflanzenanalyse" (K. Paech and M. V. Tracey, eds.), Vol. IV, p. 142. Springer, Berlin, 1955.

(9) S. Moore and W. H. Stein, in "Methods in Enzymology" (S. P. Colowick and N. O. Kaplan, eds.), Vol. 6, p. 819. Academic Press, New York, 1963.

(10) H. K. Schachman, in "Methods in Enzymology" (S. P. Colowick and N. O. Kaplan, eds.), Vol. 4, p. 70. Academic Press, New York, 1957.

(11) E. A. Kabat and M. M. Mayer, "Experimental Immunochemistry," p. 287. Thomas, Springfield, Illinois, 1948.

(12) G. Donnay, *Arch. Biochem. Biophys.* **80**, 80 (1959).

(13) M. Kasha, *Radiation Res.* **20**, 55 (1963).

(14) G. Nemethy and H. A. Scheraga, *J. Phys. Chem.* **66**, 30 (1962).

(15) H. F. Fisher, *Proc. Natl. Acad. Sci. U.S.* **51**, 1285 (1964).

(16) J. M. Olson, *Biochim. Biophys. Acta* **88**, 660 (1964).

(17) S. S. Brody and M. Brody, *Biochim. Biophys. Acta* **54**, 495 (1961).

(18) E. G. McRae and M. Kasha, *J. Chem. Phys.* **28**, 721 (1958).

(19) E. I. Rabinowitch, "Photosynthesis and Related Processes," Vol. 2, Part 2, p. 1796. Wiley (Interscience), New York, 1956.

(20) S. Silvers and A. Tulinsky, *J. Am. Chem. Soc.* **86**, 927 (1964).

(21) C. Sybesma and W. J. Vredenberg, *Biochim. Biophys. Acta* **75**, 439 (1963).

(22) C. Sybesma and W. J. Vredenberg, *Biochim. Biophys. Acta* **88**, 205 (1964).

—13—

Chlorophyll-Protein Complexes
Part III
Optical Rotatory Dispersion of Chlorophyll-Containing Particles from Green Plants and Photosynthetic Bacteria*

BACON KE

Charles F. Kettering Research Laboratory, Yellow Springs, Ohio

In 1931, Fischer *et al.* first pointed out the presence of asymmetric carbon atoms in the chlorophyll molecules (*1*). However, their initial attempt to observe optical activity was unsuccessful because of experimental difficulties arising from strong absorption of light by the pigment. In 1933, Stoll and Wiedemann measured the optical rotation of chlorophylls *a* and *b* and some of their derivatives at a wavelength beyond the main red absorption band (*2*). They used a half-shade polarimeter together with an intense monochromatic beam, and the measurement was carried out by making a series of exposures on an infrared-sensitive film. Chlorophylls *a* and *b* were both found to be levorotatory, with a specific rotation of 260 degrees at 720 mμ. In 1935, Fischer and Stern confirmed these results and extended their measurements to a number of other chlorophyll derivatives in the wavelength range 690–720 mμ (*3, 4*).

The optical-rotatory-dispersion (ORD) spectra of chlorophylls *a* and *b* in diethyl ether in the visible and near-ultraviolet regions were first measured by Ke and Miller (*5*), using a sensitive spectropolarimeter (Cary Model 60, Applied Physics Corporation). Typical Cotton effects, with marked changes in optical rotation, were found in the regions of the absorption bands, indicating that the electronic transitions corresponding to the absorption bands are optically active. The specific rotation versus wavelength is plotted in Fig. 1 for Chl *a*, Chl *b* (*5–7*), and chlorobium chlorophyll (Cchl) 660 (*7*). Negative and positive Cotton effects are present in the major red and blue absorption regions of all three chloro-

---

* Contribution No. 215 from the Charles F. Kettering Research Laboratory, Yellow Springs, Ohio.

phylls. Cotton effects associated with the minor bands are also discernible. For instance, in Chl *a*, positive Cotton effects are associated with the 615- and 575-mμ bands and a negative one with the 415-mμ violet satellite (5). While the specific rotation of Cchl 660 is greater than that of Chl *a* or Chl *b*, the rotation of Bchl (not shown in Fig. 1) is negligibly small by comparison.

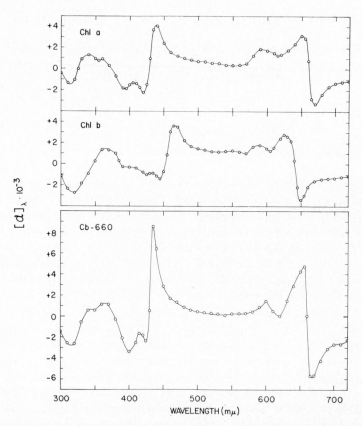

FIG. 1.   Specific rotation of Chl *a*, Chl *b*, and Cchl 660 in diethyl ether.

Subsequently, ORD spectra of chloroplast-lamellae fragments, which represent chlorophyll bound to the lipoprotein in the native state, were reported independently by Sauer (8) and Ke (9). As shown in Fig. 2, the ORD spectrum of the chlorophyll-lipoprotein shows considerably more complexity and intensity than that of free chlorophylls $(a + b)$ at the same concentration in an organic solvent. Cotton effects corresponding to almost all the absorption bands of Chl *a* and Chl *b* are retained in the lipoprotein complex. Because of overlapping absorption by the chlorophylls and carotenoids in the Soret region, the assignment of the Cotton effects to the individual absorption bands between 400 and 500 mμ is more complicated.

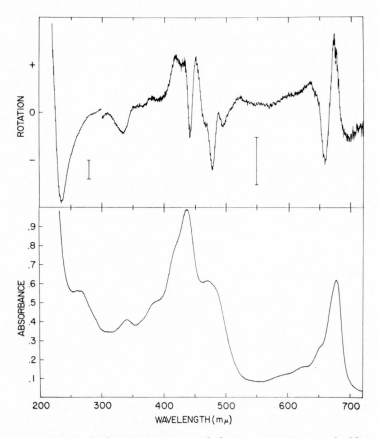

FIG. 2. ORD and absorption spectra of the same suspension of chloroplast-lamellae fragments. Suspension medium: $10^{-3} M$ phosphate buffer, pH 7.5. Chlorophyll concentrations: Chl a, 15 μg/ml; Chl b, 5 μg/ml. Vertical bars in the ORD spectra represent 10 millidegrees.

A pronounced trough with maximum depth at 235 mμ is present in the ORD spectrum of the chloroplast preparation. This Cotton effect is presumably associated with the n-π^* transition in the peptide chromophore. Assuming that the magnitude of the 235-mμ trough is representative of the helix content in a peptide or protein, and further assuming the helical protein in the chloroplast lamellae to be a right-handed α-helix, the magnitude of the 235-mµ trough measured for the chloroplast-lamellae preparation corresponds to a helical content of about 17%.

Heating a chloroplast preparation presumably causes denaturation of enzyme proteins as well as disruption of the organizational relationship between the pigment and the lipoprotein. Structural changes of this sort are not readily revealed by absorption spectra. On the other hand, ORD appears to be very sensitive to such changes. Fig. 3 shows the ORD spectra of a chloroplast solution heated at different temperatures for 5 minutes. The magnitude of the Cotton effects in the pigment ab-

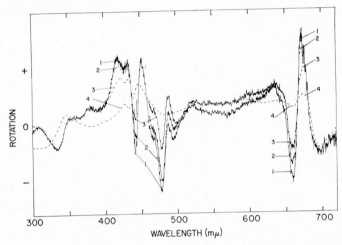

FIG. 3. ORD spectra of a suspension of chloroplast-lamellae fragments after being heated at different temperatures for 5 minutes. 1, unheated sample (reference); 2, 60°; 3, 70°; 4, 80°C. Chlorophyll concentrations are the same as listed for Fig. 2.

sorption regions steadily decreases with increasing temperature. After heating at 80°C for 5 minutes, the ORD spectrum ultimately becomes similar to that of a mixture of isolated chlorophylls *a* and *b*. In the ultraviolet region, the magnitude of the 235-mµ trough decreases slowly at first with increasing temperatures, rather abruptly between 70 and 80°C, then slowly tapers off at 100°C. It was also found that the 235-mµ trough decreased in magnitude after the lamellae fragments were suspended in saturated urea for 24 hours; however, the ORD spectrum in the visible region was largely unchanged from that of the untreated lamellae preparation.

ORD spectra of the chlorophyll-containing particles from several photosynthetic bacteria were also reported recently (7). A brief discussion of some ORD spectra together with spectral changes caused by physical or chemical treatment, or by genetic modification of the cells, will be presented below.

The absorption and ORD spectra of *Chromatium* chromatophores are shown in Fig. 4 (7). Cotton effects associated with the carotenoids can be clearly seen, since, unlike in chloroplasts, the absorption regions of Bchl and the carotenoids are well separated. Cotton effects associated

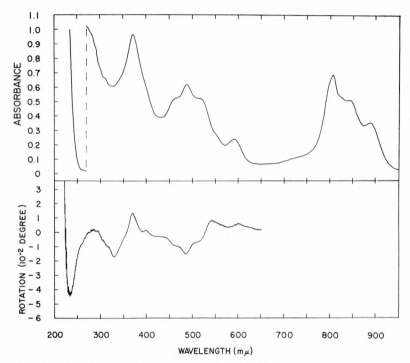

FIG. 4.   Absorption and ORD spectra of *Chromatium* chromatophores. Bchl concentration for both curves = 13.5 μg/ml.

with the Bchl absorption bands at 590 and 372 mμ are small but discernible. This is consistent with the small rotation of free Bchl in solution. The 233-mμ trough corresponds to a helical content of 10–12% in the chromatophore protein.

The absorption and ORD spectra of chromatophores from wild-type and the carotenoidless mutant *Rhodopseudomonas spheroides*, respectively, are shown in Figs. 5 and 6. In the wild-type *spheroides* chromatophores, pronounced Cotton effects for the carotenoid absorption bands at 508, 473, and 372 mμ can be seen. In the carotenoidless mutant, not only are the Cotton effects associated with the carotenoid bands absent,

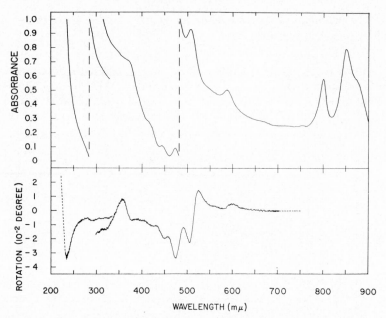

Fig. 5. Absorption and ORD spectra of chromatophores from wild-type *Rhodopseudomonas spheroides*. Bchl concentration: 9.1 µg/ml.

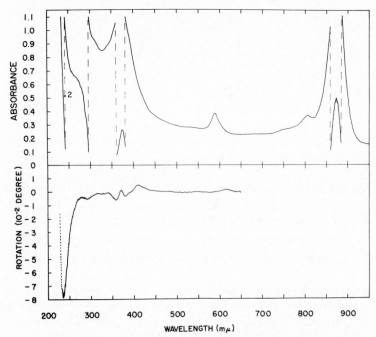

Fig. 6. Absorption and ORD spectra of chromatophores from a blue-green mutant of *Rhodopseudomonas spheroides*. Bchl concentration: 7.9 µg/ml.

but those associated with the Bchl absorption bands are also much weaker.

Figure 7 shows the absorption and ORD spectra of pigmented particles isolated from a green bacterium, *Chloropseudomonas ethylicum*. The red and blue regions of chlorobium chlorophyll show large Cotton effects, consistent with the high specific rotation of the free pigment (cf. Fig. 1). In the green bacteria, apparently the binding of Cchl to the

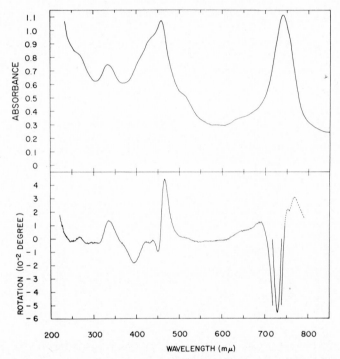

Fɪɢ. 7.   Absorption and ORD spectra of pigmented particles from *Chloropseudomonas ethylicum* (red peak folded at −5 × 10⁻² degrees).

lipoprotein carrier can be easily affected. As shown in Fig. 8, when the pigmented particles were treated with a 0.1% detergent, Triton X-100, the height of the original absorption band decreased and a new band appeared at a wavelength characteristic of the free chlorophyll in solution. Judging from the relative changes in the absorption-band heights, approximately half of the chlorophyll was converted to a more dispersed state. In the corresponding ORD spectrum, there is a proportionate decrease in the magnitude of the Cotton effects associated with the bound pigment. However, in the newly appeared absorption regions of the dispersed chlorophyll, the Cotton effects are much smaller, approaching those of the free pigments.

A Bchl-protein complex of well defined composition has recently been isolated (see Chapter 13, Part II). The absorption and ORD spectra of this complex (sample kindly furnished by Dr. J. M. Olson) are shown in Fig. 9. The major absorption bands of Bchl in the protein complex are at 810, 603, and 371 mµ. Cotton effects are associated with all absorption bands; even for the 810-mµ band (in part), despite the high noise level in the recording. In general, the Cotton effects associated with

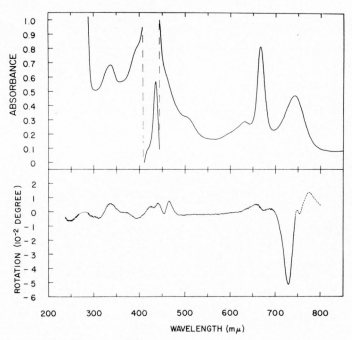

FIG. 8. Absorption and ORD spectra of *Chloropseudomonas ethylicum* pigmented particles in 0.1% Triton X-100.

the Bchl in the protein complex are much greater than those of the free pigments at an equivalent concentration.

The enhancement of the intrinsic Cotton effects of the chlorophylls and carotenoids that are complexed with the lipoprotein suggests a strong interaction either between the pigment molecules themselves or between the pigment molecules and the attached lipoprotein macromolecules. Nevertheless, a more definite theoretical interpretation of the relationship between the electronic transition of the bound pigment moieties and the observed Cotton effects still remains to be worked out. In a recent ORD study of quantasomes, Sauer (8) found that isolated Chl *a* at high concentrations in carbon tetrachloride exhibits profound changes in its ORD spectrum. Since it is known that Chl *a* under these conditions can undergo an intermolecular aggregation, Sauer suggested that a similar pigment-pigment interaction may also account for the enhanced ORD observed in quantasomes.

Alternative or additional sources of interaction such as those between pigment and protein or between pigment and lipids are also indicated by some of the results described above. For instance, spectral changes

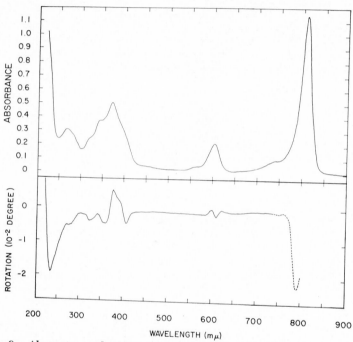

Fig. 9. Absorption and ORD spectra of a Bchl 770-protein complex isolated from *Chloropseudomonas ethylicum*. Bchl concentration = 6.9 μg/ml.

caused by heating the chloroplast-lamellae fragments and by treating the pigmented particles from *C. ethylicum* with Triton X-100 could be interpreted as due to a disruption of the Chl-Chl interaction or protein-Chl interaction. The decrease in the Cotton effects of Bchl in the carotenoidless mutant of *Rhodopseudomonas spheroides* even implicates a possible interaction between chlorophyll and carotenoid molecules. The decrease in the pigment Cotton effects by heating quantasomes also suggests a melting of a lipid matrix with which the pigment may be complexed. These preliminary ORD studies have shown that the technique is sensitive not only for detecting conformational changes in proteins, but also for detecting changes in the mode of interaction between small molecules and lipoprotein macromolecules.

REFERENCES

(1) H. Fischer, O. Süs, and G. Klebs, *Ann. Chem.* **499**, 84 (1931).
(2) A. Stoll and E. Wiedemann, *Helv. Chim. Acta* **16**, 307 (1933).
(3) H. Fischer and A. Stern, *Ann. Chem.* **519**, 58; **520**, 88 (1935).
(4) H. Fischer and A. Stern, "Die Chemie des Pyrrols," Vol. 2, Part II. Akad. Verlagsges., Leipzig, 1940.
(5) B. Ke and R. M. Miller, *Naturwissenschaften* **51**, 436 (1964).
(6) B. Ke, *Arch. Biochem. Biophys.* **112**, 554 (1965).
(7) B. Ke, in "Currents in Photosynthesis" (J. B. Thomas and J. C. Goedheer, eds.). Donker Press, Rotterdam, 1965.
(8) K. Sauer, *Proc. Natl. Acad. Sci. U.S.* **53**, 716 (1965).
(9) B. Ke, *Nature* **208**, 573 (1965).

—14—

Protochlorophyll

N. K. BOARDMAN

Commonwealth Scientific and Industrial Research Organization,
Division of Plant Industry, Canberra, Australia

I. Introduction

With a few exceptions, angiosperm seedlings if germinated in darkness do not contain chlorophyll. The etiolated seedlings are yellow in color owing to the presence of carotenoids, but they also contain a small quantity of a yellow-green pigment which is transformed to chlorophyll a in the presence of light. Monteverde (1) extracted this substance with alcohol and called it protochlorophyll. Noack and Kiessling (2) isolated protochlorophyll from cucurbit seed coats as well as from etiolated oat seedlings and, because of the similarity in their absorption spectra, concluded that the pigments were identical. Recent evidence shows, however, that the main protochlorophyll-like pigment in etiolated leaves is not protochlorophyll, but its nonphytol derivative, protochlorophyllide, or magnesium vinyl pheoporphyrin a₅, while cucurbit seed coats may contain in addition to protochlorophyll one or more closely related compounds.

In contrast to the angiosperms, many other classes of plants, e.g., gymnosperm seedlings, some ferns, and many algae, form chlorophyll in the dark, but it appears likely that the biosynthetic pathway for chlorophyll is the same as in the angiosperms, with protochlorophyllide as an intermediate (3). Granick (4) produced several mutant strains of *Chlorella vulgaris* which did not green in the dark (4); one of these accumulated protochlorophyllide. A similar mutant was described by Bryan and Bogorad (5). The green flagellate *Euglena gracilis* is devoid of chlorophyll when grown in the dark, but the cells contain small amounts of protochlorophyll (protochlorophyllide?) (6). Protochlorophyll has been detected in dark-grown roots of several species of plants from ferns to angiosperms (7). In contrast to the protochlorophyllide in etiolated angiosperm seedlings, root (8) and *Euglena* (6) protochlorophylls are transformed only slowly to chlorophyll in the light and seed coat protochlorophyll does not appear to be transformable (9).

Pigments resembling protochlorophyll in absorption spectra have been observed in the growth media of certain strains of the photosynthetic bacterium *Rhodopseudomonas spheroides* (10, 11) and when wild-type strains were grown in the presence of 8-hydroxyquinoline (12).

Protochlorophyll is in a different chemical state *in vivo* than when it is extracted into organic solvents; in the natural state, protochlorophyll and protochlorophyllide are complexed with protein. Our present knowledge of the natural state of protochlorophyll owes much to the fine work of Smith and his collaborators at the Carnegie Institution. Smith and Young (13) proposed the term protochlorophyll holochrome to designate the pigment in its natural state, and this term will be retained in this chapter, in view of its widespread use in the literature. But it should be kept in mind that the protochlorophyll holochrome in the etiolated seedling is a mixture of protochlorophyll holochrome and protochlorophyllide holochrome.

The greater part of this chapter will be devoted to the physical, biochemical, and photochemical aspects of the protochlorophyll holochrome of dark-grown seedlings, both as it exists *in vivo* and after isolation as a physiologically active particle. The transformation of protochlorophyll (and protochlorophyllide) into chlorophyll *a* (and chlorophyllide *a*) will be discussed in some detail, and the scope of the article has been extended to include some aspects of the greening process.

II. Properties of Protochlorophyll

A. Structure

The chemical relationship between protochlorophyll and chlorophyll *a* was established by the analytical and synthetic work of Fischer and

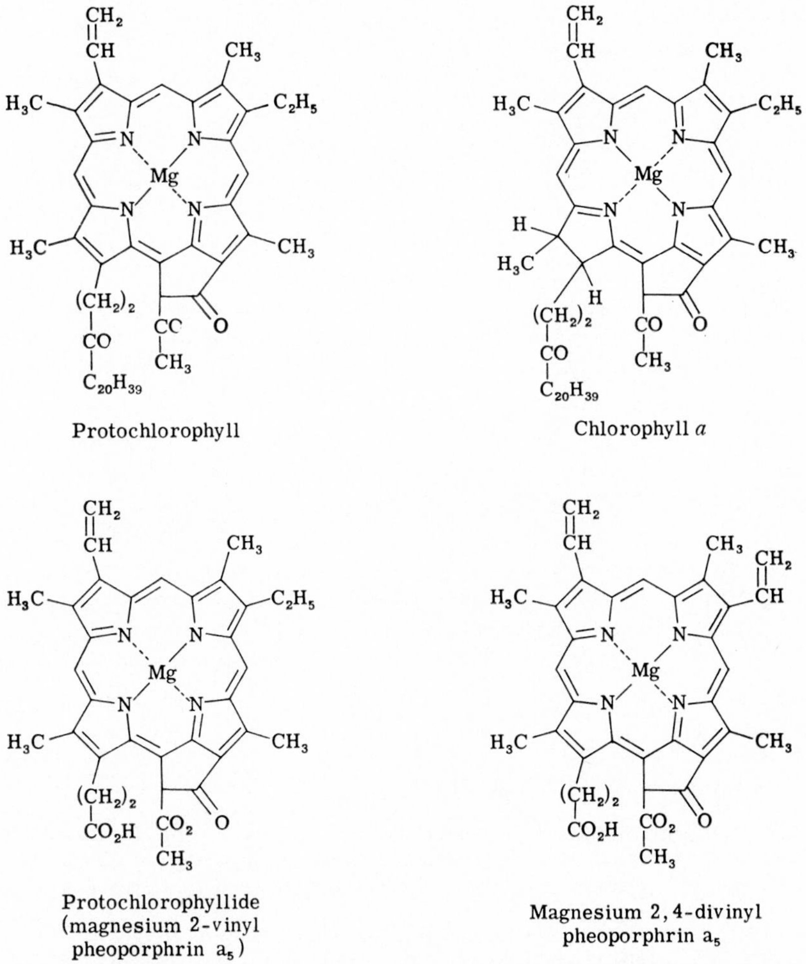

FIG. 1. Chemical structures of protochlorophyll and related compounds.

his collaborators (*14–16*). Fischer and Oestreicher (*16*) synthesized the phytyl ester of magnesium vinyl pheoporphyrin a₅ and showed that it differed from chlorophyll *a* in having two less hydrogen atoms at positions 7 and 8 of ring IV of the porphyrin nucleus (Fig. 1). The structure of chlorophyll *a* has now been confirmed by total synthesis (*17*) (cf. Chapter 5). Unlike chlorophyll *a*, protochlorophyll is optically inactive. The protochlorophyll synthesized by Fischer and Oestreicher (*16*) was compared spectroscopically with the pigment isolated from cucurbits, but not with leaf protochlorophyll. A comparison of solubility properties was not reported.

B. Leaf Protochlorophyll

1. Purification

The major pheoporphyrin pigment in etiolated leaves is protochlorophyllide (*18, 19*). Protochlorophyll is also present, but in smaller amounts, the ratio of protochlorophyllide to protochlorophyll depending on the age of the seedlings (*18, 20*). Fischer and Rudiger (*21*), using a microchemical method, reported that they were unable to detect phytyl alcohol in the hydrolysis products of barley leaf protochlorophyll, and they assumed that it was esterified with another alcohol. Godnev *et al.* (*22*) subsequently showed, however, that the small amount of protochlorophyll in etiolated seedlings was esterified with phytyl alcohol.

Koski and Smith (*23*) prepared protochlorophyll (protochlorophyllide?) in solid form from 5 kg of etiolated barley leaves. The leaves were harvested from 9-day seedlings in a dim green light to avoid transformation of the protochlorophyll, and the pigments were extracted with acetone. The protochlorophyll was purified by repeated chromatography on columns of powdered sucrose. From the magnesium content, Koski and Smith (*23*) concluded that their purified pigment contained the phytol group, but no confirmatory evidence such as a distribution coefficient between acetone and petroleum ether or a hydrochloric acid number was reported.

2. Absorption Spectrum

A quantitative absorption spectrum of protochlorophyll in ether is shown in Fig. 2 (*23, 24*). The principal absorption maxima are at 623 mμ and 432 mμ, with specific absorption coefficients of 39.9 and 325.5 liter/gm cm, respectively. Both the positions of absorption maxima and the specific absorption coefficients vary with the nature of the solvent (*24*); e.g., in 80% acetone the visible maximum is at 626 mμ (*25*). Because of the coexistence of protochlorophyll and protochlorophyllide in the etiolated leaf it would be preferable to express the absorbancy of protochlorophyll as millimolar extinction coefficients, rather than specific absorption coefficients. From the molecular weight of 891.5 for protochlorophyll and the specific absorption coefficients given above, the following millimolar extinction coefficients for protochlorophyll in ether were calculated: 35.6 at 623 mμ and 289.5 at 432 mμ.

Compared with chlorophyll *a*, protochlorophyll shows a greatly reduced absorption at the red maximum but an increased absorption in the Soret region. The position of the red maximum of protochlorophyll is shifted to a shorter wavelength by 39 mμ, a shift indicating that it is a less resonant structure than chlorophyll *a* in spite of its extra double

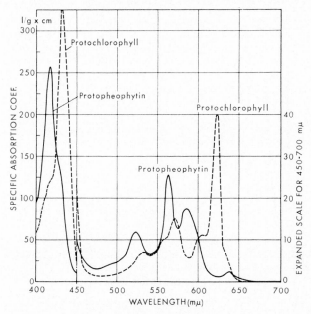

Fig. 2. Absorption spectra of protochlorophyll and protopheophytin in ether (*23, 24*).

bond. It has been assumed that protochlorophyll and protochlorophyllide have identical absorption spectra, but this is probably not so in view of the recent finding that the absorption spectra of chlorophyll *a* and chlorophyllide *a* differ slightly (*20*). The position of the visible band of chlorophyllide *a* is shifted on phytylation to a shorter wavelength by 1.5–2 mμ.

Protochlorophyll shows a strong orange-red fluorescence when dissolved in a polar solvent or a nonpolar solvent to which a trace of polar solvent has been added (*26*). The fluorescence spectra are shown in Fig. 3 (*26, 27*).

3. Estimation

Protochlorophyll (and protochlorophyllide) may be determined spectrophotometrically in 80% acetone (25) or after transfer to ether (24, 28). In the absence of the chlorophylls, absorbancy measurements at one wavelength (maximum of protochlorophyll) are sufficient, but if chlorophyll a is present, measurements at two wavelengths are necessary. If both chlorophyll a and chlorophyll b are present, absorbancies at three wavelengths are required and the concentrations of the pigments are obtained by solving simultaneous equations. Specific absorption coeffi-

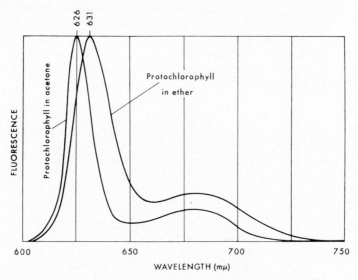

Fig. 3. Fluorescence spectra of protochlorophyll (26, 27).

cients are available for chlorophyll a, chlorophyll b, and protochlorophyll in ether (24) and in 80% acetone (25, 29), but it would seem preferable to convert these to millimolar extinction coefficients for the determination of protochlorophyll + protochlorophyllide.

In strongly acid solutions, the magnesium atom is removed from protochlorophyll to give protopheophytin. Loss of the magnesium from protochlorophyllide gives vinyl pheoporphyrin a_5. The absorption spectrum of protopheophytin (Fig. 2) is that of a typical porphyrin with one conjugated side chain (rhodo-type spectrum). In the estimation of protochlorophyll, care should be taken to avoid acidic conditions and the formation of protopheophytin. If the cell sap is acidic it is recommended that magnesium carbonate be added to the tissue at the time of grinding the tissue (24).

4. SEPARATION OF PROTOCHLOROPHYLL AND PROTOCHLOROPHYLLIDE

Protochlorophyllide with its acidic carboxyl group is readily separated from protochlorophyll. If the pigments are extracted from etiolated leaves in acetone (preferably with a trace of ammonia or dilute NaOH) and the extract shaken with petroleum ether or isooctane, protochlorophyll but not protochlorophyllide is transferred to the less polar solvent (*18*, *19*, *30*). Protochlorophyll and protochlorophyllide have been separated by paper (*18*, *19*, *31*) and thin-layer chromatography (*32*) and on columns of powdered polyethylene (*33*, *34*), but no solvent mixture has been reported which will give a distinct separation of protochlorophyll, protochlorophyllide, chlorophyll *a*, and chlorophyllide *a*. Hydrochloric acid numbers of 22 and 11 were obtained for protochlorophyll and protochlorophyllide, respectively (*18*).

C. Seed Coat and Bacterial Protochlorophylls

As mentioned earlier, Noack and Kiessling (*2*) concluded that the protochlorophylls isolated from pumpkin seed coats and etiolated leaves were identical. The same conclusion was reached by Smith and Young (*13*), but more recently a number of discrepancies were observed between the absorption spectra of the pigments, both in the positions and relative heights of the bands (*10*, *11*). The most obvious difference was in the position of the Soret band; 433 mμ for leaf protochlorophyll and 438 mμ for seed coat protochlorophyll. The spectral curve of seed coat protochlorophyll more closely resembled that of the bacterial protochlorophyll found in extracts of the "tan" mutant of *Rhodopseudomonas spheroides*, which was unable to synthesize bacteriochlorophyll (*10*, *11*).

Seybold (*35–37*) reported the isolation of two seed coat protochlorophylls which he named protochlorophyll *a* and protochlorophyll *b*, but no evidence was produced to support the view that protochlorophyll *b* was an immediate precursor of chlorophyll *b*. The recent labeling experiments of Shlyk *et al.* (*38*) support the hypothesis that chlorophyll *b* is formed from chlorophyll *a*. Neither Fischer and Oestreicher (*16*) nor Krasnovsky and Voinovskaya (*39*) could detect the two components reported by Seybold, but recently two protochlorophyll-like pigments have been isolated from marrow seeds and separated by countercurrent distribution, followed by chromatography on sugar columns (*40*). One pigment (λ_{max} in ether 432, 535, 571, and 623 mμ) was identical spectroscopically with leaf protochlorophyll (protochlorophyllide); the other (λ_{max} 438, 537, 574, and 624 mμ) closely resembled bacterial protochlorophyll, prepared either by treatment of *R. spheroides* with 8-hydroxyquinoline (*12*) or from the "tan" mutant of the same organism

(10, 11). A better separation of the seed coat pigments was achieved after acid hydrolysis of the ester groups. The resulting pheoporphyrins were identical in spectroscopic and solubility properties with the corresponding derivatives from bacterial protochlorophyll and leaf protochlorophyll (protochlorophyllide), respectively (Table I).

TABLE I

COMPARISON OF SPECTROSCOPIC PROPERTIES OF THE PHEOPORPHYRINS DERIVED
FROM PROTOCHLOROPHYLLS OF VARIOUS ORIGINS[a,b]

Origin	Band maxima (mμ)					Ratio Band III:band IV
Higher plants	638	586	565	524	417	2.1
Seed coat, fraction I	639	586	565	524	417	2.1
Mutant bacteria	644	591	567	527	421.5	1.6
Seed coat, fraction II	644	591	567	527	421.5	1.6

[a] Taken from Jones (41).
[b] Solvent was ether.

Bacterial protochlorophyll was identified as magnesium 2:4-divinylpheoporphyrin a_5 (12); thus it differs from protochlorophyllide in having a vinyl instead of an ethyl substituent at position 4 of the porphyrin ring. Jones (12, 40) has suggested that 2:4-divinylpheoporphyrin may be an intermediate, both in bacteriochlorophyll and chlorophyll synthesis, between magnesium protoporphyrin monomethyl ester and protochlorophyllide.

The second seed coat protochlorophyll differs from the bacterial pigment in being esterified, probably with phytol (41). Direct evidence for the presence of the phytol group in seed coat protochlorophyll was provided by Fischer and Rudiger (21). Protochlorophyll, however, does not appear to be a substrate for chlorophyllase (42).

Sudyina and Lozovaya (42, 43) studied the formation of protochlorophyll during the maturation of the pumpkin seed. Young, immature seeds contained small amounts of chlorophyll a and a form of protochlorophyll (protochlorophyll I). The amount of protochlorophyll increased during maturation of the seed. Just prior to maturity, chlorophyll a disappeared and a second protochlorophyll appeared (protochlorophyll II). The mature germ was found to contain two further protochlorophylls (protochlorophyll III and protochlorophyll IV). The pigments were separated by paper chromatography; their positions on the chromatogram indicated that they were esterified, and only traces of protochlorophyllide remained at the chromatographic origin. The accumulation of protochlorophyll in the mature seed after chlorophyll a synthesis has ceased bears analogy with the mutant strains of Chlorella (4, 5) and R. spher-

oides (*10*, *11*), which are unable to synthesize chlorophyll and bacteriochlorophyll, respectively.

III. Protochlorophyll *in Vivo*

A. Protochlorophyll of Etiolated Leaves

1. Spectra

The existence of two forms of protochlorophyll in etiolated leaves was demonstrated by Hill *et al.* (*44*) with the use of a Zeiss microspectroscope. The band which existed when the light was first turned on for observing the spectrum was at about 650 mμ. The absorption spectrum rapidly changed within seconds to show a weaker band at 635 mμ and an intense band between 670 and 680 mμ. Difficulty was encountered in seeing the absorption bands initially present, due to the rapid transformation of protochlorophyll, but these qualitative observations were later confirmed by quantitative measurements of the absorption spectrum of etiolated leaves (*45*). Young etiolated bean leaves gave a single absorption band at 650 mμ, which disappeared on illumination of the leaves for 1 min at 1000 foot-candles (10,800 lux) and was replaced by a strong band at 684 mμ and a band at about 635 mμ. The spectrum of old etiolated leaves (2 weeks) showed a double peak slightly below 650 mμ, suggesting two components with absorption peaks at 650 and 635 mμ. The shape of the absorption band in young leaves suggested that these leaves also contained a small amount of the component absorbing at 635 mμ. On successive illumination of old leaves with white light, only the peak at the longer-wavelength side of the double peak disappeared, with the formation of a peak near 680 mμ. (Fig. 4). Shibata (*45*) concluded from these and other experiments that etiolated leaves contain two forms of protochlorophyll absorbing at 636 mμ and 650 mμ (P636 and P650), but only one form, P650, is transformed to chlorophyll on illumination.

Similar conclusions were reached by Litvin and Krasnovsky (*46*) from measurements of fluorescence emission spectra of etiolated bean leaves after rapid freezing to −150°C. Two main bands were observed with maxima at 635 mμ and 655 mμ. Illumination of the leaves before freezing caused the disappearance of the fluorescence band at 655 mμ and the appearance of a band at 690 mμ due to a form of chlorophyll *a*. Heating the etiolated leaves to 100° destroyed both the transformation and the fluorescence band at 655 mμ. The form of protochlorophyll fluorescing at 655 mμ appears to correspond therefore to the form which absorbs at 650 mμ. More recently, Krasnovsky and Bystrova (*46a*) have reported that the form of protochlorophyll fluorescing at 635 mμ also is partly transformed to a form of chlorophyll *a*. Most of the 635-mμ form, however, is destroyed by bleaching.

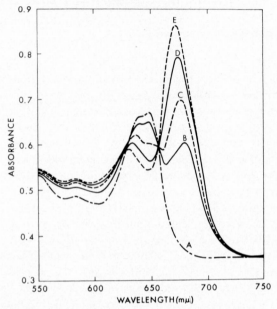

FIG. 4. Effect of successive illumination periods on the spectrum of old etiolated bean leaves. Curve *A*, two superimposed old etiolated bean leaves before illumination: Curves *B*, *C*, *D*, and *E*, after 0.5, 1.0, 2.5, and 12.0 minutes' total illumination, respectively (*45*).

2. PHYTOL CONTENT

The studies of Wolff and Price (*19*) indicated that about 20% of the protochlorophyll pigments of an etiolated bean leaf were esterified with phytol, but this proportion can vary widely for different species of plants (*47*). It also depends on the age of the seedlings (*20*). For example, the ratio of protochlorophyllide to protochlorophyll was found to vary from 7–8 in 6-day-old barley seedlings to 2–3 in 17-day seedlings (Table II). The rise in the proportion of protochlorophyll with age of seedling roughly parallels the increase in the relative content of P636 (*45*), suggesting that P636 may be protochlorophyll and P650 protochlorophyllide. This hypothesis is supported by the studies of Wolff and Price (*19*), which showed that the pigment formed when dark-grown bean seedlings were illuminated for short periods was chlorophyllide *a;* the pigment had the absorption spectrum of chlorophyll *a*, but it could not be transferred to petroleum ether or isooctane from an acetone extract of the illuminated leaves. If after illumination the leaves were allowed to stand in the dark at 25° for 1 hour before acetone extraction, the chlorophyll *a* could then be transferred to petroleum ether or isooctane. The incubation in the dark had no detectable effect on the absorption spectrum of the pigment. Lowering the temperature of the leaves to 0° did not affect the conversion of protochlorophyllide to chlorophyllide *a*, but it pre-

TABLE II

PROTOCHLOROPHYLLIDE:PROTOCHLOROPHYLL RATIO AS A FUNCTION OF
THE AGE OF ETIOLATED BARLEY SEEDLINGS[a,b]

Age (days after germination)	Ratio
6	8.7, 7.2, 7.1
15	4.3
17	2.7, 3.3, 3.6

[a] Taken from Sironval et al. (20).

[b] Each figure represents a measurement done on a batch of 60 plants. The pigments were separated by paper chromatography.

vented the esterification, which is catalyzed by an enzyme, chlorophyllase. Wolff and Price (19) therefore concluded that the two terminal steps in chlorophyll *a* biosynthesis were as follows:

$$\text{protochlorophyllide} \xrightarrow{\text{light}} \text{chlorophyllide } a \xrightarrow{\text{dark}} \text{chlorophyll } a$$

Virgin (31) reached a similar conclusion after studying the pigment transformations in etiolated wheat leaves, but in Virgin's experiments the pigments were subjected to paper chromatography and then determined spectrofluorimetrically.

3. SPECTRAL SHIFTS ON ILLUMINATION

Shibata (45) made the interesting observation that the 684-mμ absorbing form of chlorophyll (C684) which resulted from the photoconversion of P650 after a short illumination was slowly transformed in about 10 minutes in the dark to a form of chlorophyll absorbing at 672 mμ (C672). The sharp isosbestic point obtained for the C684 → C672 shift indicated conversion of one definite compound to another. A similar shift in the position of the fluorescence maximum of chlorophyll (from 690 to 680 mμ) was observed by Litvin and Krasnovsky (46) under similar conditions.

Some results of Smith et al. (48) on the transformation of protochlorophyll in the etiolated leaves of a number of corn mutants are interesting. Transformation of protochlorophyllide to chlorophyllide *a* occurred in all instances, but in several of the mutants the chlorophyllide *a* was not subsequently esterified in the dark. These mutants did not show the spectral shift from 684 to 672 mμ, suggesting that phytylation of chlorophyllide *a* may be responsible for the shift in normal seedlings. Sironval *et al.* (*20*) have recently provided strong evidence in favor of this hypothesis. Etiolated barley leaves were cut into small pieces and illuminated with an electronic flash of 1 millisecond (msec) duration. The flash transformed 80–90% of the protochlorophyllide into chlorophyllide *a* (*49*). The pigments were extracted into 80% acetone either immediately or after a period in the dark, and analyzed by paper

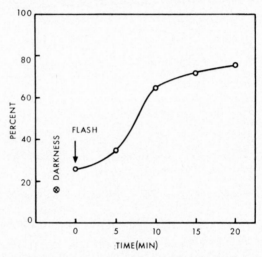

Fig. 5. Time course of the esterification of chlorophyllide in darkness after an illumination flash. Ordinate: esterified pigments (protochlorophyll plus chlorophyll *a*) as percentage of total green pigments (*20*).

chromatography. The bulk of the green pigments extracted immediately after the flash remained near the starting line of the chromatogram, but if the leaves were left in the dark for 20 minutes before extraction, most of the green pigments migrated down the paper, showing that they were then more lipophilic. The time course of the esterification of chlorophyllide *a* is shown in Fig. 5. Immediately after the flash, about 25% of the green pigments were already esterified, due presumably to the transformation of protochlorophyll. The amount of esterification increased only slightly during the first 5 minutes in darkness, but during the next 5 minutes it increased rapidly and was practically complete 15–20 minutes after the flash.

The spectrum of old etiolated barley leaves showed the presence of both the 635-mμ and 650-mμ forms of pigment, due to protochlorophyll and protochlorophyllide. Sironval *et al.* (*20*) state in the discussion of their paper that the 635-mμ form was transformed on illumination to chlorophyll *a*, absorbing at 672 mμ. In contrast, the 635-mμ form in old etiolated bean leaves is apparently inactive or only partly active (*31, 45, 46a*). The Russian investigators (*47*) have also reported that illumination of etiolated barley leaves in the cold gave the same proportion of "acidic" and "neutral" forms of chlorophyll *a* as found for protochlorophyll before illumination. The principal pathway for chlorophyll *a* formation in barley seedlings is the same as in bean seedlings, viz. protochlorophyllide → chlorophyllide *a* → chlorophyll *a*, but in addition chlorophyll *a* may be formed by the secondary pathway, protochlorophyllide → protochlorophyll → chlorophyll *a*. It seems that the percentage of protochlorophyll which is convertible to chlorophyll *a* (secondary pathway) varies with different plants.

B. Transformation of Protochlorophyll (Protochlorophyllide)

1. ACTION SPECTRUM

That protochlorophyllide is the photoreceptor for its own conversion was beautifully demonstrated by the work of Koski *et al.* (*50*). These investigators measured action spectra for the transformation of protochlorophyll to chlorophyll *a* both in normal etiolated corn seedlings and in an albino mutant which contained almost no other pigments. Their results are reproduced in Fig. 6, together with the absorption curve for protochlorophyll in methanol. The lower effectiveness of blue light for the normal seedling was attributed to the screening effect of carotenoids. The action spectra showed that the active form of protochlorophyll *in vivo* has absorption maxima at 650 mμ and 445 mμ. Light absorbed by chlorophyll in partly greened leaves was shown to be ineffective in causing the transformation.

2. EFFECT OF TEMPERATURE

The effect of temperature on the transformation of protochlorophyll (protochlorophyllide) in etiolated barley leaves was studied extensively by Smith and Benitez (*51*). The transformation was completely inhibited at −195°; it occurred at −70°, but it was not complete at this temperature. Both the rate and extent of the conversion increased as the temperature was increased to 40° (Fig. 7). At temperatures above 40° the transformation was progressively inhibited, the heat of inactivation being in the range 60,000–280,000 cal and suggestive of a protein denaturation.

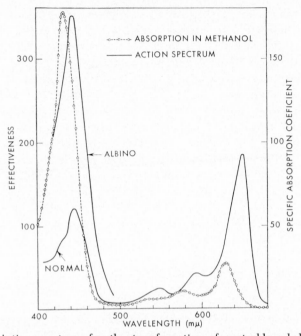

FIG. 6. Action spectrum for the transformation of protochlorophyll in dark-grown normal and albino corn seedlings (solid lines) compared with absorption spectrum of protochlorophyll in methanol (broken line). The effectiveness values for the normal and albino seedlings are equal in the long-wavelength region (50).

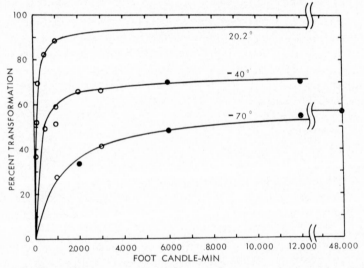

FIG. 7. Effect of temperature on the transformation of protochlorophyll in etiolated barley leaves (51).

3. KINETICS

The kinetics of the transformation are consistent with a second-order reaction, but the temperature coefficients are low for a bimolecular collision reaction (51). Calculations of energy of activation gave 5270 cal in the temperature range $-70°$ to $-40°$ and 1330 cal from $-40°$ to $20°$. The rate of the transformation was found to be directly proportional to the light intensity, a result showing that the apparent bimolecular reaction was not dependent on the collision of two photochemically excited molecules. Smith and Benitez (51) presented two hypotheses to explain the apparent bimolecular nature of the conversion with respect to the protochlorophyll; either there is reaction between excited and unexcited protochlorophyll molecules or between unexcited protochlorophyll molecules and a photodissociated product of the protochlorophyll-protein complex. These explanations now appear unlikely in view of recent work on the isolated protochlorophyll holochrome (see Section IV).

The dependency on temperature shows that the transformation is not a purely photochemical process, but that some thermochemical steps are also involved. Virgin (52) studied the transformation in intermittent as well as continuous light. The amount of transformation per unit of irradiation was the same, showing that the conversion was not dependent on a nonphotochemical reaction of long time. Since dark periods as short as 13 msec had no effect on the yield per unit of irradiation, Virgin (52) concluded that the half-life of any species produced in the light was less than one-tenth of this value, or about 1 msec. That this was indeed so was demonstrated experimentally by Madsen (49). The transformation of protochlorophyll was complete in leaves in the shortest time that could be measured, about 4 msec. In preparations of the protochlorophyll holochrome, the time resolution was reduced to 1 msec. The intensity of the flash was sufficient to produce 60% transformation, but there was no further transformation after the 1 msec illumination. The significance of these findings is considered later in relation to the structure of the protochlorophyll holochrome.

4. TRANSFORMATION OF PROTOCHLOROPHYLL IN ORGANISMS OTHER THAN ETIOLATED LEAVES

The protochlorophyll present in dark-grown roots of wheat seedlings is almost entirely esterified with the possible exception of a small amount of protochlorophyllide in the apical regions (8). Illumination of the roots leads to the formation of chlorophyll a, but correspondence between protochlorophyll disappearance and chlorophyll a formation has not been demonstrated. The rate of protochlorophyll disappearance in roots was found to be extremely slow (half-life of 5 hours at 31.7 microwatts/cm² or 210 lux) compared with the rate of transformation of protochlorophyllide in etiolated bean leaves (ca. 1 minute). The time course of protochlorophyll disappearance deviated from the second-order law observed with leaves and the rate was not proportional to light intensity (8).

Conversion of protochlorophyll (protochlorophyllide?) to chlorophyll a has been shown to occur when dark-grown cells of *Euglena* are illuminated (6). The rate of protochlorophyll disappearance roughly followed first-order kinetics, at least for the initial part of the conversion, and there was a good correspondence between rate of protochlorophyll disappearance and rate of chlorophyll formation. The half-life of the conversion was 24 minutes at 8800 lux, which was of the same order of magnitude as for root protochlorophyll. At a light intensity of 8800 lux, leaf protochlorophyllide is transformed in a few seconds. The action spectrum for chlorophyll formation in *Euglena* showed a peak around 650 mμ (6).

C. Localization of Protochlorophyll

By the use of fluorescence and phase-contrast microscopy it was demonstrated that the protochlorophyll (protochlorophyllide) in an etiolated bean leaf (10- to 14-day seedling) is localized in discrete centers, 0.7–1.3 μ in diameter, within the proplastids which are 3–5 μ in diameter (53, 54). A similar finding was reported for tomato cotyledons (55). Dark-grown *Euglena* cells also contain proplastids with fluorescing centers (56). It is usual to find one or two fluorescing centers in a single bean proplastid, although up to 4 or 5 have been observed occasionally (53, 54). The proplastids were isolated and purified by differential and density-gradient centrifugation (53, 54), and it was shown that 75–80% of the protochlorophyll (presumably protochlorophyllide) of the isolated proplastid was photoconvertible to chlorophyll a (chlorophyllide a) (54, 57).

When sections of etiolated leaves are examined by electron microscopy, it is seen that the lamellated structures normally observed in higher plant chloroplasts are absent from the proplastids. Instead the proplastids may contain one or more dense vesicular centers (58–60) which have been termed prolamellar bodies (61). Isolated proplastids also have been examined by electron microscopy and found to contain prolamellar bodies (57). Because the fluorescing centers observed by optical microscopy correspond in size and number with the prolamellar bodies (53), it would appear very likely that protochlorophyll is localized within these bodies. But whereas the prolamellar bodies rapidly disappeared when 16-day etiolated plants were illuminated and were replaced by vesicles arranged in concentric layers around the proplastid (59, 60), no change was observed in the distribution of chlorophyll for at least 3 hours when 10- to 14-day plants were illuminated (54). Plastids isolated from plants which had been illuminated for 3 hours resembled the plastids of etiolated plants when viewed by phase-contrast and fluorescence microscopy. The formation of the concentric vesicular structures occurred under conditions which allowed transformation of protochlorophyllide to chlorophyllide a, but not phytylation to chlorophyll a (60, 62).

It has been suggested that the molecules of protochlorophyll holochrome may constitute the units of the lattice of the prolamellar body (63). It appears from detailed electron micrographs, however, that the prolamellar body consists of groups of tubes or a three-dimensional lattice of tubes rather than a lattice of discrete particles (57, 64).

D. Protochlorophyll in Green Leaves of Angiosperms

Since etiolation may be considered almost as a pathological state of the normal seedling, the question arises whether protochlorophyll (protochlorophyllide) is the precursor of chlorophyll a throughout the period of chlorophyll accumulation which follows exposure of etiolated plants to light, and in normal seedlings grown under natural lighting conditions. Good experimental evidence now exists to justify the conclusion that the main pathway of chlorophyll a synthesis in green leaves is the same as in etiolated leaves on exposure to light.

If etiolated seedlings are illuminated for a brief period and then returned to darkness at room temperature, new protochlorophyll (mainly protochlorophyllide) is rapidly synthesized (10–20 minutes) to the level found in the etiolated leaf without any significant decrease in the chlorophyll level (49, 65). Lowering the temperature reduced the rate of protochlorophyllide formation. Further, the initial rate of protochlorophyllide formation paralleled the rate of chlorophyll a formation in the light (65). A similar situation was found to hold when wheat seedlings grown under normal lighting conditions were kept in darkness (66).

The presence of traces of protochlorophyll (protochlorophyllide) in bean plants grown in the light was clearly demonstrated by Litvin et al., (67) by the use of fluorescence. If the plants were then placed in darkness, there was a considerable increase in the amount of protochlorophyll; this was transformed on further exposure of the plants to light. The level of active protochlorophyll found in plants after 10 hours of darkness reached 1.6% of the chlorophyll content and was severalfold higher than the amount found in an etiolated plant. The active protochlorophyll of green leaves was shown by solubility studies and chromatography to be protochlorophyllide (30, 38). Protochlorophyllide was isolated from illuminated green leaves of a number of higher plants and also from *Chlorella* (30). Protochlorophyll may also be present in green leaves, but in amounts which are small compared with the quantities of protochlorophyllide.

Shlyk and his co-workers (38) have provided convincing evidence from labeling experiments with $^{14}CO_2$ that the terminal steps in the main biosynthetic pathway for chlorophyll *a* in the green leaf are protochlorophyllide→ chlorophyllide *a* → chlorophyll *a*. A similar conclusion was reached by Virgin (31) from pigment analyses of extracts of green wheat seedlings which had been kept in darkness for 5 hours and then subjected to light and dark conditions (31).

E. Chlorophyll Formation during Greening of Angiosperms

Several investigators (25, 65, 68–70, and others) have observed a lag phase of 2–3 hours in the formation of chlorophyll in etiolated plants during continuous illumination, after the initial protochlorophyll had been transformed. The chlorophyll *a* formed during the lag period is not effective for photosynthesis (71, 72). Thereafter the rate of chlorophyll *a* formation is accelerated, and it is accompanied by the synthesis of chlorophyll *b* (73). Several changes have been correlated with the enhanced rate of chlorophyll formation, including the initiation of photosynthesis (71, 72), the appearance of a red fluorescence in regions of the plastid other than the fluorescing centers of the proplastid (25, 54), a dramatic decrease in the fluorescence yield of the chlorophyll (70) and the commencement of a shift in the adsorption band of chlorophyll *a in vivo* from 673 mμ to 677 mμ (45). The lag phase extends over a longer period than is required for phytylation. Excitation energy absorbed by carotenoid pigments is not transferred to newly formed chlorophyll *a* in etiolated bean leaves (74). The capacity for energy transfer developed during a dark period which appeared to correspond more with the time required for phytylation than with the lag phase.

The lag phase in chlorophyll synthesis observed with continuous illumination of etiolated plants may be eliminated by a short impulse of red light, followed by several hours of darkness (75, 76). Continuous irradiation following the dark period causes an immediate rapid formation of chlorophyll. The optimum period in darkness was found to be 3–5 hours (77, 78). Virgin (77) showed that the increased rate of chlorophyll formation corresponded with an increased rate of protochlorophyllide formation. It is not known whether the effect of red light is a direct one on the stimulation of the biosynthetic pathway of protochlorophyllide or an indirect one on the growth of the plastid, but the latter appears more likely. Inhibitors of RNA synthesis (79) and protein synthesis (80) are known to inhibit chlorophyll formation in greening bean seedlings.

The action of the short impulse of red light can be partly reversed by far-red light (78, 81, 82) suggesting the participation of the photomorphogenetic pigment, phytochrome, in the elimination of the lag phase.

Madsen (49) has described what appears to be a different effect from the red, far-red phenomenon. The lag phase for chlorophyll formation was absent when etiolated barley leaves were treated with flashes of strong light of 1 msec duration, alternated with dark periods of 15 minutes. Madsen suggested that continuous illumination may cause destruction of the newly formed chlorophyllide a or a precursor of protochlorophyllide by photooxidative reactions.

F. Chlorophyll Formation in Gymnosperms

Chlorophyll formation in dark-grown pine seedlings was studied by Schmidt (83) and Bogorad (84). Embryos removed from ungerminated seeds and grown either on synthetic media or extracts of endosperm formed chlorophyll on illumination, but not in the dark. However, if the embryos were left in the dark in contact with a small piece of endosperm, they greened. The amount of chlorophyll produced in darkness by the unextirpated embryo was equal to the amount of chlorophyll formed in the extirpated embryo on illumination, suggesting that a chlorophyll precursor was present in the embryo and was converted either by the action of light or a substance transferred from the endosperm.

Sudyina (42) studied the formation of chlorophyll in seedlings of three species of gymnosperms: Scotch pine (Pinus silvestris), spruce (Picea excelsa), and Siberian larch (Larix sibirica). She reported the detection of protochlorophyllide and chlorophyllide in dark-grown seedlings and chlorophyllide in light-grown ones, suggesting that the biosynthetic pathway for chlorophyll in the gymnosperms in the dark is the same as for the angiosperms in the light. The capacity for the dark reduction of protochlorophyllide varied among the species; pine seedlings in the dark accumulated a considerable amount of chlorophyllide and a small amount of protochlorophyllide, spruce seedlings almost equal amounts of protochlorophyllide and chlorophyllide, while larch seedlings accumulated more protochlorophyllide. Thus pine, spruce, and larch differ in their ability to reduce protochlorophyllide in the dark. The observed accumulation of protochlorophyllide in the dark in gymnosperm seedlings shows that chlorophyllide formation is slower in the dark than the light and suggests that photoreduction of protochlorophyllide is the main pathway for chlorophyllide formation in the light. These experiments of Sudyina raise the question as to whether there are two protochlorophyllide holochromes in the gymnosperms, one of which is transformed to chlorophyllide in the light as in the angiosperms and the other not dependent on light for reduction; or perhaps there is only one protochlorophyllide holochrome which is convertible to chlorophyllide by two different mechanisms.

IV. Protochlorophyll Holochrome *in Vitro*

A. Spectral Properties of Homogenates

Extracts containing the protochlorophyll holochrome (protochlorophyllide holochrome) in a physiologically active state were first described by Krasnovsky and Kosobutskaya (85) and Smith et al. (86, 87). The Russian investigators ground etiolated bean leaves in a phosphate buffer, pH 7. Illumination of the extracts converted a large fraction of the protochlorophyll to chlorophyll a with an absorption maximum around 670 mμ. Smith et al. (86, 87) used glycerol as a grinding medium; the resulting brei was squeezed through a fine cloth and centrifuged for 10–15 minutes at 10,000 g. The absorption spectrum of the supernatant was measured before and after illumination. The positions of the absorption maxima of the protochlorophyll holochrome and the chlorophyll a derived by illumination were found to vary for different plants. In extracts from etiolated barley leaves, the respective maxima were at 650 and 680 mμ, in those from bean leaves at 640 and 675 mμ, and in extracts from squash cotyledons at 645 and 680 mμ. Smith et al. observed that if extracts of barley leaves were allowed to stand in the dark for 1–2 hours before illumination, the maximum shifted from 650 to 630–635 mμ. Illumination of the aged extracts gave chlorophyll a with an absorption maximum at 672 mμ, which agrees with the position found by Krasnovsky and Kosobutskaya (85) for their bean extracts.

More recently, Krasnovsky and Bystrova (88) studied the formation of chlorophyll in homogenates of etiolated leaves by spectrofluorimetry at liquid air temperatures. Homogenates made in $M/15$ phosphate buffer, pH 8.5, showed two major maxima (primary fluorescence) in the fluorescence spectrum, at 635 and 655 mμ. Lowering the pH of the grinding buffer to 6.5 reduced the relative proportion of the 635-mμ form. Illumination of a homogenate at $0°$ resulted in a rapid transformation of the protochlorophyll fluorescing at 655 mμ to chlorophyll a fluorescing at 687–689 mμ, but after a few minutes the peak shifted to a shorter wavelength, around 680 mμ. Further illumination resulted in a bleaching of the chlorophyll presumably by a photooxidative mechanism, since bleaching did not occur under anaerobic conditions. The sequence of the spectral changes was the same at pH 4.5, 6.5, and 8.5, but the rate at which the chlorophyll fluorescence band was displaced to shorter wavelengths and the extent of the shift were dependent on the pH. Both were greatest at pH 8.5.

Although the spectral changes observed with homogenates paralleled those observed with intact leaves (46), the shift from 687–689 to 680 mμ was much more rapid with the homogenates, being complete in 1–2 minutes at pH 8.5. In leaves 20–30 minutes were required to complete the spectral shift. The addition of glycerol (final concentration 60%, v/v) to the homogenate stabilized the fluorescence maximum at 687–689 mμ. The height of the 655-mμ band in the unilluminated homogenate was reduced by heating the homogenate above 40°C or by the addition of pyridine (1–2%) or alcohol (20–30%). Conditions which eliminated the 655-mμ band resulted in a complete inhibition of the transformation.

Smith et al. (87, 89) confirmed the finding of Krasnovsky and Kosobutskaya (85) that aqueous extracts of etiolated bean leaves contained active protochlorophyll holochrome, but they were not successful in obtaining active extracts from barley leaves or squash cotyledons. Barley leaf extracts in phosphate buffer pH 7 contained practically no protochlorophyll and showed no transformation. In fact, they appeared to contain an inhibitor, since an extract from barley leaves reduced the amount of transformation in a bean leaf extract.

B. Ultracentrifugal Analysis of Extracts

Smith et al. (87, 89) extracted the protochlorophyll holochrome from etiolated bean leaves with a glycine buffer pH 9.5, as the extracts appeared more stable than those in a phosphate buffer, pH 7. After filtration through a fine cloth, the extract was centrifuged at 10,000 g for 10 minutes. The supernatant was passed through a French-Milner press, centrifuged for 1 hour at 10,000 g, and dialyzed against glycine buffer. Ultracentrifugal analysis of the dialyzed extract showed two peaks with sedimentation coefficients of 3–4 S and 16–17 S. Centrifugation of the extract for 3 hours at 105,000 g gave a pellet which contained a large proportion of the active protochlorophyll, and examination of the resuspended pellet in the analytical ultracentrifuge showed that it contained a much higher proportion of the 16–17 S component than the original extract. These results led Smith and Kupke to conclude that the active protochlorophyll was associated with the 16–17 S boundary. Several tests showed that the active material consisted largely of protein.

C. Purification of the Protochlorophyll Holochrome

The protochlorophyll holochrome of dark-grown bean seedlings was extensively purified by Smith (11, 90) and by Boardman (91). These investigators independently found that the content of transformable protochlorophyll in an extract was higher if glycerol (20–30%) was added to the extracting buffer. Extracts prepared at pH 7.5 (0.05 M phosphate buffer and 30% glycerol) contained as much active protochlorophyll as those prepared at pH 9.6 (0.05 M glycine buffer and 30% glycerol), but it was found that the more alkaline conditions favored the removal of carotenoids during the ammonium sulfate fractionation and subsequent dialyses (91). The procedures used by Smith (90, 92) and by Boardman (91) to purify the holochrome were similar. A flow sheet of the author's method is shown in Fig. 8. All operations were carried out at 0–4°C, either in the dark or under a weak green light which was in-effective in transforming the protochlorophyll.

Fifteen-gram portions of etiolated bean leaves were ground in a mortar, chilled in ice, with 25 ml of a glycine buffer pH 9.6, containing 30% glycerol. The extract was filtered and centrifuged, first at 10,000 g for 15 minutes and then at 105,000 g for 1 hour. The supernatant was treated with ammonium sulfate, and the material precipitating between 20% and 50% saturation was collected. The precipitate was dissolved in glycine buffer and dialyzed; after dialysis the solution was centrifuged at 144,000 g for 20 minutes and the pellet was discarded. The supernatant was treated with calcium phosphate gel to remove nucleic acid, the gel was removed by centrifugation, and the protochlorophyll holochrome was precipitated again with ammonium sulfate. The next step in the purification procedure was zone centrifugation in a density gradient of sucrose. In some experiments the most active fractions from the zone centrifugation were purified further by zone electrophoresis in a density gradient of sucrose.

The purification was followed by measurements of specific activity and by ultracentrifugal analyses. Boardman (91) used the increase in absorbance at 675 mμ on illumination as a measure of the amount of chlorophyll formed and therefore of the amount of active protochlorophyll. The activity of a solution was defined as the product of the increase in extinction at 675 (ΔE, 675) and the volume of the solution. Thus 1 ml of a solution which gave an increase in extinction of one contained one unit of activity. The specific activity of a solution was the number of units of activity, divided by the number of milligrams of protein.

Preparations prior to density gradient centrifugation showed two main components in the analytical ultracentrifuge with sedimentation coefficients of 3–4 S and 18 S, in agreement with the findings of Smith and

Etiolated bean leaves ground in 0. 05 M glycine (pH 9. 6) and 30% glycerol

|

Filter through cloth

|

10, 000 g , 15 min

|

Supernatant
105, 000 g, 60 min

|

Supernatant
Ammonium sulfate fractionation, 20-50% saturation

|

Precipitate
Dissolve in 0. 05 M glycine (pH 9. 6); dialyze

|

144, 000 g, 20 min

|

Supernatant
Treat with calcium phosphate gel; remove gel by centrifugation

|

Supernatant
Ammonium sulfate fractionation, 20-40% saturation

|

Precipitate
Dissolve in 0. 05M glycine (pH 9. 6); dialyze

|

144, 000 g, 20 min

|

Supernatant
Density gradient centrifugation, 144, 000 g, 3 hr

Divide into fractions

FIG. 8. Flow sheet for the preparation of the protochlorophyll holochrome (91).

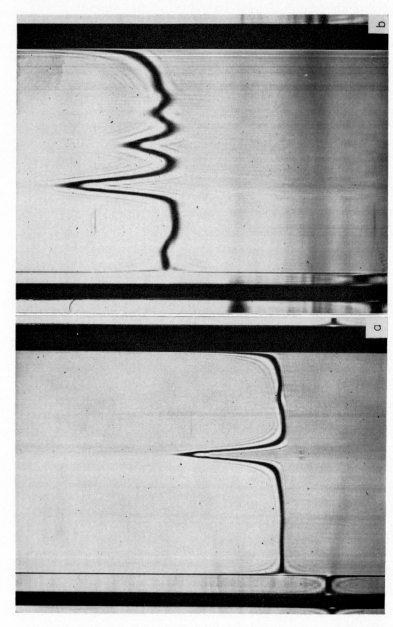

Fig. 9. Ultracentrifugal patterns of the protochlorophyll holochrome: (a) Purified by density gradient centrifugation; (b) same as (a) but after freezing and thawing (91).

Kupke (89). The areas of the peaks were in the approximate ratio 3:2. Density gradient centrifugation showed conclusively that the active protochlorophyll was associated with the 18 S component (91). Fractions obtained from near the bottom of the tube after centrifugation for 3 hours at 144,000 g had the highest activity. The specific activity of the fractions decreased from the bottom to the top of the gradient, and it paralleled the decrease in the amount of the 18 S component, as determined by ultracentrifugal analysis. In the fractions obtained from near the bottom of the tube, the 18 S component was accompanied by small but variable amounts of a component with a sedimentation coefficient of 25–27 S. The 26 S component was thought to be a dimer of the 18 S material.

A typical ultracentrifuge pattern of the protochlorophyll holochrome after density gradient centrifugation is shown in Fig. 9A. In some preparations very small amounts of higher aggregates were observed. Figure 9B shows the ultracentrifuge pattern of a purified preparation of the protochlorophyll holochrome which had been stored frozen for several months. Extensive aggregation had occurred, but there was no significant loss in activity on storage, indicating that the aggregates (dimer, trimer, and higher) also were active.

The total protochlorophyll content of preparations after purification by density gradient centrifugation varied somewhat. From 50–60% of the protochlorophyll was transformable to chlorophyll a. The minimum chemical molecular weight based on one protochlorophyll per protein molecule was calculated from the total protochlorophyll content, and it was found to vary from 1.0×10^6 to 1.7×10^6. The protochlorophyll holochrome was apparently contaminated with a colorless protein of similar size and shape because the molecular weight of the 18 S protein was shown by physical measurements to be of the order of 600,000. Boundary electrophoresis supported this conclusion. Figure 10 shows the electrophoretic pattern of a preparation of the holochrome which was reasonably homogeneous in the ultracentrifuge. The protochlorophyll appeared to be associated with the slower moving component, which constituted about 40% of the total protein in the two peaks.

Boardman (91) attempted to separate the two components on a preparative scale by electrophoresis in a density gradient of sucrose. Although there was not a separation into two discrete peaks, analysis of the fractions for protein and protochlorophyll showed that the slower-moving fractions had a higher content of protochlorophyll (Fig. 11). The concentrations of protochlorophyll in the fractions were too low for reliable absorbance measurements in the 600–700 mμ region. Measurements were made therefore at 440 mμ, before and after illumination, and the decrease in extinction at 440 mμ was taken as a measure of the active protochlorophyll (*11*). A purification factor of 3.7 was obtained for the protochlorophyll holochrome in fraction 34 (Fig. 11), compared with the preparation before electrophoresis. Using this factor, a chemical molecular weight of 400,000–500,000 was calculated for the holochrome in

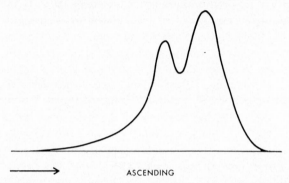

ASCENDING

Fɪɢ. 10. Boundary electrophoretic pattern of the protochlorophyll holochrome purified by density gradient centrifugation. Buffer, 0.05 M glycine (pH 9.6); potential drop 11.2 volt/cm; time 61 minutes (*91*).

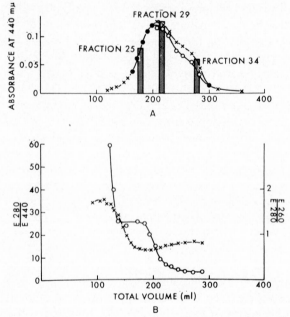

Fɪɢ. 11. Density gradient electrophoresis of the protochlorophyll holochrome (*91*). Buffer 0.025 M glycine (pH 9.6). (A) x———x, Extinction at 440 mμ before illumination; o———o after illumination. (B) Extinction ratios o———o E_{280}: E_{440}, x———x E_{260}: E_{280}.

fraction 34. This means that there was an average of one protochlorophyll molecule per protein molecule of molecular weight 600,000 (Section IV, D, 2). The active fractions gave slightly higher $E_{260}:E_{280}$ ratios (0.76–0.80) than the inactive fractions emerging from the column between 170 and 210 ml (Fig. 11). This could be due either to a different content of tyrosine or tryptophan in the protein of the active fractions, or to the presence of about 1% of some nucleotide on the weight of protein.

Smith and Coomber (92) reported chemical molecular weights of 900,000–1,000,000 for their preparations of protochlorophyll holochrome, purified by density gradient centrifugation. Kupke (93) analyzed in the ultracentrifuge aqueous extracts of the primary leaves of dark-grown bean seedlings from 3 to 20 days after planting and found a good correlation between the amount of the 18 S protein and the protochlorophyll content of the leaves. A tentative correlation of two protochlorophyll molecules per 18 S protein molecule was obtained, but it was not certain that all the 18 S protein was extracted from the leaves.

D. Properties of the Protochlorophyll Holochrome

1. ABSORPTION SPECTRUM

An absorption spectrum of the protochlorophyll holochrome after density gradient centrifugation is seen in Fig. 12 (11). It shows the typical protein absorption at 280 mμ, the Soret band of protochlorophyll at 441 mμ, a band at 485 mμ characteristic of carotenoids, and the red absorption band of protochlorophyll at 637.5 mμ. Carotenoids appear always to accompany the protochlorophyll holochrome derived from normal etiolated leaves. This does not necessarily mean that the carotenoids function in the transformation of protochlorophyll, because in albino leaves almost complete transformation occurs in the absence of carotenoids (94).

A comparison of the spectra of illuminated and unilluminated holochromes is shown in Fig. 13. In the visible region of the spectrum, illumination results in the appearance of a chlorophyll a band at 675 mμ and a decrease in the height of the protochlorophyll absorption band. The protochlorophyll which is not transformed on illumination has an absorption maximum at about 630 mμ. Boardman (94a) found that the absorption band of the purified protochlorophyll holochrome immediately after transformation was at 680–682 mμ, but it shifted rapidly to 675 mμ within 1 or 2 minutes either in the dark or the light. Smith (90) has reported that the purified holochrome contains only protochlorophyllide and this is not esterified by standing at room temperature after illumination and transformation to chlorophyllide a.

Illumination of the holochrome results in a large decrease in absorp-

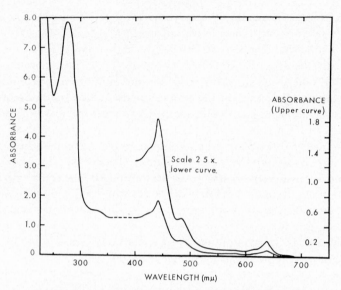

Fig. 12. Absorption spectrum of the protochlorophyll holochrome purified by density gradient centrifugation (*11*).

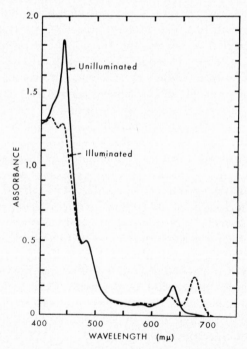

Fig. 13. Absorption spectra of illuminated and unilluminated holochrome (*11*).

tion in the Soret region at the peak for the protochlorophyll holochrome. No change is observed in either the carotenoid or protein absorption bands. After illumination the absorption band at 410 mμ is higher than the band at 440, whereas in solutions of chlorophyll *a* the 410-mμ band is relatively much lower. The spectrum of the chlorophyll holochrome resembles that of chlorophyll *a*, either when adsorbed on filter paper (95) or at an oil–water interface (11, 96).

2. FLUORESCENCE POLARIZATION

The degree of polarization of the fluorescence emitted by the initially transformed protochlorophyll in the holochrome (13.6%) was found to be much lower than that obtained (28.9%) for chlorophyll *a* in castor oil (11, 96a), where the pigment's rotation is limited. To explain these results Goedheer and Smith (96a) suggested that the chlorophyll molecules of the transformed holochrome might be able to rotate independently of the protein molecule.

3. MOLECULAR WEIGHT

Boardman (91) calculated the molecular weight of the protochlorophyll holochrome as 600,000 ± 50,000 from measurements of sedimentation coefficient (S), diffusion coefficient (D), and partial specific volume ($\overline{V}$), on a preparation purified by density gradient centrifugation. Sedimentation coefficients were measured at a number of protein concentrations (C) and the S-C plot extrapolated to zero concentration to give $S°_{20,w} = 18.0$ Svedbergs. The values obtained for the other parameters were: for $D°_{20,w}$, 2.70 10^{-7} cm²/sec and for $\overline{V}$, 0.730. The density of the holochrome as calculated from the reciprocal of the partial specific volume was 1.37, which is consistent with the densities of other pure proteins. Electron micrographs of the protochlorophyll holochrome indicated that the molecule was roughly spherical with a diameter of 100–110Å (Fig. 14). Negatively stained particles showed dense centers, indicating that the center of the molecules may have an open structure, which the phosphotungstic acid can penetrate.

4. SIMILARITY TO FRACTION 1 PROTEIN AND CARBOXYDISMUTASE

The physical properties of the protochlorophyll holochrome correspond closely with those of a colorless protein obtained from the chloroplasts of green leaves, the so-called fraction 1 protein (93, 97–101), and also with the carbon dioxide-fixing enzyme, carboxydismutase (100, 101) (Table III). Partly purified preparations of fraction 1 protein contain other enzymatic activities, e.g., phosphoribulokinase and phosphoribulo-

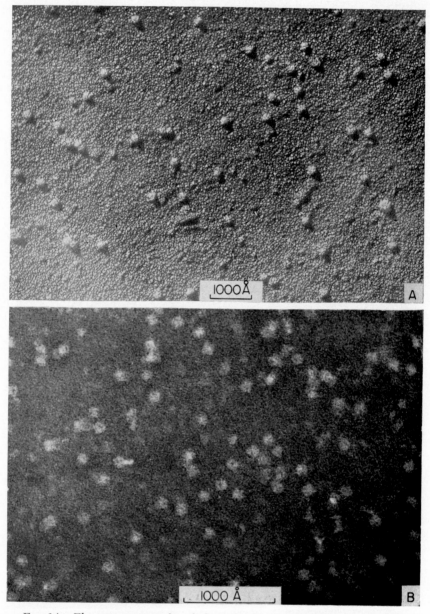

Fig. 14. Electron micrographs of the protochlorophyll holochrome (91). (A) Shadowed with platinum. (B) Stained with phosphotungstic acid.

TABLE III

COMPARISON OF PHYSICAL PROPERTIES OF PROTOCHLOROPHYLL HOLOCHROME, FRACTION I PROTEIN, AND CARBOXYDISMUTASE[a]

Sample	Sedimentation coefficient (Svedbergs)	Diffusion coefficient ($cm^2/sec \times 10^{-7}$)	Partial specific volume (ml/gm)	Molecular weight	Particle dimensions (Å)
Protochlorophyll holochrome (91)	$S^0_{20, W} = 18.0$	$D^0_{20, W} = 2.70$	0.73	$600,000 \pm 50,000$	100–110
Fraction I protein	$S^0_{20, W} = 17.9 - 19.5$ (97)	$D^0_{20, W} = 2.60 - 2.75$ (98)	—	$550,000-574,000^b$ (98)	100×200 (102)
Carboxy-dismutase (101)	$S^0_{20, W} = 18.6$	$D^0_{20, W} = 2.93$	0.73	515,000	100

[a] Taken from Trown (101).

[b] Molecular weight calculated assuming $\overline{V} = 0.73$.

isomerase, besides carboxydismutase (*103, 104*), but these may be separated from the 18 S protein either by gel filtration (*101*) or by serological methods (*105*). The other enzymes are apparently adsorbed onto the carboxydismutase.

Boardman (*91*) extracted fraction 1 protein from mature bean plants and purified it by ammonium sulfate fractionation and density gradient centrifugation. The purified preparation was indistinguishable from the protochlorophyll holochrome in the ultracentrifuge. However, it would be premature to conclude that the protochlorophyll holochrome is identical both with purified fraction 1 protein and with the enzyme carboxydismutase, particularly in view of the electrophoretic results. On present evidence one cannot exclude the other possibilities, either that the protochlorophyll holochrome is a protein of similar size and shape to fraction 1 protein and carboxydismutase, or that the protochlorophyll holochrome is a protein of smaller size which becomes associated specifically with the 18 S, fraction 1 protein either in the plant or during isolation.

E. Transformation of Protochlorophyll

1. QUANTUM YIELD OF CONVERSION

The quantum efficiency for the protochlorophyll-chlorophyll transformation was measured by Smith and French (*63, 106*). Two wavelengths were used: 642 mμ, obtained from a monochromator; and 644 mμ, from a cadmium arc. The quantum yields obtained in four experiments were 0.625, 0.478, 0.703, and 0.610, and an average of 0.60. As pointed out by Smith (*11*), it is difficult to say from these measurements whether the intrinsic quantum yield for the conversion is 1.0 or 0.5, i.e., whether one or two quanta are required for the conversion of each protochlorophyll molecule. Since the rate of transformation was directly dependent on the light intensity, it seems more likely that the transformation is a one-quantum process with an overall efficiency of 0.6 (*11*). It is obvious that light is highly efficient in converting protochlorophyll (protochlorophyllide) to chlorophyll *a* (chlorophyllide *a*).

2. EFFECT OF ENVIRONMENT

Boardman (*107*) studied the transformation of protochlorophyll in partly purified preparations of the protochlorophyll holochrome from temperatures of −55° to +25°. As observed by Smith and Benitez (*51*) with whole leaves, lowering the temperature influenced not only the rate of conversion, but also the extent of the conversion. For example, at −55° only about 50% of the active protochlorophyll was transformable. The protochlorophyll could be freeze-dried and redissolved without any significant change in the rate of transformation. The rate of transformation in D$_2$O was not significantly different from the rate in water; at pH 7.0 in phosphate buffer the rate was slightly lower than at pH 9.6 in glycine buffer (Fig. 15).

The rate of transformation, expressed as a percentage of the convertible protochlorophyll, was found to be independent of the initial concentration of the protochlorophyll holochrome and was not influenced

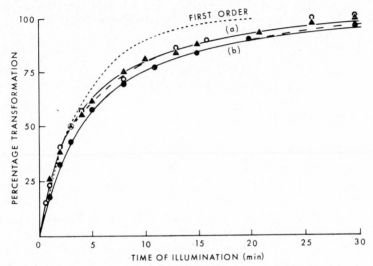

FIG. 15. Transformation of protochlorophyll in the holochrome (107), as percentage of convertible protochlorophyll. Temperature, 5°; light intensity, 8.5 μwatt/cm². The points represent the values obtained experimentally. The continuous lines show the values calculated using Eq. 2a; the broken lines show where values calculated from Eq. 2b differ from those calculated from Eq. 2a. The dotted line was calculated from a simple first-order reaction. Buffers: (a) o———o, 0.05 M glycine (pH 9.6); ▲———▲ 0.025 M glycine (pH 9.6) and 50% glycerol; (b) ●———● 0.1 M phosphate (pH 7.0). Rate constants: (a) $k_1 = 0.50$, $k_2 = 0.09$ (Eq. 2a) and $k_1 = 0.60$, $k_2 = 0.10$ (Eq. 2b). (b) $k_1 = 0.37$, $k_2 = 0.075$ (Eq. 2a) and $k_1 = 0.48$, $k_2 = 0.09$ (Eq. 2b).

by the viscosity of the medium. These results strongly suggest that the transformation does not involve a collision process either between independent protein molecules or between a protein molecule and a hydrogen donor molecule. Boardman (107) postulated that the phototransformation of protochlorophyll involves a restricted collision process between the photoactivated protochlorophyll molecule and the hydrogen- or electron-donor molecule. The protochlorophyll and donor molecules both appear to be an integral part of the holochrome. The hydrogenation reaction apparently does not involve rotation of the whole molecule of molecular weight 600,000 since this would be influenced by the viscosity of the medium. However, the rate of transformation is dependent on the temperature, and it thus seems likely that the hydrogenation involves some vibrational or rotational movement of that part of the protein molecule which is in close spatial relationship with the protochlorophyll. Of the reagents tested, the only ones which inhibited the conversion were protein denaturants. It thus seems possible that both the active protochlorophyll and the group or groups which supply the hydrogen atoms or electrons required for the conversion are located in a hydrophobic region of the protochlorophyll-protein complex, which is not readily accessible to aqueous reagents.

This model of the protochlorophyll holochrome would appear to be consistent with the findings of Virgin (52) and Madsen (49) (Section III, B, 3) that the transformation of protochlorophyll is not dependent on a nonphotochemical reaction of long life.

3. KINETICS

From their experiments with whole leaves Smith and Benitez (51) proposed that the conversion of protochlorophyll to chlorophyll *a* involved a bimolecular reaction with respect to protochlorophyll, but this seems unlikely in view of the low protochlorophyll content of purified preparations of the holochrome, and the independence of the rate of transformation on the viscosity of the medium and the initial concentration of the holochrome. If there were one protochlorophyll molecule per protein molecule, it might be expected that the transformation would obey first-order kinetics since protochlorophyll is the photoreceptor for its own conversion. However, the transformation does not obey first-order kinetics either in the leaf or in the isolated holochrome.

The kinetic data obtained by Boardman (107) for the transformation in the isolated holochrome were fitted to equations derived on the assumption that the overall reaction was composed of two first-order reactions. Boardman (107) proposed the following alternative hypotheses.

1. The protochlorophyll holochrome is a mixture of two types of protochlorophyll-protein complexes, each of which contains one protochlorophyll molecule per protein molecule.

2(a). There are two protochlorophyll molecules per protein molecule, and the rate of transformation of one protochlorophyll is different from the rate of transformation of the second protochlorophyll, owing either to some interaction between the protochlorophyll molecules or to different methods of binding to the protein.

2(b). The protochlorophyll holochrome has two protochlorophyll molecules per protein molecule, but conversion of the first protochlorophyll causes some conformational change in the protein molecule which influences the rate of transformation of the second protochlorophyll.

The following equations were derived on the assumption that both forms of protochlorophyll are transformed to chlorophyll a by first-order reactions. They show the dependence of the degree of transformation, T (expressed as a percentage of convertible protochlorophyll), on the time of illumination (t) and the first-order rate constants k_1 and k_2.

$$T = 100 \left[1 - Ae^{-k_1 t} - (1 - A) e^{-k_2 t} \right] \tag{1}$$

where A is the fraction of active molecules of type 1 in the mixture and $(1 - A)$ the fraction of type 2.

$$T = 50 \left(2 - e^{-k_1 t} - e^{-k_2 t} \right) \tag{2a}$$

$$T = \frac{50}{k_1 - k_2} \left[(k_1 - 2 k_2) (1 - e^{-k_1 t}) + k_1 (1 - e^{-k_2 t}) \right] \tag{2b}$$

Rate constants were calculated by fitting Eqs. 2a and 2b to experimental curves. Examples are given in the legend to Fig. 15. It was not possible from the kinetic data to decide between the various models. Rate constants determined at a number of temperatures enable a calculation of activation energies. Activation energies obtained for the conversion of one form of protochlorophyll (2400–5900 cal, depending on the temperature range) were low compared with those usually found for an enzymatically activated collision process, but those obtained for the second form (5700–11,900) were within the range found for enzymatic reactions.

4. Transformation in Ultraviolet Light

The studies of McLeod and Coomber (*108*) on the effect of ultraviolet light on the transformation of protochlorophyll support the hypothesis that not all the active protochlorophyll is bound to the protein in the same way (Fig. 16). Only 20–30% of the protochlorophyll was transformed by light of wavelength 250–300 mμ, a result suggesting that this amount of pigment was attached to an aromatic amino acid residue, which adsorbed at this wavelength and transferred its energy to protochlorophyll. The incomplete conversion produced by ultraviolet light could be completed by subsequent exposure to visible light.

5. The Hydrogen Donor

The protochlorophyll holochrome may be dialyzed without loss of activity (*87, 91*), thus indicating that the transformation is not dependent on a dialyzable cofactor.

Boardman (*107*) studied the inhibitory effect of a number of reagents on the transformation of protochlorophyll in an attempt to determine the nature of the hydrogen- or electron-donor. The transformation was not inhibited by the metal ion chelators ethylenediamine tetraacetate (EDTA) or 8-hydroxyquinoline, by a variety of metal ions, by the thiol-combining reagents, *p*-chloromercuribenzoate, *p*-chloromercuriphenyl sulfonate, and *N*-ethyl maleimide, by a strong reducing agent such as sodium dithionite, or by sodium arsenite, a known inhibitor of lipoic acid. Of the reagents

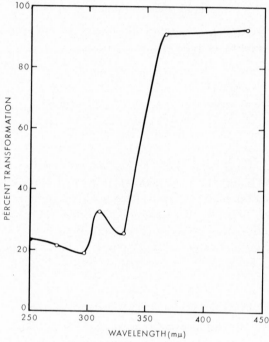

FIG. 16. Transformation of the protochlorophyll holochrome in ultraviolet light (*108*).

tested, the only ones which inhibited the conversion were protein-denaturing agents, such as 6 *M* urea, thioethanol at concentrations above 0.5 *M*, and sodium dodecyl sulfate.

In separate experiments the protochlorophyll holochrome was incubated with acetaldehyde and yeast alcohol dehydrogenase, with oxidized glutathione and glutathione reductase, and with a pyridine nucleotide pyrophosphatase, but in no instance was the subsequent transformation impaired compared with control incubations which did not contain enzyme.

These experiments indicate that freely accessible reduced pyridine nucleotides or thiol groups are not the hydrogen donor, but they do not exclude the possibility that a pyridine nucleotide or thiol group bound to the protein in some inaccessible region is the hydrogen donor. Smith (71) in his studies with etiolated barley leaves showed that it is unlikely that water is the hydrogen donor.

Röbbelen (109) studied the conversion of protochlorophyll to chlorophyll in a mutant of *Arabidopsis thaliana*. He claimed to have obtained spectrophotometric evidence which suggested the participation of reduced pyridine nucleotide in the conversion.

F. Bonding between Protochlorophyll and Protein

1. COORDINATION PROPERTIES OF MAGNESIUM

The magnesium atom may attain an inert gas configuration by the loss of two electrons; hence its primary tendency is to form electrostatic linear complexes with other ions. In many of its chemical properties magnesium resembles the atoms of group 2B, Zn and Cd, which usually have a coordination number of 4 and prefer a tetrahedral distribution of ligands (110). In protochlorophyll, however, magnesium is coordinated with the square-planar porphyrin nucleus, which distorts the usual magnesium structure. Several compounds of magnesium are known in which magnesium shows a coordination number of 6 (110).

2. COORDINATION PROPERTIES OF METALLOPORPHYRINS

In discussing the coordination properties of protochlorophyll, it is relevant to consider some general aspects of the coordination chemistry of porphyrins and metalloporphyrins and, in particular, the correlations which exist between the nature of the porphyrin side chains, the stabilities of the metal complexes, and the positions of the absorption bands (111).

The square-planar chelates of porphyrins with divalent metal ions have a Soret and two visible absorption bands. On further coordination of many of the metalloporphyrins (e.g., those of Zn and Cd) with one or two extra ligands, there is a shift of absorption bands to longer wavelengths. Increasing the electron-attracting power of the porphyrin side chains causes shifts to longer wavelengths both in the free porphyrins and the metalloporphyrins; for example, the absorption bands of protoporphyrin, which has vinyl substituents at positions 2 and 4, are shifted about 10 mμ to longer wavelengths compared with the corresponding bands of mesoporphyrin, which has ethyl substituents. The absorption bands of pheoporphyrin a_5, the porphyrin derivative of protochlorophyllide, are at longer wavelengths than those of protoporphyrin, owing apparently to an interaction of the isocyclic ring (ring V) with the aromatic porphyrin system.

Increasing the electron attracting power of the porphyrin side chains decreases the electron density at the ring nitrogen atoms. This leads, in the zinc and magnesium porphyrins, to greater affinities for the extra donor ligands and to greater stabilities of the complexes so formed.

3. FURTHER COORDINATION OF MAGNESIUM PORPHYRINS

By the use of spectrophotometric techniques, it was shown (112) that magnesium tetraphenyl porphin can form complexes with pyridine. At the lower concentrations of pyridine, one pyridine was added to the magnesium porphyrin to form a complex with a fairly high association constant, and at higher concentrations of pyridine a second ligand was added with a low association constant. The pyridine ligands are added perpendicularly to the porphyrin plane; one ligand gives a pentacoordinate complex which is essentially square pyrimidal, while the addition of two ligands gives a distorted octahedron. Dipyridine complexes of magnesium porphyrins have been reported also by Wei et al. (113).

TABLE IV

STABILITY CONSTANT VALUES FOR CHLOROPHYLL a IN BENZENE

Ligand	Log K_{s1}[a]
Dimethylaniline	+1.02
Phenol	+1.19
Aniline	+1.66
Benzyl alcohol	+3.46
Octyl alcohol	+3.66
Quinoline	+4.12
Benzylamine	+4.43
Water	+4.47
n-Heptylamine	+5.19

[a] Calculated (117) from data of Livingston et al. (114), on the assumption that one ligand molecule combines with one porphyrin molecule.

The fluorescence studies of Livingston et al. (114, 115) provide good evidence for the further coordination of magnesium in the chlorophylls. The chlorophylls were nonfluorescent in anhydrous benzene, but the addition of small amounts of various activators restored the fluorescence to the values observed in polar solvents. It was originally postulated that the activators were hydrogen bonded to the carbonyl group of ring V, but Evstigneev et al. (116) found that the magnesium-free pheophytins did not change in fluorescence as did the chlorophylls. On the assumption that the fluorescent species and the spectroscopic changes were due to the further coordination of a molecule of activator with the magnesium, stability constants were calculated (111, 117) from the activator concentrations (*114*) necessary to restore the fluorescence in pure benzene to half its limiting value (Table IV). The stabilities of the complexes increase with the electron-donor power of the activators.

From infrared and nuclear magnetic resonance spectral studies of the chlorophylls, it was postulated (*118, 119*) (cf. Chapter 7) that the chlorophylls are fairly strongly aggregated in anhydrous benzene because of the specific interaction between the carbonyl group of one molecule (in chlorophyll *a*, the carbonyl of ring V) and the magnesium atom of a second molecule. The interaction is readily broken by the addition of electronegative bases, which activate the fluorescence.

4. HYDROGEN BONDING AND FURTHER COORDINATION IN THE HOLO-CHROME

In the protochlorophyll and protochlorophyllide holochromes, it is assumed that the protochlorophyll and protochlorophyllide molecules are bonded to amino acid side chains in the protein molecules by noncovalent forces. The visible absorption bands of protochlorophyll in the holochrome are displaced about 10 mμ to longer wavelengths compared with the corresponding bands of protochlorophyll in organic solvents, and the bands of protochlorophyllide holochrome are displaced a further 15 mμ beyond the bands of protochlorophyll holochrome. On the basis of the coordination properties of the metalloporphyrins discussed above, it might be expected that the spectral properties of protochlorophyll would be influenced by protein binding to the carbonyl group of ring V and to the magnesium. Hydrogen bonding or coulombic interaction may occur between the carboxylic group at position 7 in protochlorophyllide and appropriate amino acid side chains, but this would not be expected to influence to any extent the electron density of the porphyrin ring and therefore the spectral properties.

If protochlorophyll and protochlorophyllide exist predominantly in the keto form, as does chlorophyll *a* (*117*), hydrogen bonding of a protein side chain to the isocyclic carbonyl group would tend to attract electrons away from the isocyclic ring and might lead to a spectral shift to longer wavelength. At the same time, one would expect an increased affinity of the magnesium for a protein ligand group, which in itself would cause a spectral shift to longer wavelengths, and give greater stability to the protochlorophyll-protein complex. Protein side chain substituents which have the potential ability to form hydrogen bonds with the keto group in ring V are the phenolic hydroxyl group of tyrosine and the amide groups of glutamine and asparagine. Strong hydrogen bonding between the keto group and the hydroxyl group of serine or threonine, or the ε-amino group of lysine seems unlikely. Because of its high pK, it is expected that the ε-amino group would be positively charged at neutral pH.

Potential electron donor groups for coordination to the magnesium are the phenolic group of tyrosine (phenol itself is a weak activator of chlorophyll fluorescence), the imino group in the imidazole ring of histidine, the ε-amino group of lysine, and the thiol group of cysteine.

Experimental evidence exists for believing that in 20–30% of the holochrome molecules there is an intimate association of a tyrosine side chain with the protochlorophyll or protochlorophyllide (see Section IV, E, 4). Smith and Coomber (120) have suggested from their studies on the effect of pH on the inactivation of the protochlorophyll holochrome that tyrosine, lysine, or cysteine may be involved in the binding of protochlorophyll to protein. The influence of pH on the ultraviolet spectrum of the holochrome was reported (120), but not the effect of pH on the protochlorophyll spectrum.

The binding between protochlorophyllide and protein apparently can be modified somewhat without loss of the ability to be transformed. *In vivo*, a spectral shift to a shorter wavelength occurs on phytylation of the protochlorophyllide holochrome, and, at least in barley seedlings, the phytylated pigment is transformed on illumination. Similar spectral shifts are observed *in vitro*, but these are not due to phytylation of protochlorophyllide. The modified protochlorophyllide holochrome may be transformed to the corresponding chlorophyllide *a*-protein complex, provided the modification is not too great. More drastic treatments of the protochlorophyllide holochrome result in a complete loss of activity.

REFERENCES

(1) N. A. Monteverde, *Acta Horti Petropolitani* **13**, 201 (1893-1894).

(2) K. Noack and W. Kiessling, *Z. Physiol. Chem.* **182**, 13 (1929).

(3) S. Granick and D. Mauzerall, *in* "Metabolic Pathways" (D. M. Greenberg, ed.), Vol. 2, p. 525. Academic Press, New York, 1961.

(4) S. Granick, *J. Biol. Chem.* **183**, 713 (1950).

(5) G. W. Bryan and L. Bogorad, *in* "Studies on Microalgae and Photosynthetic Bacteria" (Japan. Soc. Plant Physiol., eds.), p. 399. Univ. of Tokyo Press, Tokyo, 1963.

(6) M. Nishimura and H. Huzisige, *J. Biochem.* **46**, 225 (1959).

(7) Z. Hejnowicz, *Physiol. Plantarum* **11**, 878 (1958).

(8) L. O. Björn, *Physiol. Plantarum* **16**, 142 (1963).

(9) A. A. Krasnovsky, *Ann. Rev. Plant Physiol.* **11**, 363 (1960).

(10) R. Y. Stanier and J. H. C. Smith, *Carnegie Inst. Wash. Year Book* **58**, 336 (1959).

(11) J. H. C. Smith, *in* "Comparative Biochemistry of Photoreactive Systems" (M. B. Allen, ed.), p. 257. Academic Press, New York, 1960.

(12) O. T. G. Jones, *Biochem. J.* **89**, 182 (1962).

(13) J. H. C. Smith and V. M. K. Young, *in* "Radiation Biology" (A. Hollaender, ed.), Vol. III, p. 393. McGraw-Hill, New York, 1956.

(*14*) H. Fischer and A. Stern, "Die Chemie des Pyrrols," Vol. 2, Part II, Akad. Verlagsges., Leipzig, 1940.

(*15*) H. Fischer, *Naturwissenschaften* **28**, 401 (1940).

(*16*) H. Fischer and A. Oestreicher, *Z. Physiol. Chem.* **262**, 243 (1940).

(*17*) R. B. Woodward, W. A. Ayer, J. M. Beaton, F. Bickelhaupt, R. Bonnett, P. Buchschacher, G. L. Closs, H. Dutler, J. Hannah, F. P. Hauck, S. Ito, A. Langemann, E. Le Goff, W. Leimgruber, W. Lwowski, J. Sauer, Z. Valenta, and H. Volz, *J. Am. Chem. Soc.* **82**, 3800 (1960).

(*18*) J. E. Loeffler, *Carnegie Inst. Wash. Year Book* **54**, 159 (1955).

(*19*) J. B. Wolff and L. Price, *Arch. Biochem. Biophys.* **72**, 293 (1957).

(*20*) C. Sironval, M. R. Michel-Wolwertz, and A. Madsen, *Biochim. Biophys. Acta* **94**, 344 (1965).

(*21*) F. Fischer and W. J. Rudiger, *Ann. Chem.* **627**, 35 (1959).

(*22*) T. N. Godnev, J. L. Kaler, and R. M. Rotfarb, *Dokl. Akad. Nauk SSSR* **140**, 1445 (1961).

(*23*) V. M. Koski and J. H. C. Smith, *J. Am. Chem. Soc.* **70**, 3558 (1948).

(*24*) J. H. C. Smith and A. Benitez, *in* "Moderne Methoden der Pflanzenanalyse" (K. Paech and M. V. Tracey, eds.), p. 142. Springer, Berlin, 1955.

(*25*) J. M. Anderson and N. K. Boardman, *Australian J. Biol. Sci.* **17**, 93 (1964).

(*26*) C. S. French, *in* "Handbuch der Pflanzenphysiologie" (W. Ruhland, ed.), Vol. 5, Part 1, p. 282. Springer, Berlin, 1960.

(*27*) C. S. French, J. H. C. Smith, H. I. Virgin, and R. L. Airth, *Plant Physiol.* **31**, 369 (1956).

(*28*) V. M. Koski, *Arch. Biochem.* **29**, 339 (1950).

(*29*) G. Mackinney, *J. Biol. Chem.* **140**, 315 (1941).

(*30*) V. L. Kaler and A. A. Shlyk, *Biokhimiya* **27**, 599 (1962).

(*31*) H. I. Virgin, *Physiol. Plantarum* **13**, 155 (1960).

(*32*) J. M. Anderson, unpublished observations (1963).

(*33*) N. K. Boardman, unpublished observations (1963).

(*34*) A. F. H. Anderson and M. Calvin, *Nature* **194**, 285 (1962).

(*35*) A. Seybold, *Planta* **26**, 712 (1937).

(*36*) A. Seybold and K. Egle, *Planta* **29**, 119 (1938).

(*37*) A. Seybold, *Planta* **36**, 371 (1948).

(*38*) A. A. Shlyk, V. L. Kaler, L. I. Vlasenok, and V. I. Gaponenko, *Photochem. Photobiol.* **2**, 129 (1963).

(*39*) A. A. Krasnovsky and K. K. Voinovskaya, *Dokl. Akad. Nauk SSSR* **66**, 663 (1949) [see *Chem. Abstr.* **43**, 7092 (1949)].

(*40*) O. T. G. Jones, *Biochem. J.* **96**, 6P (1965).

(*41*) O. T. G. Jones, private communication (1965).

(*42*) E. G. Sudyina, *Photochem. Photobiol.* **2**, 181 (1963).

(*43*) E. G. Sudyina and G. G. Lozovaya, *Ukr. Botan. Zh.* **18**, No. 5 (1961).

(*44*) R. Hill, J. H. C. Smith, and C. S. French, *Carnegie Inst. Wash. Year Book* **52**, 153 (1953).

(*45*) K. Shibata, *J. Biochem.* **44**, 147 (1957).

(*46*) F. F. Litvin and A. A. Krasnovsky, *Dokl. Akad. Nauk SSSR* **117**, 106 (1957).

(*46a*) A. A. Krasnovsky and M. I. Bystrova, *Biokhimiya* **25**, 168 (1960).

(*47*) T. N. Godnev, R. M. Rotfarb, and N. K. Akulovich, *Photochem. Photobiol.* **2**, 119 (1963).

(*48*) J. H. C. Smith, L. J. Durham, and C. F. Wurster, *Plant Physiol.* **34**, 340 (1959).

(*49*) A. Madsen, *Photochem. Photobiol.* **2**, 93 (1963).

(50) V. M. Koski, C. S. French, and J. H. C. Smith, *Arch. Biochem. Biophys.* **31**, 1 (1951).
(51) J. H. C. Smith and A. Benitez, *Plant Physiol.* **29**, 135 (1954).
(52) H. I. Virgin, *Physiol. Plantarum* **8**, 389 (1955).
(53) N. K. Boardman and S. G. Wildman, *Biochim. Biophys. Acta* **59**, 222 (1962).
(54) N. K. Boardman and J. M. Anderson, *Australian J. Biol. Sci.* **17**, 86 (1964).
(55) R. Hageman, *Biol. Zentr.* **79**, 393 (1960) [see *Biol. Abstr.* **36**, 73445 (1961)].
(56) A. Gibor and S. Granick *J. Protozool.* **9**, 327 (1962).
(57) S. Klein and A. Poljakoff-Mayber, *J. Biophys. Biochem. Cytol.* **11**, 433 (1961).
(58) D. von Wettstein, *Brookhaven Symp. Biol.* **11**, 138 (1959).
(59) D. von Wettstein and A. Kahn, *Proc. 2nd Reg. Conf. (Eur.) Electron Microscopy, Delft, 1960* Vol. 2, p. 1051 (A. L. Houwink and B. J. Spit, eds.) Nederl. Ver. v. Electronenmicroscopie, Delft, 1960.
(60) Y. Eilam and S. Klein, *J. Cell Biol.* **14**, 169 (1962).
(61) A. J. Hodge, J. D. McLean, and F. V. Mercer, *J. Biophys. Biochem. Cytol.* **2**, 597 (1956).
(62) S. Klein, *Nature* **196**, 992 (1962).
(63) J. H. C. Smith, *Brookhaven Symp. Biol.* **11**, 296 (1959).
(64) H. I. Virgin, A. Kahn, and D. von Wettstein, *Photochem. Photobiol.* **2**, 83 (1963).
(65) H. I. Virgin, *Physiol. Plantarum* **8**, 630 (1955).
(66) H. I. Virgin, *Physiol. Plantarum* **14**, 384 (1961).
(67) F. F. Litvin, A. A. Krasnovsky, and G. T. Rikhireva, *Dokl. Akad. Nauk SSSR* **127**, 699 (1959).
(68) J. I. Liro, *Ann. Acad. Sci. Fennicae: Ser. A I* **1**, 1 (1909).
(69) G. Blaauw-Jansen, J. G. Kamen, and J. B. Thomas, *Biochim. Biophys. Acta* **5**, 179 (1950).
(70) J. C. Goedheer, *Biochim. Biophys. Acta* **51**, 494 (1961).
(71) J. H. C. Smith, *Plant Physiol.* **29**, 143 (1954).
(72) E. K. Gabrielsen, A. Madsen, and K. Vejlby, *Physiol. Plantarum* **14**, 98 (1961).
(73) K. Egle, in "Handbuch der Pflanzenphysiologie" (W. Ruhland, ed.), Vol. 5, Part 1, p. 323. Springer, Berlin, 1960.
(74) W. L. Butler, *Arch. Biochem. Biophys.* **92**, 287 (1961).
(75) R. B. Withrow, J. B. Wolff, and L. Price, *Plant Physiol.* **31**, Suppl., xiii (1956).
(76) H. I. Virgin, *Physiol. Plantarum* **10**, 445 (1957).
(77) H. I. Virgin, *Physiol. Plantarum* **11**, 347 (1958).
(78) K. Mitrakos, *Physiol. Plantarum* **14**, 497 (1961).
(79) L. Bogorad and A. B. Jacobson, *Biochem. Biophys. Res. Commun.* **14**, 113 (1964).
(80) M. M. Margulies, *Plant Physiol.* **37**, 473 (1962).
(81) L. Price and W. H. Klein, *Plant Physiol.* **36**, 733 (1961).
(82) H. I. Virgin, *Physiol. Plantarum* **14**, 439 (1961).
(83) A. Schmidt, *Botan. Arch.* **5**, 260 (1924).
(84) L. Bogorad, *Botan. Gaz.* **111**, 221 (1950).
(85) A. A. Krasnovsky and L. M. Kosobutskaya, *Dokl. Akad. Nauk SSSR* **85**, 177 (1952).
(86) J. H. C. Smith and A. Benitez, *Carnegie Inst. Wash. Year Book* **52**, 151 (1953).

(87) J. H. C. Smith, D. W. Kupke, J. E. Loeffler, A. Benitez, I. Ahrne, and A. T. Giese in "Research in Photosynthesis" (H. Gaffron *et al.*, eds.), p. 464. Wiley (Interscience), New York, 1957.

(88) A. A. Krasnovsky and M. I. Bystrova, *Biokhimiya* **25**, 168 (1960).

(89) J. H. C. Smith and D. W. Kupke, *Nature* **178**, 751 (1956).

(90) J. H. C. Smith, *Carnegie Inst. Wash. Year Book* **57**, 287 (1958).

(91) N. K. Boardman, *Biochim. Biophys. Acta* **62**, 63 (1962).

(92) J. H. C. Smith and J. Coomber, *Carnegie Inst. Wash. Year Book* **58**, 331 (1959).

(93) D. W. Kupke, *J. Biol. Chem.* **237**, 3287 (1962).

(94) V. M. Koski and J. H. C. Smith, *Arch. Biochem. Biophys.* **34**, 189 (1951).

(94a) N. K. Boardman, unpublished observations (1963).

(95) J. H. C. Smith, K. Shibata, and R. W. Hart, *Arch. Biochem. Biophys.* **72**, 457 (1957).

(96) H. J. Trurnit and G. Colmano, *Biochim. Biophys. Acta* **31**, 434 (1959).

(96a) J. C. Goedheer and J. H. C. Smith, *Carnegie Inst. Wash. Year Book* **58**, 334 (1959).

(97) S. J. Singer, L. Eggman, J. M. Campbell, and S. G. Wildman, *J. Biol. Chem.* **197**, 233 (1952).

(98) L. Eggman, S. J. Singer, and S. G. Wildman, *J. Biol. Chem.* **205**, 969 (1953).

(99) J. W. Lyttleton, *Biochem. J.* **64**, 70 (1956).

(100) J. W. Lyttleton and P. O. P. Ts'o, *Arch. Biochem. Biophys.* **73**, 120 (1958).

(101) P. W. Trown, *Biochemistry* **4**, 908 (1965).

(102) R. B. Park and N. G. Pon, *J. Mol. Biol.* **3**, 1 (1961).

(103) G. Van Noort, W. Hudson, and S .G. Wildman, *Plant Physiol.* **36**, Suppl., xix (1961).

(104) L. Mendiola and T. Akazawa, *Biochemistry* **3**, 174 (1964).

(105) G. Van Noort and S. G. Wildman, *Biochim. Biophys. Acta* **90**, 309 (1964).

(106) J. H. C. Smith and C. S. French, *Carnegie Inst. Wash. Year Book* **57**, 290 (1958).

(107) N. K. Boardman, *Biochim. Biophys. Acta* **64**, 279 (1962).

(108) G. C. McLeod and J. Coomber, *Carnegie Inst. Wash. Year Book* **59**, 324 (1960).

(109) G. Röbbelen, *Planta* **47**, 532 (1956).

(110) N. V. Sidgwick, "The Chemical Elements," Vol. 1, p. 241. Oxford Univ. Press, London and New York, 1950.

(111) J. E. Falk, "Porphyrins and Metalloporphyrins." Elsevier, Amsterdam, 1964.

(112) J. R. Miller and G. D. Dorough, *J. Am. Chem. Soc.* **74**, 3977 (1952).

(113) P. E. Wei, A. H. Corwin, and R. Arellano, *J. Org. Chem.* **27**, 3344 (1962).

(114) R. Livingston, W. F. Watson, and J. McArdle, *J. Am. Chem. Soc.* **71**, 1542 (1949).

(115) R. Livingston, *Quart. Rev. (London)* **14**, 174 (1960).

(116) V. B. Evstigneev, V. A. Gavrilova, and A. A. Krasnovsky, *Dokl. Akad. Nauk SSSR* **70**, 261 (1960).

(117) J. N. Phillips, *Rev. Pure Appl. Chem.* **10**, 35 (1960).

(118) J. J. Katz, G. L. Closs, F. C. Pennington, M. R. Thomas, and H. H. Strain, *J. Am. Chem. Soc.* **85**, 3801 (1963).

(119). G. L. Closs, J. J. Katz, F. C. Pennington, M. R. Thomas, and H. H. Strain, *J. Am. Chem. Soc.* **85**, 3809 (1963).

(120) J. H. C. Smith and J. Coomber, *Carnegie Inst. Wash. Year Book* **59**, 325 (1960).

—15—

The Biosynthesis of Chlorophylls

LAWRENCE BOGORAD*

Department of Botany
The University of Chicago, Chicago, Illinois

I. Pathways in the Formation of Chlorophylls

The sequence of reactions shown in Fig. 1 summarizes knowledge and conjecture regarding the biosynthesis of protochlorophyllide *a* (PChlide). The formation of chlorophyllide *a* (Chlide *a*) from this compound is discussed by Boardman in Chapter 14.

The evidence which supports or suggests each of these steps in porphyrin biosynthesis has been obtained in a variety of types of experiments including (a) tracer studies with whole animals or with animal cells *in vitro;* (b) observations on the production of various porphyrins by photosynthetic bacteria cultured under a variety of conditions; (c) identification of porphyrins formed by algal mutants in which chlorophyll forma-

* Research Career Awardee of the National Institute of General Medical Sciences, United States Public Health Service.

tion is blocked; and (d) demonstrations of individual enzymatic reactions using cell-free preparations or purified enzymes. Data from chlorophyll-forming organisms will be emphasized in discussions of each step in porphyrin biosynthesis; in some cases, however, information from experiments with such organisms is not available.

A. The Formation of δ-Aminolevulinic Acid

In 1945 Block and Rittenberg (1) found that deuteroheme was produced by rats which had been fed deuterium-labeled acetate. The following year Shemin and Rittenberg (2) described experiments designed to determine the source of nitrogen in heme: a number of amino acids labeled with ^{15}N were fed to rats and the extent of dilution of the label in the heme of the animals was determined. [15]N from glycine was clearly used most directly, judging from the dilution of the isotope in the porphyrin.

A series of tracer experiments performed during the next few years, mostly using suspensions of avian red cells, demonstrated that the α-carbon atom of glycine, as well as the nitrogen atom, is incorporated into heme, but the carboxyl carbon of the amino acid is not (3–5); eight α-carbon atoms from glycine molecules are incorporated into heme per four nitrogen atoms from this source—one carbon atom in each pyrrole ring and the carbon atoms which form the bridges between pyrrole rings arise from the α-carbon atom of glycine (5–7); the remaining carbon atoms of heme are derived from acetate (8).

Lemberg and Legge (9) have reviewed earlier, but then current, views that a porphyrin such as protoporphyrin (Proto), which has two pyrrole rings substituted with a methyl and a vinyl side chain each and two pyrrole rings with propionic acid and methyl substituents, might arise by the independent formation of two distinctly different pyrroles which would finally be condensed enzymatically to form the biologically correct Proto, i.e., isomer IX. Some of the tracer experiments (10, 11) established that the nitrogen atoms of all four pyrrole rings were derived from glycine—thus arguing against independent origins for the different types of pyrroles in Proto—and then Shemin and Wittenberg (12) analyzed, virtually carbon atom-by-atom, Proto formed by avian red cells incubated with either carboxyl- or methyl-labeled acetate [the acetate entered into the biosynthesis of heme as succinate, as was shown conclusively by Shemin and Kumin in 1952 (13)]. From the pattern of incorporation of labeled atoms they concluded that all four pyrrole rings of Proto are derived from a common pyrrole which has one acetic and one propionic acid side chain.

Shemin and Russel (14) visualized that succinate (or succinyl CoA) and glycine might condense to form α-amino β-keto adipic acid which, after loss of the carboxyl group derived from glycine, would give rise to δ-aminolevulinic acid (ALA). The latter molecule would then contain all the carbon atoms from the sources known to contribute to the formation of heme. ^{14}C-ALA was synthesized and shown to serve as the sole source of all the carbon and nitrogen atoms for the synthesis of Proto by the same avian red cell system which Shemin's group had been using in their earlier tracer experiments with glycine and succinate. These experiments were confirmed by Neuberger and Scott (15).

Pyridoxal phosphate is a cofactor for ALA synthetase, the enzyme which catalyzes the synthesis of ALA from succinyl CoA and glycine (16, 17). The reaction is inhibited by aminomalonate and by substances which complex with the aldehydic group of pyridoxal phosphate such as l-cysteine, cyanide, and penicillamine (17, 18). Burnham and Lascelles (19) have shown that ALA synthetase from Rhodopseudomonas spheroides is inhibited by Protohemin (ferric protoporphyrin IX) and iron complexes of other porphyrins. Cobalt, copper, zinc, manganese, and magnesium complexes of Proto as well as Proto and some hemoproteins were also inhibitory but considerably less effective. Protoheme (ferrous protoporphyrin IX) was about as effective an inhibitor as the ferric compound: 0.2 mM Protohemin inhibited ALA synthetase about 87% but significant inhibition was displayed by concentrations as low as 0.1 μM. Burnham and Lascelles (19) have suggested that Protoheme (or hemin) might function as a negative feedback inhibitor in regulating porphyrin biosynthesis.

The purest preparations of ALA synthetase have been made from extracts of the photosynthetic bacterium Rhodopseudomonas spheroides; Kikuchi et al. (16) obtained an eightyfold purified, and Burnham (20) a twentyfold purified, preparation. α-Amino-β-ketoadipic acid, the expected product of the condensation of succinyl CoA and glycine, has not been detected during the formation of ALA; however, Laver et al. (21) report that at pH 7 in aqueous solution spontaneous decarboxylation to δ-aminolevulinic acid is very rapid (α-amino-β-ketoadipic acid has a half-life of less than 1 minute under these conditions).

The paper by Shemin and Russel (14), in which the involvement of ALA in porphyrin synthesis was first described, included the proposal that deamination of ALA to α-ketoglutaraldehyde (γ,δ-dioxovaleric acid) might be the first step in a cycle leading to the contribution of one-carbon units which might be used, e.g., in purine metabolism and to the regeneration of succinate.

Support for this proposal came from Nemeth *et al.* (22), who studied the fate of ALA-5-^{14}C, or ALA-1,4-^{14}C, and α-ketoglutaraldehyde-5-^{14}C in ducks, pigeons, rats, and duck blood. The δ-carbon atom of ALA had the same metabolic fate as the α-carbon atom of glycine—i.e., appearing as the ureido groups of guanine, in uric acid, and as formate—while the succinyl moiety of ALA was converted to succinate. That no conversion of α-ketoglutaraldehyde to ALA could be detected indicated that the deamination was irreversible or at least was not favored. ALA transaminase, an enzyme which catalyzes the transfer of amino groups to α-keto acids has been studied in homogenates and extracts of mammalian tissues by Kowalski *et al.* (23, 24), in cell-free preparations of *Corynebacterium diphtheriae* by Bagdasarian (25), and in cell-free preparations of *R. spheroides* by Neuberger and Turner (26). The enzyme studied by Kowalski *et al.* (24) catalyzes the transfer of amino groups from ALA to pyruvate or α-oxoglutarate (to form α-ketoglutaraldehyde, alanine, and glutamate, respectively); with some clear exceptions, the enzyme from most tissues seemed to favor α-oxoglutarate over pyruvate as an amino acceptor. The equilibrium for the *R. spheroides* enzyme is well toward ALA formation; in addition, L-α-alanine and β-alanine are 5–7 times more effective as donors of amino groups to α-ketoglutaraldehyde than is glutamate (18, 26).

ALA synthetase has not yet been demonstrated in green plants, including algae, but ALA transaminase activity measured as ALA formation has been found in broken-cell preparations from *Chlorella* (27). The relative importance of these two possible enzymatic routes to ALA synthesis in green plants remains to be established.

B. Porphobilinogen: Its Utilization and Formation

Porphobilinogen (PBG) is a monopyrrole first isolated by Westall (28) from the urine of individuals afflicted with acute porphyria, a disease inherited as a Mendelian recessive trait. The trivial name of this pyrrole developed in the following way: Urine of acute porphyriacs is generally normal in color when freshly passed but it becomes red (to the color of port wine) on standing at room temperature and especially in the light. The red color is mostly attributable to porphobilin, an as yet uncharacterized compound which is probably a pyrrole polymer. In 1931, Sachs (29) observed that fresh urine of acute porphyria sufferers gave a strongly positive reaction (a red color) with *p*-dimethylaminobenzaldehyde in acid (Ehrlich's reagent) whereas the same urine after standing long enough to form large amounts of porphobilin failed to react. The precursor of the porphobilin, which gave the strong positive Ehrlich reagent, was consequently named porphobilinogen.

The Ehrlich reaction is characteristic of pyrroles with unsubstituted α-positions (the ring carbons attached to the pyrrole nitrogen). Also, in addition to porphobilin some porphyrins were later detected in acute porphyria urines. These factors led to great interest in the chemical nature of PBG.

After PBG was obtained in crystalline form in 1952 it was shown that broken-cell preparations of *Chlorella* (30) and of avian erythrocytes (31) could synthesize a number of porphyrins, including Proto, when PBG was the only substrate provided.

On the other hand, once the role of ALA was established, it was quite clear that PBG itself could be formed by the condensation of two molecules of ALA.

The enzyme ALA dehydrase, which catalyzes the condensation of two molecules of ALA to form one of PBG, with the removal of two molecules of water, has not been purified from plants but has been shown to be present in a number of species: ALA is converted to PBG by extracts of *Chlorella* and of spinach (32); protochlorophyllide accumulates in etiolated barley (33, 34) or etiolated bean leaves (35, 36) supplied with ALA. The most highly purified enzyme preparation studied to date (270-fold purification) was isolated from ox liver (37). ALA dehydrase has also been purified from avian erythrocytes (32, 38), rabbit reticulocytes (39), beef liver (40), and *Rhodopseudomonas spheroides* (19), but not from green plants. The enzyme, regardless of source, is activated by sulfhydryl compounds such as glutathione or cysteine. Potassium ions are required for activation of ALA dehydrase from *R. spheroides;* this enzyme is inhibited by Protohemin but much less than is ALA synthetase.

Uroporphyrinogens (Urogens), porphyrinogens whose constituent pyrroles each carry one propionic and one acetic acid side chain, are formed by the condensation of four molecules of PBG. Of the four possible isomers of Urogen, only isomers I and III are known biologically (Fig. 1). The latter is a precursor of Proto IX and presumably of the chlorophylls.

Uro I (Urogen I — 6H → Uro I) is found in humans and other mammals with the hereditary disease congenital porphyria, which is transmitted as a Mendelian dominant. It also occurs in the shell of the pearl oyster, *Pinctada vulgaris* (41), and in a few other organisms but has not been reported to occur in plants. However, it and the III isomer, as well as other porphyrins, are produced by broken-cell preparations of *Chlorella* incubated with PBG. When the cell-free preparations are heated to 55° for 30 minutes before incubation (30) only isomer I is produced.

Urogen I synthetase has been partially purified from extracts of spinach leaf acetone powder (42, 43) and from *R. spheroides* (44, 45).

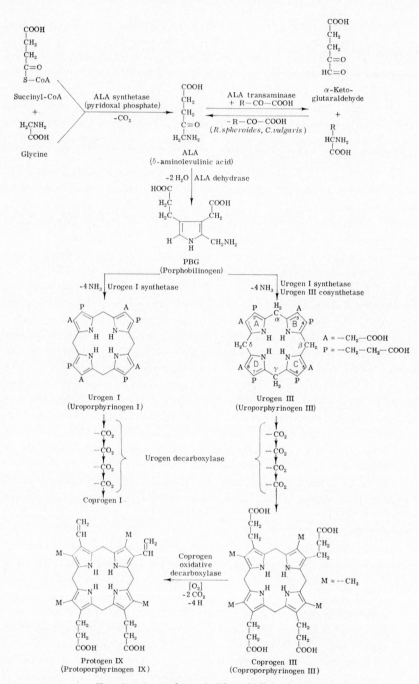

Fig. 1. An outline of chlorophyll biogenesis.

FIG. 1 (*Continued*)

Urogen I can be visualized as forming by the sequential head to tail addition, with the elimination of a molecule of ammonia at each condensation, of four molecules of PBG and the cyclization of the open-chain tetrapyrrole. Another possible mode of action of this enzyme, the formation of dipyrrylmethanes followed by the condensation of pairs of these, now appears to be eliminated. When Urogen I synthetase is incubated with a synthetic dipyrrylmethane equivalent to one-half a molecule of Urogen I but with a free aminomethyl group, i.e., a dipyrrylmethane

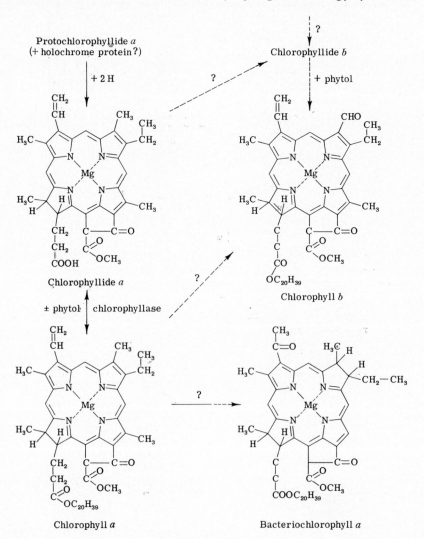

Fig. 1 (*Continued*)

corresponding to the one which would be formed by the condensation of two PBG molecules, Urogen I is not formed. However, when Urogen I synthetase is incubated with this dipyrrole *and* PBG, Urogen I is formed and the yield of tetrapyrrole is about equal to the PBG plus dipyrrylmethane included in the reaction mixture (46). Thus Urogen I synthetase is essentially a PBG polymerase; it is not known whether the cyclization of the linear tetrapyrrole to form Urogen I is enzymatic—it may be that this step has a very high probability of occurring spontaneously because of the shape of the molecule. Mauzerall (47) estimates that the maximum distance between the ends of an open-chain tetrapyrrylmethane is about 18 Å (equivalent to a concentration of about $0.5\,M$) while the mean distance is about 9 Å (roughly equivalent to a concentration of $4\,M$).

Spinach leaf Urogen I synthetase is only slightly affected by heating at 70° at pH 8.2 but is inactivated at 100° at this pH value and at lower temperature in either more acid or alkaline solutions. The enzyme is inhibited competitively by PBG analogs such as opsopyrroledicarboxylic acid, which lacks the aminomethyl substituent present on PBG, and isoPBG, a PBG isomer in which the aminomethyl substituent is on the same side of the molecule as the propionic acid side chain (48–51). Formaldehyde, p-chloromercuribenzoate, and silver and mercuric ions are noncompetitive inhibitors (43); sulfhydryl-containing compounds reverse inhibition by the three latter agents. Finally, hydroxylamine and high concentrations of ammonium ions as well as at least one dipyrrylmethane do not greatly affect the rate of enzymatic consumption of PBG, but appear to force the accumulation of open chain polypyrrolic intermediates in Urogen I synthesis (52).

For the synthesis of Urogen III from PBG, two enzymes seem to be required: Urogen I synthetase and Urogen III cosynthetase. The latter enzyme has been prepared from wheat germ (52a) but it appears to be present and active in unheated preparations of *Chlorella* (30), *R. spheroides* (44), as well as hemolyzed red cells of chickens (31, 53–55), rabbits (39), and humans (56)—all of which can catalyze the formation of Urogen III, or its derivatives, from PBG.

A large number of suggestions have been made regarding the possible modes of synthesis of Urogen III from PBG. Many of these have been discussed in detail (49, 52); the hypotheses now range from untenable to untestable or untested. This situation is likely to continue until the substrates for Urogen III cosynthetase are established. It seems reasonable at present to suggest that some product of Urogen I synthetase action short of a cyclized tetrapyrrole (perhaps a tripyrrylmethane, formed by the Urogen I synthetase catalyzed sequential condensation of three molecules of PBG) and another molecule of PBG are *somehow* reassembled by Urogen III cosynthetase to form a molecule of Urogen III (Fig. 2). The following evidence supports this position:

1. Urogen I is not a precursor of Urogen III. If this tetrapyrrole is incubated with an enzyme system capable of forming Urogen III, the latter is not produced (39, 56a, 57).

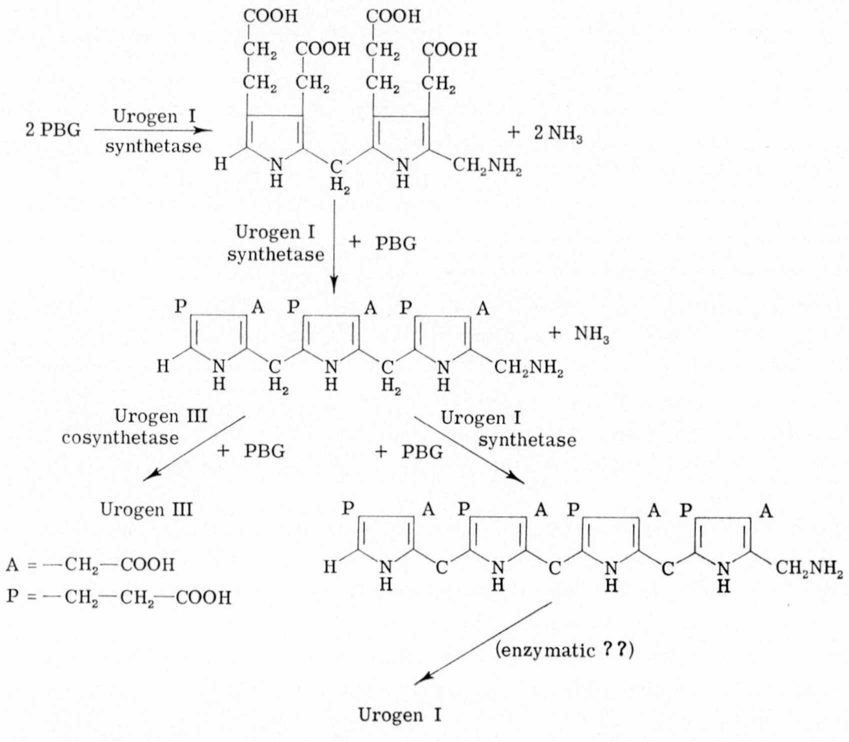

FIG. 2. Mode of action of Urogen I synthetase and its possible role in the formation of Urogen III.

2. Preparations of Urogen III cosynthetase from wheat germ are variable with regard to the degree of contamination with Urogen I synthetase, but the best preparations fail to catalyze the condensation of PBG molecules. However, the same concentration of Urogen III synthetase accelerates the rate of PBG consumption in the presence of Urogen I synthetase, suggesting that some product of Urogen I synthetase action is required for the second enzyme to be active (52a).

Unlike Urogen I synthetase, the Urogen III cosynthetase is quite heat labile.

C. The Production of Coproporphyrinogens

Each pyrrole ring of coproporphyrinogen (Coprogen) is substituted with a methyl and a propionic acid side chain; four isomers are possible although only two, isomers I and III, are formed biologically. The methyl substituents of the Coprogens are produced by the decarboxylation of the acetic acid substituents of Urogens. Uro is not attacked by the enzyme. The enzyme Urogen decarboxylase appears to decarboxylate one acetic acid residue at a time; porphyrins with three, two, or one acetic acid residue per molecule have been detected in reaction mixtures of rabbit erythrocyte Urogen decarboxylase and Urogen (58). All four Urogen isomers can serve as substrates for the rabbit enzyme, but the rate of decarboxylation varies with the type of Urogen provided as a substrate in this order: Urogen III > IV > II > I; Urogen III is decarboxylated about twice as rapidly as Urogen I.

Urogen decarboxylase has not been purified from green plants, but broken-cell preparations of *Chlorella* (56a) can catalyze the synthesis of Coprogens I or III from the corresponding Urogen isomers.

Heath and Hoare (44) have studied Urogen decarboxylase from *R. spheroides* and have also detected a heat stable cofactor for the enzymatic decarboxylation, but this material has not been characterized.

Hoare and Heath (45) found *R. spheroides* Urogen decarboxylase to be unstable in the absence of sulfhydryl compounds. The enzyme obtained from rabbit reticulocytes is inhibited by Hg^{++}, Cu^{++}, Mn^{++}, iodoacetate, and p-chloromercuribenzoate, but these effects are prevented by excess glutathione (58).

D. The Formation of Protoporphyrinogen IX

Coproporphyrinogen oxidative decarboxylase has been purified about twentyfold from beef liver mitochondria by Sano and Granick (59), who also explored the distribution of this enzyme in various rabbit and guinea pig tissues. In guinea pig liver, 80% of the Coprogen oxidative decarboxylase was estimated to be present in the mitochondria. Porra and Falk (60, 61) have studied this enzyme in extracts of acetone powder prepared from ox-liver mitochondria.

The beef liver enzyme catalyzes the formation of protoporphyrinogen (Protogen) IX from Coprogen III but does not act on Coprogen I; Porra and Falk (61) reported that Coprogen IV as well as III served as a substrate for the ox-mitochondria enzyme. This provides information about the specificity of the enzyme, however, as only Urogens I and III are made enzymatically and the discrimination against the I isomer assures the production of a single type of Proto, i.e., isomer IX.

Coprogen oxidative decarboxylase has not been purified from photosynthetic organisms, but broken-cell preparations of *Chlorella vulgaris* catalyze the formation of Proto IX from Urogen III as well as from PBG (*30, 56a*), and ruptured *Euglena gracilis* plastids have been shown to form Proto from PBG (*62*).

Oxygen is required for the formation of the Proto IX vinyl groups from the propionic acid side chains on rings A and B of Coprogen III by decarboxylation and dehydrogenation. Bénard *et al.* (*63*) had observed that blood from anemic rabbits produced mostly Copro when incubated with glycine anaerobically, but formed Proto under aerobic conditions, and Falk *et al.* (*31*) found that porphyrin synthesis from

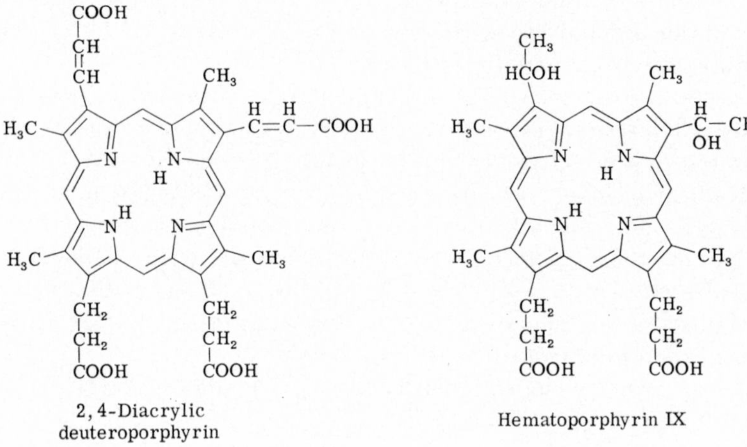

2, 4-Diacrylic
deuteroporphyrin

Hematoporphyrin IX

Fig. 3.

PBG by hemolyzed fowl red cells followed the same pattern. Attempts to substitute a wide variety of oxidants other than O_2 for the biosynthesis of Protogen IX from Coprogen III have been unsuccessful (*59, 61*). As discussed later, O_2 suppresses bacteriochlorophyll production in *R. spheroides*; an oxidant other than O_2 may serve for Protogen IX synthesis in this bacterium, but no investigations along this line have been reported.

One problem encountered in studying this enzyme is that the substrate is Coprogen III—which in air is easily oxidized to Copro III—but at the same time the reaction requires O_2. Sano and Granick (*59*) were able to circumvent the problem in part by including thioglycolate (0.05 M) in the reaction mixture. Experimentally it can also be avoided by using relatively high enzyme-to-substrate ratios (*56a*).

It is not yet clear whether any free intermediates exist between Coprogen III and a porphyrin with one vinyl and three propionic acid side chains. *C. vulgaris* mutant W-B-17 produces Uro, Copro, and copious amounts of hematoporphyrin IX (Hemato IX) (Fig. 3) (*61a, 64*). A series of reactions can be written in which Hemato IX could be a precursor of Proto IX [e.g., Bogorad (*49*)]; however, beef liver Coprogen oxidative decarboxylase preparations do not convert Hematogen to Proto (*59*); *trans*-2,4-diacrylic deuteroporphyrinogen (Fig. 3) and monoacrylic monopropionic deuteroporphyrinogen are also not converted. On the basis of results of experiments conducted in the presence of T_2O, Granick and Sano (*65*) conclude that vinyl groups are formed from two of the propionic acid side chains of Coprogen by the simultaneous decarboxylation and removal of a hydride ion from the propionic acid HCH group adjacent to the pyrrole ring.

E. The Formation of Protoporphyrin and Magnesium Protoporphyrin

The hexahydroporphyrins, or porphyrinogens which are substrates for Urogen decarboxylase and Coprogen oxidative decarboxylase, cannot chelate metals. However, Mg protoporphyrin, rather than Protogen IX is the substrate for the enzyme which catalyzes the first esterifications leading to chlorophyll production.

Porphyrinogens are colorless compounds; their absorption spectra in the ultraviolet (UV) region depend upon the nature of the pyrroles of which they are composed. The metal-free porphyrins, on the other hand, are red compounds which absorb strongly at about 400 mμ (the Soret band) as well as at other wavelengths throughout the visible region of the spectrum. For example, Urogen absorbs at about 227 mμ, the absorption maximum of PBG, but Uro in pH 8.2 Tris buffer displays absorption maxima at 398.5, 503, 538, 561, and 612 mμ (E_{mM}^{1cm} 220: 13.4, 10.7, 8.02, and 4.06 respectively) (*66*). [In 1 *N* HCl: 406, 552, and 592 mμ; E_{mM}^{1cm} 527: 17.6 and 6.11 (*66*).]

Urogen, as an example, is oxidized quantitatively to Uro by I_2 but at least one compound at an intermediate level of oxidation can be distinguished. During spectrophotometric titration of Urogen at pH 7 with I_2, a compound (or two compounds) which absorbs strongly at about 500 mμ appears prior to the development of Uro—this is most likely a tetrapyrrole with one or two *pairs* of pyrrole rings in conjugation (or a mixture of the two types), since dipyrrylmethenes absorb in this region of the spectrum (*42, 43, 57, 58*). Urogen can also be photooxidized to Uro; in 1 *M* HCl yields are about 90%, but only about 50% in 1 *M* NaOH (*58*). [Some problems regarding attacks on the vinyl side chains during the chemical oxidation of Protogen are described by Sano and Granick (*59*).] Furthermore, Urogen is autooxidized when incubated aerobically in darkness in neutral or slightly alkaline solutions, but the yield of Uro is only about 60–75% (43); under these circumstances an absorption band at about 638 mμ develops in addition to the bands of Uro. [Treatment of a solution containing 638-mμ absorbing material with sodium amalgam, which reduces porphyrins to porphyrinogens and presumably reduces the 638 mμ-absorbing substance to the same level, followed by oxidation with I_2 results in the formation of additional Uro (67).] Sano and Granick (59) found some evidence for the existence of a catalyst of Protogen oxidation in liver mitochondria, and Bogorad (43) detected a heat-labile catalyst of Urogen oxidation in extracts of spinach leaf acetone powders; however, the existence of a specific enzyme for porphyrinogen oxidation cannot be considered to be firmly established. The conversion of Protogen to Proto has not been studied in R. spheroides extracts.

The vinyl groups of partially oxidized intermediates of Protogen IX appear to be very susceptible to attack by sulfhydryl compounds. For example, a porphyrin C-type compound is formed if cysteine is present during the photooxidation of Protogen IX in acid solution (59). Porra and Falk (60) observed that a considerable fraction of porphyrin became covalently bound to protein during incubation of Coprogen with their Coprogen oxidative decarboxylase from bovine liver mitochondria. Sano and Granick (59) suggest that cytochrome c may be formed by the interaction of the —SH groups of the peptide chain with Protogen during its oxidation. The enzymatic insertion of iron into the porphyrin of such a complex has not been studied, but Porra and Jones (68) found that iron is inserted into Proto (but not porphyrin c) by an enzyme from pig liver. (The Hemato IX which accumulates in the culture medium of Chlorella vulgaris W_5-B-17 may be derived from some oxidizing Proto IX; the fact that another mutant, W_5B, produces Proto IX suggests that one difference between these mutants could be the presence of a Protogen oxidase.)

An enzyme from animal tissues which catalyzes the incorporation of ferrous ions into Proto has been studied by Labbe and his co-workers [e.g., Nishida and Labbe (69); Labbe et al. (70)], by Oyama et al. (71), Schwartz et al. (72), and Porra and Jones (68). It has also been demonstrated to occur in extracts of Chromatium strain D (68). Almost every time this enzyme has been rediscovered it has been renamed: e.g., protoporphyrin iron-chelating enzyme (PICE), ferrochelatase.

The insertion of magnesium into Proto, the next step after Protogen oxidation en route to the chlorophylls, has not been demonstrated with either a crude broken-cell system or a purified enzyme. Magnesium ions inhibit the utilization of ferrous ions by PICE.

Neuberger and Tait (73) have found an enzyme in the chromatophore fraction of R. spheroides, as well as in several mammalian tissues, which catalyzes the incorporation of zinc into protoporphyrin. Little and Kelsey (74) have described a similar enzyme in extracts of barley leaves. The enzyme preparation from R. spheroides is inhibited competitively by ferrous ions, but if ascorbate is included in the medium, PICE activity is displayed; these preparations may contain two enzymes or this may merely be the iron-chelating enzyme. In any event, neither the iron- nor the zinc-chelating enzymes, if the latter is a different catalyst, promote the incorporation of magnesium into Proto.

In 1947 Smith (75) found that an ether-soluble magnesium-containing compound disappeared during chlorophyll accumulation by dark-grown barley leaves illuminated for short periods. There is no other evidence for what may be a magnesium carrier involved in the synthesis of Mg proto.

F. The Formation of Magnesium Protoporphyrin Monomethyl Ester

Tait and Gibson (76) have shown that a chromatophore fraction from *Rhodopseudomonas spheroides* catalyzes the formation of magnesium protoporphyrin IX monomethyl ester when incubated with Mg Proto IX and S-adenosylmethionine. Mg Proto IX is esterified about fifteen times more rapidly than Proto IX, indicating that the biosynthetic sequence toward bacteriochlorophyll (Bchl)—and presumably Chl *a*—is: Proto → Mg Proto → Mg Proto monomethyl ester → chlorophyll. Attempts to solubilize the enzyme were not successful. Formate had been shown to serve as a precursor of the methyl ester group of Chl *a* in *Chlorella* by Green et al. (77).

Mutant strains of C. vulgaris which accumulated no Chl but instead produced Proto IX and Mg Proto IX were described by Granick in 1948 (78, 79). These observations served to connect the biosynthesis of Chl with that of Proto long before the biosynthetic details were known. A mutant which accumulated Mg Proto IX monomethyl ester was described by Granick in 1961 (34); in deducing the biosynthetic chain of Chl this compound appeared to be a precursor of magnesium vinyl pheoporphyrin a_5 (protochlorophyllide) which had been found to accumulate in still another *Chlorella* mutant (80).

Mg Proto monomethyl ester—presumably with the propionic acid residue on ring C being the point of esterification—has also been detected in etiolated barley leaves supplied with ALA when α,α'-dipyridyl is also given (34); in Bchl-containing illuminated anaerobic cultures of R. spheroides—particularly when grown on iron-deficient medium (81)

—and in cultures of a strain of *Rhodopseudomonas capsulata* incubated in the light in a mixture of succinate and glycine (*82*). Production of the porphyrin ester was greater at iron concentrations below those optimal for Bchl formation.

Ethionine and threonine interfere with Bchl synthesis in *R. spheroides;* inhibition by ethionine is reversed by methionine, and inhibition by threonine is overcome by administration of homoserine or methionine (*83*). Homocysteine thiolactone reverses the effect of ethionine in part and of threonine completely. Ethionine is thought to inhibit by arresting formation of the methyl ester group of methionine; threonine by interfering with the production of homoserine and consequently of methionine.

G. Magnesium Vinyl Pheoporphyrin a_5 (protochlorophyllide *a*)

The formation of chlorophyllide *a* (Chlide *a*) from protochlorophyllide a (PChlide) is discussed in Chapter 14 by Boardman.

The conversion of Mg Proto IX monomethyl ester to PChlide requires the formation of the cyclopentanone ring in which the propionic acid group at position 6 is linked to the γ-bridge carbon atom and the reduction to an ethyl group of the vinyl side chain at position 4.

It appears likely that formation of the cyclopentanone ring precedes reduction of the vinyl group at position 4. Jones (*84*) has recovered Mg 2,4-divinylpheoporphyrin a_5 monomethyl ester (i.e., 4-desethyl, 4-vinyl PChlide) from medium in which *R. spheroides* have been grown anaerobically in the light in the presence of 8-hydroxyquinoline (25 μM). Under these culture conditions bacterial growth and Bchl formation are both inhibited somewhat. [Jones suggests that the divinyl pheoporphyrin may be the compound detected in the "tan mutant" of *R. spheroides* and described as "bacterial protochlorophyll" by Stanier and Smith (*85*).]

A *possible* reaction sequence leading to the formation of 2,4-divinyl pheoporphyrin a_5 monomethylester from Mg Proto monomethyl ester is shown in Fig. 1. The significance of the esterification of Mg Proto preceding the formation of the cyclopentanone ring may lie, at least in part, in preventing decarboxylation of the propionic acid side chain after oxidation of the carbon atom proximal to the ring to a carbonyl group; the latter may be required to activate the central carbon atom of the side chain to permit condensation with the γ-bridge carbon atom.

H. The Formation of Chlorophyllide *a*

Besides the light-mediated reduction of PChlide to Chlide *a* in higher plants as well as in some algae (see Chapter 14), many organisms, particularly algae, are capable of producing large amounts of chlorophylls in complete darkness. Traces of Chl have been detected in dark-grown oats (86) and in a number of angiosperms by Godnev et al. (87). No information is available regarding the mechanism of enzymatic reduction of PChlide.

Two mutant strains of C. vulgaris are known which accumulate some PChlide in darkness but resemble wild-type cells when exposed to illumination (80, 88). The action spectrum for pigment production by one of these displays a maximum at about 650 mμ just as do higher plants (89). This situation may be more complicated than it seems, for the algae accumulate negligible amounts of photoconvertible PChlide when grown in darkness—not an amount equivalent to the amount of Chls formed by the wild-type or by the mutant strain upon illumination.

I. Esterification with Phytol

The final step in the biosynthesis of Chl a appears to be esterification with phytol of the propionic acid residue at position 7 of Chlide a.

The enzyme chlorophyllase has long been known to catalyze the hydrolysis of phytol from chlorophyll (90). This esterase is remarkable in being active in solutions containing high concentrations of organic solvents including acetone, ethanol, or methanol. In alcoholic solutions phytol is removed and the Chlide is esterified with the alcohol present in high concentration; in alcohol-free solutions, such as aqueous acetone, the products are Chlide and phytol (3,7,11,15-tetramethyl-2-hexadocen-1-ol). Böger (91) reports that the chlorophyllase he studied—from C. vulgaris—was more active in 40% than in other concentrations of acetone. It is clear from these observations that the enzyme catalyzes both hydrolysis and esterification. The reaction can be driven far toward esterification if a two-phase system is provided including a material, such as a lipid or a lipid solvent in which the Chlide is not soluble, but the Chl a is soluble—i.e., if the esterified form is removed from the reaction mixture as it is produced.

As already noted, chlorophyllase appears to display little, if any, specificity toward the esterifying alcohol. On the other hand, the enzyme affects only chlorophylls, including bacteriochlorophyll (Bchl) and chlorobium chlorophyll (Cchl) [which lacks a carboxymethyl substituent on the isocyclic ring (see Chapter 4)] and the corresponding pheophytins; it does not act on Pchl [Fischer and Stern (92); Mayer (93); Sudyina (94); also see summary by Holden (95)]. Etiolated leaves generally contain Pchl as well as PChlide despite the apparent specificity of the enzyme in vitro.

There may be another enzyme for phytolation of Chlides (and

PChlides?) but the conviction that chlorophyllase catalyzes Chl formation seems to be growing: Holden (96) has found that when etiolated
seedlings are transferred to light an increase in chlorophyllase activity
is exhibited which parallels chlorophyll accumulation; Sudyina (94) reported that "in the first minutes of illumination of etiolated plants the
activity of chlorophyllase sharply increases"; Shimizu and Tamaki (97,
98) have determined that chlorophyllase activity in tobacco reaches a
maximum and declines even before the chlorophyll content of the tissues
is maximal; and Böger (91) has observed that in C. vulgaris chlorophyllase activity increases and Chls accumulate roughly in parallel.

Sudyina (94) concludes that chlorophyllase plays no role in Chl
breakdown in vivo.

Chlorophyllase is apparently intimately associated with plastid lipids.
Until recently the enzyme has been available only in organic-aqueous
solutions, e.g., aqueous acetone or aqueous methanol, or in detergent
solutions [e.g., Ardao and Vennesland (99); Klein and Vishniac (100)],
but now purified water-soluble chlorophyllase preparations have been
made (96, 97). In addition, Böger (91) reports that about 50% of the
chlorophyllase of Chlorella vulgaris is solubilized when the cells are disintegrated at pH 7.3 in a solution containing polyethylene glycol.

J. Bacteriochlorophyll a

The observation that mutants of Rhodopseudomonas spheroides accumulate pheophorbide a (101) and the fact that pigments which spectroscopically resemble Pchl or PChlide and Chl a may be formed from
Chl a (102) support the view that Bchl a may be formed from Chl a.
(Chlorophyllase also acts on Bchl a, thus phytolation could, but may
not, be postponed until after BChlide is formed.)

Again, from cultures of R. spheroides grown in the presence of 8-
hydroxyquinoline, Jones (103) has isolated a compound which he has
identified as 2-devinyl-2-hydroxyethyl pheophorbide a. The corresponding magnesium compound, from which this might be derived, would be
a chemically reasonable intermediate between PChlide and BChlide a;
the hydroxyethyl side chain at position 2 is suggested as an intermediate
between the vinyl group of, for example, PChlide and the —CO—CH₃
at this position in Bchl a. [The chlorobium chlorophylls bear hydroxyethyl substituents at position 2 (104).] The production of this pigment
may indicate that esterification with phytol may occur later in Bchl
formation than suggested above. However, the experience with the accumulation of Hemato IX by C. vulgaris mutant W₅B-17 makes complete
acceptance of the bacterial compound as an intermediate in Bchl forma-

tion less comfortable than it might otherwise be. No information is available regarding the reduction of ring B in the formation of Bchl.

Another Bchl, named Bchl b, has recently been observed in a newly discovered *Rhodopseudomonas* (*105*).

Six different Cchls have been described and characterized structurally by Holt and his collaborators together with MacDonald's group [see Holt (*104*)]. The basic "skeleton" of the Cchls is shown in Fig. 2 of Chapter 4; the differences among these pigments are in the substituents on positions 4 and 5 and on the δ-bridge carbon atom. Formation of these compounds from Chlide a is quite possible. Species of *Chlorobium* contain Bchl as well as Cchls; the ratio is about 1:20.

K. Chlorophyll b

The immediate precursor of Chl b has not been identified. The following possible relationships between the formation of Chls a and b have been enumerated by Shlyk and Godnev (*106*) and again by Shlyk *et al.* (*107*):

1. Chls a and b are formed via independent and distinct biosynthetic paths, (which diverge at a stage prior to pyrrole formation).
2. Both pigments have a common precursor (e.g., some tetrapyrrole or even pheoporphyrin) and are formed in parallel from it.
3. Chl b is produced from Chl a.
4. Chl a is produced from Chl b.

[The Minsk groups of Godnev and of Shlyk have been very active in this field. The most accessible English summaries of their work can be found in Godnev *et al.* (*108*) and Shlyk *et al.* (*107*).]

Among the few indisputable points regarding the biogenesis of Chl b are these: (a) Chlide a is derived from PChlide a (see Boardman, Chapter 14); (b) during greening of etiolated leaves, some Chl a accumulates before any b is detectable; (c) many species of algae contain Chl a but not b; and (d) the following compounds have not been detected: Pchl b, PChlide b, Chlide b [evidence for the presence of the last of these has been obtained, but under circumstances where chlorophyllase was active (*107*)].

To evaluate the four proposed relationships in slightly more detail: proposition 4 seems unlikely in view of facts (a), (b), and (c). Proposition 1 is conceivable, but only if all the enzymes for the synthesis of Chl a are physically separate within the plastid from those used in making Chl b. Thus, by default propositions 2 and 3 have the greatest appeal at present, although 1 is not completely ruled out.

Godnev *et al.* (*108, 109*) attempted to attack the problem directly by introducing ^{14}C-Chl *a* dissolved in sunflower oil into the hollow center of an onion leaf which was just becoming green. After 2 days the Chl *b* of the leaf was isolated and found to be radioactive; in a similar experiment, in which ^{14}C-Chl *b* was introduced, no ^{14}C-Chl *a* could be detected. One of the great problems in using radioactive tracers for studies of chlorophyll interconversions and turnover is the difficulty of bringing chlorophylls to radiopurity—a problem by now well-recognized by students of chlorophyll metabolism. It is possible that the ^{14}C-Chl *a* used by Godnev *et al.* was contaminated with ^{14}C-Chl *b*. It is difficult to expect that Chl *a* would enter the chloroplasts of the living cell; it is possible, however, that an enzyme for the conversion of Chl *a* to Chl *b* exists and that it acted when onion leaves were being macerated for Chl extraction. Unfortunately no other direct experiments have been performed; the final resolution of this problem—like many others—will probably have to wait until Chl *b* can be produced enzymatically *in vitro*. Meanwhile, the available data are mostly from tracer experiments with leaves or intact plants.

In general, when [14]CO$_2$, ^{14}C-glycine, ^{14}C-acetate, or ^{14}C-ALA have been administered, either as a "pulse" or for a long period, the specific radioactivity of the Chl *a* has been found to be greater than that of the Chl *b*. Experiments with numerous species including leaves of higher plants (*106, 110, 111*) as well as with *Scenedesmus* (*112*) and *Chlorella* (*107, 112–115*) show the Chl *a* to be about 2.5 to 4.7 times higher in specific activity than the Chl *b*. On the other hand, there have also been demonstrations of Chls *a* and *b* becoming almost equally radioactive in *Chlorella* (*107, 114, 116*), and in one instance (*117*) radioactivity from glycine-2-^{14}C was observed to appear in higher concentration in Chl *b* than in Chl *a*. Finally, Michel-Wolwertz (*115*) provided etiolated barley leaves in darkness with ALA-4-^{14}C and later illuminated them; after 4 hours the specific radioactivity of Chl *b* was about the same (roughly 10% lower) as that of *a*.

Most of these data [excluding those of Brzeski and Rucker (*117*)] have been taken by Shlyk and by Michel-Wolwertz to favor the view that Chl *b* is formed from Chl *a*. Some explanation must be given to accommodate differences in specific activities of Chl *a* and Chl *b* with this position. One possibility is that, against the background of Chls *a* and *b* already present, relatively less additional Chl *b* than Chl *a* is being formed. Shlyk *et al.* (*107*) discuss this possibility and provide data from experiments with *Chlorella* that the ratio specific activity Chl *a*:Chl *b* is much smaller after the cells have made a great deal of Chl than when only small additional amounts have accumulated before analysis. When about one-third of the total Chl was formed after administration of ^{14}C-bicarbonate the ratio was about 3, but when eight-ninths was formed in the presence of labeled bicarbonate the ratio was 1.28. These data also explain some of the variable results obtained earlier. Wieckowski (118) reports that in *Phaseolus vulgaris* growing on 12 hours of light daily the total concentration of Chls *a* and *b* fluctuates with a 24-hour rhythm. In addition, as a consequence of differences in the rates of formation of the two Chls, the ratio of Chl *a* to Chl *b* decreases during the first few hours of exposure to light but later increases and continues to increase during the dark period. Another possible way to explain some of the differences rests on recent suggestions by Shlyk and Nikolayeva (119) and Michel-Wolwertz (115) that newly synthesized and "old" molecules of Chl *a* may behave differently, e.g., with regard to their availability for transformation into Chl *b*.

Finally, Shlyk *et al.* (107) report that if barley leaves are placed in darkness for 8 days after having been illuminated for 20–30 minutes in the presence of $^{14}CO_2$, the specific activity of the Chl *a* drops steadily after about the first day while that of Chl *b* increases by about 25% during the period in darkness. Shlyk *et al.* take this to mean that Chl *b* can be synthesized in darkness by barley leaves and that the "new" Chl *a* molecules (those formed during the 20–30 minutes in the light in the presence of $^{14}CO_2$) are preferentially used for Chl *b* formation.

The problem of the origin of Chl *b* has been reviewed critically by Smith (120) and more recently by Smith and French (121).

To summarize this discussion: The evidence for the origin of Chl *b* from Chl *a* now seems somewhat more acceptable than it has been. However, the data do not yet clearly discriminate between origin from Chl *a*, a special kind (or age?) of Chl *a*, or a common precursor of the two Chls.

L. Chlorophylls *c* and *d*

The structure of Chl *c* is not known. Chl *d* (2-devinyl-2-formyl Chl *a* —see Chapter 4) might, obviously, be formed from Chl *a*, although divergence from the biosynthetic path of Chl *a* at Chlide *a* or earlier is readily conceivable.

M. Summary

The enzymatic steps leading to the formation of Proto from ALA have been demonstrated *in vitro* with cell-free or partially purified enzyme preparations. Most of the enzymes have not been purified from chlorophyll-forming organisms, although cell-free preparations of *Chlorella* appear to contain all of them.

Purer preparations of ALA synthetase have been made from *Rhodopseudomonas spheroides* than any other organism, but the presence of this enzyme has not been established in any other chlorophyll-forming organisms. Succinyl thiokinase, which catalyzes the formation of succinyl CoA from succinate and coenzyme A, has been prepared from spinach (*122*) and wheat leaves (*123*) as well as from *R. spheroides* (*124*). ALA transaminase has been demonstrated in extracts of *R. spheroides* and *Chlorella vulgaris*, but what role, if any, it may play in plants *in vivo* remains to be established.

Except for the demonstration with *R. spheroides* chromatophore preparations of Mg Proto IX methyl esterase, information about intermediates between Proto IX and PChlide comes from analyses of *C. vulgaris* and *R. spheroides* mutants and culture media; the suggested sequence of compounds is based on chemical reasonableness. In this span the reactions have yet to be demonstrated, even as a group, in cell-free preparations. The sequence of reactions leading to formation of the isocyclic ring is entirely unknown. Knowledge of the enzyme catalyzing the formation of Mg Proto is conspicuously lacking; especially in view of progress with the enzyme for promoting the chelation of iron by Proto IX.

The enzymatic steps in the formation of other chlorophylls, except where they are identical to those in Chl *a* biosynthesis, are not known; even information on Chl *b* production does not provide a clear picture of its origin.

Among the reactions which have been demonstrated with purified enzymes, mechanisms are generally unresolved. Screening of isomers, i.e., blocking the production of Proto isomers other than IX, appears to occur by (a) abundance of Urogen III synthetase or, if this fails, by (b) differential rates of decarboxylation of Urogen isomers strongly favoring Urogen III, and, finally (c) the complete discrimination by Coprogen oxidative decarboxylase against Coprogen I, if it should be produced.

II. The Control of Chlorophyll Metabolism

The range of environmental effects on Chl production is great. The literature on quantitative alterations, e.g., effects of light intensity and other factors on Chl concentration as well as the ratio of Chl *a* to Chl *b*, is extensive; Egle (*125*) has discussed these data, including the classical work of Seybold and Egle, in detail. Some more recent contributions to this problem have been made by Friend (*126, 127*), Smillie and Krotkov (*128*), Bukatsch and Rudolph (*129*), Michel-Wolwertz (*115*), Sesták (*130*), and Wieckowski (*118*).

Among the most thoroughly examined situations in which Chl accumulation is affected by the environment in a more-or-less qualitative way are: (a) the effect of light on pigment production in angiosperms and some algae, and (b) the effect of oxygen and light on pigment production in *Rhodopseudomonas spheroides.*

Angiosperms germinated and grown in darkness (i.e. etiolated) contain small amounts of PChlide, and traces of Chls have been detected in dark-grown oats (*86*) and a number of other seed plants (*87*). As in much of the rest of this discussion, it will be necessary to generalize from data obtained with a few species—in this case, mostly beans, wheat, barley, and maize. The amount of Pchl + PChlide present is of the order of 1/100 to 1/300 of the amount of Chls *a* + *b* present in the same kind of leaf after prolonged illumination; the concentration of Chls in a "fully-greened" etiolated leaf may, however, be smaller than in a leaf of a plant grown in the light from germination. Etiolated leaves normally contain no detectable quantities of porphyrin precursors of PChlide. However, these leaves apparently contain active enzymes for the production of PChlide from ALA for, as pointed out above, Granick (*33, 34*) showed that when etiolated barley leaves are supplied ALA they form 10 times the normal amount of PChlide, although it has generally been observed that little, if any, of this "extra" PChlide is photoconverted to Chlide *a*. The limiting factor for PChlide formation in etiolated leaves is thus the capacity for ALA synthesis. After brief illumination of an etiolated leaf, during which much or all of the PChlide is transformed to Chlide, additional PChlide is formed even if the leaf is returned to darkness. For example, Madsen (*131*) found that within 12–19 minutes after being returned to darkness after a very brief light exposure, etiolated leaves of barley or wheat already contained about as much new PChlide as they had before illumination; after this time the rate of PChlide production was constant for about an additional 45 minutes—then no more pigment was produced. In *some* plants of *some* ages, no more pigment is formed in light than in darkness during the first 2–3 hours of illumination (*35, 132*). Under constant illumination, following this lag phase of Chl formation—and again the lag does not always occur—pigment production is accelerated about 15–20 times over the rate during the lag phase and continues at this new high speed until no further accumulation occurs [e.g., Sisler and Klein (*35*)]. However, if plants are returned to darkness during the period of rapid greening, Chl accumulation quickly slows and within 1–2 hours it ceases (*133*)—quite reminiscent of the kinetics of pigment production following initial illumination of etiolated leaves.

Massive Chl production is blocked if the protein synthesis inhibitor chloramphenicol (*134, 135*)—or if actinomycin D (*136*), an inhibitor of DNA-dependent RNA synthesis—is administered before the initial illumination of etiolated bean or maise leaves. Similarly, these agents, as well as puromycin, another inhibitor of protein synthesis, stop pigment production when bean leaves are exposed to them during the period of rapid greening (*133*). It appears that RNA and protein synthesis are required for PChlide formation (in these organisms light is, of course, also required for photoconversion of PChlide to Chlide) and that light somehow promotes these processes in the plastids. Chloroplasts contain DNA, RNA, and ribosomes and have the capacity to synthesize RNA and protein (*137–154;* also *155, 156*). The relatively rapid stoppage of PChlide formation after transfer of plants to darkness could simply indicate that, for example, inhibitors of the enzyme system for ALA synthesis accumulate quickly in darkness—feedback inhibitors would be most appropriate —but the effects of inhibitors of RNA and protein synthesis encourage a different interpretation: namely, that the informational RNA required for the synthesis of the ALA-making system and the enzyme itself (or some enzyme which is required for the production of a substrate or cofactor) have quite short half-lives, i.e., they decay relatively rapidly. Light could, perhaps act quite indirectly to promote RNA and protein synthesis; or, or course, darkness could interfere with these processes. Unfortunately the enzyme for ALA synthesis has not yet been positively identified in green plants.

On the other hand, ALA synthetase is easily demonstrable in *Rhodopseudomonas spheroides*. This bacterium forms Bchl when grown anaerobically in the light, but only traces of pigment are found in well-aerated cultures in light or darkness although the quantity of pigment produced is greater in illuminated cultures (*157–160*). Thus there seems to be a qualitative effect of O_2 and a quantitative effect of light on pigment production. ALA synthetase activity of cells actively forming Bchl is about five times that of cells from well-aerated cultures (*159*); the level of ALA dehydrase varies in the same manner. The effect of O_2 upon ALA synthetase and ALA dehydrase activities in *R. spheroides* cells is very rapid. ALA dehydrase activity fails to continue to increase, and a net decrease in ALA synthetase activity is detectable within about 50 minutes (the first measurement made) after cells are transferred from an atmosphere of 5% (v/v) CO_2 in N_2 to 5% (v/v) CO_2 in air (*161*). Upon return to anaerobic conditions, the synthesis of these two enzymes commences without delay. Bchl production also stops upon aeration. Thus in the presence of O_2, ALA synthetase activity stops and no additional detectable enzyme accumulates—in fact there is a net decrease.

Several inhibitors of protein and nucleic acid formation—chloramphenicol, *p*-fluorophenylalanine, 8-azaguanine, 5-fluorouracil (*159, 162, 163*)—as well as acriflavin, mitomycin C, and phleomycin (*164*), resemble O_2 in arresting the formation of Bchl and ALA synthetase in *R. spheroides* cultured under conditions normally conducive to pigment produc-

tion. These observations support the proposal that at least one consequence of an increase in O_2 tension is the arrest of the synthesis of
proteins, either directly or via interference with nucleic acid metabolism.
The O_2 effect, however, is very probably more specific than that of the
inhibitors. These bacteria can grow aerobically as heterotrophs; catalase,
for example, is present only in traces in cells grown anaerobically in the
light, but it accumulates rapidly upon aeration (165). On the other hand,
the production of ribulose diphosphate carboxylase is repressed by O_2 in
R. spheroides (166).

Lascelles has also shown that addition of ALA or of hemin to growing
cultures of R. spheroides interferes with the normal increase in activity
of ALA synthetase and ALA dehydrase.

In bare essentials the R. spheroides and greening etiolated plant systems have in common these features: rapid control over ALA synthesis
by enzymes present in the cell and also over the production of additional
synthetic capacity; the latter appears to involve nucleic acid and protein
synthesis. The suggestions made regarding possible devices for the control of greening in leaves, e.g., rapid decay of the enzyme and informational RNA, appear to be applicable to R. spheroides. In the bacterial
system, in addition, there is more information about the enzymes involved
and there is the knowledge that ALA itself can control the production of
ALA synthetase while hemin inhibits its action, but there is no evidence
that either ALA or hemin accumulates under aerobic conditions. (The
discussion of effects of light and O_2 seems to emphasize the need for
independent control mechanisms for early steps in the production of Chl
and of the respiratory cytochrome hemes. In higher plants, this obviously
could be related at least in part to control over plastid vs. mitochondrial
development; in bacteria, to regulation of chromatophore production. For
these moderating systems to be effective, each type of subcellular unit
must have its own equipment for ALA formation.)

Chlorophyll production in algae and higher plants is also affected by
deficiencies of a variety of mineral nutrients. Besides magnesium, the
effect of availability of iron is most notable. In none of the nutrient deficiencies do Chl precursors accumulate, although the possibility that they
are produced in small amounts and rapidly destroyed is not completely
excluded. The absence of intermediates in PChlide synthesis argues that
the capacity for ALA production may be limited. Marsh *et al.* (167)
found that chlorotic leaves from iron-deficient cowpeas formed protoporphyrin when supplied ALA in darkness; this indicated that Chl formation by these leaves is blocked by their incapacity to form ALA and to
insert Mg into Proto IX. Depending upon the degree of chlorosis, discs
from leaves of iron-deficient Swiss chard plants produce either Proto,

plus small amounts of Uro and Copro, or PChlide, Mg Proto, and Proto (*168*) when incubated with ALA in darkness. Iron deficiency chlorosis may be a manifestation of inhibition of protein or nucleic acid metabolism in plastids. Van Noort and Wallace (*169*) report that greening of iron-deficient bush beans upon supplying iron was blocked by 5-fluorouracil or 5-fluorodeoxyuridine. Perur *et al.* (*170*) found the chloroplast protein fraction of iron-deficient maize to be 82%, which is lower than normal; other leaf proteins fractions appeared to be virtually unaffected by iron deficiency.

Iron *suppresses* porphyrin production by *R. spheroides* (*171*). Burnham and Lascelles (*19*) have demonstrated that hemin inhibits ALA formation by ALA synthetase. They suggest that the availability of iron promotes the production of hemin.

The interference with Chl formation by heat and various antibiotics has long been studied, particularly in *Euglena:* e.g., by streptomycin (*172–176*), heat (*177*), chlortetracycline (*178*), acriflavin (*179*), pyrabenzamine (*180*). It is very likely that the primary effect is on plastid development.

ACKNOWLEDGMENTS

The preparation of this chapter was aided in part by research grants from the National Science Foundation and the National Institutes of Health.

REFERENCES

(*1*) K. Bloch and D. Rittenberg, *J. Biol. Chem.* **159**, 45 (1945).
(*2*) D. Shemin and D. Rittenberg, *J. Biol. Chem.* **166**, 621 (1946).
(*3*) K. I. Altman, G. W. Casarett, R. E. Masters, T. R. Noonan, and K. Salomon, *J. Biol. Chem.* **176**, 319 (1948).
(*4*) M. Grinstein, M. D. Kamen, and C. V. Moore, *J. Biol. Chem.* **179**, 359 (1949).
(*5*) N. S. Radin, D. Rittenberg, and D. Shemin, *J. Biol. Chem.* **184**, 745 (1950).
(*6*) H. M. Muir and A. Neuberger, *Biochem. J.* **47**, 97 (1950).
(*7*) J. Wittenberg and D. Shemin, *J. Biol. Chem.* **185**, 103 (1950).
(*8*) N. S. Radin, D. Rittenberg, and D. Shemin, *J. Biol. Chem.* **184**, 755 (1950).
(*9*) R. Lemberg and J. W. Legge, "Hematin Compounds and Bile Pigments," 749 pp. Wiley (Interscience), New York, 1949.
(*10*) J. Wittenberg and D. Shemin, *J. Biol. Chem.* **178**, 47 (1949).
(*11*) H. M. Muir and A. Neuberger, *Biochem. J.* **45**, 163 (1949).
(*12*) D. Shemin and J. Wittenberg, *J. Biol. Chem.* **192**, 315 (1951).
(*13*) D. Shemin and S. Kumin, *J. Biol. Chem.* **198**, 827 (1952).
(*14*) D. Shemin and C. S. Russel, *J. Am. Chem. Soc.* **75**, 4873 (1953).
(*15*) A. Neuberger and J. J. Scott, *Nature* **172**, 1093 (1953).
(*16*) G. Kikuchi, A. Kumar, P. Talmadge, and D. Shemin, *J. Biol. Chem.* **233**, 1214 (1958).
(*17*) K. D. Gibson, W. G. Laver, and A. Neuberger, *Biochem. J.* **70**, 71 (1958).
(*18*) K. D. Gibson, M. Matthew, A. Neuberger, and G. H. Tait, *Nature* **192**, 204 (1961).
(*19*) B. F. Burnham and J. Lascelles, *Biochem. J.* **87**, 462 (1963).
(*20*) B. F. Burnham, *Biochem. Biophys. Res. Commun.* **7**, 351 (1962).

(21) W. G. Laver, A. Neuberger, and J. J. Scott, *J. Chem. Soc.* p. 1483 (1959).
(22) A. M. Nemeth, C. S. Russel, and D. Shemin, *J. Biol. Chem.* **229**, 415 (1957).
(23) E. Kowalski, A. M. Dancewicz, and Z. Szot, *Bull. Acad. Polon. Sci., Ser. Sci. Biol.* **5**, 223 (1957).
(24) E. Kowalski, A. M. Dancewicz, Z. Szot, B. Lipinski, and O. Rosiek, *Acta Biochim. Polon.* **6**, 257 (1959).
(25) M. Bagdasarian, *Nature* **181**, 1399 (1958).
(26) A. Neuberger and J. M. Turner, *Biochim. Biophys. Acta* **67**, 345 (1963).
(27) M. Gassman, J. Pluscec, and L. Bogorad, unpublished data (1965).
(28) R. G. Westall, *Nature* **170**, 614 (1952).
(29) P. Sachs, *Klin. Wochschr.* **10**, 1123 (1931).
(30) L. Bogorad and S. Granick, *Proc. Natl. Acad. Sci. U.S.* **39**, 1176 (1953).
(31) J. E. Falk, E. I. B. Dresel, and C. Rimington, *Nature* **172**, 292 (1953).
(32) S. Granick, *Science* **120**, 1105 (1954).
(33) S. Granick, *Plant Physiol.* **34**, xviii (1959).
(34) S. Granick, *J. Biol. Chem.* **236**, 1168 (1961).
(35) E. G. Sisler and W. H. Klein, *Physiol. Plantarum* **16**, 315 (1963).
(36) S. Klein and L. Bogorad, *J. Cell Biol.* **22**, 443 (1964).
(37) K. D. Gibson, A. Neuberger, and J. J. Scott, *Biochem. J.* **61**, 618 (1955).
(38) R. Schmid and D. Shemin, *J. Am. Chem. Soc.* **77**, 506 (1955).
(39) S. Granick and D. Mauzerall, *J. Biol. Chem.* **232**, 1119 (1958).
(40) A. A. Iodice, D. A. Richert, and M. P. Schulman, *Federation Proc.* **17**, 248 (1958).
(41) A. Comfort, *Science* **112**, 279 (1950).
(42) L. Bogorad, *Science* **121**, 878 (1955).
(43) L. Bogorad, *J. Biol. Chem.* **233**, 501 (1958).
(44) H. Heath and D. S. Hoare, *Biochem. J.* **72**, 14 (1959).
(45) D. S. Hoare and H. Heath, *Biochem. J.* **73**, 679 (1959).
(46) J. Pluscec, L. Bogorad, and S. F. MacDonald, unpublished data (1965).
(47) D. Mauzerall, *J. Am. Chem. Soc.* **82**, 2605 (1960).
(48) L. Bogorad, *Plant Physiol.* **32**, xli (1957).
(49) L. Bogorad, *in* "Comparative Biochemistry of Photoreactive Systems" (M. B. Allen, ed.), p. 227. Academic Press, New York, 1960.
(50) A. T. Carpenter and J. J. Scott, *Biochem. J.* **71**, 325 (1959).
(51) A. T. Carpenter and J. J. Scott, *Biochim. Biophys. Acta* **52**, 195 (1961).
(52) L. Bogorad, *Ann. N. Y. Acad. Sci.* **104**, 676 (1963).
(53) J. E. Falk, E. I. B. Dresel, A. Benson, and B. C. Knight, *Biochem. J.* **63**, 87 (1956).
(54) E. I. B. Dresel and J. E. Falk, *Biochem. J.* **63**, 80 (1956).
(55) W. H. Lockwood and A. Benson, *Biochem. J.* **75**, 372 (1960).
(56) H. L. Booij and C. Rimington, *Biochem. J.* **65**, 4pp. (1957).
(57) L. Bogorad, *in* "Research in Photosynthesis" (H. Gaffron *et al.*, eds.), p. 475. Wiley (Interscience), New York, 1957.
(58) D. Mauzerall and S. Granick, *J. Biol. Chem.* **232**, 1141 (1958).
(59) S. Sano and S. Granick, *J. Biol. Chem.* **236**, 1173 (1961).
(60) R. J. Porra and J. E. Falk, *Biochem. Biophys. Res. Commun.* **5**, 179 (1961).
(61) R. J. Porra and J. E. Falk, *Biochem. J.* **90**, 69 (1964).
(62) E. F. Carell and J. S. Kahn, *Arch. Biochem. Biophys.* **108**, 1 (1964).
(63) H. Bénard, A. Gadjos, and M. Gadjos-Török, *Compt. Rend. Soc. Biol.* **145**, 538 (1951).
(64) S. Granick, L. Bogorad, and H. Jaffe, *J. Biol. Chem.* **202**, 801 (1953).

(65) S. Granick and S. Sano, Federation Proc. 20, 376 (1961).
(66) L. Bogorad, in "Methods in Enzymology" (S. P. Colowick and N. O. Kaplan, eds.), Vol. 5, p. 885. Academic Press, New York, 1962.
(67) G. S. Marks and L. Bogorad, unpublished data (1960).
(68) R. J. Porra and O. T. G. Jones, Biochem. J. 87, 186 (1963).
(69) G. Nishida and R. F. Labbe, Biochim. Biophys. Acta 31, 519 (1959).
(70) R. F. Labbe, N. Hubbard, and W. Caughey, Biochemistry 2, 372 (1963).
(71) H. Oyama, Y. Sugita, J. Yoneyama, and H. Yoshikaya, Biochem. Biophys. Acta 47, 413 (1961).
(72) H. C. Schwartz, R. L. Hill, G. E. Cartwright, and M. M. Wintrobe, Federation Proc. 18, 545 (1959).
(73) A. Neuberger and G. H. Tait, Biochem. J. 90, 607 (1964).
(74) H. N. Little and M. I. Kelsey, Federation Proc. 23, 223 (1964).
(75) J. H. C. Smith, J. Am. Chem. Soc. 69, 1492 (1947).
(76) G. H. Tait and K. D. Gibson, Biochim. Biophys. Acta 52, 614 (1961).
(77) M. Green, K. I. Altman, J. E. Richmond, and K. Solomon, Nature 179, 375 (1957).
(78) S. Granick, J. Biol. Chem. 172, 717 (1948).
(79) S. Granick, J. Biol. Chem. 175, 333 (1948).
(80) S. Granick, J. Biol. Chem. 183, 713 (1950).
(81) O. T. G. Jones, Biochem. J. 86, 429 (1963).
(82) R. Cooper, Biochem. J. 89, 100 (1963).
(83) K. D. Gibson, A. Neuberger, and G. H. Tait, Biochem. J. 83, 550 (1962).
(84) O. T. G. Jones, Biochem. J. 89, 182 (1963).
(85) R. Y. Stanier and J. H. C. Smith, Carnegie Inst. Wash. Year Book 58, 336 (1959).
(86) R. H. Goodwin and O. H. Owens, Plant Physiol. 22, 197 (1947).
(87) T. N. Godnev, A. A. Shlyk, and R. M. Rotfarb, Fiziol. Rast. 6, 36 (1959). [Plant Physiol. (USSR) (Engl. Transl.) 6, 33.]
(88) G. W. Bryan and L. Bogorad, (1963). Plant Cell Physiol. (Tokyo) p. 399 (special issue).
(89) G. W. Bryan, Ph.D. Thesis, University of Chicago (1962).
(90) R. Willstätter and A. Stoll, (1913). "Investigations on Chlorophyll" [English translation by F. M. Schertz and A. R. Mertz, (1928)], 385 pp. Science Press, Lancaster, Pennsylvania.
(91) P. Böger, Phytochemistry 4, 435 (1965).
(92) H. Fischer and A. Stern, "Die Chemie des Pyrrols" Vol. 2, Part II, 478 pp. Akad. Verlagsges., Leipzig, 1940.
(93) H. Mayer, Planta 11, 294 (1930).
(94) E. G. Sudyina, Photochem. Photobiol. 2, 181 (1963).
(95) M. Holden, Photochem. Photobiol. 2, 175 (1963).
(96) M. Holden, Biochem. J. 78, 359 (1961).
(97) S. Shimizu and E. Tamaki, Botan. Mag. (Toyko) 75, 462 (1962).
(98) S. Shimizu and E. Tamaki, Arch. Biochem. Biophys. 102, 152 (1963).
(99) C. Ardao and B. Vennesland, Plant Physiol. 35, 368 (1960).
(100) A. Klein and W. Vishniac, J. Biol. Chem. 236, 2544 (1961).
(101) W. R. Sistrom, M. Griffiths, and R. Y. Stanier, J. Cellular Comp. Physiol. 48, 459 (1956).
(102) M. Griffiths, J. Gen. Microbiol. 27, 427 (1962).
(103) O. T. G. Jones, Biochem. J. 91, 572 (1964).
(104) A. S. Holt, in "Chemistry and Biochemistry of Plant Pigments" (T. W. Goodwin, ed.), p. 3. Academic Press, New York, 1965.

(105) K. E. Eimhjellen, O. Aasmundrud, and A. Jensen, *Biochem. Biophys. Res. Commun.* **10**, 232 (1963).

(106) A. A. Shlyk and T. N. Godnev, *1st Intern. Conf. Radioisotopes Sci. Res.*, Vol. IV, pp. 479-493. Pergamon Press, Oxford, 1958.

(107) A. A. Shlyk, V. L. Kaler, L. I. Vlasenok, and V. I. Gaponenko, *Photochem. Photobiol.* **2**, 129 (1963).

(108) T. N. Godnev, R. M. Rotfarb, and N. K. Akulovich, *Photochem. Photobiol.* **2**, 119 (1963).

(109) T. N. Godnev, R. M. Rotfarb, and A. A. Shlyk, *Dokl. Akad. Nauk SSSR* **120**, 663 (1960).

(110) H. J. Perkins and D. W. A. Roberts, *Biochim. Biophys. Acta* **45**, 613 (1960).

(111) J. Duranton, J. M. Galmiche, and E. Roux, *Compt. Rend.* **246**, 992 (1958).

(112) U. Blass, J. M. Anderson, and M. Calvin, *Plant Physiol.* **34**, 329 (1959).

(113) R. S. Becker and R. K. Sheline, *Arch. Biochem. Biophys.* **54**, 259 (1955).

(114) A. A. Shlyk, S. A. Mikhailova, V. I. Gaponenko, and T. V. Kukhtenko, *Fiziol. Rast.* **10**, 275 (1963) [Plant *Physiol.* (*USSR*) (*English Transl.*) **10**, 227].

(115) M. R. Michel-Wolwertz, *Photochem. Photobiol.* **2**, 149 (1963).

(116) R. J. Della Rosa, K. I. Altman, and K. Salomon, *J. Biol. Chem.* **202**, 771 (1953).

(117) W. Brzeski and W. Rucker, *Nature* **185**, 922 (1960).

(118) S. Wieckowski, *Photochem. Photobiol.* **2**, 199 (1963).

(119) A. A. Shlyk and G. N. Nikolayeva, *Colloq. Intern. Centre Natl. Rech. Sci.* (*Paris*) **119**, 301 (1963).

(120) J. H. C. Smith, in "Comparative Biochemistry of Photoreactive Systems" (M. B. Allen, ed.), pp. 257-275. Academic Press, New York, 1960.

(121) J. H. C. Smith and C. S. French, *Ann. Rev. Plant Physiol.* **14**, 181 (1963).

(122) S. Kaufman and S. G. A. Alivisatos, *J. Biol. Chem.* **216**, 141 (1955).

(123) D. L. Nandi and E. R. Waygood, *Can. J. Biochem.* **43**, 1605 (1965).

(124) B. F. Burnham, *Acta Chem. Scand.* **17**, 123 (1963).

(125) K. Egle, in "Handbuch der Pflanzenphysiologie" (W. Ruhland, ed.), Vol. 5, Part I, p. 444. Springer, Berlin, 1960.

(126) D. J. C. Friend, *Physiol. Plantarum* **13**, 776 (1960).

(127) D. J. C. Friend, *Physiol. Plantarum* **14**, 28 (1961).

(128) R. M. Smillie and G. Krotkov, *Can. J. Botany* **39**, 891 (1961).

(129) F. Bukatsch and E. Rudolph, *Photochem. Photobiol.* **2**, 191 (1963).

(130) Z. Sesták, *Photochem. Photobiol.* **2**, 101 (1963).

(131) A. Madsen, *Photochem. Photobiol.* **2**, 93 (1963).

(132) H. I. Virgin, *Physiol. Plantarum* **8**, 630 (1955).

(133) M. Gassman and L. Bogorad, *Plant. Physiol.* **40**, lii (1965).

(134) M. M. Margulies, *Plant Physiol.* **37**, 473 (1962).

(135) M. M. Margulies, *Plant Physiol.* **39**, 579 (1964).

(136) L. Bogorad and A. B. Jacobson, *Biochem. Biophys. Res. Commun.* **14**, 113 (1964).

(137) N. K. Boardman, R. I. B. Francki, and S. G. Wildman, *Biochemistry* **4**, 872 (1965).

(138) G. Brawerman and J. M. Eisenstadt, *Biochim. Biophys. Acta* **91**, 477 (1964).

(139) E. H. L. Chun, M. H. Vaughn, and A. Rich, *J. Mol. Biol.* **7**, 130 (1963).

(140) M. F. Clark, R. E. F. Matthews, and R. K. Ralph, *Biochim. Biophys. Acta* **91**, 289 (1964).

(141) J. M. Eisenstadt and G. Brawerman, *J. Mol. Biol.* **10**, 392 (1964).

(*142*) A. Gibor and M. Izawa, *Proc. Natl. Acad. Sci. U. S.* **50**, 1164 (1963).

(*143*) A. B. Jacobson, H. Swift, and L. Bogorad, *J. Cell Biol.* **17**, 557 (1963).

(*144*) J. T. O. Kirk, *Biochim. Biophys. Acta* **76**, 417 (1963).

(*144a*)J. T. O. Kirk, *Biochem. Biophys. Res. Commun.* **16**, 233 (1964).

(*145*) J. Leff, M. Mandel, H. T. Epstein, and J. A. Schiff, *Biochem. Biophys. Res. Commun.* **13**, 126 (1963).

(*146*) J. W. Lyttleton, *Biochem. J.* **74**, 82 (1960).

(*147*) J. W. Lyttleton, *Exptl. Cell Res.* **26**, 312 (1962).

(*148*) N. Kislev, H. Swift, and L. Bogorad, *J. Cell Biol.* **25**, 327 (1965).

(*149*) D. S. Ray and P. C. Hanawalt, *J. Mol. Biol.* **9**, 812 (1964).

(*150*) H. Ris and W. Plaut, *J. Cell Biol.* **13**, 383 (1962).

(*151*) R. Sager and M. R. Ishida, *Proc. Natl. Acad. Sci. U. S.* **50**, 725 (1963).

(*152*) J. Semal, D. Spencer, Y. T. Kim, and S. G. Wildman, *Biochim. Biophys. Acta* **91**, 205 (1964).

(*153*) N. M. Sissakian, I. I. Filippovich, E. N. Svetailo, and K. A. Aliyev, *Biochim. Biophys. Acta* **95**, 474 (1965).

(*154*) C. R. Stocking and E. M. Gifford, *Biochem. Biophys. Res. Commun.* **1**, 159 (1959).

(*155*) A. Gibor and S. Granick, *Science* **145**, 890 (1964).

(*156*) L. Bogorad, in "Molecular Organization and Biological Function" (J. W. Allen, ed.), Harper and Row, New York, 1966.

(*157*) C. B. van Niel, *Bacteriol. Rev.* **8**, 1 (1944).

(*158*) G. Cohen-Bazire, W. R. Sistrom, and R. Y. Stanier, *J. Cellular Comp. Physiol.* **49**, 25 (1957).

(*159*) J. Lascelles, *Biochem. J.* **72**, 508 (1959).

(*160*) J. Lascelles, in "Bacterial Photosynthesis" (H. Gest, A. San Pietro, and L. P. Vernon, eds.), p. 35. Antioch Press, Yellow Springs, Ohio, 1963.

(*161*) J. Lascelles, *J. Gen. Microbiol.* **23**, 487 (1960).

(*162*) M. J. Bull and J. Lascelles, *Biochem. J.* **87**, 15 (1963).

(*163*) W. R. Sistrom, *J. Gen. Microbiol.* **28**, 599 (1962).

(*164*) M. Higuchi, K. Goto, M. Fujimoto, O. Namiki, and G. Kikuchi, *Biochim. Biophys. Acta* **95**, 94 (1965).

(*165*) R. K. Clayton, *J. Biol. Chem.* **235**, 405 (1960).

(*166*) J. Lascelles, *J. Gen. Microbiol.* **23**, 499 (1960).

(*167*) H. U. Marsh, Jr., H. J. Evans, and G. Matrone, *Plant Physiol.* **38**, 632 (1963).

(*168*) T. E. Treffry, Ph.D. Thesis, University of Chicago (1965).

(*169*) D. van Noort and A. Wallace, *Biochem. Biophys. Res. Commun.* **10**, 109 (1963).

(*170*) N. G. Perur, R. L. Smith, and H. H. Wiebe, *Plant Physiol.* **36**, 736 (1961).

(*171*) J. Lascelles, *Biochem. J.* **62**, 78 (1956).

(*172*) L. Provasoli, S. H. Hutner, and A. Schatz, *Proc. Soc. Exptl. Biol. Med.* **69**, 279 (1948).

(*173*) A. Lwoff, and P. Schaeffer, *Compt. Rend.* **228**, 779 (1949).

(*174*) W. G. Rosen and S. R. Gawlik, *J. Protozool.* **8**, 90 (1961).

(*175*) J. J. Wolken, *J. Protozool.* **3**, 211 (1956).

(*176*) N. C. H. Tong, J. A. Gross, and T. L. Jahn, *J. Protozool.* **12**, 153 (1965).

(*177*) E. G. Pringsheim and O. Pringsheim, *New Phytologist* **51**, 65 (1952).

(*178*) W. J. Robbins, A. Hervey, and E. Stebbins, *Ann. N. Y. Acad. Sci.* **56**, 818 (1953).

(*179*) A. Lwoff, "Problems of Morphogenesis in Ciliates." Wiley, New York, 1950.

(*180*) J. A. Gross, T. L. Jahn, and E. Bernstein, *J. Protozool.* **2**, 71 (1955).

—16—

Distribution of the Chlorophylls

M. B. ALLEN

Laboratory of Physical Biology, National Institute of Arthritis and Metabolic Diseases, Bethesda, Maryland

I. Introduction

The distribution of the different chlorophylls among the various divisions of organisms that carry out photosynthesis with evolution of oxygen follows a regular and evolutionary conservative pattern that, with a few exceptions, corresponds to that found in the classification of these organisms by other taxonomic criteria. The main lines of this pattern, summarized in Table I, have been known for some time and have been often reviewed, so that discussion of the general picture here will be brief. Some of the exceptions to the accepted outline and a few problems that remain will be mentioned.

By contrast, the photosynthetic bacteria have a much more irregular pattern of chlorophyll distribution. Different chlorophylls may even be found in different isolates of what has been considered the same species. The nature of the differences between the numerous bacterial chlorophylls that have been reported and the reasons for this irregularity in the bacteria, as contrasted with the higher forms, still pose many problems.

TABLE I

DISTRIBUTION OF CHLOROPHYLLS AMONG OXYGEN-EVOLVING
PHOTOSYNTHETIC ORGANISMS

Group	Chlorophyll			
	a	b	c	d
Algal division[a]				
Cyanophyta	+	−	−	−
Rhodophyta	+	−	−	±
Cryptophyta	+	−	+	−
Pyrrophyta	+	−	+	−
Bacillariophyta	+	−	+	−
Phaeophyta	+	−	+	−
Chrysophyta	+	−	±	−
Xanthophyta	+	−	−	−
Euglenophyta	+	+	−	−
Chlorophyta	+	+	−	−
Bryophyta	+	+	−	−
Vascular plants	+	+	−	−

[a] Various systems of classification of the algae are in existence. The form used in this table is taken from that of Silva (1). For a general discussion see, for example, Papenfuss (2).

II. Chlorophylls of Oxygen-Evolving Photosynthetic Organisms

A. Chlorophyll *a*

This chlorophyll is present as a major pigment in all the organisms of this type that have been studied. It is to be expected that it will be found in the only group not so far examined, the chloromonad flagellates, when these organisms have been successfully made available for such studies. Apparently Chl *a* was such a suitable product of evolutionary biochemical experimentation that it has persisted throughout all subsequent development of the various divisions of the photosynthetic world. Numerous organisms exist with this as their only chlorophyll, and mutant strains lacking other chlorophylls have been reported in both algae and higher plants, but no organism is known that possesses only one of the other chlorophylls and not Chl *a*.

B. Chlorophyll *b*

As indicated in Table I, this chlorophyll is generally characteristic of the green plant line, including the Chlorophyta, the Euglenophyta, and all higher plants [cf. the record of the large number of organisms examined by Strain (3)]. There have, however, been several instances recorded of organisms in this line that lack Chl *b*. It has been reported missing from one higher plant, the parasitic orchid *Neottia* (*4*), from the desmid *Closterium* (*5*), and from the small green marine alga *Nannochloris oculata* (*6*). It was, however, found by Jeffrey in a related species, *N. atomus* (*7*). *Nannochloris* is a marine "weed" that often grows abundantly in marine environments, e.g., cultures, in which abundant nutrition is provided, much as *Chlorella* does in similar freshwater situations. It is possible that, as has been suggested for *Chlorella* (*8*), a number of different organisms have been lumped together under this name.

If the unusual hot spring alga *Cyanidium caldarium* is a chlorophyte, it also is an exceptional form that lacks chlorophyll *b* (*9*). If, as is generally held (*1*), it is a cryptophyte, it is equally anomalous in lacking chlorophyll *c*.

C. Chlorophyll *c*

This important chlorophyll, which is at present the least known of the major photosynthetic pigments, deserves more study. Organisms containing Chl *c* carry out most of the photosynthesis in the oceans, playing the same dominant role as the green plants on land. Although estimates of the relative amount of photosynthesis carried out on land and in the seas vary, there can be no doubt that an appreciable fraction of the total photosynthesis on earth is carried out in the marine environment by those organisms that contain Chl *c*.

Chl *c* is, so far as is known, universally present in the Phaeophyta (brown algae), Bacillariophyta (diatoms), and Pyrrophyta (dinoflagellates), although the numbers of organisms of these types examined are far fewer than those in the green plant line. It is absent from some members of the Chrysophyta, such as *Ochromonas* (*6*), but is present in high concentrations in a number of marine flagellates of this group (*7*). Here, again, more extensive studies are needed.

Chl *c* is difficult to detect from absorption spectra of living cells because of its very weak absorption band in the red (cf. Chapter 6), and is difficult to obtain pure because of its great tendency to isomerize. Nevertheless, enough measurements of its concentration in cells have been made to indicate that it must be considered a major photosynthetic pigment. Several marine organisms have been found to contain about 25% of their total chlorophyll as *c*; *Coccolithus huxleyii*, which is an important primary producer in many ocean areas, contains almost as much Chl *c* as *a* (*10*). There are numerous measurements of total phytoplankton pigments in samples of seawater that indicate the presence of more chlorophyll *c* than *a*. Although the methods used in some of these measurements are open to criticism, it seems unlikely that they can be sufficiently in error to account for this result, so it will not be surprising if increased knowledge of oceanic phytoplankton reveals organisms with large amounts of Chl c. It must be mentioned, however, that pigment analyses on water samples measure not only the chlorophyll in living cells, but also that in dead cells and cell debris, and that there is some evidence that Chl c is more stable than a when cells are disintegrated (10). This would give higher apparent values for Chl c in water samples than those actually present in the organisms in the water.

D. Chlorophyll d

This chlorophyll is often considered to be characteristic of the Rhodophyta (red algae). However, it is by no means universally present in this group, and even when present is usually only a trace constituent [here again, cf. the extensive analysis of a large number of species by Strain (3)]. The quantities found in those algae in which it does occur are variable, apparently depending on environmental conditions. For example, it has been reported to constitute up to 33% to the chlorophyll of Rhodochorton rothii (3). However, some samples of this algae have been found to contain no Chl d at all (11).

Chl d has never been definitely detected in the absorption spectra of living red algal cells or thalli, although its absorption maximum on the long-wavelength side of that of Chl a should make such detection possible if it were present in more than trace quantities. It can be formed in vitro as an oxidation product of Chl a (11), and there is no direct evidence that it is a functional photosynthetic pigment in the red algal cell.

A pigment with the spectral properties of Chl d has recently been isolated from extracts of Chlorella pyrenoidosa (12). The authors suggest that this pigment may be responsible for the 695–700-mμ absorption band observed in living cells (cf. Chapter 11).

E. Chlorophyll e

This chlorophyll is often cited as being characteristic of the Xanthophyta (Heterokontae; yellow-green algae). However, it has only twice been found in members of this group, and has not been observed in pure cultures. Its present status is probably best summed up in the statement of Strain (3), who first reported its occurrence: "In 1943 traces of a chlorophyll (chlorophyll e), with absorption maxima at 415 and 654 mμ in methanol, were isolated from a natural growth of Tribonema bombycinum. In 1948 the same or a similar chlorophyll was found in a natural stand of *Vaucheria hamata*. This chlorophyll has not been observed in the small quantities of the species that were cultured. Whether or not it is a trace constituent of the Heterokontae algae, a secondary product, or a pigment from algal contamination has yet to be established."

F. Other Chlorophylls and Chlorophyll Derivatives

Isomers of chlorophylls *a* and *b*, as well as chlorophyll derivatives, including allomerized chlorophyll, chlorophyllides, pheophytins, and pheophorbides, are frequently found in cell extracts prepared with organic solvents. These have usually been considered to be formed from the chlorophylls during the extraction process, rather than being normal constituents of the cell (3). However, it has been shown that chlorophyllide *a* is present during the initial period of greening of etiolated leaves (13). Michel-Wolwertz and Sironval (14) have recently isolated a number of chlorophyll isomers from extracts of *Chlorella vulgaris*. Since these isomers were stable under their extraction conditions and during the preparation of derivatives, the authors consider it likely that they are present in the cells, and they have suggested that the different chlorophyll absorption bands observed in living cells, which are generally considered to be due to different modes of combination between Chl *a* and lipoprotein (cf. Chapter 11), are due to combinations of lipoprotein with these different isomeric chlorophylls (15). This is, however, difficult to reconcile with the loss of some of these bands in *Ochromonas danica* on mild heating or cell breakage (6).

During the last few years there have been several reports of pigments in various algae that absorb at longer wavelengths than the known chlorophylls (16–18). These long-wavelengths absorbing pigments are highly variable in amount and sensitive to environmental conditions, but may in some cases amount to an appreciable fraction of the concentration of chlorophyll in the cell, assuming similar absorption coefficients. They are usually more pronounced in old cultures. Brown (19) showed that the 710-mμ absorption band in old cultures of *Euglena gracilis* was due to a pheophytin-protein combination. A strong absorption at 710 mμ also appears in old cultures of certain mutants of *Chlorella pyrenoidosa* (16). Correlated with this is a marked increase in the pheophorbide content of cell extracts, so that it is possible that in this case the absorbance, which was originally thought to be due to an unusual chlorophyll complex, may be due to a pheophorbide-protein combination (20). Other cases of absorption of light at wavelengths past 700 mμ have also been ascribed to special chlorophyll-protein complexes, but the evidence for their nature is not extensive.

III. Bacterial Chlorophylls

As mentioned in the Introduction, the photosynthetic bacteria display a much more complex pattern of chlorophyll distribution than the oxygen-evolving photosynthetic organisms. In three families of photosynthetic bacteria, including among them around forty species, a larger number of chlorophylls has been found than in all other photosynthetic organisms combined. These chlorophylls have largely been identified by their spectral properties; much remains to be learned of their structure. Bacteriochlorophyll (Bchl) *a*, which was previously considered to be the only chlorophyll present in the purple bacteria (*21, 22*), is the only one for which the proposed chemical structure has been generally accepted (*23*).

Eimhjellen *et al.* (*24*) discovered another chlorophyll, Bchl *b*, in a *Rhodopseudomonas* sp., and the same chlorophyll has been found in a related, if not identical, organism isolated by Lascelles (*25*). All the Thiorhodaceae so far examined contain only Bchl *a*, as does *Rhodomicrobium vannielii*.

The green photosynthetic bacteria have recently been found to contain a number of different chlorophylls. These bacteria have long been known to have a characteristic chlorophyll, named bacterioviridin by Metzner (*26*) and renamed chlorobium chlorophyll by Larsen (*27*), who studied the two green bacteria *Chlorobium limicola* and *Chlorobium thiosulfatophilum*. Later, Stanier and Smith (*28*) found that another strain of *Chlorobium thiosulfatophilum* contained a similar yet different chlorophyll, with its principal light absorption peak at 650 mμ (in ether), compared to 660 mμ for the chlorobium chlorophyll studied by Larsen. These will be referred to as Cchl 650 and Cchl 660, respectively. Measurements on several strains of *Chlorobium limicola* have shown that a similar situation exists in this bacterium (*25*). Only Cchl 660 has been found in the photoorganotrophic green bacterium *Chloropseudomonas ethylicum*.

It is possible that the chlorobium chlorophylls are not homogeneous compounds. Holt *et al.* (*29*) obtained six pheophorbides from Cchl 650 and seven from Cchl 660. The chlorophylls themselves, however, have not been separated into such a large number of components. Jensen *et al.* (*25*) obtained three zones on chromatography of Cchl 650. The absorption spectra of all three fractions were similar, if not identical.

Olson and Romano (*30*) reported the presence of an additional chlorophyll in green bacteria, which has been shown to be identical with Bchl *a* (*29*). An additional minor chlorophyll, distinct from either of the chlorobium chlorophylls and from Bchl *a* has been isolated from *Chlorobium thiosulfatophilum* strain L (25).

The distribution of the chlorophylls in the photosynthetic bacteria is summarized in Table II. The earlier data, based on spectroscopic observations on whole cells or cell fractions, were subject to question because of the possibility that different complexes of the same pigment will

TABLE II

CHLOROPHYLLS OF PHOTOSYNTHETIC BACTERIA

Organism	Type of chlorophyll
Thiorhodaceae	
Thiocystis violacea	Bchl *a*
Chromatium sp.	Bchl *a*
Chromatium strain D	Bchl *a*
Chromatium okenii	Bchl *a*
Chromatium vinosum	Bchl *a*
Chromatium warmingii	Bchl *a*
Thiospirillum jenense	Bchl *a*
Athiorhodaceae	
Rhodopseudomonas palustris	Bchl *a*
Rhodopseudomonas spheroides	Bchl *a*
Rhodopseudomonas capsulata	Bchl *a*
Rhodopseudomonas gelatinosa	Bchl *a*
Rhodopseudomonas sp.	Bchl *b*
Rhodospirillum rubrum	Bchl *a*
Rhodospirillum photometricum	Bchl *a*
Rhodospirillum molischianum	Bchl *a*
Hyphomicrobaceae	
Rhodomicrobium vannielii	Bchl *a*
Chlorobacteriaceae	
Chlorobium limicola	Cchl 660 (and 650?), Bchl *a*
Chlorobium thiosulfatophilum, LI	Cchl 650, Bchl *a*, unknown Chl
Chlorobium thiosulfatophilum, NCIB 8346	Cchl 660, Bchl *a*, unknown Chl
Chloropseudomonas ethylicum	Cchl 660, Bchl *a*, unknown Chl

show different absorption bands. However, most of these determinations have now been verified by chromatographic separation of the pigments and determination of their absorption spectra. (25).

IV. Concluding Remarks

The principal problem at present in studies of chlorophyll distribution, both in bacteria and in higher organisms, appears to be the relation of the various isomeric forms of the chlorophylls, and of the chlorophyll derivatives, found in extracts to the functional pigments of the living cell. Since there are pronounced variations in the absorption bands of the chlorophylls in different organisms (cf. Chapter 11), careful comparative studies should make it possible to decide whether these isomers and derivatives are responsible for the variations in chlorophyll absorbance *in vivo* and for some of the functionally different chlorophyll complexes of photosynthetic organisms.

REFERENCES

(1) P. C. Silva, in "Physiology and Biochemistry of Algae" (R. A. Lewin, ed.), pp. 827-837. Academic Press, New York, 1962.

(2) G. F. Papenfuss, in "A Century of Progress in the Natural Sciences," pp. 115-224. Calif. Acad. Sci., San Francisco, California, 1955.

(3) H. H. Strain, "Chloroplast Pigments and Chemical Analysis." Penn State Univ. Press, University Park, Pennsylvania, 1958.

(4) C. S. French, in "Handbuch der Pflanzenphysiologie" (W. Ruhland, ed.), Vol. 5, p. 269. Springer, Berlin, 1960.

(5) R. Barer, Science 121, 709-715 (1955).

(6) M. B. Allen, C. S. French, and J. S. Brown, in "Comparative Biochemistry of Photoreactive Systems" (M. B. Allen, ed.), p. 39. Academic Press, New York, 1960.

(7) S. W. Jeffrey, Biochem. J. 80, 336-342 (1961).

(8) G. M. Smith, "Fresh Water Algae of the United States." McGraw-Hill, New York, 1933.

(9) M. B. Allen, Arch. Mikrobiol. 32, 270-277 (1959).

(10) S. W. Jeffrey and M. B. Allen, J. Gen. Microbiol. 36, 277-288 (1964).

(11) A. S. Holt and H. V. Morley, Can. J. Chem. 37, 507-514 (1959).

(12) M. R. Michel-Wolwertz, C. Sironval, and J. C. Goedheer, Biochim. Biophys. Acta 94, 584-585 (1965).

(13) J. B. Wolff and L. Price, Arch. Biochem. Biophys. 72, 293 (1957).

(14) M. R. Michel-Wolwertz and C. Sironval, Biochim. Biophys. Acta 94, 330-343 (1965).

(15) C. Sironval, M. R. Michel-Wolwertz, and A. Madsen, Biochim. Biophys. Acta 94, 344-354 (1965).

(16) M. B. Allen, in "Light and Life" (W. D. McElroy and B. Glass, eds.), pp. 479-480. Johns Hopkins Press, Baltimore, Maryland, 1961.

(17) Govindjee, C. Cederstrand, and E. Rabinowitch, Science 134, 391-392 (1961).

(18) E. H. Gassner, Plant Physiol. 37, 637-639 (1962).

(19) J. S. Brown, Biochim. Biophys. Acta 75, 299-305 (1963).

(20) M. B. Allen, unpublished observations (1964).

(21) C. B. van Niel in "Bergey's Manual of Determinative Bacteriology" (R. S. Breed, E. G. D. Murray, and N. R. Smith, eds.), 7th ed., p. 35. Williams & Wilkins, Baltimore, Maryland, 1957.

(22) C. B. van Niel, Bacteriol. Rev. 8, 1 (1944).

(23) H. Fischer, R. Lambrecht, and H. Mittenzwei, Z. Physiol. Chem. 253, 1 (1938).

(24) K. E. Eimhjellen, O. Aasmundrud, and A. Jensen, Biochem. Biophys. Res. Commun. 10, 232 (1963).

(25) A. Jensen, O. Aasmundrud, and K. E. Eimhjellen, *Biochim. Biophys. Acta* **88**, 466 (1964).

(26) P. Metzner, *Ber. Deut. Botan. Ges.* **40**, 125 (1922).

(27) H. Larsen, *Kgl. Norske Videnskab. Selskabs, Skrifter* No. 1, p. 199 (1953).

(28) R. Y. Stanier and J. H. C. Smith, *Biochim. Biophys. Acta* **41**, 478 (1960).

(29) A. S. Holt, D. W. Hughes, H. J. Kende, and J. W. Purdie, *Plant Cell. Physiol. (Tokyo)* **4**, 49 (1963).

(30) J. M. Olson and C. A. Romano, *Biochim. Biophys. Acta* **59**, 726 (1962).

Section IV
Photochemistry and Photophysics

—17—

Photochemistry of Chlorophylls in Vitro*

G. R. SEELY

Charles F. Kettering Research Laboratory, Yellow Springs, Ohio

I. Introduction

The earliest investigations of chlorophyll photochemistry, near the beginning of the century, were concerned with photosensitization of photographic emulsions to red light, and with photodynamic action on animals. In the 1920's, the study of chlorophyll-sensitized oxidation of various organic compounds by O_2 was begun by Noack, Meyer, Gaffron, and others. Sensitized reactions involving oxidants other than oxygen were first systematically studied by Böhi, who observed the bleaching of a number of azo dyes and other oxidants in the presence of chlorophyll, phenylhydrazine, and light (1). One azo dye, methyl red, has been studied more extensively than the others, beginning with Ghosh and Sen Gupta in 1934 (2). Because this dye is reduced irreversibly to colorless products, it has been employed extensively as an indicator of photo-sensitizing capacity of chlorophyll preparations, especially by Evstigneev and co-workers. Although the bleaching of chlorophyll in the presence

---

* Contribution No. 217 from the Charles F. Kettering Research Laboratory, Yellow Springs, Ohio. The preparation of this chapter was supported in part by National Science Foundation Grant No. GB-2089.

of light and air had been known for a long time, the discovery in 1948 by Krasnovskii of the photoreduction of chlorophyll by ascorbic acid in pyridine solution opened up a new area of interest in the photochemical changes of chlorophyll itself (3). The development of the technique of flash photolysis (4), making it possible to follow transients during photochemical reactions, has paved the way toward a more complete understanding of the basic mechanisms of photochemical reactions of chlorophylls.

The photochemistry of chlorophyll before 1948 has been reviewed by Fischer and Stern (5) and by Rabinowitch (6); reviews of later work include those of Rabinowitch (7), Livingston (8), Rosenberg (9), and Krasnovskii (10, 11). The subject of photodynamic action, important not for chlorophyll a itself but for one of its derivatives, phylloerythrin, is beyond the scope of this chapter; the interested reader is referred to the reviews of Clare (12), and of Spikes and Glad (13).

II. The Initiation of Photochemical Processes by Chlorophylls

A. Photochemical Mechanisms Applicable to Chlorophylls

1. PRIMARY PHOTOCHEMICAL REACTIONS

a. Population of Excited States. Absorption of a quantum of light with the energy hv^* of the lowest frequency absorption band of a chlorophyll raises it to the lowest vibrational level of its first singlet excited state (Chl^*). Light of higher energy excites a chlorophyll to higher

$$Chl \xrightarrow{\quad hv^* \quad} Chl^* \qquad (1)$$

vibrational and electronic singlet states, but the extra energy is very rapidly lost to the environment as the random translational energy of heat, until the lowest vibrational level of the first singlet excited state is reached. Energy may be lost from this level by *fluorescence* [Eq. (2)], *internal conversion* [for the accepted meanings of this and other photochemical terms, see Pitts *et al.* (13a)] to the ground state [Eq. (3)], *intersystem crossing* to the lowest triplet excited state Chl' [Eq. (4)], or by reactions to be listed shortly. The lifetime of the singlet excited state, defined by $\tau^* = 1/(k_2 + k_3 + k_4)$, is only about 5×10^{-9} sec for chlorophyll a (14).

$$Chl^* \xrightarrow{\quad k_2 \quad} Chl + light \qquad (2)$$

$$Chl^* \xrightarrow{\quad k_3 \quad} Chl + heat \qquad (3)$$

$$Chl^* \xrightarrow{\quad k_4 \quad} Chl' \qquad (4)$$

A chlorophyll in its triplet excited state may emit its energy as phosphorescence, but this process is unimportant for most chlorophylls, even at low temperatures. Instead, the energy is lost either by radiationless processes [Eq. (5)] to the ground state, or by reaction with other molecules. Because the lifetime of Chl′, $\tau' = 1/k_5$, ($\sim 10^{-3}$ sec for chlorophyll a), is much longer than that of Chl*, most photochemical reactions of

$$\text{Chl}' \xrightarrow{\ k_5\ } \text{Chl} + \text{heat} \tag{5}$$

chlorophyll that have been studied go through the triplet state.

b. Energy Transfer. A singlet excited chlorophyll molecule may lose energy by radiationless transfer to a nearby molecule (M), in a process described by Förster (*15, 16*), if M has a singlet excited state of the

$$\text{Chl}^* + \text{M} \xrightarrow{\ k_6\ } \text{Chl} + \text{M}^* \tag{6}$$

proper energy. The rate constant for the process, k_6, for randomly oriented molecules in a medium of refractive index n, depends on the distance R between molecules, and the overlap between the normalized fluorescence emission spectrum $\bar{f}(\lambda)$ of the energy donor and the molar extinction coefficient spectrum $\varepsilon(\lambda)$ of the acceptor, according to Eq. (7) (*17*).

$$k_6 = \frac{3000 \, k_2 \ln 10}{64\pi^5 \, N \, n^4 \, R^6} \int_0^{\infty} \varepsilon(\lambda) \, \bar{f}(\lambda) \, \lambda^4 \, d\lambda \ \sec^{-1} \tag{7}$$

(N is Avogadro's number; $1/k_2$ is called the *natural lifetime* of the excited state.)

The overlap integral in Eq. (7), and therefore the probability of energy transfer, is largest when the energy of the singlet excited state of the acceptor is just a little lower than that of the donor. Energy transfer by the Förster process can occur over distances of 50 Å or more in solution.

Of course, chlorophylls can receive energy by transfer from some other excited molecule, and energy transfer from chlorophyll to chlorophyll is possible in concentrated solution. Transfer of triplet state energy,

$$\text{Chl}' + \text{M} \xrightarrow{\ k_8\ } \text{Chl} + \text{M}' \tag{8}$$

Eq. (8), also occurs, not by the Förster process, but by an electron exchange process requiring much closer approach of the two molecules (*18*). Nevertheless, transfer of triplet state energy can be very efficient, because of the long life of the triplet state.

c. Chemical Reaction. If there are molecules present that can act as oxidants (Ox), or reductants (Red), the energy of excitation may be expended in transferring an electron by reactions (9)–(12). These reactions are much faster if not coupled with proton transfer, in accord

$$\text{Chl}^\circ + \text{Ox} \xrightarrow{k_9} \text{Chl} \cdot^+ + \text{Ox} \cdot^- \tag{9}$$

$$\text{Chl}^\circ + \text{Red} \xrightarrow{k_{10}} \text{Chl} \cdot^- + \text{Red} \cdot^+ \tag{10}$$

$$\text{Chl}' + \text{Ox} \xrightarrow{k_{11}} \text{Chl} \cdot^+ + \text{Ox} \cdot^- \tag{11}$$

$$\text{Chl}' + \text{Red} \xrightarrow{k_{12}} \text{Chl} \cdot^- + \text{Red} \cdot^+ \tag{12}$$

with the Franck-Condon principle that electronic motion is much faster than nuclear motion. Another possible process is electron ejection, [Eq. (13)], a two-quantum process for aromatic amines (19, 20), and which

$$\text{Chl}' \xrightarrow{\text{light?}} \text{Chl} \cdot^+ + e^- \tag{13}$$

has probably been observed for chlorophyll in glasses at 77°K (21, 22).

Photoexcited chlorophyll may react with some component of the solvent in a way not classifiable as oxidation or reduction, as in the photosensitized displacement of magnesium by weak acids (pheophytinization). No other clear-cut examples of this kind of reaction are known, however.

d. Rate Limitations. Unless chlorophyll forms a complex with the reagent in the dark, the rate constants for the bimolecular reactions (8)–(12) are limited by the number of collisions per second between the reactive species. The value for the collision limited rate constant, originally calculated from diffusion theory to be $k = 8RT/3000 \, \eta$ (23), has been corrected by Osborne and Porter (24) to $k = 8RT/2000 \, \eta$ to allow for "slip" between the diffusing molecule and the medium. For most common solvents at room temperature, the value of k from the latter expression is ca. $1 \times 10^{10} \, M^{-1} \, \text{sec}^{-1}$, and values in this range were in fact found for the quenching of the triplet state of naphthalene in solvents of low viscosity (24). Quite often, values of rate constants for fast reactions such as the quenching of triplet states are about tenfold lower, approximating $10^9 \, M^{-1} \, \text{sec}^{-1}$. It is not clear whether the difference between this and the theoretical value means that only one collision out of ten is effective, or whether the number of collisions has been overestimated. In this chapter, reactions with rate constants ca. $10^9 \, M^{-1} \, \text{sec}^{-1}$, as well as those with larger rate constants, will be called *diffusion controlled*.

2. SECONDARY REACTIONS

a. Quantum Yields. The *quantum yield* (ϕ) of a product of a photochemical reaction is defined as the number of moles of product formed per Einstein (N quanta) of light absorbed by the photosensitive agent. The *primary quantum yield* (ϕ_0) of a reaction such as (11) by which electronic excitation energy is annihilated, is the rate of that reaction divided by the rate of absorption of light by the photosensitive agent. For example, if the set of possible reactions includes Eqs. (2)–(6), (8), and (11), the primary quantum yield of reaction (11) is

$$\phi_0 = \frac{k_4}{k_2 + k_3 + k_4 + k_6 \, [M]} \cdot \frac{k_{11} \, [Ox]}{k_5 + k_8 \, [M] + k_{11} \, [Ox]} \tag{14}$$

b. Cage Effects. The radicals formed in a reaction such as (11) are prevented from separating very far by the resistance of the medium, and there is a large probability that they will react to regenerate starting materials. This so-called "cage" effect has long been recognized as one that may cause quantum yields to be smaller than expected.

If reaction (11), for example, is broken into the three steps (11a)–(11c), in which braces enclose caged radicals, the expression for the

$$\text{Chl}' + \text{Ox} \xrightarrow{\ k_{11a}\ } \{\text{Chl} \cdot {}^{+} + \text{Ox} \cdot {}^{-}\} \tag{11a}$$

$$\{\text{Chl} \cdot {}^{+} + \text{Ox} \cdot {}^{-}\} \xrightarrow{\ k_{11b}\ } \text{Chl} + \text{Ox} \tag{11b}$$

$$\{\text{Chl} \cdot {}^{+} + \text{Ox} \cdot {}^{-}\} \xrightarrow{\ k_{11c}\ } \text{Chl} \cdot {}^{+} + \text{Ox} \cdot {}^{-} \tag{11c}$$

quantum yield of $\text{Ox} \cdot {}^{-}$ must contain the factor $k_{11c}/(k_{11b} + k_{11c})$.

Noyes has recently reviewed in detail the processes of cage recombination, which are of great importance to photochemistry (25).

c. Follow-up Reactions. The production of radicals in a reaction such as (11c) may be only the first of a long and complex series of reactions leading to a final, stable product (e.g., OxH_2). The quantum yield of the final product may be quite different from that of reaction (11c); it is usually smaller, but for chain reactions it may be larger. Some sensitized oxidations with oxygen probably belong in the latter category.

The nature of the follow-up reactions varies according to the reagents, and it is impossible to generalize. A simple, complete set of follow-up reactions is the disproportionation of the radicals in pairs. The reduction of chlorophyllide by ascorbic acid in pyridine can be interpreted essentially in terms of this set (see Section III, A, 1, a.)

B. Direct Evidence for Primary Reactions of Chlorophylls

1. THE TRIPLET STATE

The existence of a long-lived state of chlorophyll, intermediate in energy between the ground and first excited singlet states, was originally postulated to explain the kinetics of photochemical reactions. This state is identified as the triplet state by analogy with other dyes (26). Although its magnetic susceptibility has not been measured, the enhancement of phosphorescence by the paramagnetic metal in Cu pheophytin a supports this belief. An electron spin resonance (ESR) signal belonging to the triplet state of chlorophyll b in illuminated ethanol solution has been reported by Rikhireva et al. (27). With the aid of flash photolysis, absorption spectra believed to be of the triplet states of chlorophyll a (28–31), chlorophyll b (28–32), pheophytin a (28, 29, 31), and bacteriochlorophyll (33) have been obtained independently by Livingston, Linschitz, Claesson, Witt, and their associates. In addition, triplet state spectra of tetraphenylporphin (33), protoporphyrin (34), and their zinc complexes, coproporphyrin dimethyl ester (28, 29) and Mg phthalocyanine (34), have been published. These spectra all have relatively featureless absorption extending into the infrared, the most prominent band lying just to the red of the position of the Soret band in the ground-state spectrum.

In fluid solutions (as distinct from glasses) the triplet state of a chlorophyll decays according to Eq. (15) (29).

$$-\frac{d\,[\text{Chl}']}{dt} = K_1\,[\text{Chl}'] + K_2\,[\text{Chl}']^2 + K_3\,[\text{Chl}']\,[\text{Chl}] \tag{15}$$

K_1 and K_2 are temperature dependent, but their activation energies are comparable to the "activation energy" for viscous flow of the solvent. Table I collects known values for these constants, first measured accurately by Linschitz and Sarkanen (30). Reaction (16) is evidently diffusion controlled, but it is not known whether one or both excited

$$\text{Chl}' + \text{Chl}' \xrightarrow{\;\;K_2\;\;} 2\,\text{Chl or Chl}' + \text{Chl} \tag{16}$$

molecules are quenched. It is possible that the reaction characterized by K_3 is quenching of Chl′ by a minor, perhaps carotenoid, impurity.

McCartin has listed values of K_1 for a number of porphins and phthalocyanines, in solvents of varying viscosity and at temperatures down to 77°K, where the bimolecular process [Eq. (16)] is suppressed (38).

This process [Eq. (16)] is important only at light intensities comparable to those used in flash photolysis. Unless great care is exercised to ensure purity of the chlorophyll and the solvent, Chl′ is usually deactivated largely through quenching by O_2 or other impurities.

2. ENERGY TRANSFER

a. *Singlet–singlet.* If the acceptor M in reaction (6) is fluorescent, energy transfer is manifested by an increase of the fluorescence of M at the expense of fluorescence of Chl. Thus, Watson and Livingston detected transfer of energy from chlorophyll b to chlorophyll a in solution (39), and Duysens detected it in *Chlorella* (40). The fluorescence of chlorophyll a in monolayers is quenched by energy transfer to non-fluorescent Cu pheophytin a (41).

The fluorescence of chlorophylls may be sensitized by transfer of energy from other molecules: from carotenoids to bacteriochlorophyll in bacteria, from phycobilins to chlorophyll a in *Porphyra lacineata,* and from phycoerythrin to phycocyanin to chlorophyll a in *Porphyridium cruentum* (40). Transfer from fucoxanthol or lutein to chlorophylls a or b was detected in detergent micelles (42). The extent of energy transfer is in approximate accord with the predictions of Förster's theory.

b. Triplet–triplet. The writer is not aware of reports of energy transfer from the triplet excited state of a donor to a chlorophyll. Energy transfer from a chlorophyll to an acceptor, reaction (8), has been demonstrated with reasonable certainty only for carotenoids, but transfer to O_2 and tetracene is probable, as will be shown later. Energy transfer to thioctic acid was once proposed as a mechanism of photosynthesis (43, 44), but has not received experimental support (45, 46). Very few kinds of organic molecules have triplet states of lower energy than those of the chlorophylls.

Livingston *et al.* found by flash photolysis that carotenoids quench the triplet state of chlorophylls a and b, with bimolecular rate constants k_8 equal to 1.0 to 1.6×10^9 for α- and β-carotenes, and 0.7×10^9 for retinene. Cyclooctatetraene also quenched, but with a rate constant of only 1.2×10^6 (47, 48). Claes *et al.* found that chlorophyll sensitized the *cis→trans* isomerization of the poly-*cis*-carotenoids prolycopene and *cis*-ζ-carotene (49, 50). Claes found that the photooxidation of chlorophyll a by O_2 was inhibited by the carotenoids β-carotene (eleven conjugated double bonds), lycopene (eleven), neurosporene (nine), to a lesser extent by ζ-carotene (seven), but not by phytofluene (five) (50). The photoreduction of chlorophyll by ascorbic acid was inhibited by the same carotenoids, and to some extent by phytofluene, but not by phytoene (three). The inhibition was ascribed to energy transfer from chlorophyll to carotenoids having at least five conjugated double bonds.

According to Krasnovskii and Drozdova (51), the photoreduction of chlorophyll is very effectively inhibited by carotene, lutein (ten), and violaxanthin (nine), less effectively inhibited by vitamin A acetate (five), a tetraene of structure (I), and β-ionone, and not inhibited by allocymene (2, 4, 6-octatriene) and crotonic acid. They attributed the inhibition to reoxidation of reduced chlorophyll by the carotenoid, although they were

TABLE I

RATE CONSTANTS FOR SPONTANEOUS AND INDUCED DECAY OF PORPHYRIN TRIPLET STATES

$$-d[\text{Chl}']/dt = K_1[\text{Chl}'] + K_2[\text{Chl}']^2 + K_3[\text{Chl}'][\text{Chl}]$$

Porphyrin	Solvent	K_1 (sec^{-1})	$10^{-9} K_2$ (M^{-1} sec^{-1})	$10^{-7} K_3$ (M^{-1} sec^{-1})	Reference
Chlorophyll a	Pyridine	670	1.51	2	(30)
Chlorophyll a	Pyridine	600 + 22000 × exp(−2100/RT)	—	—	(35)
Chlorophyll a	Pyridine, dry	870	—	—	(36)
Chlorophyll a	Benzene	440	2.07	5	(30)
Chlorophyll a	Toluene	600 + 13000 × exp(−2100/RT)	—	—	(35)
Chlorophyll ($a + b$)	Toluene	500	100 exp(−2600/RT)	—	(37)
Chlorophyll a	Ethyl acetate	exp(−2600/RT) 850 + 44000 ×	—	—	(35)
Chlorophyll a	Methanol	1500 + 14000 × exp(−1800/RT)	—	—	(35)
Chlorophyll a	Isobutanol	1460	—	—	(35)
Chlorophyll a	Propylene Glycol	1420	—	—	(35)
Chlorophyll a	Castor oil	750	—	—	(35)
Chlorophyll a	EPAF, 77° K	660	—	—	(38)
Chlorophyll b	Pyridine	310	1.58	2	(30)
Chlorophyll b	Benzene	330	2.19	9	(30)
Chlorophyll b	Benzene	400	2.35	3.5	(32)

		$600 + 30000 \times$ $\exp(-3500/RT)$			
Chlorophyll b	Propylene glycol	—	—	—	(35)
Chlorophyll b	EPAF, 77° K	370	—	—	(38)
Pheophytin a	EPAF, 77° K	1120	—	—	(38)
Pheophytin b	EPAF, 77° K	810	—	—	(38)
Bacteriochlorophyll	Pyridine	11810	0.9	—	(33)
Tetraphenylporphin (TPP)	Pyridine	670	2.5	—	(33)
TPP	Toluene	740	2.7	5	(33)
Zn TPP	Pyridine	870	2.4	—	(33)
Zn TPP	Toluene	800	3.0	2	(33)

unable to demonstrate reduction of any of the carotenoids. They also found inhibition by tetracene and anthracene.

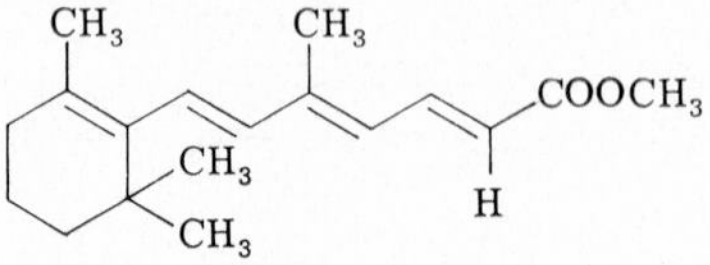

I

Oxygen under pressure enhances the intensity of ground singlet state to triplet excited-state transitions. By this technique, Evans has observed the singlet → triplet absorption bands of several polyenes and polyene aldehydes (52, 53). The position of these bands, expressed in wave numbers v'_0, and the singlet-triplet splitting, $v^*_0 - v'_0$, are listed in Table II. As the energy of triplet chlorophyll corresponds to 11,500 cm^{-1} (see below), it is apparent that transfer becomes possible to polyenes having at least five conjugated double bonds.

TABLE II

ENERGY (v'_0) OF TRIPLET EXCITED STATES OF POLYENES, AND ENERGY DIFFERENCE ($v^*_0 - v'_0$) BETWEEN FIRST EXCITED SINGLET AND TRIPLET STATES

Polyene	v'_0 (cm^{-1})	$v^*_0 - v'_0$ (cm^{-1})
Ethylene	28700	24200
Butadiene	20830	22600
Hexatriene	16450	21000
Octatetraene	13750	19400
$CH_3(CH{=}CH)_3CHO$	15210	16100
$CH_3(CH{=}CH)_4CHO$	12700	15300
$CH_3(CH{=}CH)_5CHO$	11050	14500

Platt has pointed out that the planar configuration of the triplet state of a polyene is unstable with respect to twisting about a double bond (54). Energy transfer from chlorophyll to the lower carotenoids may be coupled with a vibrational twisting motion.

The suppression of reactions of triplet chlorophyll by tetracene (51) is most interesting, in that McGlynn et al. (55, 56) have recently found its lowest triplet state to be at 10250 cm^{-1}, instead of at 18500 cm^{-1} as had previously been supposed (57). Energy transfer from triplet chlorophyll a to tetracene is therefore possible and accords with the observations of Krasnovskii and Drozdova. The much weaker quenching action ascribed to anthracene (51) may have been due to an anthraquinone or tetracene impurity.

3. OXIDATION

The most direct evidence for the occurrence of reaction (9) or (11) is the detection of ESR signals of the semiquinones of *p*-benzoquinone, *p*-chloranil, *o*-chloranil, and phenanthrenequinone, when illuminated with chlorophyll in 8:3:5 ether:isopentane:ethanol at −45°C (58). No ESR signal for the radical cation Chl·+ could be recognized, however, for reasons which are not clear. The variation with temperature of the ESR signal from chlorophyll and benzoquinone in ethanol was interpreted in terms of cage effects, and the effect of viscosity on diffusion rates (59). Similarly, a signal from the semiquinone of riboflavin was observed, when a solution of chlorophyll and riboflavin in acidified ethanol was illuminated with red light (46). Ke *et al.* ascribed to reaction (11) the rapid, transient changes produced in the spectrum of chlorophyll in ethanol, by flashing with light in the presence of pyocyanine, phenazine methosulfate, ubiquinone 6, or trimethylbenzoquinone (60). Fujimori and Livingston found a bimolecular rate constant of $2.4 \times 10^9 \, M^{-1} \, \text{sec}^{-1}$ for the quenching of the triplet state of chlorophyll by benzoquinone (47).

The several reports of reversible bleaching of chlorophylls in viscous solution or glasses at low temperatures in the presence of quinones, do not constitute evidence of a bimolecular reaction, because diffusion is slow under these conditions, and reaction with the solvent or 2-quantum electron ejection are conceivable alternatives (21, 22, 61–63). Nevertheless, Krasnovskii and Drozdova have published difference spectra which probably show the oxidized forms of chlorophyll *a*, chlorobium chlorophyll 660, and bacteriochlorophyll in ethanol-glycerol mixtures at −70°C (63). The first two show absorption increases in the 480–500 mμ region, and in the red beyond 700 mμ; oxidized bacteriochlorophyll has bands at 430–530 mμ and 840 mμ. The parent chlorophylls were regenerated on warming the glass from −60° to room temperature. Bacteriochlorophyll is photooxidized by O_2 at room temperature, apparently to the radical cation (64). The reported reduction of menadione by chlorophyll in hexane (65) would appear to have been initiated by reaction of photoexcited (365 mμ) menadione with the solvent.

4. REDUCTION

Unlike the one-electron oxidation of chlorophyll, which is usually reversed immediately upon cessation of illumination, reduction of chlorophylls leads to stable nonradical products. There is evidence, however, that the reduction is initiated by a one-electron step such as that of Eq. (12).

Evstigneev and Gavrilova found that a gold electrode in a solution

TABLE III
QUENCHING OF FLUORESCENCE OF CHLOROPHYLLS: MOLARITY OF QUENCHER, $[Q]_{1/2}$, AT WHICH THE FLUORESCENCE IS HALF QUENCHED

Chlorophyll	Quencher	Solvent	$[Q]_{1/2}{}^a$	Reference
	Oxidizing agents			
Chlorophyll a	p-Benzoquinone	Methanol	0.0096	(74)
Chlorophyll a	p-Benzoquinone	Acetone	0.0081	(74)
Chlorophyll a + b	p-Benzoquinone	Ethanol, 12°C	0.030	(84)
Chlorophyll a + b	p-Benzoquinone	Pyridine	(0.0058)	(82)
Chlorophyll a + b	p-Benzoquinone	Toluene	(0.0072)	(82)
Chlorophyll a	Duroquinone	Methanol	0.0116	(74)
Chlorophyll a	Chloranil	Acetone	0.0050	(74)
Chlorophyll a	Nitrobenzene	Methanol	0.034	(74)
Chlorophyll a + b	Nitrobenzene	Ethanol	(0.047)	(82)
Chlorophyll a + b	Nitrobenzene	Pyridine	(0.17)	(82)
Chlorophyll a + b	Nitrobenzene	Toluene	(0.25)	(82)
Chlorophyll a	Nitrobenzene	Benzene	(0.016)	(77)
Chlorophyll b	Nitrobenzene	Benzene	0.026	(77)
Pheophytin a	Nitrobenzene	Methanol	0.11	(77)
Pheophytin a	Nitrobenzene	Benzene	(0.2)	(77)
Chlorophyll a	m-Dinitrobenzene	Methanol	0.0122	(74)
Chlorophyll a + b	m-Dinitrobenzene	Ethanol	(0.015)	(82)
Chlorophyll a + b	m-Dinitrobenzene	Pyridine	(0.028)	(82)
Chlorophyll a + b	m-Dinitrobenzene	Toluene	(0.031)	(82)
Chlorophyll a	m-Dinitrobenzene	Benzene	0.00135	(76)
Chlorophyll a	p-Dinitrobenzene	Benzene	0.0008	(76)
Chlorophyll a	Trinitrotoluene	Methanol	0.0100	(74)
Chlorophyll a	m-Nitrophenol	Benzene	(0.0064)	(77)
Chlorophyll a	2, 4-Dinitrophenol	Benzene	0.0035	(77)
Chlorophyll a	2, 4, 6-Trinitrophenol	Benzene	0.0010	(77)
Chlorophyll a	β-Nitrostyrene	Methanol	0.0170	(74)
Chlorophyll a	β-Nitro-β-methyl styrene	Methanol	0.022	(74)
Chlorophyll a	β-Nitro-β, γ-hexene	Methanol	0.064	(74)
Chlorophyll a	2-Phenyl-3-nitro-bicyclo-[1, 2, 2] heptene-5	Methanol	(0.61)	(74)
Chlorophyll a	β-Nitroso-α-naphthol	Methanol	0.0140	(74)
Chlorophyll a	Nitric oxide	Ethanol	(0.0178)	(74)
Chlorophyll a	Oxygen	Ethanol	(0.023)	(74)
Chlorophyll a + b	Iodine	Ethanol	(0.0044)	(82)
Chlorophyll a + b	Triphenyltetra-zolium chloride	Ethanol	0.0040	(75)
Chlorophyll a	Methyl red	Methanol	(0.0088)	(74)
Chlorophyll a	Safranine T	30% aq. pyridine	0.0036	(86)

TABLE III (*continued*)

Chlorophyll	Quencher	Solvent	$[Q]_{1/2}{}^a$	Reference
		Reducing agents		
Chlorophyll $a + b$	Hydroquinone	Ethanol	7.0	(*84*)
Chlorophyll a	p-Dimethylaniline	Methanol	0.0138	(*74*)
Chlorophyll a	Dimethylaniline	Methanol	(0.42)	(*74*)
Chlorophyll $a + b$	Dimethylaniline	Ethanol	(0.26)	(*82*)
Chlorophyll $a + b$	Benzidine	Ethanol	7.0	(*87*)
Chlorophyll a	2, 6-Diaminopyridine	Methanol	(0.62)	(*74*)
Chlorophyll a	Phenylhydrazine	Methanol	0.31	(*74*)
Chlorophyll a	Phenylhydrazine	Ethyl ether	0.151	(*74*)
Chlorophyll a	Phenylhydrazine	Benzene	0.159	(*77*)
Chlorophyll $a + b$	Phenylhydrazine HCl	Ethanol	(0.5)	(*82*)
Chlorophyll b	Phenylhydrazine	Benzene	0.113	(*77*)
Pheophytin a	Phenylhydrazine	Benzene	0.067	(*77*)
Pheophytin a	Phenylhydrazine	Benzene	0.033	(*77*)
Pheophytin a	β-Naphthylhydrazine	Benzene	0.0084	(*77*)
Chlorophyll $a + b$	NaI	Ethanol	2.6	(*87*)

^{a} Values were taken from a curve of ϕ versus [Q] when there were sufficient data. Values listed in parentheses were calculated from the fluorescence measured at some concentration, assuming the validity of Eq. (17).

containing ascorbic acid and chlorophylls became more negative on illumination, by about 0.3 volt (*66, 67*). They regarded this as evidence for the presence of the reduced semiquinones of the pigments. Later, they were able to get spectra of primary products of photoreduction of pheophytin a and protopheophytin in pyridine solutions at $-40°C$, and some indication of primary products from hematoporphyrin and bacteriopheophytin (*68–70*). The primary product of reduction of pheophytin a has bands at 340, 470, and 620 mμ in descending order of intensity. On warming to room temperature, it disproportionates into a mixture of pheophytin and the dihydropheophytin (*67, 70*).

In the flash photolysis of solutions of chlorophyll and ascorbic acid, Livingston *et al.* (*36, 71*), and Zieger and Witt (*71a*), detected a transient with an absorption band near 470 mμ, appearing after the triplet state transient and decaying more slowly. As this transient decayed, absorption increased at 525 mμ, the position of the absorption band of dihydrochlorophyll.

Livingston and Pugh could not detect quenching of Chl' by ascorbic acid in pyridine, unless water was present (*36*). The bimolecular rate constant for quenching by ascorbic acid is $1.5 \times 10^5 \, M^{-1} \, sec^{-1}$ in 20% aqueous pyridine, but only 1×10^4 in dry pyridine (*71*), and 2 to 4×10^4

in ethanol (72). The rate constants for quenching by phenylhydrazine are 1.3×10^4 and 1.1×10^4 in dry and wet (2.4 % water) pyridine, respectively (71). These rate constants, presumably for primary reaction (12) are only about 10^{-5} to 10^{-4} of the value for diffusion controlled reactions.

The transient formation of negative ion radicals of Mg phthalocyanine and hematoporphyrin, by flash photolysis in the presence of reducing agents, has also been reported (73).

5. FLUORESCENCE QUENCHING

Because of the very short lifetime ($\sim 5 \times 10^9$ sec) of the singlet excited state of chlorophyll, there is not the kind of evidence for reactions (9) and (10) that flash spectrophotometry provides for reactions (11) and (12). Nevertheless, quenching of fluorescence is evidence of some sort of interaction between the quencher and Chl*.

Table III lists the known quenchers of chlorophyll fluorescence, and the concentration $[Q]_{\frac{1}{2}}$ at which the fluorescence intensity is reduced to one-half. If fluorescence quenching follows the Stern-Volmer equation (17), $[Q]_{\frac{1}{2}}$ is just the reciprocal of the constant K_Q. (ϕ_Q and ϕ are the fluorescence yields with and without the quencher Q.)

$$\frac{\phi}{\phi_Q} = 1 + K_Q [Q] \qquad (17)$$

Actually, quenching of chlorophyll fluorescence deviates somewhat from the Stern-Volmer equation (74, 75). The values of $[Q]_{\frac{1}{2}}$ in Table III were therefore taken from a curve of ϕ versus $[Q]$ when sufficient data existed; values listed in parentheses were calculated from the fluorescence at some concentration at which it was measured, assuming the validity of Eq. (17).

The quenchers are evidently either oxidants or reductants, and the oxidants are by far the better quenchers. Nonquenchers (at the concentrations used) included nitropropane, aniline, hydrazine, thiourea, phenol, and t-hexylmercaptan, benzaldehyde, N_2O, CO, and ascorbic acid (74). It has been proposed (76, 77) that fluorescence quenching takes place by transfer of an electron to or from the excited chlorophyll, i.e., via reactions (9) and (10), as has often been proposed for other systems (78–80). Although this seems very probable, an alternative mechanism would be formation of a charge transfer complex between Chl* and Q, in which radiationless conversion to the ground state or to the triplet state was accelerated (81).

The order of quenching ability of oxidants (Table III) is parallel to the order expected of their oxidizing ability, at least with nitro compounds (77) and quinones (74). Some of the $[Q]_{\frac{1}{2}}$ values are so low (e.g., the nitrophenols) (76, 77), that, if they are not in error, they may indicate static quenching via solvate formation in benzene solution. There is a pronounced solvent effect for nitrobenzene; it is a much better quencher in alcohols than in pyridine or toluene (82).

It is noteworthy that phenylhydrazine in small quantities activates the fluorescence of chlorophyll in benzene; in larger quantities it acts as a quencher (83). Apparently a molecule of phenylhydrazine, bound in a solvate to the Mg of chlorophyll, is not in position to quench fluorescence.

The action of quenchers on the fluorescence of Mg phthalocyanine (82, 84), mesoporphyrin, and protoporphyrin (85), is parallel to that on the fluorescence of chlorophylls.

As the participation of the singlet excited state of chlorophyll in photosynthesis is by no means excluded, it is clearly desirable to have more direct and unambiguous evidence for the electron-transfer reactions (9) and (10) than now exists, for example, the sensitized reduction of a quenching oxidant by a reductant, neither of which is able to react with chlorophyll in the triplet state.

C. Energy Requirements

1. CONDITIONS FOR THE OCCURRENCE OF PRIMARY REACTIONS

Having presented evidence for the participation of chlorophylls in the primary photochemical reactions (6) and (8)–(12), we turn to the energetic requirements for these reactions, in the hope of answering the question, what reactions does photoexcitation enable chlorophyll to undergo, that it could not undergo in the ground state? For energy transfer, reactions (6) and (8), the answer is simple: the acceptor M must have an excited electronic state of the proper multiplicity with energy less than that of the excited state of the chlorophyll. The requirements for (9) to (12) are a little more complex, however.

Consider a primary photochemical reaction such as (11a). It can occur only if the enthalpy change, ΔH_{11a}, is negative, or at most, very slightly positive. The energy excitation of chlorophyll to the triplet state (ca. 33000 cal/mole) is so much greater than ambient thermal energies (ca. 600 cal/mole), that additional activation energy is usually superfluous. The energetic requirements for this reaction can be estimated if the reaction is broken conceptually into the following steps, with the designated enthalpy and free-energy changes, or half-cell potentials.

$$\text{Chl} \xrightarrow{\nu'_0} \text{Chl}', \Delta H_{18} = Nh\nu'_0 c \text{ erg} \cdot \text{mole}^{-1} \qquad (18)$$
$$= 2.859 \, \nu'_0 \text{ cal} \cdot \text{mole}^{-1}$$

($\nu'_0 =$ wave number of light corresponding to the triplet state energy; $h =$ Planck's constant; $c =$ speed of light).

$$Chl \rightleftharpoons Chl \cdot^+ + e^-, \qquad E_{19} \text{ (volt)} \qquad (19)$$

$$Ox + e^- \rightleftharpoons Ox \cdot^-, \qquad E_{20} \text{ (volt)} \qquad (20)$$

$$Chl + Ox \rightleftharpoons Chl \cdot^+ + Ox \cdot^- \qquad \Delta F_{21} = 23060 \, (E_{19} - E_{20})$$
$$\text{cal. mole}^{-1} \quad (21)$$

$$Chl \cdot^+ + Ox \cdot^- \rightleftharpoons \{Chl \cdot^+ + Ox \cdot^-\} \qquad - \Delta H_{11c} \qquad (11c)$$

If we allow the somewhat restricting condition that there is no entropy change in reaction (21), so that $\Delta F_{21} = \Delta H_{21}$, then the requirement that ΔH_{11a} be negative leads to the condition expressed in Eq. (22).

$$\Delta H_{11a} = \Delta H_{21} - \Delta H_{18} - \Delta H_{11c} < 0 \qquad (22)$$

If ΔH_{11c} is negligible, Eq. (22) reduces to the simple condition (23).

$$E_{19} - E_{20} < 1.2398 \times 10^{-4} \, \nu'_0 \qquad (23)$$

With these simplifications it is only necessary to know the energy of the triplet state of chlorophyll, and the potential of the $Chl/Chl \cdot^+$ couple, to predict what oxidants can be reduced by chlorophyll in reaction (11a). Similar arguments apply to the primary reactions (9), (10), and (12).

2. EXCITED-STATE ENERGIES AND OXIDATION-REDUCTION POTENTIALS

Since the first absorption band and the first fluorescence band of a chlorophyll are usually only a few millimicrons apart, the energy of the lowest singlet excited state is known with precision. The energies of the lowest triplet states of chlorophylls a and b, estimated from observed low-temperature phosphorescence of these compounds, Cu pheophytin b, and Cu ethyl pheophorbide a, are about 11,500 cm^{-1}, or 33,000 cal (88–90). The energy of triplet-state bacteriochlorophyll is unknown.

The $Chl/Chl \cdot^+$ half-cell potentials are perhaps most reliably estimated by polarography. Two one-electron anodic polarographic waves, the first presumably marking reaction (19), have been reported by Stanienda for chlorophylls and pheophytins in acetonitrile (91). Goedheer et al. determined potentials, presumably for the same reaction, for chlorophylls a and b, bacteriochlorophyll, and chlorobium chlorophyll by titration with $FeCl_3$ in methanol, but were unable to oxidize the pheophytins (92).

Polarographic half-wave reduction potentials were determined by Gilman for a number of chlorophyll derivatives in ethanol, measured against an ethanolic saturated calomel electrode (E.S.C.E.), consisting of $Hg/Hg_2Cl_2/$sat. NH_4Cl in ethanol (93). The E.S.C.E. has a potential of $+$ 0.064 volt, relative to the normal hydrogen electrode (N.H.E.). The first two-electron wave (Table IV) corresponds to the electrode reaction

TABLE IV

POLAROGRAPHIC HALF-WAVE REDUCTION POTENTIALS OF CHLOROPHYLL DERIVATIVES IN ETHANOL VERSUS THE ETHANOLIC SATURATED CALOMEL ELECTRODE[a]

Substance	pH	$-E_{\frac{1}{2}}^{(1)}$	$-E_{\frac{1}{2}}^{(2)}$	$I^{(1)}$ [b]	$I^{(2)}$
Chlorophyll a [c]	unbuff.	1.07	—	1.75	—
Chlorophyll b	9.3	1.11	—	3.89	—
10-Hydroxy-chlorophyll a	7.4	1.01	—	2.30 ⎫	
	8.3	1.06	—	1.97 ⎬ $\frac{\Delta E_{\frac{1}{2}}}{\Delta(\text{pH})} = -0.52$	
	9.3	1.11	—	1.48 ⎭	
	12.1	1.23	—	1.55	—
Pheophytin a	2.3	0.24	—	2.22	—
	7.4	0.56	—	2.33 ⎫	
	8.3	0.62	—	2.14 ⎪	
	9.3	0.68	1.15(?)	1.93 ⎬ $\frac{\Delta E_{\frac{1}{2}}}{\Delta(\text{pH})} = -0.60$	
	12.1	0.88	—	2.18 ⎭	
Pheophorbide a	9.4	0.67	1.06	2.04	0.90
Mesopheophorbide a	9.4	0.71	1.12	1.91	1.92
Chlorin e_6	9.4	0.66	1.03	1.39	2.15
Pheophytin b	9.4	0.75	1.02	2.20	2.03
Bacteriopheophytin	9.4	0.58	—	2.09	—
Pheoporphyrin a_5	9.4	0.75	1.03	2.31	6.30

[a] Median values of measurements in approximate concentration range $5 \times 10^{-5}M$ to $2 \times 10^{-3}M$ taken from thesis of Gilman (93).

[b] A value for the diffusion current constant of 1.8–2.0 was typical of a 2-electron reduction.

[c] Chlorophylls a and b were rapidly allomerized in the solvent.

(24), but the potential ($E_{\frac{1}{2}}^{(1)}$) is presumably set by the one-electron reaction (25).

$$\text{Chl} + 2\,e^- + 2\,\text{H}^+ \longrightarrow \text{Chl H}_2, \qquad (24)$$
$$\text{Chl} + e^- + \text{H}^+ \rightarrow \text{ChlH} \cdot \qquad (25)$$

The second two-electron wave ($E_{\frac{1}{2}}^{(2)}$) seen with some derivatives usually merges into a rapidly rising background current of irreversible reduction. Potentials were practically independent of concentration.

Similar cathodic potentials have been reported by Felton *et al.* for chlorophyll a (-1.1 volt) and chlorophyll b (-1.05 volt) in dimethylsulfoxide, versus the (aqueous) saturated calomel electrode (S.C.E.) (94).

Half-wave potentials for several tetraphenylporphin derivatives were also reported by Gilman.

With these polarographic data, it is possible, by means of equations such as Eq. (23), to estimate the minimum value of the potential an oxidant must have, if it is to be reduced by a chlorophyll via reactions (9) or (11), and the maximum value allowed to a reductant, if it is to be oxidized by a chlorophyll via reactions (10) and (12). The results are summarized in Table V.

In this Table, Gilman's potentials have been corrected to the N.H.E. by addition of 0.064 volt; Stanienda's values are first listed versus the S.C.E., as he reported them, but corrected to the N.H.E. for subsequent calculation by addition of 0.24 volt. In general, the potentials of chlorophylls must be expected to vary somewhat with solvent; we have noted, for example, that chlorophyll a is much more easily oxidized by $FeCl_3$ in methanol than in other solvents, including ethanol. Perhaps methanol has an especially strong solvating action on $Chl \cdot^+$. The singlet and triplet excited state energies are comparatively insensitive to solvent.

It is common practice to discuss oxidizing and reducing power of chlorophyll with reference to potentials of oxidants and reductants compiled in tables such as those of Clark (95). These, however, are measured under conditions of equilibrium, often achieved with the aid of a mediator, and take into account effects of pH, semiquinone formation, etc. Consequently, tabulated half-cell potentials are germane to a photochemical reaction such as (11) only if they are for simple, one-electron reductions, as of viologens, or of ferric complexes, such as ferricyanide. In other cases, the tabulated potentials must be corrected by terms involving the acid ionization constants of Ox, $OxH \cdot$ and OxH_2, and the equilibrium constants for the formation of $OxH \cdot$ from Ox and OxH_2, etc. For many oxidants these constants are not known. Clark has discussed these problems in detail and listed some of the known values of these constants.

The conditions of polarographic oxidation and reduction more closely resemble those of photochemical reactions, and the half-wave potentials are probably more appropriate for discussion of photochemical processes than the thermodynamic oxidation-reduction potentials.

3. SOLVENT EFFECTS

The hitherto neglected term in Eq. (22), ΔH_{11c}, is the energy of dissociation of the products of reaction (11a) from the solvent cage in which they are formed. This term becomes particularly important if the chlorophyll and the oxidant are initially uncharged, so that the photoreaction produces an ion pair. The stabilization of the ion pair, relative to the separated ions, depends on the dielectric constant of the medium.

TABLE V

ENERGETICS OF PHOTOCHEMICAL REACTIONS OF CHLOROPHYLLS. SUMMARY OF EXCITED-STATE ENERGIES, HALF-CELL, OR POLAROGRAPHIC HALF-WAVE POTENTIALS, AND PREDICTED LIMITS TO OXIDIZING AND REDUCING POTENTIAL OF EXCITED CHLOROPHYLL SPECIES

Parameter	Chlorophyll a	Chlorophyll b	Pheophytin a	Pheophytin b	Chlorobium chlorophyll 660	Bacterio-chlorophyll
Potentials						
$Chl \rightleftharpoons Chl \cdot^+ + e^-$ (S.C.E.), CH_3CN (Stanienda)	+0.52	+0.65	+0.86	+0.99	—	—
$Chl \rightleftharpoons Chl \cdot^+ + e^-$ (N.H.E.), CH_3OH (Goedheer)	+0.645	+0.680	—	—	+0.55	+0.55
$Chl + e^- + H^+ \rightarrow ChlH \cdot$ (N.H.E.), C_2H_5OH (Gilman)	−1.01	−1.05	−0.62	−0.69	—	—
Singlet excited state						
Energy (cm^{-1})	15000	15500	15000	15200	15100	13000
Half-cell potentials (N.H.E.)						
$E(Ox)$,[b] CH_3CN	−1.10	−1.03	−0.76	−0.65	—	—
$E(Ox)$, CH_3OH	−1.21	−1.24	—	—	−1.32	−1.07
$E(Red)$, C_2H_5OH	+0.85	+0.87	+1.24	+1.19	—	—
Triplet excited state						
Energy (cm^{-1})	11500	11500	(11500)[a]	(11500)	(11500)	(9500)
Half-cell potentials (N.H.E.)						
$E(Ox)$, CH_3CN	−0.66	−0.53	−0.32	−0.19	—	—
$E(Ox)$, CH_3OH	−0.78	−0.74	—	—	−0.87	−0.63
$E(Red)$, C_2H_5OH	+0.41	+0.37	+0.80	+0.73	—	—

[a] Values in parentheses are estimated, assuming a singlet-triplet separation of 3500–4000 cm^{-1}.
[b] $E(ox)$ is the minimum value of the oxidation-reduction potential an oxidant must have if it is to be reduced by chlorophyll in the designated excited state; $E(Red)$ is the maximum value a reductant may have, if it is to be oxidized by photoexcited chlorophyll.

The effect of dielectric constant on the energy terms of Eq. (22) can be readily evaluated, if attention is restricted to ionic interaction.

Let the ions of the pair be represented by charges q_1 and q_2 esu in spheres of radii r_1 and r_2. The energy of dissociation of the ion pair from the separation $r_1 + r_2$ to infinity in a medium of dielectric constant D is

$$\Delta H_{11c} = \frac{-N q_1 q_2}{(r_1 + r_2) D} \, \mathrm{erg} \cdot \mathrm{mole}^{-1} \tag{26}$$

There is a further correction, however. The potentials E_{19} and E_{20}, and ΔF_{21}, are normally calculated and tabulated assuming that water is the solvent. If a different solvent is used, ΔF_{21} must be corrected by the free energy of transport of the ions from water $(D = 78.5)$ to the solvent,

$$\Delta F_{\mathrm{transport}} = N \left(\frac{q_1^2}{2 r_1} + \frac{q_2^2}{2 r_2} \right) \left(\frac{1}{D} - \frac{1}{78.5} \right) \mathrm{erg} \cdot \mathrm{mole}^{-1} \tag{27}$$

With these additional terms, Eq. (24) becomes

$$\Delta H_{11a} = 23060 \, (E_{19} - E_{20}) - 2.859 \, v'_0 + N \left(\frac{q_1^2}{2 r_1} + \frac{q_2^2}{2 r_2} \right) \left(\frac{1}{D} - 0.0127 \right)$$
$$\times 2.39 \times 10^{-8} + N \frac{q_1 q_2}{(r_1 + r_2) D} \times 2.39 \times 10^{-8} < 0 \tag{28}$$

For unit charges $q_1 = - q_2 = 4.8 \times 10^{-10}$ esu, and $r_1 = r_2 = 5 \times 10^{-8}$ cm, Eq. (28) reduces to

$$23060 \, (E_{19} - E_{20}) < 2.859 \, v'_0 - 66400 \left(\frac{1}{2D} - 0.0127 \right) \tag{29}$$

The second term on the right is negligible for alcohols and water (Table VI) but becomes appreciable for solvents less polar than pyridine, and in nonpolar solvents, amounts to almost half the energy of the chlorophyll triplet state. It is predictable that there will be reactions that

TABLE VI

ELECTROSTATIC STABILIZATION OF ION PAIRS IN VARIOUS SOLVENTS

Solvent	D	$66400 \left(\dfrac{1}{2D} - 0.0127 \right)$
Water	78.5	−420 cal
Methanol	32.6	+170
Ethanol	24.3	520
Pyridine	12.3	1860
Dioxane	2.2	14250

go in polar solvents and not in nonpolar solvents, but there has been little verification of this with chlorophylls.

III. Photochemical Reactions of Chlorophylls

A. Reduction

1. THE "KRASNOVSKII REACTION"

a. Chlorophyll a. In 1948, Krasnovskii reported that chlorophyll *a* was photoreduced by ascorbic acid in pyridine to a red compound ($ChlH_2$), with an absorption band at 525 mμ (*3*). $ChlH_2$ is reoxidized to chlorophyll in the dark by dehydroascorbic acid, O_2, and such oxidants as safranine T, riboflavin (*96*), Fe^{+++}, quinone, thionine, NO_3^-, NO_2^-, and neutral red (*97*). Phenylhydrazine, cysteine, H_2S, and dihydroxymaleic acid also effect reduction (*98*). There is little reduction in very dry pyridine; the reaction rate is considerably increased by the presence of at least 5% water or alcohol (*71a, 99, 100*). On the other hand, there is little reduction in pure ethanol or acetone; a base such as pyridine, imidazole, histidine, ammonia, or piperidine is required (*3, 98, 101*). Ammonia also much accelerates the reoxidation of $ChlH_2$ to chlorophyll by dehydroascorbic acid (*102*). Both the photoreduction and reoxidation are retarded by the presence of D_2O instead of H_2O in the pyridine (*103*).

The red product is also formed by chemical reduction of chlorophyll with Zn/CH_3COOH (*104*) or Zn amalgam (*105*). It is not a radical, or at least no ESR signal can be detected from it (*106, 107*). It is most probably a dihydrochlorophyll, reduced at two methine bridge positions. However, Zieger and Witt detected a short-lived transient species with an absorption band at 475 mμ, which probably is the radical intermediate ChlH· (*71a*). Further evidence for radical intermediates is the initiation of polymerization of methyl methacrylate during photoreduction of chlorophyll (*108*).

$ChlH_2$ loses Mg to form a reduced pheophytin, the more readily, the higher the concentration of ascorbic acid and water or ethanol in the pyridine (*71a, 99, 100, 109, 110*). Reduced pheophytin ($PheH_2$) thus formed is reoxidized to pheophytin by O_2 and other oxidants. $PheH_2$ is also formed directly by photoreduction of pheophytin *a* (*101, 111*). The solvent requirement is somewhat relaxed, in that a base is not needed for reduction by ascorbic acid in ethanol (*112*). Pheophytin is reduced by the same agents that are active for chlorophyll; in addition, pheophytin hydrochloride is reduced by $SnCl_2$ (*113*) or $FeCl_2$ (*114*) in acetone.

Pheophytin has been reduced by H_2S in aqueous pyridine in quantities large enough to permit taking infrared spectra (115).

The spectrum of $PheH_2$ varies with the solvent (71a, 99, 112, 116); there may be two isomeric forms in equilibrium, with characteristic absorption bands near 620 and 525 mμ, differing in the location of the added hydrogens (116). There is also an acid form of $PheH_2$ with an absorption band in the 510–515 mμ region, prepared by adding acid to solutions of $PheH_2$ in ethanol or acetone (112, 114), or by reduction of pheophytin in formic acid (112).

The radical intermediate, PheH$\cdot$, is stable in pyridine at $-40°$; on warming, it disproportionates into Phe and $PheH_2$ (67, 68, 70, 117). Even at $-120°$ the intermediate is an active reducing agent (68).

The primary photochemical reaction in the reduction of chlorophyll by ascorbic acid is between the solvates Chl $(ethanol)_2$ and Chl ($ethanol$) ($pyridine$), and the pyridinium ascorbate ion pair, {pyrH$^+$ AH$^-$} (116). The requirements for base, and for water or ethanol, and the inactivity of sodium or tetrabutylammonium ascorbate, are all explained this way (99, 100, 116). The transfer of an electron from ascorbate to Chl' is probably coupled with transfer of a proton from the pyridinium ion, which might explain the rather low rate constant found for the reaction, $k_{30} = 1.8 \times 10^5 M^{-1} sec^{-1}$.

The following sequence of reactions adequately accounts for events during the photoreduction of chlorophyll a by ascorbic acid.

$$\text{Chl}' + \text{pyrH}^+ \text{AH}^- \rightarrow \text{ChlH} \cdot + \text{AH} \cdot + \text{pyr}$$
$$k_{30} = 1.8 \times 10^5 M^{-1} sec^{-1} \tag{30}$$

$$2\,\text{AH} \cdot \longrightarrow \text{A} + \text{AH}_2 \tag{31}$$

$$\text{AH} \cdot + \text{ChlH} \cdot \longrightarrow \text{AH}_2 + \text{Chl} \tag{32}$$

$$2\,\text{ChlH} \cdot \longrightarrow \text{ChlH}_2 + \text{Chl} \tag{33}$$

$$\text{ChlH}_2 + 2\,\text{H}^+ \longrightarrow \text{PheH}_2 + \text{Mg}^{++} \tag{34}$$

$$\text{ChlH}_2 + \text{A} \longrightarrow \text{Chl} + \text{AH}_2 \tag{35}$$

$$\text{PheH}_2 + \text{A} \longrightarrow \text{Phe} + \text{AH}_2 \tag{36}$$

The quantum yield for $ChlH_2$ production approaches 0.02 in ethanol-pyridine (116), and a yield of 0.07 has been reported in water-pyridine mixtures (99, 100). The yield is much higher with diazabicyclooctane as base instead of pyridine (118).

b. *Reduction with Phenylhydrazine.* In polar solvents (ethanol, pyridine), chlorophyll is photoreduced to the usual red product, $ChlH_2$ (3, 98, 101). In toluene, ether, and other nonpolar solvents (99, 109, 119), there appears in addition a band at 585 mμ, belonging to another reduction product. Production of the 585 mμ material is relatively greater at

—45° (*109*). There is much conversion to pheophytin on reoxidation, and it is possible that the 585 mμ compound lacks Mg. Addition of pyridine or ethanol to a solution containing the 585 compound results in its immediate conversion to pheophytin.

On reduction in 5:5:2 ether:isopentane:ethanol, there is formed, in addition to ChlH$_2$ and the 585 mμ compound, a third reduction product absorbing at 615 mμ (*119*). Fluorescence spectra of various reduction products of the chlorophylls and pheophytins have been published (*119, 120*).

Unlike ascorbic acid, phenylhydrazine quenches the fluorescence of chlorophyll at concentrations often used for photoreduction (0.1 *M*), probably by electron transfer to the singlet excited state. However, the monosolvate of chlorophyll with phenylhydrazine is fluorescent, and apparently it is not photosensitive in the absence of excess phenylhydrazine (*83*). The exact nature of photochemical interactions of chlorophyll and phenylhydrazine is still very unclear.

c. Other Chlorophylls. Chlorophyll *b* is reduced irreversibly by ascorbic acid in pyridine (*98*), but reversibly by phenylhydrazine in toluene to two products, absorbing at 565 mμ and 635 mμ (*109*), which appear to be analogous to the 520 mμ and 585 mμ reduction products of chlorophyll *a*. Pheophytin *b* is reduced by ascorbic acid in pyridine to a red product absorbing at 520 mμ (*117*). An intermediate radical was stable enough at —65°C for detection.

Chlorobium chlorophyll is reduced by ascorbic acid or Na$_2$S in pyridine to a red product absorbing at 510 mμ (*121*) and by phenylhydrazine in toluene to this and another product absorbing at 560 mμ (*122*). Chlorobium pheophytin, like pheophytin *a*, is reduced to a primary radical form at —40° (*121*).

Chlorin e$_6$, rhodin g$_7$, and their Mg complexes are photoreduced by ascorbic acid in aqueous pyridine and other media to products absorbing near 550 mμ (*123, 124*), but chlorin e$_6$-6-butylcarboxamide is reduced slowly and irreversibly, no red product being visible (*110*).

Bacteriochlorophyll and bacteriopheophytin are reduced by ascorbic acid or Na$_2$S in aqueous pyridine to compounds absorbing near 660 mμ, which rapidly revert to the parent pigment (*64, 70, 125*). It is not clear whether these are one-electron or two-electron reduction products.

If a solution containing two chlorophylls and a reducing agent is illuminated, one of them is reduced more rapidly than the other, no matter which receives the light. This apparently comes about through reduction of one chlorophyll by the intermediate reduced radical of the other. Thus, chlorophylls *a* and *b* sensitize the photoreduction of pheophytins *a* and *b* (Eq. (37)), bacteriochlorophyll, and chlorobium chlorophyll by ascorbic acid or phenylhydrazine, but chlorobium chlorophyll cannot sensitize the reduction of bacteriochlorophyll (126). Sensitized reductions of this sort are probably quite general among the porphyrins (127).

$$\text{ChlH} \cdot + \text{Phe} \longrightarrow \text{Chl} + \text{PheH} \cdot \qquad (37)$$

d. *Photopotentials.* Evstigneev and Gavrilova measured the change in the potential of a gold electrode immersed in the solution, brought about by photoreduction of chlorophylls and pheophytins by ascorbic acid in pyridine (128, 129). During photoreduction of chlorophyll *a*, chlorophyll *b*, and pheophytin *a* + *b*, the electrode potentials dropped from their dark values of −0.37, −0.34, and −0.36 volt (S.C.E.), to −0.66, −0.59, and −0.70 volt, respectively. When the light was turned off, the dark values were slowly restored. The photopotentials approximate to the equilibrium potentials for the couple Chl/ChlH_2, −0.35 volt, determined by chemical means (97). Reversible potential drops were also noted on illuminating chlorophyll *a*, bacteriochlorophyll, and chlorobium chlorophyll with Na_2S in pyridine, but to lower values, −0.85, −0.90, −0.80 volt (S.C.E.) (125). Changes in photopotential have been useful for following reactions qualitatively. Conductivity changes have also been noted during reduction by phenylhydrazine in pyridine, indicating increase of ionic species during reduction (128, 130).

2. REDUCTION OF OTHER PORPHYRINS

a. *Phlorins and Isophlorins.* It has become clear recently that the normal product of photoreduction of the free base porphin [uroporphyrin (131–133), hematoporphyrin (102, 134, 135), protoporphyrin (135, 136), mesoporphyrin (137), tetraphenylporphin (137, 138)] is a phlorin (II) (139, 140). Compounds of this class are dihydroporphins, in which hydrogens have been added to one pyrrole nitrogen and one methine bridge [cf. also (XVIII) of Chapter 3]. Their spectra have two prominent band systems in the red and in the blue, which show a pronounced acid-base shift. They are reduced further to porphomethenes (III) (131, 132, 138) and are readily reoxidized by air to porphins.

Photoreduction of metal complexes of porphins cannot give phlorins unless the metal is lost, because the pyrrole nitrogens are blocked. Instead, dihydroporphins are formed which resemble the acid form of phlorins spectrally in having an intense band in the blue, but which lack a prominent broad band in the red. These have been observed in the reduction of Zn porphin (127), Zn tetraphenylporphin (141), and Zn protoporphyrin (142). The structure (IV) has been proposed for this sort of dihydroporphin (127). It is convenient to refer to these dihydroporphins as "isophlorins," to distinguish them from phlorins. As yet, they are known only in the form of metal complexes.

Like phlorins, isophlorins are readily reoxidized to porphins by O_2 and other oxidants. Unlike phlorins, the isophlorins of the above-mentioned porphins rearrange to chlorins in photosensitized reactions with ascorbic acid and aliphatic amines.

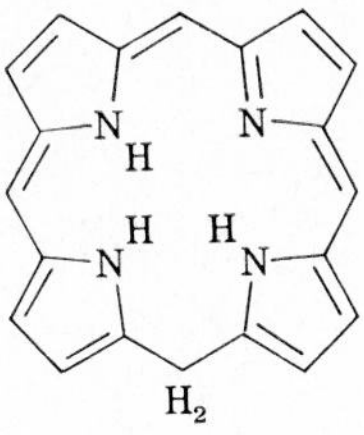

Phlorin

II

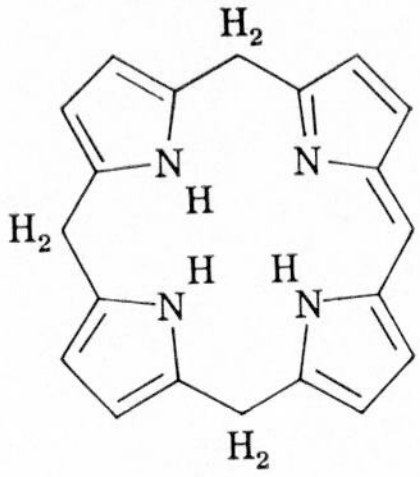

Porphomethene

III

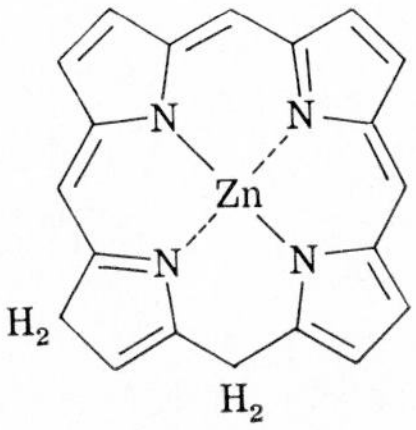

Zinc "isophlorin"

IV

Protochlorophyll is photoreduced by ascorbic acid in pyridine to a compound with an intense absorption band at 470 mμ, which is probably an isophlorin (143). The same compound is formed by reduction with Zn/CH_3COOH (144). It is readily oxidized back to protochlorophyll, but is not isomerized to a chlorin (chlorophyll) under photoreducing conditions.

b. Hypochlorins, Bacteriochlorins, and Hexahydroporphins. Photochemical reduction of Zn chlorin (127) and Zn tetraphenylchlorin (145) by ascorbic acid gives tetrahydroporphins of a kind distinct from bacteriochlorins [structure (XVII) of Chapter 3], for which structures like (V) have been suggested. Similar tetrahydroporphins have been prepared, along with bacteriochlorins and other reduced porphyrins, by catalytic hydrogenation of tetraphenylporphin (146), and by Na/isoamyl alcohol reduction of ferric complexes of porphin, octaethylporphin, octamethylporphin, and etioporphyrin II (*147*). Their reduction level has been confirmed by quantitative dehydrogenation (*147*); assignment of the four hydrogens to adjacent pyrrole rings is supported by molecular orbital calculations (*148–150*). Their spectra closely resemble those of the corresponding chlorins, but the bands lie somewhat more to the blue. Metal complexes and free bases are known (*145, 147*), but their salts with mineral acids are rapidly oxidized to chlorins (*147*).

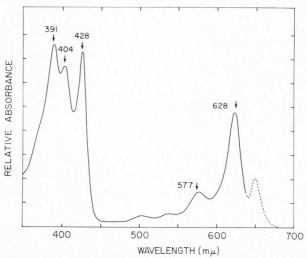

Fig. 1. Spectrum of ethyl hypochlorophyllide in ether. The extinction coefficient at 625 mμ is roughly $6 \times 10^4 M^{-1} cm^{-1}$. The dotted band belongs to a by-product, ethyl mesochlorophyllide.

Chlorophyll and some of its derivatives also are photoreduced by ascorbic acid in ethanolic pyridine to a tetrahydroporphin, apparently of this type, if a small amount of a tertiary aliphatic amine, such as diazabicyclooctane, is included (*118*). The function of the amine is probably to isomerize $ChlH_2$ or a product of reduction of $ChlH_2$, to the compound the spectrum of which is shown in Fig. 1. This compound is dehydrogenated photochemically by quinones, but the regenerated chlorin is mesochlorophyll, not chlorophyll. Pyrochlorophyll, chlorophyll *b*-3-methanol, and probably Mg chlorin e_6-6-butylcarboxamide are reduced to similar compounds under the same conditions, but chlorophyll *b*, Mg chlorin e_6, and ethyl-9-deoxo-9-hydroxychlorophyllide *a* are not. The compound with a band at 615 mμ prepared from chlorophyll and phenylhydrazine by Brody (*119*) may belong to this class.

There has been some confusion in the literature between these tetrahydroporphins and bacteriochlorins. It has therefore been proposed (*118*) that they be called *hypochlorins*, to emphasize their derivation by reduction of chlorins. *Hypochlorophyll* is therefore dihydromesochlorophyll. If structure (V) is correct for simple hypochlorins, the two extra hydrogens must either be at the 5,6-positions or at the 1,2-positions, probably the latter (VI).

Hexahydroporphins of unknown structures and differing spectra have been prepared from Zn tetraphenylporphin (*145*), octaethylporphin (*147*), bacteriochlorophyll, bacteriopheophytin (*64, 125*), and hypochlorophyll (*118*). They are all readily oxidized by O_2 back to the tetrahydroporphin level.

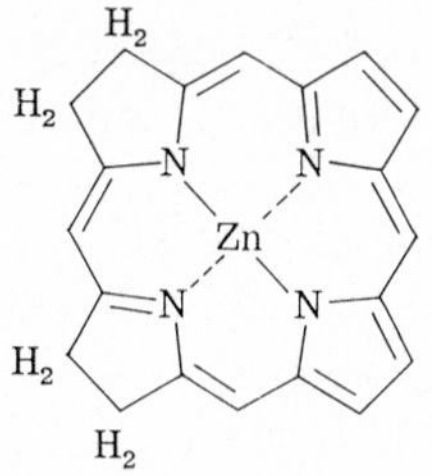

V

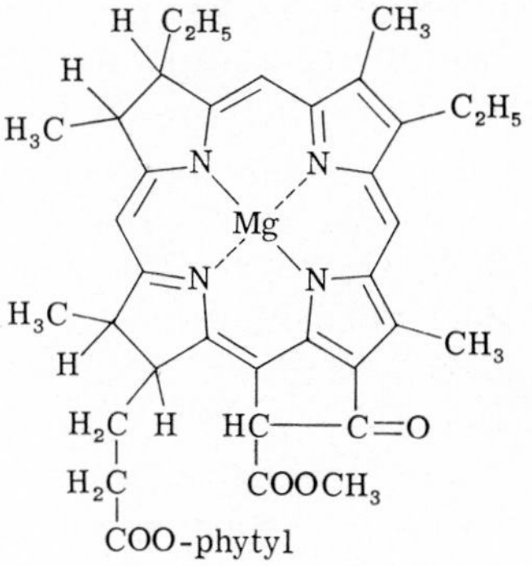

VI

B. Oxidation

Simple chlorins and tetrahydroporphins are dehydrogenated photochemically by quinones, in benzene and other solvents. Thus, Zn and Mg tetraphenylchlorin (*151, 152*), and Zn tetrahydrotetraphenylporphin (tetraphenylhypochlorin) (*145*), are oxidized to porphins by 1,2-naphthoquinone and other o- and p-quinones, and the free bases tetraphenylbacteriochlorin and tetraphenylhypochlorin are oxidized to tetraphenylchlorin (*146*). The quantum yield of dehydrogenation is rather low (10^{-3} to 10^{-2}), and depends on the type (o- or p-) and oxidation potential of the quinone (*152*). In view of the ability of quinones to quench the triplet state of chlorophyll, probably by electron transfer (*47, 60*), a plausible reaction sequence for the dehydrogenation of a hydroporphin PH_2 is the following, in which the quantum yield is determined by the relative values of the rate constants k_{39} and k_{40}.

$$PH_2' + Q \longrightarrow \{PH_2 \cdot^+ Q \cdot^-\} \rightleftharpoons PH_2 \cdot^+ + Q \cdot^- \qquad (38)$$

$$\{PH_2 \cdot^+ Q \cdot^-\} \xrightarrow{k_{39}} PH_2 + Q \qquad (39)$$

$$\{PH_2 \cdot^+ Q \cdot^-\} \xrightarrow{k_{40}} PH \cdot + QH \cdot \qquad (40)$$

$$2 PH \cdot \longrightarrow P + PH_2 \qquad (41)$$

The reaction is probably the photochemical equivalent of the thermal dehydrogenation of chlorins, bacteriochlorins, hypochlorins, and hexa-hydroporphins by dichlorodicyanoquinone and other quinones (147, 153).

Chlorophylls and most of their derivatives are not dehydrogenated this way; apparently reaction (40) is sterically hindered or otherwise blocked. Bacteriochlorophyll is oxidized, even in the dark, by o-benzoquinone to a chlorophyll derivative with an absorption band at 680 mμ, but not by quinones of lower potential (64). Hypochlorophylls are photooxidized to mesochlorophylls quantitatively, but with low quantum yield, by benzoquinone, 1,4-naphthoquinone, 2,5-dimethylbenzoquinone, and even phenosafranine (118).

On prolonged exposure to light in the presence of O_2, chlorophyll is oxidized to yellow products which have not been identified, but which include a small amount of a porphin (71a, 154–158). Chlorophyll is not regenerated by the action of reducing agents. Simpler chlorins are oxidized by O_2 to porphins, but in poorer yield than by quinones (159).

Bacteriochlorophyll is photooxidized by O_2, probably to Bchl·$^+$; the reaction is reversed by addition of ascorbic acid or H_2S (64). Chlorophylls a and b, and pheophytin a sensitize the oxidation of bacteriochlorophyll and chlorobium chlorophyll by O_2, being thereby protected, in accord with the oxidation potentials listed in Table V (160).

The photooxidations with O_2 are probably not initiated by reaction (38), but as in other sensitized oxidations by O_2 (Section III, F).

The equilibrium established in the dark between chlorophyll a and Fe^{+++} in methanol solution is displaced to the right by light (161).

$$Chl + Fe^{+++} \rightleftharpoons Chl·^+ + Fe^{++} \tag{42}$$

C. Pheophytinization

Bona fide examples of primary photochemical reactions of chlorophyll, not classifiable as oxidation or reduction, are not abundant. The reaction of chlorophyll a with phenylhydrazine at $-193°C$ to form a compound absorbing at 640 mμ (119), might be a photosensitized aminolysis of the cyclopentanone ring. Light accelerates the aggregation of chlorophyllin in aqueous buffer of pH 7, and in the presence of polyvinylpyrrolidone (162).

That light accelerates pheophytinization has been suspected for some time [Rabinowitch (6), p. 493]. Bannister has verified this reaction between succinic, maleic, and malonic acids and chlorophyll a in 30% aqueous pyridine, but the quantum yields are low ($< 10^{-3}$) (99, 100). Krasnovskii and Pakshina studied pheophytinization of chlorophyll a and b, protochlorophyll, bacteriochlorophyll, and chlorobium chlorophyll by oxalic and hydrochloric acids in aqueous acetone and pyridine (*163*). There was acceleration by light in every case except that of chlorobium chlorophyll, where the loss of Mg was quite rapid in the dark. Photo-pheophytinization was inhibited by carotene, methyl red, and tetracene at low concentration, clearly indicating that it goes via the triplet state of the chlorophyll.

D. Reversible Photobleaching

Porret and Rabinowitch discovered that deoxygenated solutions of chlorophyll in methanol were perceptibly bleached when exposed to intense light (*164*). Livingston *et al.* confirmed this *reversible photobleach-*

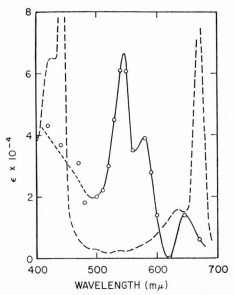

FIG. 2. Spectra of chlorophyll *a* (broken line) and its "photobleached" product (solid line) in pyridine. Reprinted from Livingston and Stockman (*170*).

ing or *phototropy* of chlorophyll, and studied the reaction intensively (*165–170*). The extent of bleaching is proportional to the square root of light intensity, and amounts to about 1% at their highest light intensities. The recovery of chlorophyll in the dark is complete in one second if the solvent is pure. The back reaction in methanol is second order in bleached product concentration (*165*), but in viscous solvents tends toward first order (*170*). The quantum yield is only about 5×10^{-4} (*165*), but the extent of bleaching is enhanced by traces of oxalic acid or I_2 (*166*). Bleaching occurs in polar solvents, but not in dry benzene unless "activated" with about 1% of methanol (*170*).

Photobleached chlorophyll is not identical with the triplet state, but is produced from it, because carotene and O_2 inhibit the reaction (170). Chlorophyll b and Mg phthalocyanine are photobleached to about the same extent as chlorophyll a, but pheophytin a, protoporphyrin, and mesoporphyrin are bleached to a much smaller extent or not at all. Cu pheophytin, Cu protoporphyrin, and ferric protoporphyrin chloride (hemin) are bleached, the last to a greater extent than chlorophyll a. The spectrum of photobleached chlorophyll a is shown in Fig. 2; the spectrum for photobleached chlorophyll b is quite similar.

Electron transfer from the solvent to the porphyrin has been suggested as the cause of photobleaching (168). It is difficult to reconcile this explanation with the fact that although the quantum yield is very low, it is practically independent of solvent (ethyl acetate, methanol, pyridine, cyclohexanol, castor oil) (170), and not enhanced in the presence of reductants like allylthiourea, hydroquinone, and phenylhydrazine (166). However, no alternative is readily apparent, consistent with the established kinetic data. Cu pheophytin and hemin are probably bleached by electron transfer from the ligand to the central metal.

Possibly related to reversible photobleaching of chlorophyll are the irreversible photobleaching of pheophytin hydrochloride in ethanol solution reported by Tollin and Green (46), and weak transient absorption changes in the spectrum of Cu pheophytin a reportedly evoked by flash photolysis (171).

E. Sensitized Oxidation-Reduction Reactions Not Involving Oxygen

1. DETERMINATION OF THE MECHANISM

Photosynthesis may be regarded as a gigantic sensitized reaction in which electrons are transferred from H_2O to CO_2 through many intermediate steps. Recognizing this, many investigators have studied reactions sensitized by chlorophylls in vitro, first those involving O_2 ("reverse of photosynthesis"), then those involving other oxidants. We treat the former separately, as they probably go by an entirely different mechanism.

Böhi discussed the results of his first systematic study of chlorophyll sensitized reduction of dyes (1) in terms of Baur's theory of photosensitized reactions (172), according to which a photoexcited molecule is polarized into a sort of intramolecular cathode and anode, which can react, respectively, with cathodic depolarizers (oxidants) and anodic depolarizers (reductants). Although the language of batteries is now superseded by the language of molecular orbitals, two insights of value remain: (1) the photoexcited molecule has suffered a charge redistribution, altering its susceptibility to electrophilic and nucleophilic attack, and (2) the ability to oxidize and the ability to reduce have *both* been increased by photoexcitation.

Chlorophylls can therefore sensitize oxidation-reduction reactions in two general ways, which will be called (A) and (B):

Mechanism A

$$Chl' + Ox \xrightarrow{k_{11}} Chl \cdot^+ + Ox \cdot^- \qquad (11)$$

$$Chl \cdot^+ + Red \longrightarrow Chl + Red \cdot^+ \qquad (43)$$

Mechanism B

$$Chl' + Red + H^+ \xrightarrow{k_{44}} ChlH \cdot + Red \cdot^+ \qquad (44)$$

$$ChlH \cdot + Ox \longrightarrow Chl + Ox \cdot^- + H^+ \qquad (45)$$

These mechanisms have been written for triplet excited state chlorophyll, but similar mechanisms could be written for Chl*.

It is not a matter of indifference which mechanism operates, because the energy requirements are different. According to Table V, $Chl \cdot^+$ is a better oxidant than Chl', and $ChlH \cdot$ a better reductant. That is, more difficult oxidations can be performed by $Chl \cdot^+$ via (A) than by Chl' via (B), and more difficult reductions can be performed by $ChlH \cdot$ via (B) than by Chl' via (A). This appears to have been verified experimentally (*173*).

Krasnovskii's observations that many dyes and other oxidants reoxidize $ChlH_2$ led him to conclude that all sensitized reductions of dyes go through $ChlH_2$ or $ChlH \cdot$ via (B) (*174*). Others have favored (A), (B), or some intermediate, noncommittal mechanism, often without decisive experimental evidence. Flash spectrophotometry has made it possible to get the most basic data needed for determination of mechanism, namely, the rate constants for reactions (11) and (44). Knowing these, it has become possible to decide in many cases whether (A) or (B) is operating.

From an analysis of the kinetics, mechanism (A) was assigned to the chlorophyllide-sensitized reductions of methyl red (*175*) and phenosafranine (*176*) by ascorbic acid, hydrazobenzene, and mercaptosuccinic acid, with the rate constants $k_{11} = 1.4 \times 10^9$ and $0.9 \times 10^9\ M^{-1}\ \text{sec}^{-1}$, respectively. However, mechanism (B) was assigned to the chlorophyllide-sensitized reduction of azobenzene by ascorbic acid (*173*).

Sensitized reactions going via (B) may be distinguished from those going via (A), in that they require photoreducing conditions for the chlorophyll. For example, chlorophyll very actively sensitizes the reduction of methyl red and phenosafranine in alcohols, whereas chlorophyll itself is hardly at all reduced by ascorbic acid in alcohols unless an organic base is present. On the other hand, the sensitized reduction of azobenzene in ethanol is sluggish unless pyridine is present, in which case there is simultaneous reduction of the sensitizer chlorophyllide.

The comparative activity of hydrazobenzene and ascorbic acid as reductants is a criterion of mechanism. Hydrazobenzene can be oxidized by Chl·$^+$, but apparently cannot be oxidized by Chl'. It is therefore an effective reducing agent only in mechanism (A). Thus, ethyl chlorophyllide sensitizes the reduction of triphenyltetrazolium chloride in ethanol by ascorbic acid and hydrazobenzene with comparable rapidity via (A), whereas it sensitizes the reduction of NAD in aqueous pyridine by ascorbic acid, but not by hydrazobenzene, via (B) (*173*).

In a survey of chlorophyll-sensitized oxidation-reduction reactions reported in the literature, there was a relation between the mechanism assigned on kinetic grounds and the polarographic half-wave reduction potential, measured in a solvent approximating those of the photochemical reactions. The oxidants reduced via (A) all had half-wave potentials greater than −0.60 volt (S.C.E.), whereas those reduced via (B) had potentials less than −0.90 volt (*173*). Reductions are possible via (B) that are forbidden via (A) on energetic grounds. However, when both (A) and (B) are energetically allowed, the rate constant for Eq. (11) so much exceeds that for Eq. (44) in the cases studied so far, that mechanism (A) dominates unless the reducing agent is in huge excess.

2. NOTES ON VARIOUS SENSITIZED REDUCTIONS

a. Methyl Red. The reduction of this azo dye has been reported with chlorophyll *a*, ethyl chlorophyllide *a*, and pheophytin *a* as sensitizers, and with ascorbic acid, phenylhydrazine, mercaptosuccinic acid, semicarbazide hydrochloride, *tert*-hexyl mercaptan, and phenylhydroxylamine as reductants in methanol or ethanol (*2, 175, 177–179*). The kinetics have been studied repeatedly; the quantum yield is usually ca. 0.2 or less, but goes up to 0.5 at concentrations of phenylhydrazine large enough to quench chlorophyll fluorescence (*178*). Only the neutral (zwitterionic) form of dye (VIIb) was said to be reducible (*178*), but Evstigneev *et al.* report reduction of acid form (VIIc) by ascorbic acid and polyphenols, apparently sensitized by pheophytin hydrochloride (*180, 181*); the

VII a   VII b   VII c

yellow, basic form (VIIa) may be reduced reversibly by mechanism (B) (*173, 175*). Reduction consumes four electron equivalents and is irreversible; for this reason, the reaction has been used regularly as a qualitative test of photochemical activity of chlorophyll preparations. Other chlorophylls and porphyrins [hematoporphyrin and protoporphyrin (*135*), porphin (*182*), chlorobium chlorophyll (*121*), chlorophyll *b*, and pheophytin *b* (*183*)] also sensitize the reduction.

b. *Other Azo Dyes.* Böhi reported reduction of azo dyes of every color but yellow, with chlorophyll and phenylhydrazine in methanol (*1*). The following yellow azo dyes were found to have low polarographic half-wave potentials, and mechanism (B) was assigned to their reduction (*173*):azobenzene ($E_{\frac{1}{2}} = -0.91$), methyl orange (basic form) (−1.05), butter yellow (−1.09), methyl red (basic form) (−1.10). These polarographic reduction potentials are much lower than those obtainable under more nearly reversible conditions (*184*). Reduction of butter yellow by pheophytin and ascorbic acid in methanol is catalyzed by a product, dimethylaminoazobenzene (*185*), probably acting as a basic catalyst in reaction (44). The quantum yield for the chlorophyll-sensitized reduction of butter yellow by ascorbic acid is much lower in dioxane than in methanol (*186*). Chlorophyll sensitizes reduction of azobenzene by phenylhydrazine in methanol, but pyridine is required for reduction by ascorbic acid (*173, 179*). Fast red S is reduced by chlorophyllin and ascorbic acid in water (*124*), and by ethyl chlorophyllide *a* and ascorbic acid or hydrazobenzene in ethanol (*173*), by mechanism (A) in the latter system if not in the former.

c. *Quinones.* These are of great interest because of their possibly direct role in photoreactions of photosynthesis. Transfer of an electron from chlorophyll to quinones has been followed by flash spectrophotometry (*60*) and ESR (*58, 59*). Chlorophyll sensitizes the reduction of benzoquinone by NADH in aqueous ethanol (*46*), and of plastoquinone by ascorbic acid in aqueous pyridine (*51*). Bacteriochlorophyll in solution sensitizes the reduction of vitamin K_3 by Na_2S in aqueous pyridine (*125*), and in suspension, of ubiquinones 2 and 6 by reduced phenazine methosulfate in a reversible reaction (*187*). The reduction of cytochrome *c* by trimethylhydroquinone, a reaction that is accelerated by chlorophyll suspensions, is initiated by reaction of Chl′ with small amounts of trimethylquinone (*188*).

d. *Azine Dyes.* Thionine and methylene blue form an equilibrium with ascorbic acid (*97*) that is displaced by reaction with chlorophyll in the light (*189*). Chlorophyll (*86, 174*), protochlorophyll (*143*), and bacteriochlorophyll (*190*) sensitize the reversible reduction of Safranine T by ascorbic acid in aqueous pyridine; chlorophyll sensitizes its reduction by phenylhydrazine in methanol (*1*), and by NADH in ethanol (*191*). A complex between Chl·+ and hydrazobenzene with a lifetime of at least 10^{-3} sec was postulated to account for retardation of the chlorophyllide-sensitized reduction of phenosafranine by hydrazobenzene in ethanolic pyridine (*176*). Phenazine and acridine yellow are reduced by chlorophyll and ascorbic acid in aqueous pyridine (*51*); pyocyanine quenches Chl', apparently by electron transfer (*60*); bacteriochlorophyll sensitizes reduction of phenazine methosulfate by Na_2S in aqueous pyridine (*125*). Janus green is reduced in two steps; first the azo linkage is reduced irreversibly, then the safranine that remains is reduced reversibly (*1*). Mechanism (A) is probable for all of these reductions, except perhaps those of phenazine and acridine yellow.

e. Other Dyes. Riboflavin is reduced reversibly by chlorophyll (*174*) or hematoporphyrin (*134*) and ascorbic acid in ethanol or pyridine, by bacteriochlorophyll and Na_2S in aqueous pyridine (*125, 190*), and by pheophytin hydrochloride in ethanol (*46*). Reductions of thiocarbocyanine dyes (*51*), thioindigo and triphenylmethane dyes (*1*), and triphenyltetrazolium chloride (*75, 173*) have been reported. The ability of chloroplasts to reduce certain viologens of low oxidation potential suggests that chlorophyll should be able to do so *in vitro* (*192, 193*). Mechanism (A) seems likely for all these dyes.

f. Pyridine Nucleotides. Chlorophyll-sensitized reduction of NAD by ascorbic acid in aqueous pyridine was first reported by Krasnovskii and Brin (*96*) and, though doubted (*194*), has been amply confirmed (*195, 196*). It is best seen if some ammonia is added to the aqueous pyridine, to prevent accumulation of ChlH₂, which otherwise would be further reduced in competition with NAD (*195*). Chlorophyllin, chlorin e₆, and hematoporphyrin sensitize reduction of NAD and NADP by ascorbic acid in water, but only in the presence of transhydrogenase (*197*). Bacteriochlorophyll and Na_2S cannot reduce NAD (190). NADP (*195*) and N-benzylnicotinamide (*196*) are photoreduced as is NAD. Mechanism (B) is assigned because there is no reduction with hydrazobenzene in place of ascorbic acid (*173*), and because reduction of NAD is normally accomplished by reduction of chlorophyll (*194*).

3. PHOTOSENSITIZATION BY AGGREGATES

Interest in the photochemistry of various aggregated forms of chlorophyll derives from their possible resemblance to the state of chlorophyll in the plant. Suspensions, emulsions, and adsorbates have been tested.

Chlorophyll suspensions, prepared by diluting acetone or ethanol solutions with water (*121, 198*), or by dissolving adsorbates of chlorophyll on sugar (*199*) or magnesia (*200*), are able to sensitize reduction of methyl red (*121, 198, 199, 201*) and tetrazolium blue (*199*) by ascorbic acid, thionine by Fe^{++} (*202*), ubiquinones by reduced phenazine methosulfate (*187*), and pyridine nucleotides by ascorbic acid in the presence of NADP reductase (*197, 203*). Chlorophyll adsorbed onto polycaprolactam (*204*), alumina, and a number of other materials (*201*) can sensitize reduction of methyl red by ascorbic acid. In these cases, activity is usually only a fraction of that realized in true solution.

Better yields are obtained if surface active agents such as deoxycholate (*197*) or Triton X-100 (*188*) are added to the suspensions, or the chlorophyll is adsorbed to polymers in a molecularly dispersed form. Phospholipid stimulates the bacteriochlorophyll-sensitized reduction of ubiquinones by reduced phenazine methosulfate (*187*), and chlorophyll adsorbed to egg yolk lipoprotein is active in the reduction of methyl red (*205*). Polyvinylpyrrolidone increases the rate of chlorophyllin-sensitized reduction of fast red S by ascorbic acid (*124*). Chlorin e_6 attached to a water soluble anionic copolymer through a 6-amide linkage was active in the reduction of phenosafranine by hydrazobenzene (*206*). Finally, chlorophyll in "coacervates" of oleic acid and serum albumin shows enhanced activity over that in solution toward methyl red reduction, probably because the dye is distributed preferentially into the "coacervate" phase (*207*).

The conditions under which chlorophylls are most active in aggregated states are nevertheless those under which the pigment molecules are somewhat separated from each other (micelles, emulsions). There is yet to be demonstrated an enhanced photochemical activity definitely owing to the existence of interaction between chlorophyll molecules in an aggregate.

However, the retention of some activity in suspensions that are usually nonfluorescent suggests that the following singlet excited state quenching process may occur in the aggregate:

$$Chl^* + Chl \longrightarrow Chl' + Chl \tag{46}$$

4. IRON CHLOROPHYLL COMPLEXES

Ashkinazi *et al.* have studied the photoreduction of ferric complexes of chlorophyll derivatives, a sort of oxidation-reduction reaction in which the oxidant is the central metal ion, and the reductant may be the solvent. Ferric complexes of pheophytin *a*, pheophorbide *a* (*208, 209*), and chlorin e_6 (*208, 210*) in ethanol or chloroform solution are reduced to the ferrous complexes on prolonged illumination. Reducing agents that form complexes with the iron in the chlorophyll (glycerol, glycol, phenylhydrazine) accelerate the reaction (*211*). Indeed, reduction goes in the dark with the last, unless its concentration is very low. Reduction is apparently initiated by electron transfer from the photoexcited ligand to the iron, and completed by oxidation of the solvent. Polymerization of acrylonitrile or methyl methacrylate can be initiated in the process (212).

The reaction is accompanied by notable changes in the spectrum (209). In dry ethanol, the principal red band of ferric pheophorbide is at 620 mμ, but in wet ethanol, near 675 mμ. Both forms were reduced (after 12 and 3 hours of intense illumination, respectively) to the ferrous complex, with a band at 650 mμ. By an even slower reaction, ferrous, zinc, and copper pheophytins are bleached photochemically, via a derivative with a band at 635 mμ (213).

Butsko and Dain investigated a rather complicated reaction between chlorophyll, $FeCl_3$, and acetone (114). In 365 mμ light, $FeCl_3$ reacted with acetone to give compounds which in red light reduced photoexcited chlorophyll to $ChlH_2$ (525 mμ). In the dark, $ChlH_2$ reacted with the (ferrous) iron to form a reduced ferrous pheophytin (510 mμ) which then could be reoxidized by air to ferric protopheophytin or some similar compound.

F. Sensitized Oxidations Involving Oxygen

The chlorophylls share with many other dyes the ability to sensitize the oxidation of organic substances by molecular oxygen. The only apparent requirement of a sensitizer is the existence of an excited state of sufficiently long lifetime to react with O_2; Cu pheophytin, for example, is inactive.

Among the compounds whose oxidation by chlorophyll, its derivatives, and other porphyrins has been studied are benzidine (214, 215), p-toluenediamine (124, 158), aliphatic amines such as isoamylamine (216), phenylhydrazine (177), diphenylamine and p-phenylenediamine (217), benzyl alcohol (218), pyruvic acid (219), ascorbic acid (220–222), cysteine, polyphenols, and cytochrome (223), NADH (195), dihydrodichlorophenolindophenol (224), ergosterol (225), pyrethrum (226), terpenes (227), serum proteins, casein, tyrosine, phenol, and uric acid (228). α-Terpinene (229) and rubrene (230) are oxidized to endoperoxides. The rate of oxidation of carotenes increases with the extent of saturation (50) (i.e., with the number of allylic hydrogens?). Zinc tetraphenylporphin oxidizes trimethylene disulfide, an analog of thioctic acid, to the sulfoxide (44). A cysteine-pheophorbide b adduct is oxidized by O_2, with evolution of CO_2 (231). Chlorophylls a and b sensitize the oxidation of bacteriochlorophyll and chlorobium chlorophyll, implying lower oxidation potentials for the bacterial pigments (*160*), a conclusion supported by the data of Table V. Uroporphyrin sensitizes its own formation by oxidation of uroporphyrinogen (*232*).

The sensitized oxidations of thiourea and of allylthiourea have received the most quantitative study (*218, 233, 234*), and have been developed as actinometers in which O_2 uptake is measured manometrically (*234–236*). Allylthiourea takes up $1\frac{1}{2}$moles of O_2, and gives off SO_2 (*233*); thiourea is oxidized to H_2SO_4 and cyanamide in pyridine, but 2.5 moles of O_2 are consumed per mole (*234*).

$$NH_2CSNH_2 + 2\,O_2 \xrightarrow{\text{Chl}'} NH_2C{\equiv}N + H_2SO_4 \qquad (47)$$

At high O_2 and substrate concentrations, the limiting quantum yield for oxidation of thiourea by ethyl chlorophyllide is 0.98 (*234*), and by pheophorbide, 0.7 (*235*), independent of light intensity. The limiting quantum yield of oxidation of allylthiourea by chlorophyll is 0.78 in methanol; in acetone it increases from 0.99 at 1°C to 1.3 at 35°C (*218*). These yields are based on O_2 consumption and are therefore greater than the yields of substrate consumption. For allylthiourea, the following equation expressed the quantum yield (*218*).

$$\phi = \phi_0 \cdot \frac{[O_2]}{B + [O_2]}\frac{[R]}{C + [R]} \qquad (48)$$

$\phi_0 =$ limiting quantum yield; $[R] =$ allylthiourea concentration; $B = 2 \times 10^{-6}$ in methanol or acetone; $C = 0.100$ in methanol and 0.0117 in acetone.

The mechanism of sensitized oxidations has been paid an enormous amount of attention; the earlier experiments and thought on the subject have been reviewed in detail by Terenin (*237*), and so far as chlorophyll is concerned, by Rabinowitch (*6*).

Four distinct mechanisms have been offered. Kautsky *et al.*, on the basis of their finding that trypaflavin sensitized the oxidation of leuco malachite green at low O_2 pressure, even when the dye and substrate were adsorbed onto different silica gel particles, concluded that the active oxidizing agent had to be volatile and proposed that the photoexcited dye transferred energy to O_2, raising it to the $^1\Sigma_g{}^+$ (37.5 kcal) or the $^1\Delta_g$ (22.5 kcal) state (*238, 239*). Rosenberg and Shombert, from similar observations, postulated vibrationally excited O_2 (*240*). Gaffron favored a primary photochemical reaction between the sensitizer (Chl′) and the reductant (allylthiourea) (*230*). Weiss argued for electron transfer from the photoexcited dye to O_2, as the initial step in sensitized oxidations as well as in fluorescence quenching (*79*). In a variant of this, Franck and Livingston proposed hydrogen atom transfer from their postulated long-

lived tautomer of the photoexcited sensitizer to O_2, making $HO_2 \cdot$ the
active oxidant (241). Finally, Schenck championed a proposal of Schönberg (242) that photoexcited dyes form loose complexes with oxygen
("moloxides") which were able to react with oxidizable substrates, applied it to the chlorophyll-sensitized synthesis of ascaridol (IX) from
α-terpinene (VIII) (229, 243), and subsequently to many other reactions
(222, 244).

$$\text{Chl}' \; + \; O_2 \longrightarrow [\text{Chl} \cdots O_2] \tag{49}$$

VIII   IX

Until recently, the most generally accepted mechanism has probably
been Schenck's. It is supported by the existence of endoperoxides of
rubrene and anthracene, and by the formation of ascaridol. It is usually
consistent with detailed analyses of the kinetics (244). However, peroxides of most of the sensitizers have not been isolated, and the mechanism cannot explain convincingly why the autosensitized formation of
anthracene peroxide should be second order in anthracene (245, 246),
or why rubrene should sensitize the oxidation of allylthiourea (230).

Kautsky's mechanism has recently been revived by Foote and Wexler
(247), on grounds that the products of reaction of chemically generated
O_2 ($^1\Delta_g$) are the same as those of photosensitized oxidations (248). Furthermore, O_2 ($^1\Delta_g$) generated by electrodeless discharge forms the same
endoperoxides from a α-terpinene, anthracene, and other unsaturated
compounds as those formed photochemically (249). In view of this evidence, it now appears most probable that the primary step in chlorophyll-
sensitized oxidations by O_2 is transfer of energy from Chl' to O_2, exciting
the latter to the $^1\Delta_g$ state. Even triplet excited bacteriochlorophyll has
enough energy (27 kcal according to the estimate of Table V) to excite
O_2 ($^1\Delta_g$) (22.5 kcal). Gaffron found that bacteriopheophytin sensitizes
the oxidation of allylthiourea (250).

Egerton has reasserted the case for primary electron transfer (251),
by analogy to ZnO sensitized reactions, which apparently start this way
(252). However, a primary energy transfer process does not exclude
subsequent electron transfer to O_2 ($^1\Delta_g$) from the sensitizer, as well as
the substrate, in a solvent cage or by a diffusion controlled reaction. The
quenching (rather than prolongation) by O_2 of transient absorption

spectral changes in chlorophyll solutions subjected to high intensity flashes (*169*), favors energy transfer over electron transfer as the primary photochemical reaction.

Gaffron's mechanism may be eliminated for want of evidence of interaction between Chl′ and most of the substrates of sensitized oxidation, including allylthiourea. The probable mechanism of sensitized oxidation of a substrate RH is then essentially as follows:

$$\text{Chl}' + O_2 \longrightarrow \text{Chl} + O_2\,(^1\Delta_g) \tag{51}$$

$$O_2\,(^1\Delta_g) \longrightarrow O_2\,(^3\Sigma_g^- \text{ ground state}) \tag{52}$$

$$O_2\,(^1\Delta_g) + RH \longrightarrow HO_2 \cdot + R \cdot \tag{53}$$

The limiting quantum yield ϕ_0 is determined by the nature of the reactions subsequently undergone by $HO_2 \cdot$ and $R \cdot$.

The kinetic expression for the quantum yield is of the same form as in Schenck's mechanism, and as that found by Livingston and Owens.

$$\phi = \phi_0 \cdot \frac{[O_2]}{(k_5/k_{51}) + [O_2]} \cdot \frac{[RH]}{(k_{52}/k_{53}) + [RH]} \tag{54}$$

The Kautsky mechanism of energy transfer as the initiating reaction in sensitized oxidations would appear intuitively to have the advantage over electron transfer of being faster. However, the rate of energy transfer depends on the overlap of energy levels, which may be large for some sensitizers and small for others. With the latter, electron transfer may predominate. It is not known whether the reactivities of $O_2\,(^1\Sigma_g^+)$ and $O_2\,(^1\Delta_g)$ are the same; if they are not, different results with different sensitizers could be explained. For example, the rate and products of the benzil-sensitized oxidation of Zn tetraphenylporphin are very different from those of its autosensitized oxidation (*145*). The various products of reduction of O_2 ($OO \cdot^-$, $HOO \cdot$, $HO \cdot$, H_2O_2, etc.) are themselves oxidants and will have different reactivities toward different reductants and sensitizers. The detailed course of a given photosensitized oxidation will therefore be the complex resultant of all these factors.

References

(*1*) J. Böhi, *Helv. Chim. Acta* **12**, 121 (1929).
(*2*) J. C. Ghosh and S. B. Sen Gupta, *J. Indian Chem. Soc.* **11**, 65 (1934).
(*3*) A. A. Krasnovskii, *Dokl. Akad. Nauk SSSR* **60**, 421 (1948).
(*4*) G. Porter, *Proc. Roy. Soc.* **A200**, 284 (1950).
(*5*) H. Fischer and A. Stern, "Die Chemie des Pyrrols," Vol. 2, Part II, p. 360. Akad. Verlagsges., Leipzig, 1940.
(*6*) E. I. Rabinowitch, "Photosynthesis and Related Processes," Vol. 1, p. 483. Wiley (Interscience), New York, 1945.

(7) E. I. Rabinowitch, "Photosynthesis and Related Processes," Vol. 2, Part 2, p. 1487. Wiley (Interscience), New York, 1956.

(8) R. Livingston, Quart. Rev. (London) 14, 174 (1960).

(9) J. L. Rosenberg, Ann. Rev. Plant Physiol. 8, 115 (1957).

(10) A. A. Krasnovskii, Ann. Rev. Plant Physiol. 11, 363 (1960).

(11) A. A. Krasnovskii, Usp. Khim. 29, 736 (1960).

(12) N. T. Clare, in "Radiation Biology" (A. Hollaender, ed.), Chapter XV, p. 693. McGraw-Hill, New York, 1956.

(13) J. D. Spikes and B. W. Glad, Photochem. Photobiol. 3, 471 (1964).

(13a) J. N. Pitts, Jr., F. Wilkinson, and G. S. Hammond, Advan. Photochem. 1, 1, (1963).

(14) L. A. Tumerman, Dokl. Akad. Nauk SSSR 117, 605 (1957).

(15) T. Förster, "Fluoreszenz Organischer Verbindungen," p. 83. Vandenhoeck & Ruprecht, Göttingen, 1951.

(16) T. Förster, Discussions Faraday Soc. 27, 7 (1959).

(17) I. Ketskeméty, Z. Naturforsch. 17a, 666 (1962).

(18) B. Smaller, E. C. Avery, and J. R. Remko, J. Chem. Phys. 43, 992 (1965).

(19) Kh. S. Bagdasar'yan, Z. A. Sinitsyna, and V. I. Muromtsev, Dokl. Akad. Nauk SSSR 153, 374 (1963).

(20) Kh. S. Bagdasar'yan, V. I. Muromtsev, and Z. A. Sinitsyna, Dokl. Akad. Nauk SSSR 152, 349 (1963).

(21) H. Linschitz and J. Rennert, Nature 169, 193 (1952).

(22) A. A. Kachan and B. Ya. Dain, Dokl. Akad. Nauk SSSR 80, 619 (1951).

(23) P. Debye, Trans. Electrochem. Soc. 82, 265 (1942).

(24) A. D. Osborne and G. Porter, Proc. Roy. Soc. A284, 9 (1965).

(25) R. M. Noyes, Progr. Reaction Kinetics 1, 129 (1961).

(26) G. N. Lewis and M. Kasha, J. Am. Chem. Soc. 66, 2100 (1944).

(27) G. T. Rikhireva, Z. P. Gribova, L. P. Kayushin, A. V. Umrikhina, and A. A. Krasnovskii, Dokl. Akad. Nauk SSSR 159, 196 (1964).

(28) R. Livingston, G. Porter, and M. Windsor, Nature 173, 485 (1954).

(29) R. Livingston, J. Am. Chem. Soc. 77, 2179 (1955).

(30) H. Linschitz and K. Sarkanen, J. Am. Chem. Soc. 80, 4826 (1958).

(31) G. Zieger and H. T. Witt, Z. Physik. Chem. (Frankfurt) [N.S.] 28, 273 (1961).

(32) S. Claesson, L. Lindqvist, and B. Holmström, Nature 183, 661 (1959).

(33) L. Pekkarinen and H. Linschitz, J. Am. Chem. Soc. 82, 2407 (1960).

(34) R. Livingston and E. Fujimori, J. Am. Chem. Soc. 80, 5610 (1958).

(35) R. Livingston and P. J. McCartin, J. Phys. Chem. 67, 2511 (1963).

(36) R. Livingston and A. C. P. Pugh, Nature 186, 969 (1960).

(37) A. K. Chibisov and A. V. Karyakin, Dokl. Akad. Nauk SSSR 153, 1132 (1963).

(38) P. J. McCartin, Trans. Faraday Soc. 60, 1694 (1964).

(39) W. F. Watson and R. Livingston, J. Chem. Phys. 18, 802 (1950).

(40) L. N. M. Duysens, Nature 168, 548 (1951).

(41) A. G. Tweet, W. D. Bellamy, and G. L. Gaines, Jr., J. Chem. Phys. 41, 2068 (1964).

(42) F. W. J. Teale, Nature 181, 416 (1958).

(43) M. Calvin and J. A. Barltrop, J. Am. Chem. Soc. 74, 6153 (1952).

(44) J. A. Barltrop, P. M. Hayes, and M. Calvin, J. Am. Chem. Soc. 76, 4348 (1954).

(45) R. B. Whitney and M. Calvin, *J. Chem. Phys.* 23, 1750 (1955).
(46) G. Tollin and G. Green, *Biochim. Biophys. Acta* 66, 308 (1963).
(47) E. Fujimori and R. Livingston, *Nature* 180, 1036 (1957).
(48) R. Livingston and A. C. Pugh, *Discussions Faraday Soc.* 27, 144 (1959).
(49) H. Claes and T. O. M. Nakayama, *Nature* 183, 1053 (1959).
(50) H. Claes, *Z. Naturforsch.* 16b, 445 (1961).
(51) A. A. Krasnovskii and N. N. Drozdova, *Biokhimiya* 26, 859 (1961).
(52) D. F. Evans, *J. Chem. Soc.* p. 1735 (1960).
(53) D. F. Evans, *J. Chem. Soc.* p. 2566 (1961).
(54) J. R. Platt, *in* "Radiation Biology" (A. Hollaender, ed.), Vol. III, Chapter 2, p. 71. McGraw-Hill, New York, 1956.
(55) S. P. McGlynn, T. Azumi, and M. Kasha, *J. Chem. Phys.* 40, 507 (1964).
(56) S. P. McGlynn, F. J. Smith, and G. Cilento, *Photochem. Photobiol.* 3, 269 (1964).
(57) C. Reid, *J. Chem. Phys.* 20, 1214 (1952).
(58) G. Tollin and G. Green, *Biochim. Biophys. Acta* 60, 524 (1962).
(59) G. Tollin, K. K. Chatterjee, and G. Green, *Photochem. Photobiol.* 4, 593 (1965).
(60) B. Ke, L. P. Vernon, and E. R. Shaw, *Biochemistry* 4, 137 (1965).
(61) A. A. Krasnovskii and N. N. Drozdova, *Dokl. Akad. Nauk SSSR* 150, 1378 (1963).
(62) A. A. Krasnovskii and N. N. Drozdova, *Dokl. Akad. Nauk SSSR* 153, 721 (1963).
(63) A. A. Krasnovskii and N. N. Drozdova, *Dokl. Akad. Nauk SSSR* 158, 730 (1964).
(64) A. A. Krasnovskii and K. K. Voinovskaya, *Dokl. Akad. Nauk SSSR* 81, 879 (1951).
(65) R. J. Marcus and L. K. Moss, *Nature* 183, 990 (1959).
(66) V. B. Evstigneev and V. A. Gavrilova, *Dokl. Akad. Nauk SSSR* 92, 381 (1953).
(67) V. B. Evstigneev and V. A. Gavrilova, *Dokl. Akad. Nauk SSSR* 95, 841 (1954).
(68) V. B. Evstigneev and V. A. Gavrilova, *Dokl. Akad. Nauk SSSR* 114, 1066 (1957).
(69) V. B. Evstigneev and V. A. Gavrilova, *Dokl. Akad. Nauk SSSR* 115, 530 (1957).
(70) V. B. Evstigneev and V. A. Gavrilova, *Dokl. Akad. Nauk SSSR* 118, 1146 (1958).
(71) R. Livingston and P. J. McCartin, *J. Am. Chem. Soc.* 85, 1571 (1963).
(71a) G. Zieger and H. T. Witt, *Z. Physik. Chem.* (Frankfurt) [N.S.] 28, 286 (1961).
(72) D. T. Holmes, *Photochem. Photobiol.* 4, 631 (1965).
(73) P. A. Shakhverdov and A. N. Terenin, *Dokl. Akad. Nauk SSSR* 150, 1311 (1963).
(74) R. Livingston and C.-L. Ke, *J. Am. Chem. Soc.* 72, 909 (1950).
(75) E. Fujimori, *J. Am. Chem. Soc.* 77, 6495 (1955).
(76) I. I. Dilung and I. N. Chernyuk, *Dokl. Akad. Nauk SSSR* 140, 162 (1961).
(77) I. I. Dilung and I. N. Chernyuk, *Zh. Fiz. Khim.* 37, 1100 (1963).
(78) J. Weiss and H. Fischgold, *Z. Physik. Chem.* B32, 135 (1936).
(79) J. Weiss, *Trans. Faraday Soc.* 35, 44 (1939).
(80) G. K. Rollefson and R. W. Stoughton, *J. Am. Chem. Soc.* 63, 1517 (1941).

(81) N. Christodouleas and S. P. McGlynn, *J. Chem. Phys.* **40**, 166 (1964).

(82) V. B. Evstigneev, V. A. Gavrilova, and A. A. Krasnovskii, *Dokl. Akad. Nauk SSSR* **74**, 315 (1950).

(83) R. Livingston, W. F. Watson, and J. McArdle, *J. Am. Chem. Soc.* **71**, 1542 (1949).

(84) V. B. Evstigneev and A. A. Krasnovskii, *Dokl. Akad. Nauk SSSR* **60**, 623 (1948).

(85) R. Livingston, L. Thompson, and M. V. Ramarao, *J. AM. Chem. Soc.* **74**, 1073 (1952).

(86) T. T. Bannister, *Photochem. Photobiol.* **2**, 519 (1963).

(87) J. Franck and H. Levi, *Z. Physik. Chem.* **B27**, 409 (1934).

(88) R. S. Becker and M. Kasha, *J. Am. Chem. Soc.* **77**, 3669 (1955).

(89) J. Fernandez and R. S. Becker, *J. Chem. Phys.* **31**, 467 (1959).

(90) I. S. Singh and R. S. Becker, *J. Am. Chem. Soc.* **82**, 2083 (1960).

(91) A. Stanienda, *Naturwissenschaften* **50**, 731 (1963).

(92) J. C. Goedheer, G. H. Horreus de Haas, and P. Schuller, *Biochim. Biophys. Acta* **28**, 278 (1958).

(93) S. Gilman, Ph.D. Dissertation, Syracuse University (1957).

(94) R. Felton, G. M. Sherman, and H. Linschitz, *Nature* **203**, 637 (1964).

(95) W. M. Clark, "Oxidation-Reduction Potentials of Organic Systems." Williams & Wilkins, Baltimore, Maryland, 1960.

(96) A. A. Krasnovskii and G. P. Brin, *Dokl. Akad. Nauk SSSR* **67**, 325 (1949).

(97) A. A. Krasnovskii and G. P. Brin, *Dokl. Akad. Nauk SSSR* **73**, 1239 (1950).

(98) A. A. Krasnovskii, G. P. Brin, and K. K. Voinovskaya, *Dokl. Akad. Nauk SSSR* **69**, 393 (1949).

(99) T. T. Bannister, in "Photochemistry in the Liquid and Solid States" (L. J. Heidt *et al.*, eds.), p. 110. Wiley, New York, 1960.

(100) T. T. Bannister, *Plant Physiol.* **34**, 246 (1959).

(101) A. A. Krasnovskii and G. P. Brin, *Dokl. Akad. Nauk SSSR* **89**, 527 (1952).

(102) A. A. Krasnovskii and E. V. Pakshina, *Dokl. Akad. Nauk SSSR* **120**, 581 (1958).

(103) A. A. Krasnovskii and G. P. Brin, *Dokl. Akad. Nauk SSSR* **96**, 1025 (1954).

(104) V. B. Evstigneev and V. A. Gavrilova, *Dokl. Akad. Nauk SSSR* **108**, 507 (1956).

(105) A. V. Umrikhina and A. A. Krasnovskii, *Dokl. Akad. Nauk SSSR* **155**, 904 (1964).

(106) H. Linschitz and S. I. Weissman, *Arch. Biochem. Biophys.* **67**, 491 (1957).

(107) S. S. Brody, G. Newell, and T. Castner, *J. Phys. Chem.* **64**, 554 (1960).

(108) A. A. Krasnovskii and A. V. Umrikhina, *Dokl. Akad. Nauk SSSR* **104**, 882 (1955).

(109) V. B. Evstigneev and V. A. Gavrilova, *Dokl. Akad. Nauk SSSR* **91**, 899 (1953).

(110) B. Rackow and H. König, *Z. Elektrochem.* **62**, 482 (1958).

(111) V. B. Evstigneev and V. A. Gavrilova, *Dokl. Akad. Nauk SSSR* **74**, 781 (1950).

(112) M. S. Ashkinazi, I. A. Dolidze, and V. E. Karpitskaya, *Biofizika* **6**, 294 (1961).

(113) I. I. Dilung and B. Ya. Dain, *Zh. Fiz. Khim.* **33**, 2740 (1959).

(114) S. S. Butsko and B. Ya. Dain, *Ukr. Khim. Zh.* **27**, 314 (1961).

(115) A. N. Sidorov and A. N. Terenin, *Dokl. Akad. Nauk SSSR* **145**, 1092 (1962).

(116) G. R. Seely and A. Folkmanis, *J. Am. Chem. Soc.* **86**, 2763 (1964).

(*117*) V. B. Evstigneev and V. A. Gavrilova, *Dokl. Akad. Nauk SSSR* **96**, 1201 (1954).
(*118*) G. R. Seely, submitted for publication.
(*119*) S. S. Brody, *J. Am. Chem. Soc.* **82**, 1570 (1960).
(*120*) Yu. E. Erokhin and A. A. Krasnovskii, *Biofizika* **6**, 392 (1961).
(*121*) A. A. Krasnovskii and E. V. Pakshina, *Dokl. Akad. Nauk SSSR* **127**, 913 (1959).
(*122*) V. B. Evstigneev and O. D. Bekasova, *Dokl. Akad. Nauk SSSR* **154**, 946 (1964).
(*123*) I. G. Savkina and V. B. Evstigneev, *Biofizika* **8**, 335 (1963).
(*124*) G. Oster, J. S. Bellin, and S. B. Broyde, *J. Am. Chem. Soc.* **86**, 1313 (1964).
(*125*) A. A. Krasnovskii and E. V. Pakshina, *Dokl. Akad. Nauk SSSR* **135**, 1258 (1960).
(*126*) V. B. Evstigneev and V. A. Gavrilova, *Dokl. Akad. Nauk SSSR* **141**, 477 (1961).
(*127*) G. R. Seely and K. Talmadge, *Photochem. Photobiol.* **3**, 195 (1964).
(*128*) V. B. Evstigneev and V. A. Gavrilova, *Dokl. Akad. Nauk SSSR* **103**, 97 (1955).
(*129*) V. B. Evstigneev and V. A. Gavrilova, *Dokl. Akad. Nauk SSSR* **127**, 198 (1959).
(*130*) V. B. Evstigneev and I. G. Savkina, *Biofizika* **6**, 30 (1961).
(*131*) D. Mauzerall, *J. Am. Chem. Soc.* **82**, 1832 (1960).
(*132*) D. Mauzerall, *J. Am. Chem. Soc.* **84**, 2437 (1962).
(*133*) D. Mauzerall, *J. Phys. Chem.* **66**, 2531 (1962).
(*134*) A. A. Krasnovskii and K. K. Voinovskaya, *Dokl. Akad. Nauk SSSR* **96**, 1209 (1954).
(*135*) A. A. Krasnovskii and A. V. Umrikhina, *Dokl. Akad. Nauk SSSR* **122**, 1061 (1958).
(*136*) V. B. Evstigneev, V. A. Gavrilova, and I. G. Savkina, *Dokl. Akad. Nauk SSSR* **124**, 691 (1959).
(*137*) G. P. Gurinovich, A. M. Shulga, and A. N. Sevchenko, *Dokl. Akad. Nauk SSSR* **153**, 703 (1963).
(*138*) A. N. Sidorov, V. G. Vorob'ev, and A. N. Terenin, *Dokl. Akad. Nauk SSSR* **152**, 919 (1963).
(*139*) R. B. Woodward, *Angew. Chem.* **72**, 651 (1960); *Pure Appl. Chem.* **2**, 383 (1960).
(*140*) R. B. Woodward, *Ind. Chim. Belge* **27**, 1293 (1962).
(*141*) A. N. Sidorov, *Dokl. Akad. Nauk SSSR* **158**, 973 (1964).
(*142*) G. R. Seely, unpublished observations (1964).
(*143*) A. A. Krasnovskii and K. K. Voinovskaya, *Dokl. Akad. Nauk SSSR* **66**, 663 (1949).
(*144*) V. B. Evstigneev and I. G. Savkina, *Biofizika* **4**, 289 (1959).
(*145*) G. R. Seely and M. Calvin, *J. Chem. Phys.* **23**, 1068 (1955).
(*146*) G. D. Dorough and J. R. Miller, *J. Am. Chem. Soc.* **74**, 6106 (1952).
(*147*) U. Eisner, *J. Chem. Soc.* p. 3461 (1957).
(*148*) J. R. Barnard and L. M. Jackman, *J. Chem. Soc.* p. 1172 (1956).
(*149*) G. R. Seely, *J. Chem. Phys.* **27**, 125 (1957).
(*150*) M. Gouterman, G. H. Wagniere, and L. C. Stryer, *J. Mol. Spectry.* **11**, 108 (1963).
(*151*) M. Calvin and G. D. Dorough, *J. Am. Chem. Soc.* **70**, 699 (1948).
(*152*) F. M. Huennekens and M. Calvin, *J. Am. Chem. Soc.* **71**, 4024 (1949).

(*153*) U. Eisner and R. P. Linstead, *J. Chem. Soc.* p. 3749 (1955).
(*154*) S. Aronoff and G. Mackinney, *J. Am. Chem. Soc.* **65**, 956 (1943).
(*155*) A. A. Krasnovskii, *Dokl. Akad. Nauk SSSR* **58**, 617 (1947).
(*156*) I. I. Dilung, *Ukr. Khim. Zh.* **24**, 202 (1958).
(*157*) I. I. Dilung and V. E. Karpitskaya, *Dokl. Akad. Nauk SSSR* **152**, 367 (1963).
(*158*) T. T. Bannister and J. E. Bernardini, *Photochem. Photobiol.* **2**, 535 (1963).
(*159*) F. M. Huennekens and M. Calvin, *J. Am. Chem. Soc.* **71**, 4031 (1949).
(*160*) V. B. Evstigneev and V. A. Gavrilova, *Dokl. Akad. Nauk SSSR* **154**, 714 (1964).
(*161*) E. Rabinowitch and J. Weiss, *Nature* **138**, 1098 (1936).
(*162*) G. Oster, S. B. Broyde, and J. S. Bellin, *J. Am. Chem. Soc.* **86**, 1309 (1964).
(*163*) A. A. Krasnovskii and E. V. Pakshina, *Dokl. Akad. Nauk SSSR* **148**, 935 (1963).
(*164*) D. Porret and E. Rabinowitch, *Nature* **140**, 321 (1937).
(*165*) R. Livingston, *J. Phys. Chem.* **45**, 1012 (1941).
(*166*) J. J. McBrady and R. Livingston, *J. Phys. Chem.* **52**, 662 (1948).
(*167*) R. Livingston, *J. Phys. Colloid Chem.* **52**, 527 (1948).
(*168*) J. D. Knight and R. Livingston, *J. Phys. Chem.* **54**, 703 (1950).
(*169*) R. Livingston and V. A. Ryan, *J. Am. Chem. Soc.* **75**, 2176 (1953).
(*170*) R. Livingston and D. Stockman, *J. Phys. Chem.* **66**, 2533 (1962).
(*171*) D. T. Holmes and R. Livingston, *Photochem. Photobiol.* **4**, 629 (1965).
(*172*) E. Baur, *Helv. Chim. Acta* **1**, 186 (1918).
(*173*) G. R. Seely, *J. Phys. Chem.* **69**, 2779 (1965).
(*174*) A. A. Krasnovskii, *Dokl. Akad. Nauk SSSR* **61**, 91 (1948).
(*175*) G. R. Seely, *J. Phys. Chem.* **69**, 821 (1965).
(*176*) G. R. Seely, *J. Phys. Chem.* **69**, 2633 (1965).
(*177*) R. Livingston, D. Sickle, and A. Uchiyama, *J. Phys. Chem.* **51**, 775 (1947).
(*178*) R. Livingston and R. Pariser, *J. Am. Chem. Soc.* **70**, 1510 (1948).
(*179*) R. Livingston and R. Pariser, *J. Am. Chem. Soc.* **78**, 2948 (1956).
(*180*) V. B. Evstigneev, V. A. Gavrilova, and I. G. Savkina, *Dokl. Akad. Nauk SSSR* **151**, 227 (1963).
(*181*) V. B. Evstigneev, *Photochem. Photobiol.* **4**, 171 (1965).
(*182*) A. A. Krasnovskii and A. V. Umrikhina, *Dokl. Akad. Nauk SSSR* **155**, 691 (1964).
(*183*) I. G. Savkina and V. B. Evstigneev, *Dokl. Akad. Nauk SSSR* **138**, 958 (1961).
(*184*) S. Wawzonek and J. D. Frederickson, *J. Am. Chem. Soc.* **77**, 3985 (1955).
(*185*) R. Livingston and R. Pariser, *J. Am. Chem. Soc.* **78**, 2944 (1956).
(*186*) J. W. Weigl and R. Livingston, *J. Am. Chem. Soc.* **74**, 4211 (1952).
(*187*) W. S. Zaugg, L. P. Vernon, and A. Tirpack, *Proc. Natl. Acad. Sci. U.S.* **51**, 232 (1964).
(*188*) L. P. Vernon and E. R. Shaw, *Biochemistry* **4**, 132 (1965).
(*189*) K. G. Mathai and E. Rabinowitch, *J. Phys. Chem.* **66**, 954 (1962).
(*190*) A. A. Krasnovskii and K. K. Voinovskaya, *Dokl. Akad. Nauk SSSR* **87**, 109 (1952).
(*191*) A. A. Krasnovskii and G. P. Brin, *Dokl. Akad. Nauk SSSR* **153**, 212 (1963).
(*192*) R. F. Homer, G. C. Mees, and T. E. Tomlinson, *J. Sci. Food Agr.* **11**, 309 (1960).
(*193*) C. C. Black, *Science* **149**, 62 (1965).
(*194*) T. T. Bannister and J. E. Bernardini, *Biochim. Biophys. Acta* **59**, 188 (1962).
(*195*) G. P. Brin and A. A. Krasnovskii, *Biokhimiya* **24**, 1085 (1959).

(*196*) A. A. Krasnovskii, G. P. Brin, and N. N. Drozdova, *Dokl. Akad. Nauk SSSR* **150**, 1157 (1963).

(*197*) L. P. Vernon, A. San Pietro, and D. A. Limbach, *Arch. Biochem. Biophys.* **109**, 92 (1965).

(*198*) V. B. Evstigneev and V. A. Gavrilova, *Dokl. Akad. Nauk SSSR* **126**, 410 (1959).

(*199*) L. P. Vernon, *Acta Chem. Scand.* **15**, 1639 (1961).

(*200*) V. B. Evstigneev and V. A. Gavrilova, *Biofizika* **4**, 641 (1959).

(*201*) V. B. Evstigneev and V. A. Gavrilova, *Biofizika* **5**, 599 (1960).

(*202*) S. Ichimura and E. Rabinowitch, *Science* **131**, 1314 (1960).

(*203*) L. P. Vernon, *Acta Chem. Scand.* **15**, 1651 (1961).

(*204*) G. G. Komissarov, V. A. Gavrilova, L. I. Nekrasov, N. I. Kobozev, and V. B. Evstigneev, *Dokl. Akad. Nauk SSSR* **150**, 174 (1963).

(*205*) V. B. Evstigneev and V. A. Gavrilova, *Biofizika* **6**, 563 (1961).

(*206*) R. G. Jensen, Ph.D. Dissertation, Brigham Young University (1965).

(*207*) K. B. Serebrovskaya, V. B. Evstigneev, V. A. Gavrilova, and A. I. Oparin, *Biofizika* **7**, 34 (1962).

(*208*) M. S. Ashkinazi, I. P. Gerasimova, and B. Ya. Dain, *Dokl. Akad. Nauk SSSR* **102**, 767 (1955).

(*209*) M. S. Ashkinazi, I. P. Gerasimova, and B. Ya. Dain, *Dokl. Akad. Nauk SSSR* **108**, 655 (1956).

(*210*) M. S. Ashkinazi and A. I. Kryukov, *Ukr. Khim. Zh.* **23**, 448 (1957).

(*211*) M. S. Ashkinazi and A. I. Kryukov, *Ukr. Khim. Zh.* **26**, 600 (1960).

(*212*) A. I. Kryukov and M. S. Ashkinazi, *Ukr. Khim. Zh.* **25**, 309 (1959).

(*213*) M. S. Ashkinazi and V. E. Karpitskaya, *Dokl. Akad. Nauk SSSR* **96**, 785 (1954).

(*214*) K. Noack, *Naturwissenschaften* **14**, 385 (1926).

(*215*) K. Noack, *Biochem. Z.* **183**, 153 (1927).

(*216*) H. Gaffron, *Chem. Ber.* **60B**, 2229 (1927).

(*217*) V. E. Karpitskaya, I. A. Dolidze, and M. S. Ashkinazi, *Dokl. Akad. Nauk SSSR* **146**, 844 (1962).

(*218*) R. Livingston and K. E. Owens, *J. Am. Chem. Soc.* **78**, 3301 (1956).

(*219*) K. Meyer, *J. Biol. Chem.* **103**, 39 (1933).

(*220*) A. A. Krasnovskii and G. P. Brin, *Dokl. Akad. Nauk SSSR* **58**, 1087 (1947).

(*221*) J. C. Goedheer and J. R. Vegt, *Nature* **193**, 875 (1962).

(*222*) G. O. Schenck, *Z. Elektrochem.* **64**, 997 (1960).

(*223*) G. P. Brin and A. A. Krasnovskii, *Biokhimiya* **22**, 776 (1957).

(*224*) L. P. Vernon and E. D. Ihnen, *Biochim. Biophys. Acta* **24**, 115 (1957).

(*225*) K. Meyer, *J. Biol. Chem.* **103**, 607 (1933).

(*226*) G. D. Glynne Jones, *Ann. Appl. Biol.* **48**, 352 (1960).

(*227*) K. Meyer, *J. Biol. Chem.* **103**, 597 (1933).

(*228*) H. Gaffron, *Biochem. Z.* **179**, 157 (1926).

(*229*) G. O. Schenck and K. Ziegler, *Naturwissenschaften* **32**, 157 (1944).

(*230*) H. Gaffron, *Biochem. Z.* **264**, 251 (1933).

(*231*) E. Tyray, *Ann. Chem.* **556**, 171 (1944).

(*232*) D. Mauzerall and S. Granick, *J. Biol. Chem.* **232**, 1141 (1958).

(*233*) H. Gaffron, *Chem. Ber.* **60B**, 755 (1927).

(*234*) O. Warburg and V. Schocken, *Arch. Biochem.* **21**, 363 (1949).

(*235*) D. Burk and O. Warburg, *Z. Naturforsch.* **6b**, 12 (1951).

(236) O. Warburg, G. Krippahl, W. Buchholz, and W. Schröder, Z. Naturforsch. **8b**, 675 (1953).

(237) A. N. Terenin, "Photochemistry of Dyes and Related Organic Compounds" (translated by Kresge-Hooker Scientific Library), Chapters 7 and 8. Acad. Sci., Press, Moscow, 1947.

(238) H. Kautsky, H. de Bruijn, R. Neuwirth, and W. Baumeister, Chem. Ber. **66**, 1588 (1933).

(239) H. Kautsky, Trans. Faraday Soc. **35**, 216 (1939).

(240) J. L. Rosenberg and D. J. Shombert, J. Am. Chem. Soc. **82**, 3527 (1960).

(241) J. Franck and R. Livingston, J. Chem. Phys. **9**, 184 (1941).

(242) A. Schönberg, Ann. Chem. **518**, 299 (1935).

(243) G. O. Schenck, Naturwissenschaften **35**, 28 (1948).

(244) G. O. Schenck and E. Koch, Z. Elektrochem. **64**, 170 (1960).

(245) E. J. Bowen and D. W. Tanner, Trans. Faraday Soc. **51**, 475 (1955).

(246) R. Livingston and V. Subba Rao, J. Phys. Chem. **63**, 794 (1959).

(247) C. S. Foote and S. Wexler, J. Am. Chem. Soc. **86**, 3880 (1964).

(248) C. S. Foote and S. Wexler, J. Am. Chem. Soc. **86**, 3879 (1964).

(249) E. J. Corey and W. C. Taylor, J. Am. Chem. Soc. **86**, 3881 (1964).

(250) H. Gaffron, Chem. Ber. **68B**, 1409 (1935).

(251) G. S. Egerton, Nature **204**, 1153 (1964).

(252) A. Bernas, J. Phys. Chem. **68**, 2047 (1964).

—18—

Photochemistry of Chlorophyll in Vivo*

LEO P. VERNON AND BACON KE

Charles F. Kettering Research Laboratory, Yellow Springs, Ohio

I. Introduction

The importance of chlorophyll in nature lies in its role as the photocatalyst for photosynthesis. Chlorophyll functions in photosynthesis by virtue of its ability to produce and maintain a charge separation in the highly ordered lamellar structure of the chloroplast. This charge separation initially involves chlorophyll and some other molecule complexed to it (donor or acceptor molecule), but is subsequently evidenced in discrete biochemical entities other than chlorophyll which become either reduced (accept electrons or gain negative charge) or oxidized (give up electrons or gain positive charge) after reaction with photoexcited chloro-

---

* Contribution No. 206 from the Charles F. Kettering Research Laboratory.

phyll. The following chapter discusses this initial charge separation in detail. We will not attempt to discuss the electronic nature of the excited chlorophyll which initiates the chain of electron transfer reactions peculiar to photosynthesis. We will discuss in general terms the involvement of photoexcited chlorophyll as a partner in the initial charge-separation process. We will discuss at greater length the nature of the closely coupled biochemical redox compounds which react directly with the photoactive chlorophyll, thereby changing their redox state and initiating in their own local environments the sequential electron transfer reactions leading ultimately to carbon dioxide fixation and oxygen evolution in plants and the analogous reactions in bacteria.

It is impossible to consider in this chapter in a comprehensive way the experimental evidence which supports the modern concept of the photosynthetic process which serves as the framework for this presentation. Please consult Chapter 19 for brief discussions of the photosynthetic unit, reaction-center chlorophylls, and the phenomenon of enhancement. More detailed information on the physical aspects of photosynthesis are available (1–4), and several reviews on the biochemical aspects of photosynthesis have recently appeared (5–11).

Plant cells are heavily pigmented for the efficient capture of light energy for photosynthesis. In addition to chlorophyll, such cells contain various carotenoids and phycobilins which transfer excitation energy to chlorophyll. The major chlorophylls are Chl a and Chl b. Some algae do not contain Chl b, but all photosynthetic plant cells contain Chl a, which sets it apart as the essential photoactive pigment in plant photosynthesis. Different forms of Chl a exist in plants and manifest different absorption properties, as discussed in Chapter 11. Complexing to different proteins and lipids would suffice to alter the absorption and photochemical properties of Chl a. One group, however, has reported that the different forms observed in the intact chloroplasts are due to chemically different chlorophyll molecules, which can be physically separated in the purified state (12, 13).

Not all chlorophyll molecules are photochemically active in the sense that they interact with adjoining redox components to initiate photochemical reactions. The early flashing-light experiments of Emerson and Arnold (14) and later experiments of Kok (15) have shown that the primary photochemical event of photosynthesis in plants is initiated by one chlorophyll which is structurally linked to approximately 300 other chlorophyll molecules which serve as light-harvesting members of the unit. Light energy absorbed by any chlorophyll molecule in the unit is transferred with high efficiency to the reactive chlorophyll (reaction-center chlorophyll) which initiates the photochemistry (see Chapter 19).

Since extraction of chloroplasts with organic solvents yields only one Chl *a* [however, see Michel-Wolwertz *et al.* (*12, 13*)], the reaction-center chlorophyll must differ only in its environment in the chloroplast, being more available to adjoining redox agents.

II. Photoreactions Associated with Pigment System I of Plants

Pigment system I (PSI)* of green plants includes the long-wavelength forms of Chl *a* which absorb light maximally at 683 mμ and beyond (*11*). A special form of long-wavelength chlorophyll, discussed in Chapter 19, has been described by Kok and designated P700 because of its absorbancy change in this region upon illumination. This change corresponds to an oxidation of a chlorophyll having an E'_0 of 430 mv at neutral pH. This chlorophyll is present to the extent of 1 per 400 total chlorophyll molecules (*16*). These properties coincide with those required of the reaction-center chlorophyll.

The difference spectrum associated with P700 photooxidation has been constructed from absorption-change transients induced in chloroplasts by light flashes. The P700 reactions can be experimentally isolated either by making PSII inoperative or by uncoupling PSII from PSI and at the same time coupling P700 with an external electron donor (e.g., $PMSH_2$ at a concentration $> 5 \times 10^{-5} M$). Specifically, the P700 spectra have been obtained with the following: (a) aged chloroplasts (*17, 18*); (b) digitonin-treated chloroplasts (*19, 20*); (c) chloroplasts poisoned with DCMU (*17, 18, 20*); (d) chloroplasts extracted with petroleum ether to remove plastoquinone (*17*); (e) sonicated chloroplasts (*20a*); (f) chloroplasts in the presence of $PMSH_2$ and at $-150°C$ (*19*); (g) chloroplasts or algae illuminated with a far-red background light (*19, 21*).

The far-red light presumably maintains P700 in the oxidized state, and a subsequent flash causes its re-reduction and a positive absorption change. The difference spectrum constructed from absorption-change

---

* The abbreviations used are the following: PSI, pigment system I (long wavelength); PSII, pigment system II (short wavelength); DCMU, 3-(3,4-dichlorophenyl)-1,1-dimethylurea; PMS (PMSH₂), phenazine methosulfate; DPIP (DPIPH₂), 2,6-dichlorophenolindophenol; FMN, flavin mononucleotide; TMPD, *N,N,N',N'*-tetramethyl-*p*-phenylenediamine; TMQ (TMQH₂), trimethyl-*p*-benzoquinone; P700, reaction center chlorophyll for PSI; ATP, adenosine triphosphate; NAD, nicotinamide adenine dinucleotide; NADP, nicotinamide adenine dinucleotide phosphate. The reduced forms of the redox compounds are in parentheses.

transients under these conditions appears inverted from the usual light-minus-dark spectrum obtained by all other methods (19).

A. Electron Acceptors for PSI

1. PHYSIOLOGICAL ELECTRON ACCEPTORS (FERREDOXIN)

The reducing power generated by PSI is utilized by the chloroplast for the reduction of carbon dioxide. Since $NADPH_2$ is the reductant involved in the enzymatic reactions leading to carbon dioxide fixation in the cytoplasm, we can consider this as the terminal compound in terms of photosynthetic electron transfer reactions *in vivo*. The reduction of NADP is not direct, however, but requires at least two enzymes to couple with the photochemical system as shown in Fig. 1.

Ferredoxin is the name given a class of nonheme iron proteins which occur in plants and bacteria. The compounds contain iron bound in a unique manner to labile sulfur which is present in equimolar amounts (5, 22). Addition of acid ruptures the iron-sulfur linkages and liberates hydrogen sulfide. The activity of this protein was demonstrated in 1952 by Davenport et al. (23), who were investigating the photoreduction of methemoglobin coupled to oxygen evolution by chloroplasts. They called the protein "methemoglobin reducing factor." In 1958 San Pietro and Lang independently discovered a soluble protein from leaves which catalyzed the photoreduction of NADP in the presence of spinach chloroplasts or grana (24). They designated the protein "photosynthetic pyridine nucleotide reductase." Somewhat later Tagawa and Arnon described the same protein under the name of spinach ferredoxin (25). A protein isolated from *Chlorella* in Warburg's laboratory by Gewitz and Völker (26) carries out the same reaction. All four proteins appear to be the same, with slight differences related to the different plant sources. In view of the chemical similarities of these proteins to the bacterial ferredoxins, it appears logical to call these non-heme iron proteins chloroplast ferredoxins. For a more detailed discussion of the ferredoxins and other enzymes related to photosynthesis see the review by San Pietro and Black (5).

A change in the valency of the iron is associated with the photoreduction of spinach ferredoxin (27). In the dark both iron atoms are in the ferric state, and illumination causes the formation of one ferrous iron when the photoreduction is complete, showing that spinach ferredoxin functions as a one-electron redox agent. When ferredoxin photoreduction is coupled to oxygen evolution the stoichiometry observed is four ferredoxins reduced per oxygen evolved (28). Whatley et al. (29) were able to temporally separate the photoreduction of spinach ferredoxin from its reoxidation by NADP, which clearly shows its function in the photoreduction of NADP by chloroplasts. Furthermore, the kinetic measurements of Chance and San Pietro show that ferredoxin is photo-reduced at a rate which is consistent with its postulated function in NADP photoreduction (*30*).

Ferredoxin is the first detectable compound reduced by illuminated chloroplasts in terms of its temporal proximity to the initial photo-

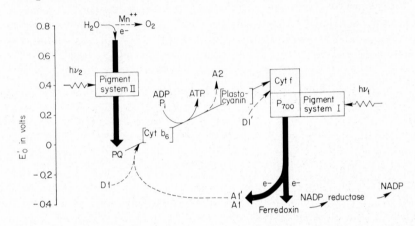

FIG. 1. Light-induced electron flow in chloroplasts. All arrows indicate the direction of electron transfer. Heavy arrows indicate the photochemical steps, starting with water oxidation by PSII. The electron acceptor for PSII, plastoquinone (PQ), transfers the electron to plastocyanin and thence to cytochrome f which serves as donor for PSI. There may be intermediate enzymatic steps between plastoquinone and plastocyanin (perhaps cytochrome b_6 whose redox state is affected in opposite ways by PSI and PSII). The other photochemical step is catalyzed by PSI, with P700 as the reaction center of the pigment system. Two fates are shown for the transferred electrons; a physiological one leading to NADP reduction and a non-physiological one leading to reduction of added electron acceptors (denoted A). Ferredoxin also catalyzes the reduction of cytochrome c and nitrite. $A1$ includes quinones, indigo carmine, benzyl viologen, ferricyanide, and manganipyrophosphate. $A1'$ includes those redox agents (DPIP, PMS, FMN, vitamin K_3, ferredoxin, and quinones having an E'_0 below zero volts) which in the reduced form are also capable of donating electrons to some redox agent (perhaps plastoquinone) prior to the ATP-forming site, and therefore catalyze cyclic electron flow via PSI with an associated phosphorylation. Donors (denoted D) feed electrons into the system at two sites. $D1$ includes the reduced forms of the compounds listed above for $A1'$. To study the noncyclic flow of electrons through PSI, the donors ordinarily used are $PMSH_2$ (below $5 \times 10^{-4} M$) and $DPIPH_2$ (greater than $10^{-5} M$). $D1'$ includes $PMSH_2$ (greater than $5 \times 10^{-4} M$), TMPD and probably $DPIPH_2$ at low concentration. The $D1'$ donors serve as electron feeders for photoreductions, but no associated phosphorylation is observed. $DPIPH_2$ reacts via plastocyanin, but TMPD oxidation is not stimulated by plastocyanin and may react directly with cytochrome f or P700. The acceptors listed for $A2$ include DPIP, ferricyanide, toluylene blue, thionine, and methylene blue. Their reduction via PSII would support ATP formation, as has been shown for the first two listed. See text for additional information.

chemical act of PSI. If ferredoxin does react directly with chlorophyll, two important questions remain concerning this photoreduction: (a) does ferredoxin react directly with the photoexcited chlorophyll to gain an electron from photoexcited chlorophyll in a charge transfer reaction, or (b) does it react with a reduced chlorophyll molecule produced in another charge transfer reaction?

Some indirect evidence for the existence of a primary electron acceptor (designated Z) other than ferredoxin for PSI has been reported by Rumberg and Witt from the absorption-change transients associated with P700 photooxidation (19). When chloroplasts are illuminated with a far-red background light, P700 is maintained in the oxidized state (19, 21), and a subsequent flash causes its re-reduction and a positive absorption change. It was found by Rumberg and Witt (19) that for such a positive change to occur in chloroplasts, an oxidant such as indigo carmine or benzyl viologen has to be present (cf. Section II, A, 2). On the other hand, in *Chlorella* such an added electron acceptor is not required (19).

The externally added oxidants are believed to serve as an electron trap for the primary acceptor Z in these reactions. As quoted in reference (19), NADP plus ferredoxin (but not NADP alone) can serve a similar function. These experiments imply that the primary electron acceptor should have a redox potential more negative than -440 mv.

In the following section evidence is presented that illuminated chloroplasts are able to photoreduce dyes having an E'_0 of -700 mv. It was originally felt that substances with an E'_0 below zero could not be reduced in a Hill reaction (serve as an oxidant for oxygen evolution) (31). The subsequent photoreduction of NADP (24) and ferredoxin (25) by chloroplasts showed that compounds with an E'_0 as low as -432 mv for spinach ferredoxin (25) can be photoreduced. The ability of chloroplasts to reduce dyes having E'_0 values as low as -700 mv means that reduced ferredoxin need not be the initial compound produced photochemically by PSI. The experiments of Zweig and Avron (32) indicate that the E'_0 of the endogenous redox agent lies between -490 and -530 mv. If true, this would rule out ferredoxin as the first product, and some other unknown agent could fill this role. A likely candidate is a reduced form of chlorophyll. Kamen [(2), p. 155] has discussed the chemical properties of reduced chlorophyll in this context, emphasizing the capability of chlorophyll to function as either an oxidizing or reducing agent in photochemical reactions.

2. NONPHYSIOLOGICAL ELECTRON ACCEPTORS

PSI in isolated chloroplasts reacts under appropriate conditions with added electron acceptors. When both PSI and PSII are operative and coupled, oxygen evolution from PSII is coupled to the reduction by PSI of added acceptors, represented as *A1* and *A1'* in Fig. 1. This reaction, which is readily performed by isolated chloroplasts and chloroplast fragments, is known as the Hill reaction after its discoverer (*33*). In his original experiments Hill used ferric potassium oxalate as the acceptor. Other compounds which serve this purpose are ferricyanide (*34, 35*), *p*-benzoquinone (*36*), other quinones (*37*), DPIP (*38*), chromate (*39*), nitrite and nitrate (*36, 40*), FMN (*41*), indigo carmine (*42*), and manganipyrophosphate (*43*). For a more detailed description of the Hill reaction the reader is referred to the review by Clendenning (*44*). Cytochrome *c* and methemoglobin are reduced by illuminated chloroplasts at a slow rate (*45*), but the addition of ferredoxin markedly stimulates the reaction (*46, 47*). Other compounds such as PMS and quinones, which themselves serve as Hill reagents, are capable of stimulating the photoreduction of cytochrome *c* (*47, 48*), indicating that cytochromes are only indirectly photoreduced in a Hill reaction. Nitrate and nitrite also do not react directly, but via FMN (or FAD) and ferredoxin, respectively.

The Hill reagents listed above could in principle be photoreduced by either PSI or PSII. It is generally held that they are photoreduced by PSI. Certainly those compounds which catalyze cyclic photophosphorylation are photoreduced by PSI. These include PMS, DPIP, FMN, vitamin K_3, and quinones with an E'_0 below zero volts. From their effect on the flash-induced absorption-change transient of P700, Rumberg and Witt list the following as oxidants for PSI (*19*): ferricyanide, *p*-benzoquinone, PMS, indigo carmine, benzyl viologen, methyl viologen, and ferredoxin. There is mounting evidence that both ferricyanide (*49–51*) and DPIP (*17, 52–54*), however, may be reduced by PSII as well as by PSI.

A number of viologen dyes and related dipyridyl compounds have been used to determine the maximal reducing power which chloroplasts are capable of producing upon illumination. Zweig and Avron determined the extent of reduction of such dyes, and calculated an E'_0 of −490 to −530 mv at neutral pH for the endogenous factor which couples to the dye systems at equilibrium (*32*). They reported appreciable photoreduction of triquat (1,1'-trimethylene-2,2'-dipyridylium dibromide), which has an E'_0 of −550 mv. Black (*55*) also studied the photoreduction of this dye, and showed that in the presence of air there is no net reduction of the dye, but it turns over very rapidly, resulting in oxygen uptake in a Mehler reaction (reduced dye is reoxidized by oxygen to produce hydrogen peroxide). The formation of ATP was noted in these experiments. In continuing these experiments Black has found (personal communication) that 1,1'-trimethylene-2,2'-dipyridylium dibromide, which has an E'_0 of -656 mv, catalyzes photophosphorylation under aerobic conditions at a rate comparable to that observed during NADP photoreduction. These data show that the dye is being reduced by the chloroplast system. Kok and Rurainski (56) have determined, with oxygen isotopes, the rates of oxygen evolution (photosynthetic oxygen) and oxygen uptake (an indirect measure of dye reduction since the reduced dye reacts with oxygen). They report equal rates for two viologen dyes having E'_0 values of -318 and -740 mv.

Assuming the redox properties of the dipyridyl dyes do not change in the presence of chloroplasts, which is certainly open to question, the data indicate that chloroplasts have a natural redox agent which is photoreduced endogenously and has an E'_0 at least as low as -500 mv. Ferredoxin, as a physiological oxidant, would react with this compound to become reduced, later transferring electrons to NADP via the NADP reductase contained in the chloroplasts. When isolated ferredoxin is added back to a chloroplast system to promote NADP photoreduction, it appears to react at a site (or sites) which is common for the other Hill oxidants. Keister and San Pietro (47) report that in a phosphorylating system pyocyanin competes successfully with ferredoxin for available reducing power. Since ferredoxin is readily extracted from broken chloroplasts, it is likely that its functional site becomes exposed upon removal of the ferredoxin, and is thus available as a general reducing site which can interact with a wide variety of oxidants in a nonspecific manner. The nature of the endogenous reductant having an E'_0 of -500 mv is not known. It could be some compound which has been undetected to date, or could be a chlorophyll molecule which is reduced photochemically. The possibility of chlorophyll serving as the primary electron acceptor for photoexcited P700 is discussed in more detail in Section V, A.

B. Electron Donors for PSI

1. Physiological Donors

In agreement with the commonly accepted scheme of photosynthesis upon which Fig. 1 is based, the immediate electron donor (or donors, since there could be more than one compound reacting directly with the photoexcited chlorophyll at the reaction center) for PSI is the terminal member of the electron transfer chain which connects the two pigment systems. As such it should have a high oxidation potential in relation to the other members of the electron transfer system, and by occupying a site between PSI and PSII, it should be photooxidized by light absorbed by PSI and photoreduced by PSII. Two compounds which satisfy these criteria are cytochrome f and plastocyanin.

a. Cytochrome f. Cytochrome *f* was first detected by Scarisbrick and Hill (57) and further purified by Davenport and Hill (58). A similar cytochrome is found in algae (59). Cytochrome *f* has an E'_0 of 365 mv at neutral pH values and is readily photooxidized by illuminated leaves or chloroplasts. Duysens first demonstrated a photooxidation of this type with the algal cytochrome in the red alga *Porphyridium cruentum* (60). The photooxidation of cytochrome *f* in green leaves and chloroplasts has also been detected from absorption changes, which proceed even at temperatures as low as 77°K (19, 61). These reactions are discussed in more detail in Section V, A. Cytochrome *f* is tightly bound in the chloroplast, presumably to the chlorophyll, and is extracted in the reduced form by means of organic solvents or detergents. Although isolated chloroplasts do not readily photooxidize added cytochrome *f* in the presence of air, treatment with detergents allows this activity to appear (62, 63). In this case oxygen is the electron acceptor. In the experiments of Duysens and Amesz (64) on the red alga *Porphyridium cruentum* and the more recent experiments of Ke and Ngo (21) on *Plectonema boryanum*, the analogous algal cytochrome was oxidized by PSI and reduced by PSII, which places it functionally between these two systems.

b. Plastocyanin. A blue protein has been isolated from a variety of plant tissues by Katoh *et al.* (65, 66), who have shown that it occurs only in photosynthetic tissues, is reduced by chloroplasts in a regular Hill reaction, and is photooxidized by digitonin-treated chloroplasts (62). The pigment is a copper protein with two copper atoms per molecule; it is present to the extent of one plastocyanin to 30 chlorophyll molecules, and is not readily oxidized by oxygen. Its ability to serve as a Hill oxidant does not mean that it functions physiologically in this fashion, since a wide variety of natural and nonphysiological redox agents undergo the same reaction. The recent experiments of de Kouchkovsky and Fork (67) show that its redox state in the alga *Ulva lobata* can be determined spectrophotometrically and is influenced by both red and far-red light, being reduced by the former and oxidized by the latter.

Further evidence for the participation of plastocyanin close to the primary photochemical reaction of PSI is found in its direct oxidation by detergent-treated chloroplasts (in which only PSI remains functional) and by its stimulation of or requirement for photooxidation of other compounds of lower oxidation potential. Chloroplasts treated with digitonin (62) or with the detergent Tween 20 (63) catalyze a rapid photooxidation of plastocyanin. Examination of the kinetics of plastocyanin oxidation with Tween 20-treated chloroplasts (68) shows that the initial rate is dependent upon the percentage reduction of plastocyanin, which is consistent with plastocyanin existing in a complex of some type with the reaction-center chlorophyll (P700), which would assign plastocyanin the role of a primary donor to the photoexcited chlorophyll. Similar kinetics were observed for cytochrome f photooxidation (63), and similar reasoning would apply to this compound.

Plastocyanin is easily removed from the chloroplast structure, and in such cases electron-transfer reactions which ordinarily proceed via this compound are greatly depressed or are absent. Sonication of chloroplasts causes a loss of bound plastocyanin, resulting in loss of NADP photoreduction activity, both with and without added electron feeder systems such as ascorbate-DPIP (69). Addition of catalytic amounts of plastocyanin restores both activities. Recent experiments of Vernon and Shaw (70) included a study of various photoreactions catalyzed by spinach chloroplasts in the presence of the detergent Triton X-100. At high detergent concentrations the addition of plastocyanin stimulated the following reactions: photoreduction of NADP by ascorbate-DPIP, NADP photoreduction coupled to ferrocytochrome c oxidation, and ferrocytochrome c photooxidation by molecular oxygen. These data show that both cytochrome c and $DPIPH_2$ are photooxidized via plastocyanin in the treated chloroplasts, which would also be the case in the intact chloroplasts, since mere sonication removes plastocyanin and prevents oxidation of $DPIPH_2$ (69).

2. NONPHYSIOLOGICAL DONORS

A number of nonphysiological redox agents are capable of feeding electrons into PSI, and thus support the photoreduction of ferredoxin or the other agents listed as A1 or A1' in Fig. 1. For our purposes the main question is whether such donors react directly with the chlorophyll system or employ some endogenous redox compound as an intermediary. Three criteria which may be used to assess the directness of the interaction are the following: (a) Does the compound support cyclic photophosphorylation, and does the coupled photoreduction of NADP result in ATP formation? Such data indicate whether the donor acts prior to or following the ATP-forming reaction. (b) Is plastocyanin required for the reaction? A plastocyanin requirement for the photooxidation of some donors can be induced by sonication (69) or by treatment with detergents (62, 63, 70). Those donors not requiring plastocyanin for photooxidation would react more directly with the chlorophyll systems. (c) What effect does the donor have on the rapid absorption changes related to chlorophyll photochemistry? A compound reacting directly should profoundly affect such changes.

$DPIPH_2$ which is kept in the reduced form by excess ascorbate

(DPIP-ascorbate couple) supports the photoreduction of NADP when PSII is inoperative due to aging or poisoning (71). This reaction is accompanied by ATP formation (72, 73). Furthermore, low concentrations of DPIPH$_2$ catalyze cyclic photophosphorylation (73). These data show that DPIPH$_2$ interacts with some component of the electron transport chain prior to the ATP-forming site. However, at low concentrations (10^{-5} M) the photooxidation of DPIPH$_2$ is not associated with a corresponding formation of ATP, indicating that at these concentrations it reacts beyond the ATP site (73–75). PMSH$_2$ (76, 77), vitamin K$_3$ (75), FMN (78), ferredoxin (78), and various quinones of redox potential below zero volts (79) also catalyze cyclic photophosphorylation and thus interact before the ATP site in the electron transfer chain. PMSH$_2$, however, also reacts directly with the chlorophyll system, as shown by the spectroscopic evidence reported by several investigators (20, 80, 81).

The photoreduction of NADP by chloroplasts exposed to high concentrations of Triton X-100 is supported by either ascorbate-DPIP or ferrocytochrome c (70). Both reactions are markedly stimulated by the addition of plastocyanin, which shows that DPIPH$_2$ interaction with PSI is not direct, and further includes ferrocytochrome c in this category. Ferrocytochrome c photooxidation by molecular oxygen with detergent-treated chloroplasts also is markedly stimulated by plastocyanin (63, 70).

Photooxidation of TMPD (with excess ascorbate) coupled to NADP photoreduction is not accompanied by ATP formation (81, 82), an indication that it acts beyond the ATP site. With high concentrations of Triton X-100, photooxidation of either TMQH$_2$ or TMPD is not stimulated by added plastocyanin (70), a result indicating that these compounds react beyond the plastocyanin site and more directly with the chlorophyll system.

III. Photoreactions Associated with Pigment System II

Considerably less is known about PSII, both in terms of chlorophyll photochemistry and reactions directly related to it. This system contains Chl b and the short wavelength forms of Chl a absorbing maximally around 673 mμ. In addition carotenoids and phycobilins, when present, serve as additional light-absorbing molecules which transfer their excitation energy to the chlorophyll. According to the scheme shown in Fig. 1, PSII produces a reductant of intermediate potential which eventually serves as the source of electrons for PSI. The ultimate electron donor is water, leading to water oxidation and oxygen production.

A. Electron Acceptors for PSII

The primary evidence concerning acceptors for PSII comes from spectrophotometry of intact systems. On this basis, plastoquinone is generally regarded as an acceptor for PSII, and there is some biochemical evidence to support this concept. Quinones constitute one of the major classes of compounds in the chloroplasts (83). Chloroplasts contain four compounds of the plastoquinone type, three tocopherylquinones and vitamin K_1 (84). Illumination causes a reduction of endogenous quinones as determined by analysis of extracted quinone (85) or by spectroscopic analysis of the chloroplast (86, 87). Spectroscopic examination of *Anacystis nidulans*, a blue-green alga, has shown that a compound spectroscopically similar to plastoquinone is reduced upon illumination with 620 mµ light, but is oxidized in 680 mµ light (88). The reduction, but not the oxidation, is sensitive to added DCMU. The efficiencies for reduction and oxidation are high, approaching 2 quanta per equivalent reduced or oxidized. All these data are consistent with plastoquinone (or some other quinone) acting as the acceptor molecule for PSII.

Duysens and Sweers have proposed a compound designated "Q" as a primary electron acceptor for PSII (89). From fluorescence measurements on various algae and spinach chloroplasts excited with different colored lights, they found that Chl *a* fluorescence is increased by light absorbed by PSII and decreased when light absorbed by PSI is superposed on a background of PSII exciting light. Excitation spectra indicate the fluorescence comes from a short-wavelength form of Chl *a* in PSII. The change in fluorescence yield is explained by a hypothetical compound Q, closely associated with PSII and acting as quencher for Chl *a* fluorescence. The reduced form of Q, QH, which does not quench the fluorescence, is formed when PSII is illuminated. Upon illumination of PSI, QH passes on an electron to PSI, becomes oxidized to Q, and becomes active again as a quencher of fluorescence from the Chl *a* of PSII.

From fluorescence kinetics, there is also evidence suggesting that QH undergoes an alternate reaction, QH → Q′, outside the main photosynthetic electron-transfer path. This reaction may serve to protect the system from radiation damage by dissipating excess light energy. Q′ is slowly converted back to Q in a dark reaction. The suggested Q could be a quinone-type compound, but it is different from plastoquinone or the quinone described by Amesz (88), since its response to DCMU is exactly the opposite of the reaction of these quinones.

Bishop (90) first showed that chloroplasts extracted with organic solvents regained their activity for various Hill reactions if plastoquinone was added to the system. Subsequent experiments with lyophilized chloroplasts have shown that readdition of plastoquinone restored activity for NADP reduction coupled to oxygen evolution (*84, 91, 92*), cyclic photophosphorylation catalyzed by PMS, vitamin K_3, or FMN (*93, 94*), and ferricyanide or cytochrome *c* reduction (*84, 95*). The restoration of ferricyanide, NADP, and cytochrome *c* photoreduction upon the readdition of tocopherylquinones has been reported (*84, 95, 96*).

Trebst and Eck performed a graded extraction of plastoquinone from chloroplasts, and reported that after 80% of the plastoquinone had been removed, ferricyanide photoreduction was abolished while NADP photoreduction activity was unimpaired (*92, 97*). These authors postulated two sites for plastoquinone action, one as acceptor for PSI and another for PSII. In view of the mounting evidence that ferricyanide reduction may also occur via PSII (*49–51*), another interpretation is more plausible. There is likely only one plastoquinone site, that of acceptor for PSII. The effects noted above could be due to the fact that interaction between plastoquinone and ferricyanide is chemical in nature and requires a higher quinone concentration. NADP photoreduction, on the other hand, occurs via enzymatic transfer reactions, which probably saturate at lower plastoquinone concentrations.

A number of hydroquinones of high potential (particularly *o*-hydroquinones) are oxidized by chloroplasts in the light. This photooxidation is inhibited by DCMU, indicating that PSII is involved. Although the mechanism is not clear, it could mean that *o*-hydroquinones act as electron donors for PSII and are involved in some fashion in oxygen evolution (*98, 99*).

Rumberg has recently assigned the newly observed absorption change at 648 mμ together with the usually observed changes at 515 and 475 mμ to a Chl *b* reaction, presumably photoreduction (*100*). Absorption-change transients at these wavelengths were obtained for *Chlorella* cells with brief flashes of light under conditions where both P700 and cytochrome *f* were initially in the oxidized state.

The 515/475 mμ absorption changes were first observed by Duysens in 1954 (*101*) and subsequently by many other investigators (*8*) employing steady illumination of chloroplasts and *Chlorella*, and were first observed with flash illumination by Witt in 1955 (*102*). Witt and co-workers have conducted extensive studies on the physical and chemical nature of these changes by means of flash spectrophotometry (*17*). The interpretation of their flash-spectroscopic results has evolved through several stages as newer information becomes available. Besides the suggestion of Chl *b* involvement, other possible interpretations of these absorption changes have been advanced by various workers on previous occasions (8, 103). More evidence appears needed before a definite assignment of these absorption changes can be made.

Evidence that the 515/475 mμ changes are associated with PSII was given by Witt et al., who reported that these changes can be induced by flashes at 670 mμ, but very ineffectively by flashes at 710 mμ (104). The action spectrum for the 515/475 mμ changes appears identical with that for oxygen evolution; it has a peak at 674 mμ with a shoulder at 650 mμ and a far-red limit at 705 mμ (105). These results suggest that Chl b and a short-wavelength form of Chl a are the light absorbers for these reactions.

The possible connection of plastoquinone with the 515/475 mμ changes is based on the finding that the changes are absent in chloroplasts extracted with petroleum ether, and that the changes can be restored upon readdition of the petroleum-ether extract or purified plastoquinone to the extracted chloroplasts (17). The main difficulty of associating the 515/475 mμ changes to plastoquinone is that the latter does not absorb at these wavelengths. The current concept adopted by Rumberg is that the 475/515/648 mμ changes reflect a physical change induced by a redox reaction of an unknown substance (possibly plastoquinone) in close association with Chl b, and do not reflect redox changes of compounds in the main electron transfer pathway (106).

Other workers have reported complicated kinetics for the 515 mμ change, influenced by such experimental conditions as preillumination or degree of anaerobicity. Contrary to the finding of Müller et al. (105), Rubinstein and Rabinowitch reported a long-wavelength sensitization of the 520 mμ change, suggesting a PSI rather than a PSII reaction (107; cf. 105). Kok et al. also concluded that the 520 mμ change is at least partially caused by PSI and is not inhibited by DCMU (108). More recently, Rubinstein demonstrated that opposite effects on the 520 mμ change can be produced by varying the experimental conditions, and suggested a scheme to reconcile the conflicting reports in the literature. In an aerobic Chlorella suspension the 520 mμ absorption change is partially inhibited by DCMU, but is completely inhibited by DCMU when the suspension is also illuminated with a weak background of white light. When the suspension is made anaerobic, the 520 mμ change is unaffected by DCMU, and it is sensitized by the far-red light. These observations are explained by postulating that the changes are due to a compound capable of undergoing either a photoreduction (DCMU sensitive) by short-wavelength light or a photooxidation sensitized by far-red light (109).

B. Electron Donors for PSII

We can speak only in general terms about the donor for PSII. Ultimately the electron transferred through PSII must come from water (7), but little or nothing is known about the specific intermediates or enzymes involved. It is known that manganese is required for the oxygen-evolution process, since manganese deficiencies in algae inhibit those photoreactions which require an intact oxygen-evolution system (110–113). Manganese deficiency has no effect on the partial reactions which are independent of the oxygen evolution system, such as photoreduction in adapted algae (114) and photophosphorylation catalyzed by pyocyanin, (110). The manner in which manganese reacts in the oxygen-evolving system is not known, but it most likely involves a change of valency for manganese and could be related to its ability to react in a catalytic fashion in peroxidative reactions catalyzed by peroxidase (115–118).

Various inhibitors of photosynthesis are known to act somewhere on PSII. These include DCMU (119), phthiocol, o-phenanthroline, hydroxylamine (120), and some amino triazines (121). Chloride ion is also involved in some manner in the oxygen-evolution system.

IV. Photoreactions Associated with the Bacterial Systems

The general features of bacterial photosynthesis are given in Chapter 19, and in more detail in Clayton (1), Vernon (6), and Gest *et al.* (122). Only the salient features relevant to this discussion will be given here.

The purple photosynthetic bacteria (both sulfur and nonsulfur types) contain bacteriochlorophyll as the photoactive pigment. It occurs in different forms *in vivo*, showing absorption maxima at about 800, 850, and 880 mμ in different organisms. When extracted, however, only one absorption band is shown in the infrared, indicating the difference *in vivo* is caused by absorption shifts due to association of the bacteriochlorophyll. The green photosynthetic bacteria contain one of two general types of chlorobium chlorophyll which are chemically different (cf. Chapter 4). The green bacteria also contain minor amounts of bacteriochlorophyll. The bacteria also contain some specialized pigment which responds to light in a manner consistent with its being reaction-center chlorophyll. This component is called P870 for *Rhodopseudomonas spheroides* and P890 for *Rhodospirillum rubrum* and *Chromatium*, designating the wavelength of maximal absorption change. The behavior patterns of P870 and P890 are similar to P700 of plants, but P870 is present in the ratio of only one P870 per 50 regular bacteriochlorophyll molecules.

Photosynthetic bacteria do not evolve oxygen during their photosynthesis. In lieu of water oxidation, they oxidize (primarily) either organic acids or inorganic sulfur compounds. They do incorporate carbon dioxide in a manner similar to green plants, and thus in some environments have the need for producing a low potential reductant for this reaction. The bacterial chromatophores (fragments of the photosynthetic apparatus) contain specific cytochromes which are readily oxidized at high quantum efficiencies upon illumination. They do not contain plastocyanin, nor do they require manganese at the high levels needed in plants. To date no enhancement with light of different wavelengths has been observed; this finding indicates that only one pigment system is involved, or that only one primary photoact takes place in the photosynthetic bacteria. For these reasons, investigators generally compare the photochemical system of the photosynthetic bacteria with PSI of plants, and find the following similarities: the presence of detectable reaction-center chlorophyll, the ready photooxidation of endogenous cytochromes of the c type, the reduction of low potential compounds (NAD with bacteria, NADP with plants) leading to carbon dioxide fixation, the ability to perform cyclic photophosphorylation, ready interaction with similar redox agents, the presence of a photosynthetic unit (although it is smaller in the bacterial system), and the presence of ferredoxins, which couple to carbon dioxide fixation. The bacterial system differs from PSI of plants in at least three important ways: the bacterial system does not contain plastocyanin, it contains and photoreduces ubiquinone (not plastoquinone), and it photoreduces NAD instead of NADP.

A. Electron Acceptors for the Bacterial Photochemical System

The two compounds to consider as acceptors for reaction center chlorophyll in photosynthetic bacteria are ferredoxin and quinone, the physiological acceptors for PSI and PSII, respectively, of plants. Ubiquinone-9 and ubiquinone-10 are found in the purple bacteria in appreciable amounts (123, 124), and recently a new quinone "rhodoquinone" has been reported (125). Clayton (126) has demonstrated spectrophotometrically that a light-induced oxidation of endogenous cytochromes in *Rhodopseudomonas spheroides* and *Chromatium* is coupled to the reduction of quinone in the cell. Also, Bales and Vernon (127) have shown that oxidation of DPIPH$_2$ by *Rhodospirillum rubrum* chromatophores is coupled to the photoreduction of endogenous quinones in the particles. Zaugg et al. (128, 129) have shown that photoreduction of added ubiquinone can be coupled to photooxidation of either added ferrocytochrome c or PMSH$_2$. These reactions are very rapid and are stimulated by the addition of detergents. Thus, there is sufficient evidence that both endogenous and added ubiquinones function as efficient acceptors for the photosystem of purple bacteria. Isolated bacteriochlorophyll will catalyze the transfer of electrons from $PMSH_2$ to ubiquinone (129), showing the ease with which photoexcited chlorophyll reacts with quinones.

Clayton (130) has shown that the bulk of the bacteriochlorophyll can be changed to bacteriopheophytin by illumination of old cultures of *Rhodopseudomonas spheroides* in the presence of detergent, but such cells still carry out the primary reactions associated with the reaction center, which is unchanged. Sistrom has isolated a mutant of *R. spheroides* which has an apparently normal complement of bacteriochlorophyll, yet it will not photosynthesize. This mutant lacks all the properties associated with reaction-center chlorophyll (131) and is inactive in the $PMSH_2$-ubiquinone reaction which proceeds so readily with chromatophores of all photosynthesizing bacteria. These data indicate that the reaction-center chlorophyll of bacteria is involved in the photochemistry of the system, and further that quinones are intimately connected with this photochemistry.

Ferredoxin appears to be a normal constituent of photosynthetic bacteria. It has been detected in *Rhodospirillum rubrum* (132), has been isolated and crystallized from *Chromatium* (133), and has been purified from *Rhodopseudomonas palustris* (134) and *Chlorobium thiosulfatophilum* (135). The ferredoxin from *Rhodospirillum rubrum* does not stimulate the photoreduction of NAD by chromatophores of that organism (136), and only a slight stimulation of questionable significance is observed from *Chromatium* chromatophores with added *Chromatium* ferredoxin (137). The ferredoxin from *Rhodopseudomonas palustris* was not tested in a photochemical system (134). The ferredoxin from *Chlorobium thiosulfatophilum*, however, is photoreduced photochemically and supports the fixation of carbon dioxide into pyruvate (135). There is evidence, therefore, for the participation of ferredoxin in the photochemistry of the green photosynthetic bacteria, but no direct convincing evidence is available for a similar role in the purple bacteria. It should be noted, however, that the rates for NAD photoreduction observed for chromatophores of the purple bacteria are low in comparison to the rates in intact cells (138), which may indicate that some other condition prevents the attainment of an activation by the ferredoxin. In view of the absence of any other known endogenous redox agent capable of coupling NAD photoreduction to the chromatophore of the purple bacteria, and in view of the conclusion that NAD is photoreduced via a direct electron transfer from photoexcited chlorophyll (in opposition to a reversed electron flow from substrate to NAD powered by light-generated ATP) (6) it is logical to assume that ferredoxin does function in this manner and the demonstration of this function awaits further investigation on the chromatophores of purple bacteria.

As is the case with PSI of green plants, isolated bacterial chromatophores are capable of interacting with a number of artificial electron acceptors. In these cases the electrons are supplied by succinate or a feeder system such as ascorbate-DPIP. The acceptors which have been examined to date include oxygen (139, 140), quinones (128, 129), methyl viologen (coupled to a disulfide as the final acceptor) (141), methyl red (142), and sulfate (143). Space does not permit discussing each acceptor; suffice it to say that little is known about the mechanism of the reactions, the photoreduction of the low potential methyl viologen has significance in the mechanism of the photoact, and the ease of reduction of these compounds indicates a general lack of specificity for the chromatophore with potential acceptors.

B. Electron Donors for the Bacterial Photochemical System

Since the initial detection of cytochrome c_2 in *Rhodospirillum rubrum* (144, 145) and the observation by Duysens that light produces an oxidation of cytochrome c_2 in *R. rubrum* (146), numerous cytochromes have been isolated from photosynthetic bacteria (147). The predominant cytochromes found in photosynthetic bacteria are of the c type and have a high E'_0. However, cytochromes of the b type are also present. Illumination of whole cells or of chromatophores uniformly leads to an oxidation of the endogenous c type cytochromes. Cytochrome photooxidation has been shown for *R. rubrum* (146, 148, 149), *Chromatium* (150–152), *Rhodopseudomonas spheroides* (152), *Chlorobium* and *Chloropseudomonas* (153), and *Rhodomicrobium vannielii* (154). The cytochromes are photooxidized with quantum requirements of 0.7 to 1.4 for *Chromatium* (155) and 3–4 for *Rhodospirillum rubrum* (156).

The data given above indicate a very efficient reaction for endogenous cytochrome photooxidation. Evidence that the bacteriochlorophyll and cytochrome form a complex which functions to transfer an electron from the cytochrome to the reaction center chlorophyll was obtained by Chance and Nishimura (157), who showed that with *Chromatium* the photooxidation of cytochrome 423.5 (one of four cytochromes oxidized) proceeds at $-196°C$ with an efficiency about equal to that at room temperature. The cytochrome was not reduced during a subsequent dark period. On this basis it was thought that the primary photoreaction in bacterial photosynthesis was the transfer of an electron from the complexed cytochrome to the photoexcited chlorophyll. A subsequent study by Vredenberg (*158*) has shown that with all other bacteria studied the cytochrome photooxidation decreases as the temperature is lowered. The temperature at which the photooxidation stops varies for the different bacteria, and is between $-110°$ and $-120°C$ for cytochrome 422 of *Chromatium*, at $-25°C$ for *R. rubrum* and $-50°C$ for *Rhodopseudomonas spheroides*. Thus there is a marked difference in the ease of transfer of electrons from cytochrome to the bacteriochlorophyll system, which makes it less sure that the cytochrome photooxidation is the primary reaction.

The photochemical system of bacteria is able to interact with many of the artificial donors which are active in the plant system via PSI. The donor system most commonly used is ascorbate-DPIP, in which case $DPIPH_2$ is the donor for the photochemical system (*136, 139, 141, 142*) and can be coupled to the photoreduction of oxygen (*139, 159, 160*), methyl red (*142*), methyl viologen (*141, 159*), ubiquinone (*129*), and NAD (*136, 138*). A donor which reacts much as does DPIP is TMPD (*129, 138, 159*), which reacts at a faster rate with the chromatophore system. Other donors include $PMSH_2$ (*129*) and cytochrome *c* (*128, 139*). These reactions again point out the relative nonspecificity of the bacterial system for interaction with nonphysiological redox agents.

At least two of the simple electron transfer reactions catalyzed by bacterial chromatophores are also carried out by isolated chlorophylls, both chlorophyll *a* and bacteriochlorophyll. These are the photoreduction of methyl red supported by ascorbate-DPIP (*142, 160*) and the photoreduction of ubiquinone supported by $PMSH_2$ (*129*). The latter reaction has been investigated in some detail (*160*a) and related to the ATP-forming system of *Rhodospirillum rubrum* chromatophores. Since PMS is a cofactor for cyclic photophosphorylation in *R. rubrum* chromatophores, and since the reaction studied by Zaugg *et al.* (*129*) covers a span from $PMSH_2$ to ubiquinone, it was thought that there might be a phosphorylation linked to this reaction. However, it was concluded that the cyclic electron transport system responsible for ATP formation is a tight, efficient system, which is not intercepted by the added $PMSH_2$-ubiquinone system. The $PMSH_2$-ubiquinone reaction, which proceeds at rates of 6–8 mmoles per hour per mg bacteriochlorophyll, was not light saturated under the conditions employed and appeared to be a secondary reaction superimposed on the more physiological electron-transfer reactions which are responsible for ATP formation.

V. Mechanism of Photochemical Reactions *in Vivo*

According to the accepted principles of photochemistry, chlorophyll *in vivo* catalyzes specific electron transfer reactions by means of the photoexcited chlorophyll itself entering into the reaction as a chemical reactant. This could be accomplished by either of the mechanisms given below, where D represents an electron donor and A an electron acceptor:

(1) a. Chl + light ⟶ Chl*
     b. Chl* + A ⟶ Chl+ + A−
     c. Chl+ + D ⟶ Chl + A−

Sum: D + A + light $\xrightarrow{\text{(Chl)}}$ D+ + A−

(2) a. Chl + light ⟶ Chl*
     b. Chl* + D ⟶ Chl− + D+
     c. Chl− + A ⟶ Chl + A−

Sum: D + A + light $\xrightarrow{\text{(Chl)}}$ D+ + A−

Both mechanisms provide for the transfer of an electron from a donor to an acceptor, and in theory either would be satisfactory to initiate the process of photosynthesis. It is not yet possible to positively identify either mechanism as the correct one for any of the photosynthetic systems, since there is evidence in favor of both. In the organized system of the chloroplast and chromatophore the reaction-center chlorophyll is considered to be complexed to a donor cytochrome molecule and perhaps also to an acceptor molecule, such as ferredoxin. In this situation, where the reactants are not free to move about, the distinction between the two mechanisms becomes less clear since some of the reactions may proceed in concert. But in any event, the same question of primacy remains: does the photoexcited chlorophyll react first to donate an electron or to gain an electron?

A. Pigment System II of Plants

To date a reaction-center chlorophyll for PSII has not been detected. Kok *et al.* (*108*) made an extensive investigation of the absorption changes of spinach chloroplasts, but could find no change identifiable specifically with PSII in the manner that P700 is identified with PSI. Broyde and Brody (*161*) have detected a chlorophyll form which shows a fluorescence maximum at 698 mμ in acetone at −77°C, but does not fluoresce at room temperature. These investigators feel that this compound is responsible for the 698-mμ fluorescence which is observed *in vivo,* and further propose that it is involved in the photoact of PSII.

Considerably more information must be available, however, before such an assignment could be made. As the authors point out, the pigment could be an artifact of preparation.

For the reasons given in Section III, A, plastoquinone is thought to be the acceptor for PSII. Quinones are excellent acceptors for photochemical redox reactions involving chlorophyll *in vitro* (*48, 129, 162–165*). That quinones can function as electron acceptors for photoexcited chlorophyll is shown by absorption changes (*163*) and electron spin resonance responses (*164, 165*) of chlorophyll and quinones in solution. This agrees with the ability of quinones to quench fluorescence of chlorophyll in solution. For these reasons the primary photoreaction of PSII could well be the transfer of an electron to a quinone from some photoexcited chlorophyll molecule whose only architectural requirement would be that it be in close proximity to the quinone. In view of the high quinone content of the chloroplast, this would make the reaction a very probable one. Accordingly, we suggest mechanism 1 for the operation of PSII in plant chloroplasts. The oxidized chlorophyll so produced could then in some manner regain its electron from water in a reaction involving manganese ion and chloride ion. Nothing is known about a possible mechanism for this reaction.

B. Pigment System I of Plants

Illumination of PSI causes the oxidation of three compounds: cytochrome f, plastocyanin, and P700. The experiments of Fork and Urbach indicate that cytochrome f is located closer to the photochemical site of PSI, and plastocyanin serves to transfer electrons from PSII to cytochrome f (*166*). This conclusion stems from the fact that absorption changes related to cytochrome f oxidation are still observed when the plastocyanin changes are prevented by the addition of inhibitors. Therefore, we should consider cytochrome f as the donor for the chlorophyll of PSI, and the sequence as plastocyanin $\rightarrow$ cytochrome $f \rightarrow$ P700.

The first detectable reduced compound formed via PSI is reduced ferredoxin. Although there may be some other intermediate compound, there is no direct evidence for one. The experiments in which compounds with lower potentials than ferredoxin are reduced by chloroplasts (*32, 55, 56*) are not conclusive, since it must be shown that the dyes have the same redox potential in the chloroplast system as in solution. In view of this we must consider three compounds when considering the primary photochemistry of PSI, viz. cytochrome f, P700, and ferredoxin.

There is some evidence available concerning the kinetics of cytochrome f and P700 photooxidation, but the literature reports are contradictory. Witt *et al.* reported that the decay time for oxidized chlorophyll (P700) corresponds to the rise time for cytochrome oxidation, which is consistent with P700 being a photochemical donor molecule (*167*). In aged chloroplasts coupled with DPIP or PMS ($< 3 \times 10^{-5} M$) in the presence of excess ascorbate, and more recently in whole cells of the blue-green alga *Plectonema boryanum*, Ke observed transient absorption changes at wavelengths characteristic of both P700 and cytochrome *f* (*18, 21*). At a 5×10^{-5} sec time resolution of the instrument, there was no evidence of a sequential reaction between the two components. The light-minus-dark spectrum constructed from absorption-change transients in *Plectonema boryanum* induced with 20 µsec flashes has a profile similar to that reported earlier by Kok and Gott (*168*) for another blue-green alga, *Anacystis nidulans*. Both difference spectra correspond to P700 and cytochrome *f* oxidation. Chance and Bonner (*61*), in investigating the kinetics of cytochrome *f* and P700 oxidation in intact leaves, both at room temperature and 77°K, report that the rates of cytochrome *f* and P700 oxidation are similar. Furthermore, there was no noticeable lag period for cytochrome oxidation, as would be expected if cytochrome oxidation was sequential to P700 oxidation. Chance and Bonner concluded that both cytochrome *f* and P700 could be oxidized in primary photochemical events, but the two moieties would have to react at different sites and with different chlorophyll molecules.

Chance *et al.* (*169*) have studied the kinetics of oxidation of endogenous cytochrome *f* and reduction of added ferredoxin by illuminated spinach chloroplasts. The kinetics of cytochrome *f* oxidation were independent of whether endogenous or added ferredoxin was present. Experiments with pulsed light (Q-switched laser) showed that cytochrome *f* oxidation proceeded in ferredoxin-free chloroplasts. When present, however, ferredoxin was reduced at approximately the same rate as that of cytochrome *f* oxidation. All these data indicate a coupled reaction between cytochrome *f* and ferredoxin, with the cytochrome *f* reaction being primary and the ferredoxin reduction being secondary. These data show that ferredoxin probably does not react as an acceptor molecule for photoexcited chlorophyll via mechanism 1. This agrees with experiments performed with chlorophyllin in aqueous media, in which ferredoxin does not serve as an electron acceptor to this chlorophyll derivative under illumination (*170*).

Regardless of the data advanced by Chance *et al.* (*169*), it is still not possible definitely to state the mechanism for PSI action in photosynthesis. Experiments showing that cytochrome *f* is photooxidized prior to P700 are not conclusive, since in a closely coupled situation reduced cytochrome *f* could transfer an electron to photooxidized P700 so rapidly that one would not detect any oxidized P700 until all the cytochrome f was oxidized. Various states of reduction and tightness of coupling of the cytochrome f to the P700 would give responses in which the observer would notice either cytochrome f oxidation, P700 oxidation, or both. Mechanism 1 requires some acceptor (and this would not be ferredoxin) to be reduced in the primary reaction. Such a reaction has not been observed. Mechanism 2 requires that the reaction center chlorophyll becomes reduced. The available evidence indicates that the reaction center chlorophyll is P700, and there is no evidence for its reduction in a primary reaction. Therefore, it seems expedient to withhold judgment on this important question and await more refined experiments, which hopefully will give information about the primary acceptor for PSI.

C. Photochemical System of the Photosynthetic Bacteria

As described above, the photosynthetic bacteria contain one functional pigment system which resembles in many respects PSI of plants, and the same difficulties are encountered in trying to specify a mechanism for these reactions. It is less certain that ferredoxin is photoreduced by the bacteriochlorophyll system, although it is photoreduced by the chlorobium chlorophyll system of *Chlorobium thiosulfatophilum*. Since methyl viologen supports a rapid photoreduction of the disulfide DTNB (5,5'-dithiobis (2-nitrobenzoic acid)) (*141, 159*) the dye itself must be photoreduced, meaning that the chromatophores of the purple bacteria are capable of reducing a compound with a redox potential of $-470\,mv$ (*32*). Again, there is little specificity in terms of the redox agents with which the bacterial chromatophore will react, as is the case with PSI of plants.

The bacterial system differs from PSI in that the former interacts with endogenous quinone, reducing it upon illumination of the chromatophore. For the reasons given above, quinones can be considered as excellent acceptors of electrons from photoexcited chlorophyll. Therefore, we propose that activation of the bacterial reaction-center chlorophyll can lead to two alternate reactions: (1) the reduction of ubiquinone or (2) the reduction of ferredoxin. Ubiquinone reduction would function in the internal, cyclic electron transport sequence responsible for ATP formation, which is an efficient and stable system. Ferredoxin reduction would be involved in the reactions leading to NAD reduction and eventually carbon dioxide fixation. The controlling factors for these two reactions could be the availability of ADP for the cyclic flow, and the redox level of exogenous organic compounds for the ferredoxin system. Excess reducing power, which could not be funneled into NAD reduction, would be dissipated as molecular hydrogen. A scheme outlining these suggestions is given in Fig. 2. If some other low-potential compound is involved in the system leading to NAD reduction, it could be either a reduced bacteriochlorophyll or some as yet undetected compound.

Bacterial chromatophores are capable of photoreducing endogenous quinone, but nothing is known concerning the kinetics of this reaction relative to cytochrome or chlorophyll changes. Even less is known about

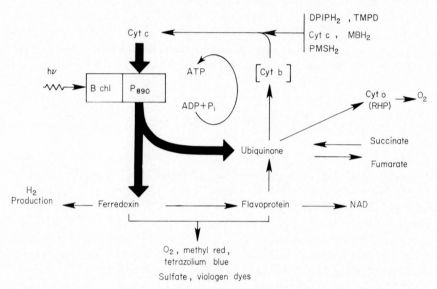

FIG. 2. Pathway of light-induced electron transfer reactions of *Rhodospirillum rubrum* chromatophores. Only one pigment system is operative in the bacterial process, which funnels excitation energy to P890 for initiation of the photochemical events. Two acceptors are portrayed for the photochemical act, ferredoxin and ubiquinone. This scheme resembles one previously published (6) except for the bifurcation of the light-induced transfer and the fact that cytochrome o is considered as the dark oxidase. RHP would be a derivative of this compound (171).

the role of ferredoxin in the bacterial system. Therefore, the same situation applies here as for PSI of plants, viz. there is kinetic information available concerning the photooxidation of cytochrome c and P870 (or P890), but nothing is known about the photoreduction reactions.

The studies which have been made concerning the kinetics of cytochrome photooxidation and P890 oxidation in *Chromatium* show that the cytochrome is oxidized preferentially (152, 172). At first glance this would appear to favor mechanism 2 for the bacterial system. However, the same situation applies as for PSI of plants; a rapid transfer of an electron from cytochrome to oxidized P890 would not allow the detection of oxidized P890 until the cytochrome was first oxidized. Therefore, these data can be fit into either scheme.

In this context, it is important to recall the experiments reported by Chance and De Vault (*173*) in which the use of a ruby laser for excitation of the *Chromatium* chromatophore showed that cytochrome 423.5 photooxidation proceeded with a half-life of 18 μsec. This leads to the conclusion that if P890 were oxidized first in the primary act, and cytochrome 423.5 oxidation was subsequent to this primary act, the P870 would of necessity have to turn over at a faster rate than that observed for the cytochrome. Existing experimental methods are incapable of detecting such rapid rates. It is also possible, that this particular cytochrome in *Chromatium*, whose photooxidation is unique in its temperature insensitivity down to −196°C, is oxidized in a photoreaction which is minor in terms of the overall photosynthetic process. The low rate of reduction of oxidized *Chromatium* cytochrome 423.5 in the dark would indicate that this cytochrome may not be important in the overall process, since a rapid turnover, not just a rapid oxidation, would be required for significant transfer in a continuing process. Our conclusion, like that of Clayton in Chapter 19, is that it is not yet possible to rule out either mechanism.

D. Comparison of *in Vivo* and *in Vitro* Photoreactions of Chlorophyll

There are a few photochemical redox reactions which are catalyzed by chlorophyll *in vivo* and by detergent-solubilized purified Chl *a* in aqueous systems *in vitro*. Those observed for Chl *a* are the following: (a) photoreduction of methyl red supported by ascorbate alone or ascorbate-DPIP (*142, 160*); (b) photoreduction of cytochrome *c* supported by TMQH$_2$ plus TMQ (*48*); (c) photoreduction of pyocyanin supported by TMQH$_2$ (*162*); (d) photoreduction of quinones (ubiquinone and TMQ) supported by PMSH$_2$ (*162, 163*); (e) photooxidation of cytochrome *c* by oxygen (*160*). From the information available, reactions (a)–(d) appear to proceed via mechanism 1, in which photoexcited chlorophyll donates an electron to the acceptor. This is shown for reaction (a) by the quantum yields for ascorbate-supported reactions and for methyl red photoreduction in ethanol (cf. Chapter 17). Analysis by flash spectrophotometry of the kinetics of absorption changes for reactions (b)–(d) (*163*) also support mechanism 1. The mechanism of reaction (e) is not known.

In a recent investigation, Vernon and Shaw examined the effect of increasing Triton X-100 concentrations on the various photoreactions of spinach chloroplasts (*70*). The Hill reaction was stimulated at low concentrations of detergent because of an uncoupling of phosphorylation. At higher concentrations (0.01%), an inhibition was observed due to destruction of the oxygen evolving system and other disruptive effects on the electron transfer enzymes. At concentrations above 0.02% Triton X-100, a few of the simpler photoreactions reappeared, apparently catalyzed by small fragments produced from the chloroplasts by the action of the detergent. These reactions included NADP photoreduction by ascorbate-DPIP, ferrocytochrome c photooxidation by either NADP or oxygen, cytochrome c photoreduction coupled to $TMQH_2$ oxidation, and methyl red reduction coupled to ascorbate-DPIP (174). All reactions involving $DPIPH_2$ and cytochrome c were stimulated by adding plastocyanin to the system. The cytochrome c–$TMQH_2$ system was not stimulated by plastocyanin. With the exception of NADP photoreduction, the same reactions were catalyzed by Chl a which was solubilized by Triton X-100 (for cytochrome c oxidation an intermediate carrier to oxygen such as methyl red must be added). With purified Chl a, no effect of added plastocyanin was observed.

NADP is photoreduced in aqueous systems using hematoporphyrin (175) or chlorophyllin a (170) as the photocatalyst. In both cases spinach ferredoxin is not required, and only the flavoprotein NADP reductase is needed. The enzymatic requirements of this system, then, are less strict than are observed for the chloroplast. The possible reason for this is discussed by these investigators (170). Bacteriochlorophyll *in vivo* and *in vitro* supports several of the reactions listed for chlorophyll a. Zaugg *et al.* have shown that isolated BChl will support reaction d (129).

Appendix 1. Absorption Spectra of Some Redox Components in the Photosynthetic Electron-Transport Chain in Plants

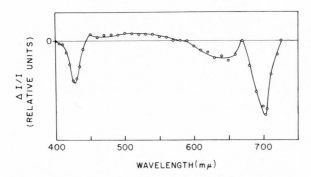

FIG. 3. Difference spectrum (oxidized-minus-reduced) of P700 (in digitonin-treated chloroplasts) (20). Experimental values are shown by o.

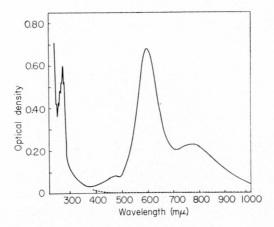

FIG. 4. Absorption spectra of spinach plastocyanin (176). Oxidized form (solid line), reduced form (broken line).

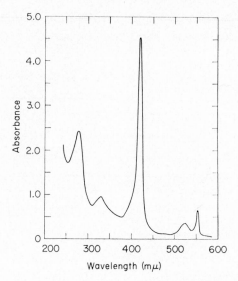

Fig. 5A. Absorption spectrum of reduced cytochrome f (58).

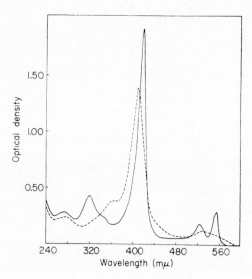

Fig. 5B. Absorption spectra of crystalline *Porphyra tenera* cytochrome 553 (59). Oxidized form (broken line), reduced form (solid line).

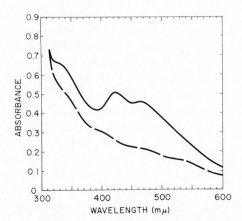

FIG. 6. Absorption spectra of spinach ferredoxin (*50*). Oxidized form (solid line), reduced form (broken line).

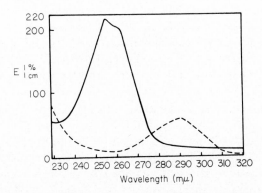

FIG. 7. Ultraviolet absorption spectra of plastoquinone (*177*). Oxidized form (solid line), reduced form (broken line).

Appendix II.   Flash-Spectrophotometric Techniques for Photosynthesis Research

In the development of the modern concept of the mechanism of photosynthesis, much of the experimental evidence has been derived by the useful technique of difference spectrophotometry first introduced by Duysens in 1952 (*178*). As the name implies, the technique measures the difference in absorbance between the active sample under illumination and the control maintained in darkness. Duysen's technique uses a monochromatic beam split either by a rotating disk or a vibrating mirror. Two cuvettes containing the cell suspension to be analyzed are placed singly in the beam paths, and both beams fall on the same photomultiplier surface. The changes in absorption in the sample cuvette are induced by an actinic light of high intensity, usually incident at right angle to the measuring beam. Since the measuring beam is modulated, the difference signal generated by the change in absorption can be amplified by a tuned circuit and recorded. Although the unmodulated actinic beam does not cause deflection, it is usually necessary to place an appropriate cutoff filter in front of the photomultiplier to prevent transients caused by the scattered actinic light.

The recorded deflection represents the change in transmission, $\Delta I/I$. For small changes, the linear relationship $\Delta A = 1/2.303\ (\Delta I/I)$ is valid and may be used for calculating the absorbance change. The light-induced absorbance change may be measured at different wavelengths of the measuring beam. The deflection plotted as a function of the measuring wavelength is called a difference spectrum (light-minus-dark), which represents the total changes in the absorption spectra of the photosynthetic cells under illumination. Absorbance changes of 10^{-4} can be measured by difference spectrophotometry with a time response of 10^{-1} to 10^{-2} sec, depending on the modulation frequency. Other investigators have adapted commercial split-beam spectrophotometers for similar measurements (*179*).

The total process of photosynthesis consists of light absorption, stabilization and conversion of the electronic energy into chemical free energy, and finally its utilization in biosynthesis. The study of the rapid events taking place at the junction bordering photochemistry and chemistry is of prime importance for the understanding of the basic process of photosynthesis. To differentiate the primary and secondary effects, to measure the onset of rapid reactions and the lifetimes of transitory intermediates by means of difference spectrophotometry, both its sensitivity and its time response must be of a high level. In 1949, Norrish and Porter introduced the technique of flash photolysis for the study of photochemical problems of a similar nature, namely, rapid reactions affected by free radicals and intramolecular processes of excited molecules (*180*). The basic operating principle of flash photolysis will be briefly described here as an introduction to the flash methods applied specifically to photosynthesis research, since the latter are actually derivations or modifications of the basic flash-photolysis technique.

In conventional flash photolysis (shown schematically in Fig. 8A), an intense actinic flash with duration of 10^{-6} to 10^{-4} sec is used to ini-

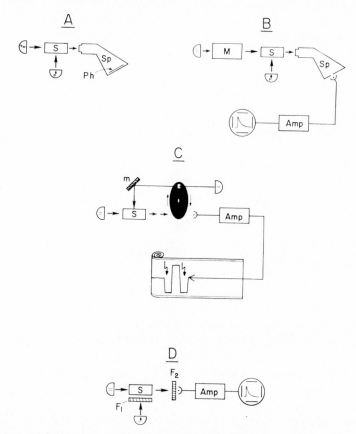

FIG. 8. Schematic presentation of several experimental techniques for measuring rapid absorption changes. (A) Conventional flash photolysis method of Norrish and Porter (*180*); (B) Modification of the flash-photolysis apparatus for kinetic measurements at individual wavelength (*181*); (C) Kok's method of "time separation" (*182*); (D) Witt's method of "color separation" (*183*). Abbreviations are: S, sample; Sp, spectrograph; Ph, photographic plate; M, monochromator; m, mirror; Amp, amplifier; F_1, F_2 are complementary filters.

tiate a photochemical reaction. At a given instant after the actinic flash, a weak measuring flash is sent through the solution cuvette and received by a spectrograph, and the entire absorption spectrum of the activated sample is recorded photographically. The photographic method is convenient in preliminary investigations since the entire absorption spectrum is recorded. Although the lifetime of the reaction intermediates can be determined from a series of spectra obtained at different times after the actinic flash, it is less convenient for kinetic investigations.

For kinetic studies, the above procedure may be modified by replacing the photographic plate with a photomultiplier positioned at a particular wavelength of interest and by replacing the measuring flash with a steady measuring beam (181). This arrangement, as shown in Fig. 8B, allows the study of the time course of absorption changes upon the application of an actinic flash.

The application of flash spectrophotometry to photosynthesis research was pioneered by Kok and by Witt, each group using a slightly different approach. In Kok's method, as shown schematically in Fig. 8C, the actinic flash isolated by a slotted rotating disk is applied at a rate of 10–20 flashes per second. The slots in the disk and the speed of disk rotation can be varied to provide light and dark periods of desired durations. Kok's original papers should be consulted for more details (182, 184). As seen from Fig. 8C, the actinic light source and the measuring light source are arranged in such a manner that at the time the sample is illuminated the rotating disk cuts off the measuring beam from the photomultiplier. This "time-separation" technique has the advantage that measuring light with the same wavelength as the actinic light can be used and that the measurement is not influenced by fluorescence caused by the actinic light. These features were probably instrumental in enabling Kok to measure the P700 absorption changes in various photosynthesizing organisms (182). The change in absorption is measured between the time immediately before the flash and the time immediately after the completion of the flash. The difference is plotted by averaging over 1 minute of periodic flashes. Kok's method has a sensitivity of 10^{-4} in absorbance, but changes with a lifetime shorter than several milliseconds or longer than one second cannot be recorded.

Witt's method is similar to the modified flash photolysis (183). A single beam version of Witt's apparatus is shown schematically in Fig. 8D. The sample cuvette is monitored with a steady monochromatic beam. The photomultiplier output is balanced by a direct current voltage, and the absorption-change transient caused by an intense flash is displayed on the oscilloscope screen directly. In Witt's arrangement, it is necessary to provide the photomultiplier with cutoff filters which will pass only the measuring beam but not the actinic flash. Hence Witt's measurement is of the "color-separating" type, namely, the measuring wavelength and the actinic wavelength must be different. The sensitivity of Witt's method has been claimed to be 0.1% in transmission change, and the time response is 10^{-5} sec.

In flash spectrophotometry the dual requirements of high sensitivity and rapid time response pose an inherent difficulty because the two requirements are mutually exclusive. The wide bandpass of the amplifier and the recording system necessary for rapid time response also decrease the signal-to-noise ratio and consequently the sensitivity. One way to circumvent the problem is to increase the total energy falling on the photomultiplier, since the signal-to-noise ratio improves as the square root of the intensity of the incident measuring beam. This can be done by either using a cuvette with a larger surface area and therefore utilizing a larger active surface of the photomultiplier, or by increasing the intensity of the measuring beam. The latter approach may not be desirable, because an intense measuring beam often causes changes in absorption, especially in the wavelength region where the system strongly absorbs.

To circumvent these difficulties, Ke *et al.* have recently introduced the use of a commercial Computer of Average Transient (CAT) to improve the signal-to-noise ratio without resorting to the use of an intense measuring light (*185*). The CAT computer functions by extracting the signal from a high amplitude of background noise. The computer samples the signal at 400 intervals, and the amplitude of the signal is converted into digital count and stored in a memory "address." The repetitive stimulus pulses synchronize the averaging process with the result that the desired response is reinforced. The random noise that is unrelated to the signal will sum out of phase and thus tend to average to zero. The improvement is proportional to the square root of the number of signal accumulations. Thus, very minute absorption changes can be measured with a reasonable number of repetitive flashes. The computer calculates and simultaneously displays on a cathode-ray tube the data as they appear. The output from the computer can also be transcribed onto an *xy* recorder.

A schematic diagram of the flashing light spectrophotometer used by Ke *et al.* is shown in Fig. 9. The basic design is similar to the color-separation method of Witt. Repetitive flashes with half-peak duration of 20 μsec have been produced with xenon flash lamps by condenser discharge. A polychromatic flash with wavelength limit between 645 and 745 mμ has an instantaneous peak intensity of 2.5×10^7 erg cm^{-2} sec^{-1}. The typical intensity of the 515-mμ measuring beam is approximately 200 ergs cm^{-2} sec^{-1}. A Schott continuous-running interference filter slide plus appropriate Corning filters are used in front of the photomultiplier for cutting off the actinic light. The double-beam arrangement shown here mainly serves the function of eliminating intensity fluctuations in the measuring light sources. By using a highly stabilized power supply for the light source, this instrument can also be operated in the single-beam mode.

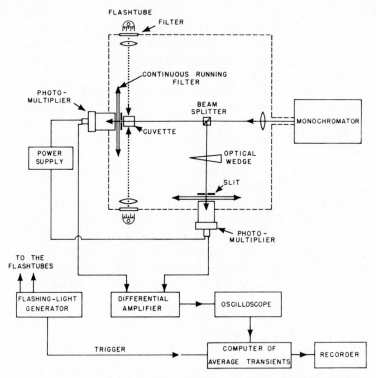

FIG. 9. Block diagram of a flashing-light spectrophotometer (185).

Rüppel et al. recently have also reported the use of a sampling technique for extracting the signal from noise by repetitive cycling (186). The authors have reported that improvement as much as 200-fold can be obtained. They have called this sampling technique "periodic chemical relaxation."

More recently, Chance and De Vault have reported the use of a "Q-switched" ruby laser as an excitation flash for studying cytochrome reactions in Chromatium (173). These pulses last 10–20 nsec and have a total energy of about 1 joule per pulse.

REFERENCES

(*1*) R. K. Clayton, "Molecular Physics in Photosynthesis." Ginn (Blaisdell), Boston, Massachusetts, 1965.
(*2*) M. D. Kamen, "Primary Processes in Photosynthesis." Academic Press, New York, 1963.
(*3*) G. W. Robinson, *Ann. Rev. Phys. Chem.* **15**, 311 (1964).
(*4*) L. N. M. Duysens, *Progr. Biophys. Biophys. Chem.* **14**, 1 (1964).
(*5*) A. San Pietro and C. C. Black, *Ann. Rev. Plant Physiol.* **16**, 155 (1965).
(*6*) L. P. Vernon, *Ann. Rev. Plant Physiol.* **15**, 73 (1964).
(*7*) L. P. Vernon, *Ann. Rev. Biochem.* **34**, 269 (1965).
(*8*) G. Hoch and B. Kok, *Ann. Rev. Plant Physiol.* **12**, 155 (1961).
(*9*) C. B. van Niel, *Ann. Rev. Plant Physiol.* **13**, 1 (1962).
(*10*) R. K. Clayton, *Ann. Rev. Plant Physiol.* **14**, 159 (1963).
(*11*) J. H. C. Smith and C. S. French, *Ann. Rev. Plant Physiol.* **14**, 181 (1963).
(*12*) M. R. Michel-Wolwertz and C. Sironval, *Biochim. Biophys. Acta* **94**, 330 (1965).
(*13*) C. Sironval, M. R. Michel-Wolwertz, and A. Madsen, *Biochim. Biophys. Acta* **94**, 344 (1965).
(*14*) R. Emerson and W. Arnold, *J. Gen. Physiol.* **15**, 391 (1932).
(*15*) B. Kok, *Biochim. Biophys. Acta* **21**, 245 (1956).
(*16*) B. Kok, *Biochim. Biophys. Acta* **48**, 527 (1961).
(*17*) B. Rumberg, P. Schmidt-Mende, J. Weikard, and H. T. Witt, *Natl. Acad. Sci. —Natl. Res. Council, Publ.* **1145**, 18 (1963).
(*18*) B. Ke, *Biochim. Biophys. Acta* **88**, 289 (1964).
(*19*) B. Rumberg and H. T. Witt, *Z. Naturforsch.* **19b**, 693 (1964).
(*20*) B. Ke, *Biochim. Biophys. Acta* **88**, 297 (1964).
(*20a*) B. Ke, S. Katoh, and A. San Pietro, unpublished experiments (1966).
(*21*) B. Ke and E. Ngo, *Biochim. Biophys. Acta,* **109**, 431 (1965).
(*22*) K. T. Fry and A. San Pietro, *Biochim. Biophys. Res. Commun.* **9**, 218 (1962).
(*23*) H. E. Davenport, R. Hill, and F. R. Whatley, *Proc. Roy. Soc.* **B139**, 346 (1952).
(*24*) A. San Pietro and H. M. Lang, *J. Biol. Chem.* **231**, 211 (1958).
(*25*) K. Tagawa and D. I. Arnon, *Nature* **195**, 537 (1962).
(*26*) H. A. Gewitz and W. Völker, *Z. Physiol. Chem.* **330**, 124 (1962).
(*27*) K. T. Fry, R. A. Lazzarini, and A. San Pietro, *Proc. Natl. Acad. Sci. U.S.* **50**, 652 (1963).
(*28*) D. I. Arnon, H. Y. Tsujimoto, and B. D. McSwain, *Proc. Natl. Acad. Sci. U.S.* **51**, 1274 (1964).
(*29*) F. R. Whatley, K. Tagawa, and D. I. Arnon, *Proc. Natl. Acad. Sci. U.S.* **49**, 266 (1963).
(*30*) B. Chance and A. San Pietro, *Proc. Natl. Acad. Sci. U.S.* **49**, 633 (1963).
(*31*) J. S. C. Wessels and E. Havinga, *Rec. Trav. Chim.* **72**, 1076 (1953).
(*32*) G. Zweig and M. Avron, *Biochem. Biophys. Res. Commun.* **19**, 397 (1965).
(*33*) R. Hill, *Nature* **139**, 881 (1937).
(*34*) R. Hill and R. Scarisbrick, *Proc. Roy. Soc.* **B129**, 238 (1940).
(*35*) J. D. Spikes, *Arch. Biochem. Biophys.* **35**, 101 (1952).
(*36*) O. Warburg and W. Lüttgens, *Naturwissenschaften* **32**, 161 (1944).
(*37*) S. Aronoff, *Plant Physiol.* **21**, 393 (1946).
(*38*) S. L. Chen, *Plant Physiol.* **27**, 35 (1952).

(39) A. S. Holt, in "Photosynthesis in Plants" (J. Franck and W. E. Loomis, eds.), p. 277. Iowa State Coll. Press, Ames, Iowa, 1949.

(40) J. M. Ramirez, F. F. Del Campo, A. Paneque, and M. Losada, Biochem. Biophys. Res. Commun. 15, 297 (1964).

(41) B. Vennesland, H. W. Gattung, and E. Birkicht, Biochem. Biophys. Acta 66, 285 (1963).

(42) L. P. Vernon and W. S. Zaugg, J. Biol. Chem. 235, 2728 (1960).

(43) A. D. Swensen and L. P. Vernon, Biochim. Biophys. Acta 102, 349 (1965).

(44) K. A. Clendenning, in "Handbuch der Pflanzenphysiologie" (W. Ruhland, ed.), Vol. 5, Part 1, p. 736. Springer, Berlin, 1960.

(45) A. S. Holt, U.S. At. Energy Comm., Doc. ORNL-752 (1950).

(46) H. Davenport and R. Hill, Biochem. J. 74, 493 (1960).

(47) D. L. Keister and A. San Pietro, Arch. Biochem. Biophys. 103, 45 (1963).

(48) L. P. Vernon and E. R. Shaw, Biochemistry 4, 132 (1964).

(49) Z. Gromet-Elhanan and M. Avron, Biochem. Biophys. Res. Commun. 10, 215 (1963).

(50) T. Horio and A. San Pietro, Proc. Natl. Acad. Sci. U.S. 51, 1226 (1964).

(51) D. C. Fork, Plant Physiol. 38, 323 (1963).

(52) P. R. Levine and R. M. Smillie, Proc. Natl. Acad. Sci. U.S. 48, 417 (1962).

(53) Y. de Kouchkovsky and J. Briantais, Natl. Acad. Sci.—Natl. Res. Council, Publ. 1145, 362 (1963).

(54) D. I. Arnon, M. Losada, F. R. Whatley, H. Y. Tsujimoto, D. O. Hall, and A. A. Horton, Proc. Natl. Acad. Sci. U.S. 9, 1314 (1961).

(55) C. C. Black, Science 149, 62 (1965).

(56) B. Kok and H. J. Rurainski, Ann. Rep. Res. Inst. Advanced Study, Baltimore, Maryland p. 4 (1964).

(57) R. Scarisbrick and R. Hill, Biochem. J. 37, xxii (1943).

(58) H. E. Davenport and R. Hill, Proc. Roy. Soc. B139, 327 (1952).

(59) S. Katoh, Plant Cell Physiol. (Tokyo) 1, 91 (1960).

(60) L. N. M. Duysens, Science 121, 210 (1955).

(61) B. Chance and W. D. Bonner, Jr., Natl. Acad. Sci.—Natl. Res. Council, Publ. 1145, 66 (1963).

(62) S. Katoh and A. Takamiya, Plant Cell Physiol. (Tokyo) 4, 335 (1963).

(63) B. Kok, H. J. Rurainski, and E. A. Harmon, Plant Physiol. 39, 513 (1964).

(64) L. N. M. Duysens and J. Amesz, Biochim. Biophys. Acta 64, 243 (1962).

(65) S. Katoh, Nature 186, 533 (1960).

(66) S. Katoh, I. Suga, I. Shiratori, and A. Takamiya, Arch. Biochem. Biophys. 94, 136 (1961).

(67) Y. de Kouchkovsky and D. C. Fork, Proc. Natl. Acad. Sci. U.S. 52, 232 (1964).

(68) B. Kok and H. J. Rurainski, Biochim. Biophys. Acta 94, 588 (1965).

(69) S. Katoh and A. Takamiya, Biochim. Biophys. Acta 99, 156 (1965).

(70) L. P. Vernon and E. R. Shaw, Plant Physiol. (1966) (in press).

(71) L. P. Vernon and W. S. Zaugg, J. Biol. Chem. 235, 2729 (1960).

(72) M. Losada, F. R. Whatley, and D. I. Arnon, Nature 190, 14 (1961).

(73) D. L. Keister, J. Biol. Chem. 240, 2673 (1965).

(74) Z. Gromet-Elhanan and M. Avron, Biochem. Biophys. Res. Commun. 10, 215 (1963).

(75) A. Trebst and H. Eck, Z. Naturforsch. 16b, 455 (1961).

(76) A. T. Jagendorf and M. Avron, Arch. Biochem. Biophys. 80, 246 (1959).

(77) A. R. Krall, N. E. Good, and B. C. Mayne, *Plant Physiol.* 36, 44 (1961).

(78) K. Tagawa, H. Y. Tsujimoto, and D. I. Arnon, *Proc. Natl. Acad. Sci. U.S.* 50, 544 (1963).

(79) A. Trebst and H. Eck, *Z. Naturforsch.* 16b, 44 (1961).

(80) H. T. Witt, A. Müller, and B. Rumberg, *Nature* 192, 967 (1961).

(81) A. Trebst, *Z. Naturforsch.* 19b, 418 (1964).

(82) J. S. C. Wessels, *Biochim. Biophys. Acta* 79, 640 (1964).

(83) H. K. Lichtenthaler and R. B. Park, *Nature* 198, 1070 (1963).

(84) R. A. Dilley, M. D. Henninger, and F. L. Crane, *Natl. Acad. Sci.—Natl. Res. Council, Publ.* 1145, 273 (1963).

(85) J. Friend and E. R. Redfearn, *Phytochemistry* 2, 397 (1963).

(86) J. Weikard, A. Müller, and H. T. Witt, *Z. Naturforsch.* 18b, 139 (1963).

(87) M. Klingenberg, A. Müller, P. Schmidt-Mende, and H. T. Witt, *Nature* 194, 379 (1962).

(88) J. Amesz, *Biochim. Biophys. Acta* 79, 257 (1964).

(89) L. N. M. Duysens and H. E. Sweers, *Plant Cell Physiol. (Tokyo)* 4, 353 (1963).

(90) N. I. Bishop, *Proc. Natl. Acad. Sci. U.S.* 45, 1696 (1959).

(91) D. I. Arnon and A. A. Horton, *Acta Chem. Scand.* 17, S135 (1963).

(92) A. Trebst and H. Eck, *Z. Naturforsch.* 18b, 694 (1963).

(93) F. R. Whatley and A. A. Horton, *Acta Chem. Scand.* 17, S140 (1963).

(94) D. W. Krogmann and E. Olivero, *J. Biol. Chem.* 237, 3292 (1962).

(95) M. D. Henninger, R. A. Dilley, and F. L. Crane, *Biochem. Biophys. Res. Commun.* 10, 237 (1963).

(96) M. D. Henninger and F. L. Crane, *Biochim. Biophys. Acta* 75, 144 (1963).

(97) A. Trebst, H. Eck, and S. Wagner, *Natl. Acad. Sci.—Natl. Res. Council, Publ.* 1145, 174 (1963).

(98) A. Trebst and S. Wagner, *Z. Naturforsch.* 17b, 396 (1962).

(99) A. Trebst, *in* "La Photosynthese," p. 499. C.N.R.S., Paris, 1963.

(100) B. Rumberg, *Nature* 204, 860 (1964).

(101) L. N. M. Duysens, *Science* 120, 353 (1954).

(102) H. T. Witt, *Naturwissenschaften* 42, 72 (1955).

(103) H. T. Witt and R. Moraw, *Z. Physik. Chem. (Frankfurt)* [N.S.] 20, 283 (1959).

(104) H. T. Witt, A. Müller, and B. Rumberg, *Nature* 191, 194 (1961).

(105) A. Müller, D. C. Fork, and H. T. Witt, *Z. Naturforsch.* 18b, 142 (1963).

(106) B. Rumberg, *Biochim. Biophys. Acta* 102, 354 (1965).

(107) D. Rubinstein and E. Rabinowitch, *Biophys. J.* 4, 107 (1964).

(108) B. Kok, B. Cooper, and L. Yang, *Plant Cell Physiol. (Tokyo)* p. 373 (1963) (special issue).

(109) D. Rubinstein, *Biochim. Biophys. Acta* 109, 41 (1965).

(110) D. Spencer and J. V. Possingham, *Biochim. Biophys. Acta* 52, 379 (1961).

(111) E. Kessler, W. Arthur, and J. E. Brugger, *Arch. Biochem. Biophys.* 71, 326 (1957).

(112) E. Kessler, R. Moraw, B. Rumberg, and H. T. Witt, *Biochim. Biophys. Acta* 43, 134 (1960).

(113) R. W. Treharne, T. E. Brown, H. C. Eyster, and H. A. Tanner, *Biochem. Biophys. Res. Commun.* 3, 119 (1960).

(114) E. Kessler, *Arch. Biochem. Biophys.* 59, 527 (1955).

(115) T. Akazawa and E. E. Conn, *J. Biol. Chem.* 232, 403 (1958).

(*116*) S. J. Klebanoff, *J. Biol. Chem.* **234**, 2480 (1959).
(*117*) S. J. Klebanoff, *Biochim. Biophys. Acta* **56**, 460 (1962).
(*118*) H. M. Habermann and H. Gaffron, *Photochem. Photobiol.* **1**, 159 (1962).
(*119*) N. I. Bishop, *Biochim. Biophys. Acta* **27**, 205 (1958).
(*120*) H. Gaffron, *Plant Physiol.* **1B**, 3 (1960).
(*121*) N. I. Bishop, *Biochim. Biophys. Acta* **57**, 186 (1962).
(*122*) H. Gest, A. San Pietro, and L. P. Vernon, eds., "Bacterial Photosynthesis." Antioch Press, Yellow Springs, Ohio, 1963.
(*123*) R. L. Lester and F. L. Crane, *J. Biol. Chem.* **234**, 2169 (1959).
(*124*) R. C. Fuller, R. M. Smillie, N. Rigopoulos, and V. Yount, *Arch. Biochem. Biophys.* **95**, 197 (1961).
(*125*) J. Glover and D. R. Threlfall, *Biochem. J.* **85**, 14P (1962).
(*126*) R. K. Clayton, *Biochem. Biophys. Res. Commun.* **9**, 49 (1962).
(*127*) H. Bales and L. P. Vernon, *in* "Bacterial Photosynthesis" (H. Gest, A. San Pietro, and L. P. Vernon, eds.), p. 269. Antioch Press, Yellow Springs, Ohio, 1963.
(*128*) W. S. Zaugg, *Proc. Natl. Acad. Sci. U.S.* **50**, 100 (1963).
(*129*) W. S. Zaugg, L. P. Vernon, and A. Tirpack, *Proc. Natl. Acad. Sci. U.S.* **51**, 232 (1964).
(*130*) R. K. Clayton, *Photochem. Photobiol.* **1**, 201 (1962).
(*131*) W. R. Sistrom and R. K. Clayton, *Biochim. Biophys. Acta* **88**, 61 (1964).
(*132*) K. Tagawa and D. I. Arnon, *Nature* **195**, 537 (1963).
(*133*) D. I. Arnon, *Natl. Acad. Sci.—Natl. Res. Council, Publ.* **1145**, 195 (1963).
(*134*) T. Yamanaka and M. D. Kamen, *Biochem. Biophys. Res. Commun.* **18**, 611 (1965).
(*135*) M. Evans and B. Buchanan, *Proc. Natl. Acad. Sci. U.S.* **53**, 1420 (1965).
(*136*) M. Nozaki, K. Tagawa, and D. I. Arnon, *in* "Bacterial Photosynthesis" (H. Gest, A. San Pietro, and L. P. Vernon, eds.), p. 175. Antioch Press, Yellow Springs, Ohio, 1963.
(*137*) S. L. Hood, *Biochim. Biophys. Acta* **88**, 461 (1964).
(*138*) L. P. Vernon, *in* "Bacterial Photosynthesis" (H. Gest, A. San Pietro, and L. P. Vernon, eds.), p. 235. Antioch Press, Yellow Springs, Ohio, 1963.
(*139*) L. P. Vernon and M. D. Kamen, *Arch. Biochem. Biophys.* **44**, 298 (1953).
(*140*) E. S. Lindstrom, *Plant Physiol.* **37**, 127 (1962).
(*141*) J. W. Newton, *J. Biol. Chem.* **237**, 3282 (1962).
(*142*) O. K. Ash, W. S. Zaugg, and L. P. Vernon, *Acta Chem. Scand.* **15**, 1629 (1961).
(*143*) M. L. Ibanez and E. S. Lindstrom, *Biochem. Biophys. Res. Commun.* **1**, 224 (1960).
(*144*) L. P. Vernon, *Arch. Biochem. Biophys.* **43**, 492 (1953).
(*145*) L. P. Vernon and M. D. Kamen, *J. Biol. Chem.* **211**, 643 (1954).
(*146*) L. N. M. Duysens, *Nature* **173**, 692 (1954).
(*147*) R. G. Bartsch, *in* "Bacterial Photosynthesis" (H. Gest, A. San Pietro, and L. P. Vernon, eds.), p. 475. Antioch Press, Yellow Springs, Ohio, 1963.
(*148*) B. Chance and L. Smith, *Nature* **175**, 803 (1955).
(*149*) L. Smith and J. Ramirez, *Brookhaven Symp. Biol.* **11**, 310 (1958).
(*150*) J. M. Olson and B. Chance, *Arch. Biochem. Biophys.* **88**, 26 (1960).
(*151*) J. M. Olson and B. Chance, *Arch. Biochem. Biophys.* **88**, 40 (1960).
(*152*) R. K. Clayton, *Photochem. Photobiol.* **1**, 313 (1962).
(*153*) J. M. Olson and C. Sybesma, *in* "Bacterial Photosynthesis" (H. Gest, A. San

Pietro, and L. P. Vernon, eds.), p. 413. Antioch Press, Yellow Springs, Ohio, 1963.

(*154*) J. M. Olson and S. Morita, *in* "Bacterial Photosynthesis" (H. Gest, A. San Pietro, and L. P. Vernon, eds.), p. 433. Antioch Press, Yellow Springs, Ohio, 1963.

(*155*) J. M. Olson, *Science* **135**, 101 (1962).

(*156*) J. Amesz, *Biochim. Biophys. Acta* **66**, 22 (1963).

(*157*) B. Chance and M. Nishimura, *Proc. Natl. Acad. Sci. U.S.* **46**, 19 (1960).

(*158*) W. J. Vredenberg, Ph.D. Thesis, University of Utrecht (1965).

(*159*) J. W. Newton, *J. Biol. Chem.* **239**, 3038 (1964).

(*160*) L. P. Vernon, *Acta Chem. Scand.* **15**, 1639 (1961).

(*160a*) W. S. Zaugg, unpublished experiments (1965).

(*161*) S. B. Broyde and S. S. Brody, *Biochem. Biophys. Res. Commun.* **19**, 444 (1965).

(*162*) L. P. Vernon, W. S. Zaugg, and E. R. Shaw, *Natl. Acad. Sci.—Natl. Res. Council, Publ.* **1145**, 509 (1963).

(*163*) B. Ke, L. P. Vernon, and E. R. Shaw, *Biochemistry* **4**, 137 (1965).

(*164*) G. Tollin and G. Green, *Biochim. Biophys. Acta* **60**, 524 (1962).

(*165*) G. Tollin and G. Green, *Biochim. Biophys. Acta* **66**, 308 (1963).

(*166*) D. C. Fork and W. Urbach, *Proc. Natl. Acad. Sci. U.S.* **53**, 1307 (1965).

(*167*) H. T. Witt, A. Müller, and B. Rumberg, *in* "La Photosynthese," p. 43. C.N.R.S., Paris, 1963.

(*168*) B. Kok and W. Gott, *Plant Physiol.* **35**, 802 (1960).

(*169*) B. Chance, A. San Pietro, M. Avron, and W. W. Hildreth, *in* "Non-Heme Iron Proteins" (A. San Pietro, ed.), p. 225. Antioch Press, Yellow Springs, Ohio, 1965.

(*170*) L. P. Vernon, A. San Pietro, and D. Limbach, *Arch. Biochem. Biophys.* **109**, 92 (1965).

(*171*) T. Horio and C. P. S. Taylor, *J. Biol. Chem.* **240**, 1772 (1965).

(*172*) T. Beugling and L. N. M. Duysens, *in* "Currents in Photosynthesis" (G. B. Thomas and G. C. Goedheer, eds.). *Donker*, Rotterdam, 1965.

(*173*) B. Chance and D. De Vault, *Ber. Bunsenges. Physik. Chem.* **68**, 722 (1964).

(*174*) L. P. Vernon and E. R. Shaw, unpublished data (1965).

(*175*) A. San Pietro, L. P. Vernon, and D. Limbach, *Natl. Acad. Sci.—Natl. Res. Council, Publ.* **1145**, 504 (1963).

(*176*) S. Katoh, I. Shiratori, and A. Takamiya, *J. Biochem. (Tokyo)* **51**, 32 (1962).

(*177*) F. L. Crane and R. A. Dilley, *Methods Biochem. Anal.* **11**, 279 (1963).

(*178*) L. N. M. Duysens, Thesis, University of Utrecht (1952).

(*179*) L. P. Vernon, *Plant Cell Physiol. (Tokyo)* **4**, 309 (1963).

(*180*) R. G. W. Norrish and G. Porter, *Nature* **164**, 658 (1949).

(*181*) R. Livingston and V. A. Ryan, *J. Am. Chem. Soc.* **75**, 2176 (1953).

(*182*) B. Kok, *Acta Botan. Neerl.* **6**, 316 (1957).

(*183*) H. T. Witt, R. Moraw, and A. Müller, *Z. Physik Chem. (Frankfurt)* [N.S.] **20**, 193 (1959).

(*184*) B. Kok, *Plant Physiol.* **34**, 184 (1959).

(*185*) B. Ke, R. W. Treharne, and C. McKibben, *Rev. Sci. Instr.* **35**, 296 (1964).

(*186*) H. Rüppel, V. Büttemann, and H. T. Witt, *Ber. Bunsenges. Physik. Chem.* **68**, 340 and 727 (1964).

—19—

Physical Processes Involving Chlorophylls in Vivo

RODERICK K. CLAYTON*

Charles F. Kettering Research Laboratory, Yellow Springs, Ohio

Photosynthetic tissues contain aggregates of pigment molecules (chlorophylls, carotenoids, and phycobilins) that absorb light quanta and consequently deliver energy to sites where the photochemistry of photosynthesis is initiated. This chapter will deal with mechanisms for the transfer of energy in the pigment aggregate and the trapping of this energy at the photochemical reaction centers. The characteristics of light re-emitted by chlorophylls *in vivo* will be examined in detail because this light indicates the fate of excitation energy absorbed by photosynthetic tissues.

* Contribution No. 203 from the Charles F. Kettering Research Laboratory, Yellow Springs, Ohio.

I. Photosynthetic Units: Light-Harvesting Pigments and Photochemical Reaction Centers

A. Photosynthetic Units and Reaction Centers

Emerson and Arnold (*1*), studying photosynthetic oxygen evolution in the green alga *Chlorella* exposed to brief flashes of light, concluded that many molecules of chlorophyll (Chl) cooperate in harvesting the energy of light quanta and delivering this energy to a smaller number of reaction sites. Their conclusion was based on two observations. First, the maximum yield of oxygen evolution in response to a single flash of light amounted approximately to one O_2 molecule for every 2500 Chl molecules in the algae. Second, the quantum requirement in these experiments was no greater than about 8 quanta per molecule of O_2 evolved. Eight quanta absorbed anywhere in an ensemble of 2500 Chl molecules could therefore bring about the evolution of one O_2 molecule. The necessary conclusion was that energy converges from many molecules of light-harvesting pigment to a small number of reaction sites.

This conclusion was reached in a different way by Gaffron and Wohl (*2*). These investigators computed that in a suspension of *Chlorella* cells exposed to dim light a single Chl molecule would require about an hour to absorb the 8 quanta needed for the evolution of one molecule of O_2 (and the concomitant assimilation of one molecule of CO_2 into carbohydrate). It had been observed that the maximum rates of O_2 evolution and CO_2 assimilation become established within a few seconds after the start of illumination, even in a thoroughly dark-adapted suspension of *Chlorella* cells exposed to dim light. Again it was clear that many Chl molecules must cooperate in harvesting light energy to initiate the photochemistry of photosynthesis.

At the time of these experiments van Niel (*3, 4*) was formulating a view of the primary photochemistry of photosynthesis that is now thoroughly substantiated: the primary photochemical act is one which produces a separation of oxidizing and reducing entities. This conception, coupled with the foregoing conclusions about the convergence of energy from Chl to reaction sites, has laid the basis for contemporary thought about physical mechanisms in photosynthesis. Photosynthetic tissues contain aggregates of Chl and other light-harvesting pigments. Energy absorbed by these pigments is delivered to reaction centers where photochemical oxido-reductions take place. The primary oxidizing and reducing entities then initiate the metabolic processes of photosynthesis.

A single reaction center, plus the amount of light-harvesting pigment associated with it, is called a photosynthetic unit. There is as yet little evidence to indicate whether the photosynthetic units are distinct morphological entities or whether an extended aggregate of pigment is studded with many reaction centers.

With van Niel's formulation of the primary photochemical act, it was hoped that the quantum efficiency of photosynthesis could have concrete significance for the photochemical mechanism. The evolution of O_2 from water and the reduction of CO_2 to the level of sugar are processes that require the transfer of four reducing equivalents (hydrogen atoms or electrons) per molecule. If the primary photoproducts carry one oxidizing and one reducing equivalent respectively, the photochemical act must occur four times to afford the evolution of one molecule of O_2 and the assimilation of one molecule of CO_2. The established quantum requirement of $8\,hv/O_2$ then implied that the primary photochemical act involves the cooperative action of two quanta, and physical mechanisms for this kind of cooperation were entertained by Franck (5) and others.

In the 1950's it became clear that light has two qualitatively different effects in the photosynthesis of algae and higher plants. Emerson and collaborators (6, 7) found that in *Chlorella* and other algae the efficiency of photosynthesis is abnormally low under light, absorbed by Chl *a*, of wavelength greater than about 680 mμ. The efficiency under far-red light could be enhanced to a "normal" level (about $8\,hv/O_2$) by superimposing light of shorter wavelengths. Myers and French (8) then observed that this enhancement occurred even when the two qualities of light were presented alternately, separated by dark intervals of several seconds. At least one of the two light effects is therefore stable for a time much longer than the lifetimes of excited states in Chl.

This background lent great interest to the finding of Duysens *et al.* (9) that a cytochrome (Cyt) in the red alga *Porphyridium* becomes oxidized under far-red light and is then partly reduced when shorter-wave light is superimposed.

These observations were supplemented by many other spectrophotometric observations, especially those of Witt and Kok and their collaborators* (10–13). Kok's chief contribution was the discovery and characterization of P700, a Chl-like pigment in green plants, that becomes oxidized and thus bleached reversibly by illumination. The oxidation is favored by far-red light, and the return to the reduced (unbleached) form is accelerated by shorter-wave illumination.

All of the foregoing observations are consistent with a formulation of photosynthesis in green plants that has become accepted almost uni-

---

* An adequate annotation and description of research in the differential spectrophotometry of photosynthetic tissues would be beyond the scope of this chapter.

versally as a working hypothesis. Photosynthesis requires the cooperation of two distinct photochemical systems, linked in series through a chain of electron carriers. One system (system I) produces the strong reductant needed for the assimilation of CO_2, and a weak oxidant. The other system (system II) produces a strong oxidant (the precursor of O_2) and a weak reductant. The weak oxidant of system I and the weak reductant of sys-

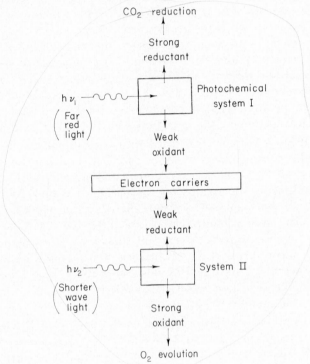

FIG. 1. An outline showing the interaction of two photochemical systems in green plant photosynthesis. This figure also appears in Clayton (12a).

tem II interact through the electron carriers linking the two systems; this interaction is perhaps coupled with formation of adenosine triphosphate. Shorter-wave light can drive both of these systems, but far-red light is effective for system I alone. This model (9, 11, 12) for photosynthesis in green plants is outlined in Fig. 1.

In this framework the quantum requirement of 8 hv/O_2 can be understood simply on the basis that the primary photochemistry in each system involves the transfer of one electron at the expense of one quantum. Eight quanta, four in each system, are needed in order to transfer four reducing equivalents from the level of strong oxidant to that of strong reductant.

There is no need, then, to invoke a concerted two-quantum reaction at a single photochemical site.

The photosynthetic bacteria do not evolve oxygen, and their growth depends on the presence of oxidizable substances such as organic acids, H_2S, or H_2. There is no convincing evidence that bacterial photosynthesis involves the cooperation of two photochemical systems. These bacteria appear to have a single photochemical system analogous to system I, with the obligatory oxidizable substrate taking the place of the weak reductant formed in green plants by system II.

Returning to the idea of a photosynthetic unit, the size of the unit can be computed by analyzing the observations (such as those of Emerson and Arnold) on the yield of the photosynthetic products in flashing light. The stoichiometry of 2500 Chl molecules per O_2 molecule corresponds, in view of the quantum requirement, to about 300 Chl molecules per quantum absorbed in a single flash of light. Then if the photochemical act at a reaction center is driven by the energy of one quantum, that reaction center is associated with 300 molecules of light-harvesting Chl. This simple approach is confused by the evidence that photosynthesis involves two photochemical systems, each with its own kind of reaction center and its own set of light-harvesting pigments. Even so there is abundant evidence that each photochemical system possesses an antenna of many light-harvesting molecules (chlorophylls and accessory pigments) for each reaction center. Dichlorodimethylurea, which represses the operation of system II, is effective at a concentration of one molecule for every 200 Chl molecules (14). Oxygen evolution, a function of system II, diminishes sharply in fragments of spinach chloroplasts when these fragments are made so small as to contain fewer than about 150 Chl molecules (15). Constituents of system I associated with photochemical reaction centers, such as Cyt and P700, are present to the extent of one molecule for approximately 300 Chl molecules. The light-induced oxidation of P700, driven by energy absorbed by the Chl, occurs with a quantum efficiency close to 100% (13).

In purple photosynthetic bacteria the size of a photosynthetic unit can be estimated, from experiments with flashing light (4, 16), to be about 50–100 molecules of bacteriochlorophyll (Bchl) for every reaction center (assuming that the photochemical act is a single electron transfer driven by the energy of one quantum). This result is supported by the molecular ratio of Bchl to minor components that react photochemically with high quantum efficiency. One of these components is P870, a pigment analogous to P700 of green plants, present in *Rhodopseudomonas spheroides* to the extent of one molecule for about 50 Bchl molecules (17).

The morphologies of photosynthetic tissues are at least consistent with photosynthetic units of the foregoing sizes. In plant chloroplasts the smallest structures that have been resolved with the electron microscope are the quantasomes, small bumps on the surfaces of lamellae. These bumps have a diameter of about 200 Å, enough to accommodate a few hundred Chl molecules (*18*). In photosynthetic bacteria the pigment and all the photochemical activity is associated with chromatophores, subcellular particles that probably arise from invaginations of the cytoplasmic membrane. These chromatophores are large enough to accommodate several hundred molecules of Bchl.

B. Chlorophylls and Accessory Pigments

In purple photosynthetic bacteria the major pigments are Bchl and carotenoids. Energy absorbed by the carotenoids can be transferred, with efficiencies ranging from about 30 to 70% (*19, 20*), to Bchl and on to the photosynthetic reaction centers. A more important function of the carotenoids appears to be their ability to protect the bacteria from the harmful effects of photooxidations sensitized by Bchl. Mutants lacking colored carotenoids are killed, and their Bchl is destroyed, in the presence of oxygen and light absorbed by Bchl (*21*). In such mutants there is a tendency for the Bchl to lose its Mg atoms and thus to become converted to bacteriopheophytin during prolonged incubation in the light (*17*). Excitation spectra show that light absorbed by this pheophytin can promote the photochemistry of photosynthesis.

In green photosynthetic bacteria the major pigment is chlorobium chlorophyll (Cchl). Two types of this pigment have been discovered, one absorbing maximally at 725 mμ and the other at 747 mμ *in vivo*. In addition these bacteria contain a trace of Bchl absorbing at 810 mμ (*22*). These pigments, together with carotenoids, make up the light-harvesting system of the green bacteria. Despite its low abundance, the trace of Bchl (one molecule for about 70 of Cchl) does not appear to be the ultimate repository for excitation energy. There is evidence (see later) that a pigment absorbing at 840 mμ is the terminal energy sink at a reaction center (*23*).

The main pigments of green plants and algae are Chl *a*, Chl *b*, carotenoids, and phycobilins. The Chl *a* exists in at least two forms (two states of aggregation or of interaction with other substances) having red absorption maxima at 672 and 683 mμ, respectively. These forms have been called Chl *a* 670 and Chl *a* 680. The association of these pigments with photochemical system I or II has been elucidated through excitation spectra for the enhancement phenomenon and for partial reactions exhibited by one or the other system (*8, 24–26*). System II is sensitized by phycobilins, by Chl *b*, and by Chl *a* 670. The light-harvesting apparatus for system I consists of Chl *a* 680 and carotenoids.

C. Specialized Pigments at Reaction Centers

The universal presence of chlorophylls in photosynthetic organisms indicates that Chl in one form or another serves a specific photochemical function in addition to the light-harvesting function that is shared with the carotenoids and phycobilins. The presence of specialized, photochemically active chlorophylls was inferred, by Duysens (27) and Kok (28) and their collaborators among others, from reversible light-induced absorbancy changes in photosynthetic tissues. The changes represent a bleaching at or near the principal long-wave absorption band of Chl or Bchl. Their magnitudes are small, as though one Chl molecule out of about 300, or one Bchl molecule out of 50, were altered. From these observations it could not be decided whether the entire set of Chl or Bchl molecules is altered slightly or a specific minor component suffers a gross alteration. The latter conclusion has been supported through techniques for removing the major Chl or Bchl component without destroying the minor, photochemically active component. The existence of photochemical reaction centers, containing specialized (photochemically active) chlorophylls, is thereby demonstrated (17, 29).

In spinach chloroplasts Kok succeeded in extracting (with mixtures of acetone and water) about 85% of the light-harvesting Chl *a* while leaving intact the component responsible for the light-induced bleaching (29). The ratio of light-induced bleaching to total absorbancy by Chl *a* was thus improved about sevenfold. The bleachable component was called P700; the maximum of bleaching is at 705 mμ in contrast to the major components of Chl *a*, which absorb maximally at 672 and 683 mμ. A light-induced bleaching at 432 mμ, accompanying the bleaching at 705 mμ, indicates that P700 has a Soret band coincident with that of Chl *a*. It is likely, therefore, that P700 is nothing but Chl *a* in a specialized environment that endows it with photochemical reactivity. The bleaching of P700 can be effected by chemical oxidants (e.g., ferricyanide) as well as by light; the oxidation has been shown to involve transfer of a single electron at a potential of 430 mv (29). The response of P700 to far-red and shorter-wave light, and the effect of dichlorodimethylurea on the light response, show that P700 is associated with system I. The light-induced oxidation of P700 and a concomitant reduction of pyridine nucleotide can proceed with quantum efficiencies greater than 80% (13, 30). The light reaction of P700 occurs, with a reversible and an irreversible component, at 77°K.

From these properties it appears likely that P700 engages in the primary photochemical act of system I and that oxidized P700 is the primary oxidizing entity (the weak oxidant made by system I). Because the absorption maximum of P700 (at 705 mμ) is at a greater wavelength than that of the light-harvesting Chl *a* 680 it was natural to suggest that P700 acts, at the reaction center of system I, as a sink for singlet excitation energy absorbed by the major pigment. Excited P700 then transfers an electron to an unspecified electron acceptor which becomes the primary strong reductant. The reducing power is eventually stabilized in the form of reduced pyridine nucleotide. Oxidized P700 is restored to its reduced form by the flow of electrons from system II.

Photosynthetic bacteria contain pigments that behave like counterparts of P700; they are called P870, P890, etc. according to the wavelength at which their reversible bleaching is maximal. A complete separation can be effected between these pigments and the major Bchl component (*31*). If chromatophores from blue-green (carotenoidless) mutant *Rhodopseudomonas spheroides* are illuminated in the presence of O_2 and a detergent such as Triton X-100, the light harvesting Bchl disappears (apparently as a result of photooxidation) but the P870, as measured by the reversible light-induced bleaching, survives the treatment. The resulting preparation has an absorption band at 870 mμ that is bleached completely and reversibly by light. Absorption spectra showing the effect of this treatment are shown in Fig. 2. The amount of reversible bleaching, about 0.2 unit of optical density, was the same before photooxidation as after. In both Figs. 2A and 2B the band at 760 mμ is due to a trace of bacteriopheophytin. The band at 800 mμ in Fig. 2B shows a reversible light-induced shift to shorter wavelengths; this band accompanies the main absorption band of P870. The 800-mμ band might be part of the spectrum of P870, or it might be another pigment associated closely with P870.

The material giving the spectrum of Fig. 2B can be precipitated with ammonium sulfate and then extracted with methanol. The extracted residue, when redispersed in water, shows no absorption bands in the near-infrared. The extract shows only the spectrum of Bchl plus a trace of bacteriopheophytin. It is therefore probable that both P870 and the substance absorbing at 800 mμ are forms of Bchl, endowed with special properties by their environment *in vivo*.

Cells and chromatophores of photosynthetic bacteria exhibit several light reactions associated with their photochemistry. First there are the light-induced absorbancy changes signifying oxidation of P870 (or P890, etc.), oxidation of one or more cytochromes, and reduction of ubiquinone (*20, 32–34*). In addition there is a light-induced electron spin resonance signal (35). Chromatophores can also catalyze photochemical electron transfer involving externally added reagents such as mammalian Cyt *c*, phenazine methosulfate, and quinones (36, 37). All these light reactions are exhibited by preparations of *R. spheroides* of the kind shown in Fig. 2B (38). The photochemistry associated with photosynthesis can therefore be accomplished by a system containing P870 without the help of the major Bchl component. On the other hand, the absence of P870 is attended by a complete failure of photochemical activity. Sistrom *et al.* (39) have isolated a mutant of *R. spheroides* that cannot grow photosynthetically even though the cells (grown aerobically in darkness)

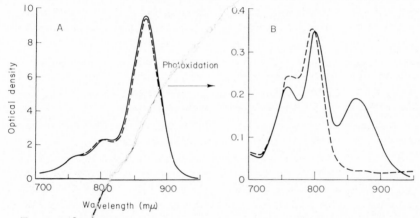

FIG. 2.   Absorption spectra of a suspension of chromatophores from carotenoidless mutant *Rhodopseudomonas spheroides*, before (A) and after (B) photooxidation in the presence of a detergent (Triton X-100, 1% v/v). The difference between the broken (light) and solid (dark) curves shows the reversible light-induced bleaching of P870 and blue-shift of a band at 800 mµ. Note the different scales of the ordinate for curves (A) and (B). This figure also appears in Clayton (12a).

possess a normal complement of light harvesting Bchl. This mutant lacks P870 (40) and also lacks the associated component absorbing at 800 mµ (40a). Cells and chromatophores of the mutant show none of the foregoing light reactions; their Bchl serves no useful photochemical function (38, 40).

In green photosynthetic bacteria the presence of a "P840" pigment has been inferred from a reversible light-induced bleaching centered at 840 mµ (23). This pigment cannot be seen as a band in absorption spectra of the bacteria because of the great preponderance of Cchl.

All photosynthetic bacteria exhibit light-induced oxidation of one or more cytochromes. In cells under anaerobic conditions the oxidation of Cyt proceeds in dim light with a quantum efficiency close to 100% (41);

under these conditions the appearance of oxidized P870 or P890 is suppressed. Light-induced Cyt oxidation can be observed in *Chromatium* at
77°K; the reaction is irreversible at this temperature (*42*) [the photochemical oxidation of P870 occurs reversibly at temperatures as low as
1°K (*43*)].

There are two major alternatives that can account for these and other
observations. The primary photochemical reaction for photosynthesis
might be one in which P870 is oxidized and an electron acceptor is reduced. In that case the Cyt, coupled closely with P870, donates an electron to oxidized P870 and becomes oxidized in turn. On the other hand,
the oxidation of Cyt might be part of the primary light reaction, coupled
with reduction of an unspecified electron acceptor (the acceptor might
even be P870). In that case the oxidation of P870 is an aberrant event
that occurs only when the Cyt is already oxidized and hence unable to
participate in the normal photochemical act of photosynthesis. Both these
positions find experimental support (*44, 45*), and neither can be ruled
out by the available evidence. Whether or not the oxidation of P870 is
part of the normal photochemistry of photosynthesis, the importance of
this pigment for photosynthesis is shown by the photochemical failure
attending its absence in Sistrom's mutant of *R. spheroides.*

The reservations regarding the significance of the oxidation of P870
can be applied also to the oxidation of P700 in green plants. An interesting possibility that cannot be discounted is the following: P870 (or P700)
serves in a photochemical reaction center as a trap for excitation energy
and as an initiator of a photochemical act in which Cyt is oxidized. The
primary reducing entity may then be either P870 itself or a closely coupled electron acceptor. If the Cyt is already oxidized, the photochemistry
is different: P870 then becomes oxidized and an acceptor becomes reduced. This reaction acts as a safety valve (*46*) for dissipating excess
excitation energy and thus preventing deleterious photooxidations.

II. Mechanisms for the Transfer and Utilization of Energy Absorbed in Photosynthetic Tissues

A. Possible Fates of Primary Singlet Excitation Energy

The primary absorption of light in photosynthetic tissues raises the
light-harvesting pigments to singlet excited states. Eventually this energy
reaches photochemical reaction centers that contain specialized forms of
Chl or Bchl (P700, P870, etc.). To consider this process of energy transfer let us begin by examining various possible fates of the singlet excitation energy produced in the initial act of light absorption.

First it is clear that higher singlet states, such as the one responsible for the Soret absorption band of Chl, are degraded rapidly to the lowest excited singlet state as exemplified by the long-wave band of Chl. This degradation ("internal conversion") proceeds through vibrational substates; its efficiency is shown by the absence of fluorescence from all but the lowest excited singlet state. In consequence of this process, quanta absorbed in the Soret band of Chl are no more effective for photochemistry than the quanta of lower energy absorbed in the long-wave band.

Energy in the lowest excited singlet state can be dissipated in a transition to the ground state. The transition may be direct and accompanied by emission of a light quantum (fluorescence), or it may proceed through substates generated by molecular interactions (vibrations, collisions, etc.). In the latter case the transition is radiationless.

A molecule in a singlet state may undergo a transition to a metastable state involving electron transfer or spin conversion. In this way the singlet energy may give rise to charge transfer states, ionized states, or triplet states. Processes of this kind could be involved in the transfer and utilization of energy for photosynthesis. Alternatively such processes, occurring in the light-harvesting pigment aggregate, could be parasitic (preventing the flow of energy to reaction centers) and could lead to injurious photochemical reactions that have nothing to do with photosynthesis.

If the foregoing events do not supervene, the singlet energy may be transferred as such from one molecule to another, or delocalized in an ensemble of similar molecules. This transfer or delocalization depends on interactions between electric dipole moments of two or more molecules. The dipole moments arise from the redistribution of charge that accompanies an electronic transition between the ground state and the excited state.

These considerations show that several mechanisms can be entertained for the transfer of energy in the light harvesting aggregate and its utilization at photochemical reaction centers. The various possibilities will be evaluated in the ensuing pages.

B. Energy Transfer from Accessory Pigments to Chlorophylls

Engelmann first showed in 1884 (47) that light absorbed by accessory pigments can promote photosynthesis. In recent years it has been confirmed repeatedly (19, 48, 49) that carotenoids and phycobilins are able to sensitize photosynthesis. In every case that has been studied, the singlet energy absorbed by the accessory pigment is transferred as such to one of the chlorophylls. This is shown by the fact that excitation spectra for photosynthesis, featuring peaks due to accessory pigments as well as to Chl, have the same form as corresponding excitation spectra for fluorescence of the Chl (*19*). The efficiency of energy transfer from carotenoids to Chl or Bchl is generally about 30–70%; the transfer from phycobilins to Chl and from Chl *b* to Chl *a* has an efficiency approaching 100% (*19, 20*).

The transfer of singlet excitation energy between dissimilar molecules has been treated theoretically by Förster (*50, 51*). The efficiency of transfer is proportional to the square of the dipole interaction energy, and hence to the sixth power (approximately) of the distance separating the molecules. The weakness of the interaction allows thermal relaxation in the vibrational substates of the donor molecule before a transfer is effected. The donor thus reaches a state, before transfer, similar to the state from which fluorescence would occur. Because of this the efficiency of transfer depends on the amount of overlap between the fluorescence band (corresponding to the lowest excited singlet state) of the donor and the absorption band of the acceptor. When the excitation donor is phycocyanin and the acceptor is Chl *a* or Chl *b* this overlap is large and the transfer is highly efficient. With carotenoids as donors and chlorophylls as acceptors the overlap is less. The efficiency of transfer should depend on temperature to the degree that the overlap between donor fluorescence and acceptor absorption is influenced by the temperature; this point has not been tested experimentally.

C. Energy Transfer in the Major Chlorophyll Aggregate

Singlet energy absorbed by Chl or Bchl promotes the oxidation of P700 or P870, and of cytochromes, with quantum efficiencies of 50–100%. Transfer to the latter substances is therefore the principal fate of light energy absorbed by photosynthetic organisms. A mechanism based on the diffusion of atoms, radicals, or molecules can be ruled out because the transfer is effected within a few microseconds at most (*45*) and because the light reaction of P870 continues to occur with high quantum efficiency at 1°K (*52*). The diffusion of electrons and holes, following local photoionization in the major pigment aggregate, is also an unlikely mechanism for this energy transfer. Arnold and others (*43, 53*) have obtained evidence that photoionization occurs in photosynthetic tissues, but there is no indication that the efficiency of this process in the major pigment aggregate is more than about 0.1%. Arnold himself now suggests (*54*) that the primary flow of energy to reaction centers occurs in the singlet excited state, with photoionization at the reaction centers playing a role in the ensuing photochemical processes.

We are left with the possibility that energy is transferred to the reaction centers in the form of excitation quanta. These quanta could be singlet or triplet; let us consider first whether the triplet excited state is involved in photosynthetic energy transfer.

Robinson, McGlynn, and others (55, 56) have shown that triplet excitons can be delocalized in molecular crystals at least as extensively as singlet excitons. In the triplet state the much weaker interaction between molecules is more than offset by the greater lifetime of the excited state (milliseconds to seconds, as contrasted with about 10^{-8} sec for the singlet excited state). Quanta of triplet excitation in the Chl of photosynthetic units could in principle migrate extensively during their lifetime and become trapped efficiently at reaction centers. There is indirect evidence, however, that the transfer of energy in photosynthesis does not proceed by way of the triplet excited state. Fluorescence of Chl and Bchl *in vivo* shows variations in its intensity related to photochemical activities at reaction centers. Typically the fluorescence yield rises as the reaction centers become saturated with incoming energy during the onset of photosynthesis. At least in the purple photosynthetic bacteria (see later), this fluorescence is due to the decay of primary singlet excitation in the light-harvesting Bchl. Such variations in the fluorescence should not be expected if the primary singlet energy is converted locally (in the light-harvesting aggregate) to triplet and then transferred to reaction centers. Saturation of the reaction centers might lead reflexively to an increase in the population of triplet states in the light-harvesting aggregate, but this should have little effect on the primary singlet state lifetime which determines the yield of fluorescence. At any rate the fraction of light-harvesting pigment in the triplet state is inappreciable even under intense illumination. Absorption spectra corresponding to the triplet state of Chl and Bchl are detected easily in illuminated solutions of these pigments (57, 58), but efforts to see the same thing *in vivo* have been negative (59, 59a).

There is evidence that a small fraction of the primary singlet energy is converted to the triplet state in the light-harvesting Bchl aggregate of purple bacteria, and that this triplet energy can sensitize harmful photooxidative reactions. Carotenoidless mutants of purple bacteria are killed, and their Bchl is destroyed, through exposure to light in the presence of oxygen (21). These processes occur in carotenoidless forms of Sistrom's nonphotosynthetic mutant of *Rhodopseudomonas spheroides;* indeed, the photooxidative self-destruction of Bchl is the only photochemical activity displayed in these organisms (40). These harmful reactions are suppressed in phenotypes that contain colored carotenoids in association with Bchl in the light-harvesting aggregate. The protection afforded by various carotenoids can be seen in a sequence of "carotenoid mutants" of *R. spheroides* (*60*); forms that contain carotenoids having nine or more conjugated double bonds are fully protected. The protection offered by ζ-carotene, with seven conjugated double bonds, is questionable, and more fully saturated compounds (phytofluene and phytoene) give no protection. Exactly the same pattern of activity has been reported by Claes (*61*) for the ability of carotenoids to interfere with photooxidations sensitized by Chl *in vitro*. These results were interpreted to mean that the triplet state of Chl is an intermediate in the photochemistry, and that carotenoids with nine or more conjugated double bonds quench the triplet state effectively enough to prevent its photochemical utilization. These findings suggest that only a little triplet excitation is deposited in the light-harvesting system *in vivo*, and this small amount is deleterious.

There remains one reasonable mechanism for photosynthetic energy transfer, the migration of quanta of singlet excitation energy. This process can be described in two distinct ways. In the treatment formulated by Perrin and extended by Förster, an excitation quantum jumps from one molecule to another through a resonant interaction between electric dipole oscillations in the two molecules. By successive jumps the quantum executes a random walk in the molecular aggregate; at any instant the energy is conceived as being localized in one molecule. This is a satisfactory description for transfer between pairs of dissimilar molecules (see earlier in connection with accessory pigments) and for transfer between similar molecules in the limit of very weak coupling, as in a dilute solution. With stronger coupling, as in a molecular crystal, the uncertainty principle dictates that the excitation must be regarded as a property of the entire set of molecules. The excitation quantum is described as a delocalized exciton (*62*). Any description in which the quantum is concentrated in one part of the ensemble must then be supported by an observable localizing action such as transfer to a molecule that has special properties.

The dipole coupling between Chl or Bchl molecules in a photosynthetic unit is strong enough (*63*) to warrant the requirement of a delocalized treatment of singlet energy transfer. The conclusions drawn from such a treatment can be translated into a language of localization, and one can then speak of a random walk, a frequency or rate of jumping of the quantum, and so forth. In this way the delocalized treatment can be blended with the localized description that arises when one considers trapping of a quantum at a reaction center.

Bay and Pearlstein (*64*) have made a careful computation of the rate of singlet energy migration in the Chl of a photosynthetic unit, using a delocalized formulation. They estimated the strength of dipole interaction from morphological evidence (*65*) as to the density of Chl molecules in the light-harvesting aggregate. The reaction center was assumed to act as an irreversible sink for excitation quanta. The computation led to the conclusion that energy absorbed by the light-harvesting Chl could be trapped at reaction centers with efficiency greater than 97%. (Some details of this computation will be considered later in connection with the yield and lifetime of Chl fluorescence *in vivo*.) The delocalization of singlet excitation energy is therefore an acceptable mechanism for the operation of a photosynthetic unit.

D. Trapping of Energy at Reaction Centers

We shall now consider ways in which singlet excitation energy in photosynthetic units can become localized at photochemical reaction centers. It will be assumed that the localization is mediated by a specialized Chl or Bchl. The specialized pigments could be P700, P870, etc., but our considerations will not depend on this specific identification.

Because the singlet excited state of P700 is at a lower energy than that of the light harvesting Chl *a* 680 (the difference is 0.04 ev or 1300 cal/mole), it was natural to imagine that P700 acts as a sink, and therefore a trap, for singlet energy. In this view the trapping act is a localizing transfer from Chl *a* 680 to P700 by the Perrin-Förster mechanism. On closer examination it becomes clear that a trapping by this mechanism is by no means irreversible; the result is simply a concentration of singlet energy in P700 relative to Chl *a* 680. The degree of concentration can be estimated by comparing two overlap integrals: the overlap between the fluorescence band of Chl *a* 680 and the absorption band of P700, which governs the rate of transfer into P700, and the overlap between P700 fluorescence and Chl *a* 680 absorption, which governs the flow of energy from P700 back into the Chl *a* 680 aggregate. From reasonable assumptions about the shapes of fluorescence bands it can be estimated that the two overlap integrals differ by a factor of about five; this then is the factor by which energy could be concentrated in P700, relative to the energy density in the major Chl aggregate.

In purple bacteria there is no basis whatever for assuming that the singlet states of P870, P890, etc., can act as sinks for singlet energy in the Bchl aggregate. The absorption bands of these pigments coincide with the long-wave absorption band of the major Bchl component in *Rhodopseudomonas spheroides*, *Rhodospirillum rubrum*, and *Chromatium*. The specialized pigments can therefore be expected to share delocalized singlet energy with the major pigment, to a degree dictated simply by the ratio of minor to major pigments. In one purple bacterium, Eimhjellen's *Rhodopseudomonas* sp. NHTC 133, the specialized pigment

(P985) actually has its absorption maximum on the short-wave side of the main absorption band at 1012 mμ (66). Here the partition of energy density between Bchl and P985 favors the former by a factor of about three on the basis of overlap integrals. The ratio of Bchl to P985 appears (from absorption and difference spectra) to be 100:1, but with the added factor of three it is as though the ratio were 300:1.

Having discounted the singlet excited state of a specialized Chl as an effective energy trap, we must invoke an efficient transition from the singlet state to a metastable state at the reaction center. The metastable state might be a triplet state of the specialized Chl. Alternatively it might be a charge transfer state involving a complex between the specialized Chl and a neighboring electron acceptor or donor. Neither of these possibilities has experimental support; both are compatible with what is known about the photochemistry of Chl *in vitro*.

In a model of this kind the environment of the specialized Chl facilitates a transition from the excited singlet state to a metastable state. By way of example, a transition to the triplet state in the specialized Chl could be facilitated by a neighboring Cyt molecule through the proximity of a heavy atom (Fe) and an extensive π-electron orbital with a shape matching that in the Chl. Alternatively the electron affinity of a molecule adjacent to the specialized Chl might favor the formation of a charge transfer state, as in the phthalocyanin-chloranil system studied by Kearns *et al.* (67). Such a state could lead directly to a separation of oxidized and reduced molecules in their ground states.

It is instructive to examine the foregoing model mathematically in order to see what parameters govern the efficiency of energy trapping. Let us assume that singlet energy delocalized in the major Chl aggregate can be transferred to a specialized Chl, and can then be used in forming a metastable state or else be transferred back into the aggregate. The singlet energy can also be lost through transition to the ground state, in the specialized Chl as well as in the aggregate. The aggregate contains N Chl molecules of which n are nearest neighbors to the specialized Chl (n might be 6 in a three-dimensional array and 4 in a two-dimensional system). The flow of energy is shown in Fig. 3. The density of singlet energy in the aggregate is x, and that in the specialized Chl is y; these quantities can be regarded as probabilities that a Chl molecule is in the excited state. Transfer between the aggregate and the specialized Chl is given first-order rate constants α and β; the rate constants for dissipation and metastable trapping are k and Q, respectively. Excitation quanta are being formed, through absorption of light, at a constant rate I. In formulating the differential equations for energy flow one must be careful to distinguish between the extensive variable, number of excitation quanta, and the intensive one, density or probability of excitation at any one Chl molecule. It must also be recognized that the transfer between the aggregate and the specialized Chl can involve any of the n nearest neighbors in the aggregate. The differential equations are

$$N(dx/dt) = I - kxN - \alpha\,xn + \beta\,yn \qquad (1)$$

and

$$dy/dt = \alpha\,xn - \beta\,yn - ky - Qy \qquad (2)$$

In the steady state the trapping efficiency is the rate of trapping Qy di-

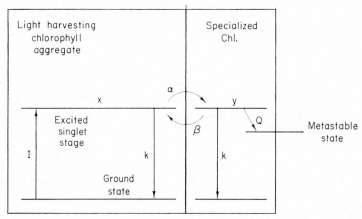

FIG. 3. Energy flow in a model of the photosynthetic unit. The quantities x and y are densities (or probabilities) of singlet excitation energy residing in a Chl molecule. The constants α and β are first-order rate constants for transfer of singlet energy between the major Chl aggregate and the specialized (reaction center) Chl. I is the rate at which quanta are absorbed, and k and Q are rate constants for dissipation of singlet energy and trapping in a metastable state.

vided by the rate of absorption I. The steady-state solution of Eqs. (1) and (2) gives

$$\text{efficiency} = Qy/I = Q\alpha n/[(\alpha n + kN)(k + Q) + kN\beta n] \qquad (3)$$

As a minimum requirement for efficient trapping the entry into a metastable state must predominate over dissipation of the singlet quantum in the specialized Chl; i.e., $Q >> k$. Equation (3) then reduces to

$$\frac{Qy}{I} = \frac{Q}{Q(1 + kN/\alpha n) + (\beta/\alpha)\,kN} \qquad (4)$$

This expression becomes simpler in special cases.

Consider first that metastable trapping is slow compared with the exchange of energy between the aggregate and the specialized Chl: $Q << an$ and $Q << \beta n$. Assume further that transfer from the aggregate to the specialized Chl is much faster than the summed probability of singlet dissipation: $an >> kN$. This assumption is necessary to account for the high efficiency of trapping. The trapping efficiency is then given approximately by

$$\frac{Qy}{I} = \frac{Q}{Q + (\beta/\alpha)\, kN} \tag{5}$$

The quantity β/α can be evaluated from the fluorescence-absorption overlap integrals for transfer in each direction between the Chl aggregate and the specialized Chl. In Eq. (5) it acts as a factor modifying the ability of dissipation (kN) to compete with metastable trapping (Q). Among the known systems of light-harvesting chlorophylls and presumptive traps, the value of β/α ranges from about 0.2 for P700 surrounded by Chl a 680, through unity for P870 and P890, to about 3 for P985 in Eimhjellen's *Rhodopseudomonas*. If the efficiency is high, it is relatively insensitive to the value of β/α. Suppose that $Q = 30\, kN$. Then the efficiency is 99% with β/α equal to 0.2, 97% with $\beta/\alpha = 1$, and 91% with $\beta/\alpha = 3$. It matters little whether the singlet state of the specialized Chl is a little lower or higher in energy than that of the light-harvesting Chl.

As a second case consider that a singlet quantum reaching the specialized Chl is overwhelmingly likely to be trapped in the metastable state and has little chance to be dissipated or to return to the aggregate. In that case $Q >> k$ and $Q >> \beta n$, and Eq. (3) reduces to

$$\frac{Qy}{I} = \frac{1}{1 + kN/an} \tag{6}$$

independent of Q and β. Now the efficiency is high if the summed rate of dissipation kN is small compared with the rate of transfer of singlet energy into the specialized Chl from any of its nearest neighbors (an). The effect of raising or lowering the singlet level of the specialized Chl relative to the aggregate is even less than before because the rate of reverse transfer (expressed by β) is no longer involved.

In summary, the high efficiency that is known to prevail in photosynthetic energy transfer shows that the reaction centers must be equipped to transform singlet excitation energy rapidly into some other form. With an efficient mechanism of this kind it makes little difference whether the singlet state of the specialized Chl is slightly above or below that in the light-harvesting Chl aggregate.

Numerical values for the rates of energy transfer and dissipation have already been explored by Bay and Pearlstein (*64*), who assumed that the trap is irreversible; i.e., that singlet energy reaching the trap never returns to the aggregate. They computed a rate of transfer greater than 10^{12}/sec between neighboring Chl molecules in the aggregate. If the rate of transfer to the specialized Chl is appreciably less than this, the present model of completely delocalized excitation in the aggregate is appropriate. If not, Eq. (6) should be replaced by a more complicated expression that shows rate-limiting diffusion in the aggregate. Let us take the conservative value of 10^{11}/sec for the transfer rate α from aggregate to specialized Chl. Based on the natural lifetime of Chl fluorescence, 15×10^{-9} sec, the constant k is $(1/15) \times 10^9$/sec. Then with $n = 6$ and $N = 300$, $kN/\alpha n$ is 0.03, and the trapping of energy according to Eq. (6) is 97% efficient.

III. The Significance of Light Emitted by Chlorophylls *in Vivo*

A. Delayed Light Emission

Strehler and Arnold, while attempting to measure photosynthetic phosphorylation using the firefly luminescence technique, discovered that illuminated *Chlorella* cells emit a feeble afterglow that persists long after the primary fluorescence of Chl has died out (*68*). This delayed light emission has been observed in all classes of photosynthetic organisms (*68, 69*). Its spectrum is like that of Chl or Bchl fluorescence (*70, 71*), but it can be detected as long as an hour after a flash of exciting light (*72*). It therefore reflects a regeneration of singlet excited states in Chl or Bchl at the expense of metastable energy.

The intensity of delayed light emission is not a simple function of exciting intensity or of time after an exciting flash (*71, 72*). Very roughly the emitted light intensity varies as the square of the exciting intensity over a limited range and becomes saturated at higher exciting intensities. The emitted intensity decays with time after a flash of exciting light; with a sufficiently weak exciting flash (to avoid saturation) the decay is described roughly by the reciprocal of the time from about 10^{-5} sec to 1 hour. The longer-lived components of emission are saturated more easily than the shorter-lived ones. From these characteristics, Arnold (*73*) has concluded that the singlet state giving rise to delayed emission is populated not from a distinct metastable energy level, but from a manifold of states having different energies below the singlet level. He suggested that the metastable energy resides in separated electrons and holes (electron vacancies) in the Chl aggregate. The separated charges, generated from primary singlet energy, recombine and singlet quanta are reconstituted.

The delayed light emitted during 1 msec, 3 msec after an exciting flash, is about 100–1000 times weaker than the primary fluorescence in most green plants (71). Extrapolating the decay curve for delayed emission to shorter times, Arnold and Davidson have suggested (72) that the faster components of delayed light, emitted 10^{-6} sec or less after an exciting flash, may have an intensity comparable to that of the primary fluorescence. Further evidence for this is found in the characteristics of fluorescence and delayed light emitted by green photosynthetic bacteria (see later). In purple bacteria the delayed light is usually about 100 times weaker than in green plants, relative to the fluorescence (71).

An intimate connection between delayed light emission and the photochemistry of photosynthesis is shown by the sensitivity of the delayed light to inhibitors of photosynthesis (74). In green plants the response to light of different wavelengths and the effects of inhibitors indicate that delayed light is associated mainly with photochemical system II (69, 74, 75). The mechanism of delayed light emission, at least in purple bacteria, involves photochemical reaction centers. This is shown by the absence of delayed light emission (76) in Sistrom's mutant of *Rhodopseudomonas spheroides*, which possesses light-harvesting Bchl but lacks functioning reaction centers.

Although reaction centers are implicated in the mechanism of delayed light emission, the light is emitted from Chl or Bchl in the light-harvesting aggregate. In green photosynthetic bacteria light energy absorbed by Cchl at 747 mμ is transferred to the Bchl component absorbing at 810 mμ, and probably ultimately to P840 (23, 77). The spectrum of delayed light emission shows that it comes from singlet excited states in Cchl and Bchl (71). It appears therefore that primary singlet energy flows from light harvesting pigments to reaction centers, where it is converted to metastable energy. Most of this energy is used photochemically, but to some extent the conversion of energy is reversed and singlet excited states reappear in the light harvesting system.

In view of the connection between delayed light emission and the functioning of photochemical reaction centers, it can be hoped that further characterization of the delayed light will elucidate the physical and photochemical processes of photosynthesis.

B. The Yield and Lifetime of Chlorophyll Fluorescence *in Vivo*

The natural lifetime of the excited singlet state of a molecule is that which prevails when fluorescence is the only avenue of de-excitation;

that is, when the quantum yield of fluorescence is 100%. The natural lifetime reflects the probability of spontaneous radiative de-excitation; this in turn is proportional to the probability of excitation from the ground state. The natural lifetime can therefore be computed from the area under the absorption band corresponding to transitions into the lowest excited singlet state. The relationship is approximately

$$1/\tau_0 = 3 \times 10^{-9} \, k^2 \, \Delta k \varepsilon_{max} \tag{7}$$

where τ_0 is the natural lifetime, ε_{max} is the extinction coefficient (optical density, $M^{-1} \, cm^{-1}$) at the peak of absorption, k is the wavenumber in cm^{-1} at the peak, and Δk is the band width at half maximum.

If the excited state has other avenues for de-excitation, the quantum yield of fluorescence is reduced and the lifetime is shortened in proportion. This relationship is displayed by chlorophylls dissolved in ether (*78–81*). For Chl *a* the natural lifetime computed from the red absorption band is 15 nsec. The observed lifetime is 5 nsec, in nice agreement with the observed quantum yield of 33%. For Chl *b* the lifetime is 3.9 nsec and the yield is 16%. From these figures the natural lifetime should be 24 nsec, in agreement with the area under the red absorption band of Chl *b*.

In *Chlorella* cells the yield of Chl *a* fluorescence in very dim exciting light is about 2% (*79*). This yield should correspond to a lifetime 2% of the natural lifetime, or 0.3 nsec. Actually the lifetime is about 1 nsec; estimates by the pulse and phase techniques, for various photosynthetic organisms, have ranged from about 0.6 to 2 nsec (*78, 82, 83*). The observed yield is thus about three times too low to match the observed lifetime. This discrepancy can be resolved (*81*) by recognizing that there are two forms of Chl *a in vivo*, associated with systems I and II. The Chl of system I has a very low fluorescence yield (*44*) and makes up about three-fourths of the total Chl *a*. The yield of fluorescence from the Chl *a* of system II is about 8%, in harmony with the lifetime of about 1 nsec. When this yield of 8% is averaged with the low yield of the system I Chl, the overall yield is about 2%.

Pearlstein (*84*) has suggested that the fluorescence yield of Chl *a* 680 is less than that of Chl *a* 670 because the molecules of the former, being more numerous, are closer to one another. The transfer of energy to a reaction center is correspondingly more efficient and the energy lost as fluorescence is less. Assuming that 400 molecules of Chl *a* 680, serving system I, and 100 of Chl *a* 670 (system II) are packed into equal volumes, Pearlstein computed rates of energy transfer to reaction centers and thus obtained predictions of the yield and lifetime of fluorescence from the two types of Chl. He concluded that the 400 molecules of system I Chl and the 100 molecules of system II Chl contribute equally[*] to a total fluorescence of average yield 2.7%, with 1.1 nsec for the lifetime in system II and 0.27 nsec in system I. These computations, which are an extension of earlier ones by Bay and Pearlstein (64), provide an adequate accounting for the observed yield and lifetime of Chl fluorescence *in vivo.*

C. Interpretations of Variations in the Fluorescence

The fluorescence of Chl and Bchl *in vivo* shows interesting variations with exciting light intensity and with time during the onset of photosynthesis. The temporal variations, or induction effects, have typically the form of an increase in the yield of fluorescence during the first few seconds of illumination, often followed by a decline to a lower value in the steady state. The fluorescence yield during steady illumination is independent of exciting intensity in dim light, increases a little as the exciting intensity reaches the compensation point (where photosynthetic oxygen evolution is just balanced by respiration), and increases again by a factor of two or more when the exciting intensity exceeds saturation for photosynthesis. All these effects, both temporal and in the steady state, can be attributed to changes in the efficiency with which energy is trapped at reaction centers. When the photochemistry cannot keep pace with the influx of light energy, less of the energy is trapped and more emerges from Chl or Bchl as fluorescence.

Variations in the fluorescence yield have received extensive study in the hope that these changes could reveal the nature of physical mechanisms in photosynthesis [see Gaffron (85) and Franck (86)]. The doubling in fluorescence yield that accompanies light saturation for photosynthesis in *Chlorella* was the basis for several interesting hypotheses advanced by Franck and collaborators (46, 87) in recent years. To explain this doubling, Franck proposed an obligatory alternation of two photochemical reactions in a single reaction center. In one reaction, involving "reaction center" Chl complexed to reduced Cyt, a photochemical electron transfer oxidizes the Cyt and produces a strong reductant. This reaction, equivalent to the photochemical act of system I in the more traditional series formulation, is driven by the triplet excited state of the specialized Chl. Now that the Cyt is oxidized a second reaction can be mediated directly by the singlet excited state of the Chl. In this second reaction, equivalent to that performed by system II, the Cyt is reduced and a strong oxidant (a precursor of O_2) is formed. The splitting of a water molecule was visualized in this process.

---

[*] The yield of fluorescence in the larger component is one-fourth that in the smaller, but this is balanced by the greater number of molecules.

The foregoing hypothesis is like a series formulation in which the two photochemical systems have been allowed to merge in a single reaction center and the chain of electron carriers between the systems has been eliminated. The doubling in fluorescence yield under light saturation was explained as follows: In dim light half of the quanta are used in a photochemistry driven directly by the singlet excited state. This process is so efficient that practically none of the singlet energy is lost as fluorescence. The remaining half of the absorbed singlet energy must be converted to triplet before it can be used photochemically. This conversion is less efficient, and some fluorescence is allowed. Under light saturation, with absorption of light exceeding the capacity for photochemical utilization, nearly all the absorbed quanta (instead of half of them) can give rise to fluorescence in competition with quenching by way of the triplet state.

While the doubling of the fluorescence yield under light saturation is convincingly precise in *Chlorella*, the change is not an exact doubling in other systems. In chloroplasts of higher plants for example, the yield of the steady state fluorescence increases by a factor as great as five as the light intensity exceeds saturation for photosynthetic oxygen evolution (*88*). These changes in fluorescence can probably be explained in the framework of the series formulation described earlier, by assuming either that the reaction centers fail to trap singlet energy when they become saturated or that the apparent increment in fluorescence is actually a fast component of delayed light emission associated with the photochemistry (*71, 88*).

Fluorescence induction effects, occurring during the first few seconds of steady illumination, could be understood more clearly when it was recognized that green plant photosynthesis involves two distinct light reactions; indeed, the induction effects provided one basis for the series formulation (*89*). The initial rise in fluorescence yield is associated with saturation of the photochemical reaction center of system II (*44, 90–92*). Apparently the primary electron acceptor becomes reduced and thus cannot participate in photochemical utilization of singlet energy. The oxidizing influence of system I alleviates this condition, restoring some of the primary "system II" electron acceptor to its active (oxidized) form. This can account for the decline that comes after the initial fluorescence rise in intact cells of green plant tissues. With chloroplasts the fluorescence rises initially but does not decline subsequently. Here the oxidizing effect of system I appears to be circumvented, but this effect can be mimicked by adding Hill oxidants such as ferricyanide. Addition of the oxidant before illumination delays the initial rise in fluorescence. If oxidant is added after the rise has already occurred, the fluorescence drops momentarily to a lower value, rising once more when the oxidant is exhausted (*88,*

92*a*). From these experiments, and others with algae exposed to qualities of light absorbed by systems I and II, respectively (*44, 90–93*), it is clear that an increase in the yield of light emission is associated with the accumulation of reductant in system II. Comparable experiments with photosynthetic bacteria will be discussed in Section III, F.

D. Delayed Light Emission and Fluorescence in Photosynthetic Bacteria

It was mentioned in Section III, A that delayed light emission depends on the functioning of photosynthetic reaction centers and that this light is emitted, at least in green photosynthetic bacteria, from singlet states regenerated in the light-harvesting pigment aggregate.

The green bacteria show a fluorescence induction effect (an initial rise in yield followed by a decline), and in several respects the time-varying part of the fluorescence resembles the delayed light measured with a delay of several milliseconds. The resemblance suggests that the varying part of the fluorescence is actually a fast, intense part of the delayed emission, distinct from the initial component of fluorescence that can be seen as soon as the exciting light is turned on. The two components of fluorescence in *Chloropseudomonas ethylicum*, and also the delayed light emission, are shown in Fig. 4. This experiment was performed with an instrument (*71*) that could measure both fluorescence and delayed light emission from a single sample under identical conditions of excitation. The delayed light was measured in a succession of intervals 3 msec after each of a succession of exciting flashes. The fluorescence was measured during the flashes, which lasted 1 msec and were spaced 6 msec apart. The delayed light shown in Fig. 4 was 1500 times weaker than the fluorescence. It can be seen that the time-varying part of the fluorescence had the same kinetics as the delayed light emission during a 10-sec succession of exciting flashes.

The spectrum of delayed light emission in *C. ethylicum* also matches that of the time-varying part of the fluorescence and differs from that of the initial fluorescence (Fig. 5). Finally the delayed light and the time-varying fluorescence show similar responses (distinct from those of the initial fluorescence) to environmental conditions including the exciting light intensity. The initial fluorescence is linear with exciting light intensity; the other emissions vary roughly with the square of the exciting intensity in dim light and become saturated in bright light. The least trace of oxygen quenches both the delayed light and the time-varying part of the fluorescence.

From these experiments it appears that the so-called fluorescence induction effect in *C. ethylicum* can be attributed to a fast component of delayed emission. Similar experiments with algae and leaves of higher plants suggest that here also the fluorescence induction effects might be the result of "fast" delayed emission. The same could be true for the increased yield of fluorescence under light saturation in the steady state.

It remains possible that the time-varying fluorescence is authentic and resembles the delayed emission simply because the environment and the photochemical state of the tissues have parallel effects on the two kinds

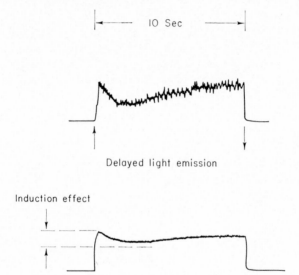

FIG. 4. Fluorescence and delayed light emission from *Chloropseudomonas ethylicum* cells during 10 sec of intermittent illumination. The exciting light was a sequence of 1-msec flashes spaced 6 msec apart. Reprinted by permission of the Rockefeller Institute Press; see Clayton (71).

of emission. Both emissions are reflections of the fate of singlet excitation energy: they differ in the way in which the singlet energy is formed and consequently in the partition of this energy among the various pigments. For a complete and fruitful understanding of the variations in fluorescence it must be established whether these variations are associated with primary or secondary (delayed) singlet excitation.

In purple photosynthetic bacteria, with the possible exception of Eimhjellen's *Rhodopseudomonas*, the delayed light bears little resemblance to observable variations in the fluorescence. First of all, the intensity of delayed light emission, compared with the fluorescence intensity, is about 100 times weaker than in green bacteria and green plants. Second, the kinetics of the delayed light emitted during several seconds of intermittent illumination bear no resemblance, in the purple bacteria, to the kinetics of the fluorescence induction effect. Finally the

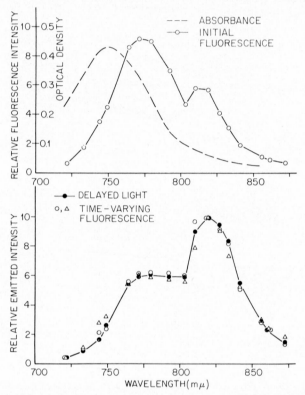

FIG. 5. Spectra of absorption and light emission by *Chloropseudomonas ethylicum* cells. The initial and time varying components of fluorescence can be distinguished in Fig. 4. Reprinted by permission of the Rockefeller Institute Press; see Clayton (*71*). This figure also appears in Clayton (*12a*).

delayed light, but not the time-varying fluorescence, is quenched by oxygen. The fluorescence in purple bacteria thus shows no evidence that it is contaminated appreciably with delayed light.

E. The Source of Fluorescence in Purple Photosynthetic Bacteria

In evaluating fluorescence as related to the transfer and utilization of energy most investigators, with the exception of Franck, have assumed tacitly or explicitly that the fluorescence comes from the light-harvesting aggregate and not from a specialized Chl at a reaction center. In green bacteria most of the fluorescence comes from Cchl and from the Bchl absorbing at 810 mμ, indicating that the trap (presumably P840) is not fluorescent.

In at least one of the purple bacteria, carotenoidless mutant *Rhodopseudomonas spheroides,* it can be shown that P870 is not fluorescent and that the fluorescence therefore originates either in the light-harvest-

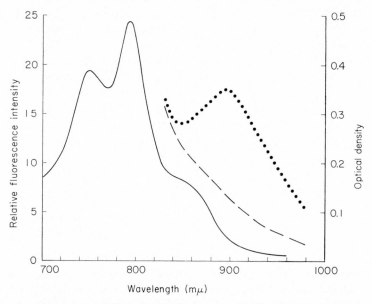

Fig. 6.   Absorption (solid line) and fluorescence (broken lines) spectra of a dry film of chromatophores from carotenoidless mutant *Rhodopseudomonas spheroides.* The cells had been incubated in the light until the major component of bacteriochlorophyll had been destroyed or converted to the pheophytin. The dashed curve shows fluorescence of the film exposed to the atmosphere. The dotted curve shows fluorescence after the film had been desiccated by enclosing it in a chamber with anhydrous calcium sulfate.

ing Bchl or in an unknown specialized Bchl. Chromatophores of this organism, treated so as to remove the light-harvesting Bchl while preserving the P870 (Fig. 2B), show no fluorescence from the P870. There is no fluorescence band near 870 mμ; only the tail of a band at 770 mμ attributable to bacteriopheophytin. The weak fluorescence that can be observed from 850 to 950 mμ does not diminish during illumination, as it should if some of it comes from P870 (the P870 is bleached reversibly during illumination).

In cells of carotenoidless mutant *R. spheroides* the light-harvesting

Bchl is partly lost and partly converted to bacteriopheophytin during prolonged incubation in the light. In chromatophores of such cells the P870 is preserved, and most of the light-harvesting Bchl has been lost, without recourse to photooxidation in the presence of a detergent. Dried films of these chromatophores, prepared on glass plates, show many photochemical activities including the reversible bleaching of the P870 (43). Figure 6 shows absorption and fluorescence spectra of such a film (solid and dashed curves respectively). There is little fluorescence attributable to P870; only the tail of a shorter-wave fluorescence band due to the pheophytin. The reversible bleaching of P870 and other photochemical activities can be suppressed by desiccating the film thoroughly, for example by enclosing it overnight in a chamber containing anhydrous calcium sulfate. A fluorescence band at about 900 mμ then makes its appearance, as shown by the dotted curve in Fig. 6. When the film is returned to moist air this fluorescence band vanishes and photochemical activity reappears. Apparently the P870 becomes fluorescent when it loses its ability to channel singlet energy into photochemical pathways.

In the absence of any evidence for a specialized Bchl other than P870 (and the associated P800), these experiments show that the fluorescence of Bchl *in vivo* comes from the light-harvesting component.

F. Variations in the Fluorescence of Bacteriochlorophyll *in Vivo*

We have seen that in purple photosynthetic bacteria the fluorescence probably comes from the major component of Bchl and contains very little contamination by delayed light. These organisms therefore provide the best opportunity to make meaningful interpretations of the variations in the yield of fluorescence.

Vredenberg and Duysens (94) first reported that the intensity of Bchl fluorescence rises in *Rhodospirillum rubrum* cells during constant illumination. The increase parallels the bleaching of P890, suggesting that P890 is an effective trap for singlet excitation energy. As P890 becomes bleached it loses its trapping property and more energy emerges from the light-harvesting aggregate as fluorescence. Vredenberg and Duysens assumed that fluorescent de-excitation, radiationless de-excitation, and trapping by P890 all are first-order processes; that is, their rates are proportional to the number of singlet excitation quanta. Furthermore the rate of trapping by P890 was assumed to be proportional to the concentration of unbleached P890. With these assumptions the fluorescence yield is predicted to be

$$\phi_f = k_f/(k_f + k_d + k[P]) \tag{8}$$

where k_f, k_d, and k are first-order rate constants for fluorescence, radiationless dissipation, and trapping, respectively, and [P] is the concentration of unbleached P890. According to this equation the reciprocal of the fluorescence yield should change during illumination to an extent proportional to the absorbancy change at 890 mμ. Adherence to this relationship was reported, at least for the difference between light and dark steady states. The data do not appear to be smooth enough to afford a reliable test during the actual change.

The greatest trapping efficiency, and hence the maximum possible quantum efficiency of P890 oxidation, should prevail at the start of illumination before any P890 has become bleached. This efficiency should approach zero in very bright light such that all of the P890 is bleached; under this condition the fluorescence yield will have its maximum value ϕ_f^{max}. From Eq. (8) one can derive an expression for the maximum efficiency of P890 oxidation, namely

$$\phi_P^{max} = 1 - \phi_f^0/\phi_f^{max} \tag{9}$$

where ϕ_f^0 is the fluorescence yield at the start of illumination. The data of Vredenberg and Duysens show that ϕ_P^{max} is at least 0.6, corresponding to a requirement fewer than two quanta per molecule of P890 oxidized.

These phenomena can be studied in chromatophores with the advantage that the data are less noisy because the preparations are not turbid. The changes in fluorescence and absorbancy can then be measured accurately throughout the transition between light and dark steady states. Experiments with chromatophores from *Rhodospirillum rubrum*, *Rhodopseudomonas spheroides*, and *Chromatium* (R. K. Clayton, unpublished) tend to confirm the conclusion of Vredenberg and Duysens that a fluorescence induction effect is related to the bleaching of P870 or P890. In fresh preparations from *Rhodospirillum rubrum* the relationship of Eq. (8) is sometimes obeyed throughout the transition from dark to light steady states. The slightest mistreatment, including exposure to oxygen, leads to a departure from this relationship. Often the data are described better by a simple proportionality between the change in fluorescence intensity and the change in absorbancy. In most cases neither the fluorescence nor its reciprocal is related in a simple way to the absorbancy at 890 mμ.

Equation (8) is appropriate only if the quanta of singlet energy in the light-harvesting aggregate are delocalized over a region embracing many molecules of P890. If each molecule of P890 and its share of light-harvesting pigment forms a distinct photosynthetic unit, isolated from neighboring units with respect to energy transfer, a correct formulation predicts that the change in fluorescence intensity is directly proportional to the amount of P890 bleached. Failure of Eq. (8) can also be expected if the population of chromatophores is not homogeneous with respect to the efficiency of energy transfer. The earliest increments of P890 bleaching will occur in the most efficient chromatophores and will contribute a disproportionately large share of the total change in fluorescence. These two effects, a restriction in the range of energy transfer and a loss of efficiency in some chromatophores, are sufficient to account for all the observed departures from Eq. (8) in *Rhodospirillum rubrum* chromatophores that have been mistreated slightly. The adherence to Eq. (8) in fresh chromatophores and intact cells carries an implication as to the structure of the photosynthetic tissue: the photosynthetic units are not discrete entities, but exist in an extended matrix of light-harvesting pigment with energy transfer extending over many reaction centers.

In *Rhodopseudomonas spheroides* and *Chromatium* a second fluorescence induction effect can be distinguished from the one associated with bleaching of P870 or P890. This other effect, also an increase during the first few seconds of illumination, can be observed under conditions where the bleaching of P870 or P890 is inappreciable, such as in anaerobic preparations exposed to dim light. This effect shows some correlation with the photochemical accumulation of reducing power. Its significance is obscure; possibly it reflects a second mechanism for energy trapping distinct from that associated with the photochemical oxidation of P870 or P890 (see Section I, C).

IV. Conclusions

On the basis of available evidence a picture can be drawn for the physical events leading to photosynthesis. Light produces singlet excited states in an aggregate of Chl, either directly or by transfer from accessory pigments. The singlet energy is trapped in some metastable form at photochemical reaction centers containing specialized chlorophylls. In green plants there are two kinds of reaction center linked through electron carriers. One kind (system II) produces a weak reductant and a strong oxidant that leads to oxygen evolution. The other kind (system I) forms a weak oxidant and a reductant strong enough to reduce pyridine nucleotide. Adenosine triphosphate is formed in reactions coupled with the interaction of photochemically produced oxidants and reductants. The photosynthetic bacteria contain just one kind of reaction center analogous to that in system I of green plants. The photochemical reactions are all oxidoreductions; they might be mediated by a triplet state or a charge transfer state involving the specialized chlorophyll.

The quantum yield of fluorescence gives an indication of the rate at which singlet energy is trapped at photochemical reaction centers and the emission of delayed light is in some way a manifestation of the functioning of the reaction centers.

This simple picture is already challenged by evidence that suggests alternatives; for one thing the roles of specialized chlorophylls such as P700 and P870 are in some doubt. A more extensive and careful study of the relationships between emitted light and photochemical events offers hope for a detailed elucidation of the physical and chemical processes involved in photosynthesis.

REFERENCES

(1) R. Emerson and W. Arnold, *J. Gen. Physiol.* **16**, 191 (1932).

(2) H. Gaffron and K. Wohl, *Naturwissenschaften* **24**, 81 (1936).

(3) C. B. van Niel, *Cold Spring Harbor Symp. Quant. Biol.* **3**, 138 (1935).

(4) C. B. van Niel, *Advan. Enzymol.* **1**, 263 (1941).

(5) J. Franck, *in* "Research in Photosynthesis" (H. Gaffron *et al.*, eds.), p. 142. Wiley (Interscience), New York, 1957.

(6) R. Emerson and C. M. Lewis, *Am. J. Botany* **30**, 165 (1943).

(7) R. Emerson, R. V. Chalmers, and C. Cederstrand, *Proc. Natl. Acad. Sci. U.S.* **43**, 133 (1957).

(8) J. Myers and C. S. French, *Plant Physiol.* **35**, 963 (1960).

(9) L. N. M. Duysens, J. Amesz, and B. M. Kamp, *Nature* **190**, 510 (1961).

(10) H. T. Witt, R. Moraw, A. Müller, B. Rumberg, and G. Zieger, *Z. Physik. Chem.* (*Frankfurt*) [N.S.] **23**, 133 (1960).

(11) H. T. Witt, A. Müller, and B. Rumberg, *Nature* **192**, 967 (1961).

(12) B. Kok and G. Hoch, *in* "Light and Life" (W. D. McElroy and B. Glass, eds.), p. 397. Johns Hopkins Press, Baltimore, Maryland, 1961.

(12a) R. K. Clayton, *Science* **149**, 1346 (1965).

(13) B. Kok, *Natl. Acad. Sci.—Natl. Res. Council, Publ.* **1145**, 35 (1963).

(14) N. I. Bishop, *Biochim. Biophys. Acta* **27**, 205 (1958).

(15) J. B. Thomas, O. H. Blaauw, and L. N. M. Duysens, *Biochim. Biophys. Acta* **10**, 230 (1953).

(16) M. Nishimura, *Biochim. Biophys. Acta* **59**, 183 (1962).

(17) R. K. Clayton, *Biochim. Biophys. Acta* **75**, 312 (1963).

(18) E. Rabinowitch, *Plant Physiol.* **34**, 213 (1959).

(19) L. N. M. Duysens, Thesis, University of Utrecht (1952).

(20) R. K. Clayton, *Photochem. Photobiol.* **1**, 313 (1962).

(21) W. R. Sistrom, M. Griffiths, and R. Y. Stanier, *J. Cellular Comp. Physiol.* **48**, 459 (1950).

(22) J. M. Olson and C. A. Romano, *Biochim. Biophys. Acta* **59**, 726 (1962).

(23) C. Sybesma and W. J. Vredenberg, *Biochim. Biophys. Acta* **88**, 205 (1964).

(24) Govindjee and E. Rabinowitch, *Science* **132**, 355 (1960).

(25) D. C. Fork, *Natl. Acad. Sci.—Natl. Res. Council, Publ.* **1145**, 355 (1963).

(26) L. W. Jones and J. Myers, *Plant Physiol.* **39**, 938 (1964).

(27) L. N. M. Duysens, W. J. Huiskamp, J. J. Vos, and J. M. van der Hart, *Biochim. Biophys. Acta* **19**, 188 (1956).

(28) B. Kok, *Biochim. Biophys. Acta* **22**, 399 (1956).

(29) B. Kok, *Biochim. Biophys. Acta* **48**, 527 (1961).
(30) G. Hoch and I. Martin, *Arch. Biochem. Biophys.* **102**, 430 (1963).
(31) R. K. Clayton, in "Bacterial Photosynthesis" (H. Gest, A. San Pietro, and L. P. Vernon, eds.), p. 377. Antioch Press, Yellow Springs, Ohio, 1963.
(32) R. K. Clayton, *Photochem. Photobiol.* **1**, 201 (1962).
(33) J. M. Olson and B. Chance, *Arch. Biochem. Biophys.* **88**, 26 (1960).
(34) R. K. Clayton, *Biochem. Biophys. Res. Commun.* **9**, 49 (1962).
(35) P. B. Sogo, N. G. Pon, and M. Calvin, *Proc. Natl. Acad. Sci. U.S.* **43**, 387 (1957).
(36) W. S. Zaugg, *Proc. Natl. Acad. Sci. U.S.* **50**, 100 (1963).
(37) W. S. Zaugg, L. P. Vernon, and A. Tirpack, *Proc. Natl. Acad. Sci. U.S.* **50**, 232 (1963).
(38) R. K. Clayton, W. R. Sistrom, and W. S. Zaugg, *Biochim. Biophys. Acta* **102**, 341 (1965).
(39) W. R. Sistrom, B. M. Ohlsson, and J. Crounse, *Biochim. Biophys. Acta* **75**, 285 (1963).
(40) W. R. Sistrom and R. K. Clayton, *Biochim. Biophys. Acta* **88**, 61 (1964).
(40a) W. R. Sistrom, verbal communication (1965).
(41) J. M. Olson, *Science* **135**, 101 (1962).
(42) B. Chance and M. Nishimura, *Proc. Natl. Acad. Sci. U.S.* **46**, 19 (1960).
(43) W. Arnold and R. K. Clayton, *Proc. Natl. Acad. Sci. U.S.* **46**, 769 (1960).
(44) L. N. M. Duysens, *Natl. Acad. Sci.—Natl. Res. Council, Publ.* **1145**, 1 (1963).
(45) B. Chance and D. deVault, *Ber. Bunsenges.* **68**, 722 (1964).
(46) J. Franck and J. L. Rosenberg, *J. Theoret. Biol.* **7**, 276 (1964).
(47) T. W. Engelmann, *Botan. Ztg.* **42**, 81 (1884).
(48) H. J. Dutton and W. M. Manning, *Am. J. Botany* **28**, 516 (1941).
(49) W. Arnold and J. R. Oppenheimer, *J. Gen. Physiol.* **33**, 423 (1950).
(50) T. Förster, *Discussions Faraday Soc.* **27**, 7 (1959).
(51) T. Förster, *Ann. Physik* [6] **2**, 55 (1948).
(52) R. K. Clayton, *Photochem. Photobiol.* **1**, 305 (1962).
(53) W. Arnold and H. K. Maclay, *Brookhaven Symp. Biol.* **11**, 1 (1959).
(54) W. Arnold, *J. Phys. Chem.* **69**, 788 (1965).
(55) G. W. Robinson, *Proc. Natl. Acad. Sci. U.S.* **49**, 521 (1963).
(56) T. Azumi and S. P. McGlynn, *J. Chem. Phys.* **39**, 1186 (1963).
(57) H. Linschitz and K. Sarkanen, *J. Am. Chem. Soc.* **80**, 4826 (1958).
(58) L. Pekkarinen and H. Linschitz, *J. Am. Chem. Soc.* **82**, 2407 (1960).
(59) J. L. Rosenberg, S. Takashima, and R. Lumry, in "Research in Photosynthesis" (H. Gaffron et al., eds.), p. 85. Wiley (Interscience), New York, 1957.
(59a) H. Linschitz, unpublished experiments (1964).
(60) J. B. Crounse, R. P. Feldman, and R. K. Clayton, *Nature* **198**, 1227 (1963).
(61) H. Claes, *Z. Naturforsch.* **16b**, 445 (1961).
(62) A. S. Davydov, "Theory of Molecular Excitons" (translated by M. Kasha and M. Oppenheimer, Jr.). McGraw-Hill, New York, 1962.
(63) Z. Bay and R. M. Pearlstein, *Proc. Natl. Acad. Sci. U.S.* **50**, 962 (1963).
(64) Z. Bay and R. M. Pearlstein, *Proc. Natl. Acad. Sci. U.S.* **50**, 1071 (1963).
(65) R. B. Park and N. G. Pon, *J. Mol. Biol.* **6**, 105 (1963).
(66) A. S. Holt and R. K. Clayton, *Photochem. Photobiol.* **4**, 829 (1965).
(67) D. R. Kearns, G. Tollin, and M. Calvin, *J. Chem. Phys.* **32**, 1013 (1960).
(68) B. L. Strehler and W. Arnold, *J. Gen. Physiol.* **34**, 809 (1951).
(69) W. F. Bertsch, *Proc. Natl. Acad. Sci. U.S.* **48**, 2000 (1962).

(70) W. Arnold and J. B. Davidson, *J. Gen. Physiol.* **37**, 677 (1954).

(71) R. K. Clayton, *J. Gen. Physiol.* **48**, 633 (1965).

(72) W. Arnold and J. B. Davidson, *Natl. Acad. Sci.—Natl. Res. Council, Publ.* **1145**, 698 (1963).

(73) W. Arnold, *in* "Research in Photosynthesis" (H. Gaffron *et al.*, eds.), p. 128. Wiley (Interscience), New York, 1957.

(74) W. F. Bertsch, J. B. Davidson, and J. R. Azzi, *Natl. Acad. Sci.—Natl. Res. Council, Publ.* **1145**, 701 (1963).

(75) J. C. Goedheer, *Biochim. Biophys. Acta* **66**, 61 (1963).

(76) R. K. Clayton and W. F. Bertsch, *Biochem. Biophys. Res. Commun.* **18**, 415 (1965).

(77) C. Sybesma and J. M. Olson, *Proc. Natl. Acad. Sci. U.S.* **49**, 248 (1963).

(78) S. S. Brody and E. Rabinowitch, *Science* **125**, 555 (1957).

(79) P. Latimer, T. T. Bannister, and E. Rabinowitch, *Science* **124**, 585 (1956).

(80) G. Tomita and E. Rabinowitch, *Biophys. J.* **2**, 483 (1962).

(81) E. Rabinowitch, *J. Phys. Chem.* **61**, 870 (1957).

(82) W. L. Butler and K. H. Norris, *Biochim. Biophys. Acta* **66**, 72 (1963).

(83) A. B. Rubin and L. K. Osnitskaya, *Mikrobiologiya* **32**, 200 (1963).

(84) R. M. Pearlstein, *Proc. Natl. Acad. Sci. U.S.* **52**, 824 (1964).

(85) H. Gaffron, *in* "Plant Physiology" (F. C. Steward, ed.), Vol. 1B, p. 3. Academic Press, New York, 1960.

(86) J. Franck, *in* "Handbuch der Pflanzenphysiologie" (W. Ruhland, ed.), Vol. 5, p. 689. Springer, Berlin, 1960.

(87) J. Franck, *Proc. Natl. Acad Sci. U.S.* **44**, 461 (1958).

(88) R. Lumry, B. C. Mayne, and J. D. Spikes, *Discussions Faraday Soc.* **27**, 149 (1959).

(89) H. Kautsky, W. Appel, and H. Amann, *Biochem. Z.* **332**, 277 (1960).

(90) B. Kok, *Natl. Acad. Sci.—Natl. Res. Council, Publ.* **1145**, 45 (1963).

(91) W. L. Butler and N. I. Bishop, *Natl. Acad. Sci.—Natl. Res. Council, Publ.* **1145**, 91 (1963).

(92) J. L. Rosenberg and T. Bigat, *Natl. Acad. Sci.—Natl. Res. Council, Publ.* **1145**, 122 (1963).

(92a) S. Malkin, verbal communication (1964).

(93) Govindjee, S. Ichimura, C. Cederstrand, and E. Rabinowitch, *Arch. Biochem. Biophys.* **89**, 322 (1960).

(94) W. J. Vredenberg and L. N. M. Duysens, *Nature* **197**, 355 (1963).

Author Index

Numbers in parentheses are reference numbers and are included to assist in locating references when the authors' names are not mentioned in the text. Numbers in italic refer to the pages on which the references are listed.

A

Aagaard, J., 336, *341*

Aasmundrud, O., 5(18a), *19*, 22(40), 23(40), 30(40), 36(40, 169, 170), 37(40, 169), 38(40, 170), 62, *66*, 116(32, 33), *118*, 150(16), 152(16), 153(16), *181*, 321(31), *340*, 384(19), 389(19), 397, 415(5), *425*, 499(105), *509*, 516(24, 25), 517(25) *518*, *519*

Abdul Majid, C. M., 22(26), *62*

Abraham, R. J., 219(41), 221, 222(41, 42, 45, 45a), 222(45), 223(45), *250*, *251*

Abrahamson, E. W., 180(106), *184*

Abram, D., 328(52), *341*

Adams, D. M., 190(9), *249*

Afzelius, B. A., 328(50), *340*

Agarwala, S. C., 22(23), *62*

Aghion, J., 407(38), *411*

Ahrne, I., 456(87), 457(87), 471(87), *479*

Ahrne, J., 407(41), 408(41), *411*

Airth, R. L., 161(44), *182*, 441(27), 442(27), *477*

Akazawa, T., 468(104), *479*, 583(115), *605*

Akimov, I., 275(40), *279*

Akulovich, N. K., 23(58), 36(160, 161, 162), 47(162), *63*, *65*, 446(47), 449(47), *477*, 499(108), *509*

Albers, V. M., 70(39), *105*, 180(105), *184*

Albert, A., 82(122), 93(171), 99(122), *107*, *108*

Albrecht, H. O., 180, *184*

Alexander, A. E., 254, 255(4), 257(4), *278*

Alivisatos, S. G. A., 502(122), *509*

Aliyev, K. A., 504(153), *510*

Allen, M. B., 22(39), 23(39), 30(39), 32(39), 36(39), *62*, 235(57), *251*, 285(4), *309*, 356, 378, 399(6), 402, *410*, 513(6, 9, 10), 514(10), 515(6, 16, 20), *518*

Altman, K. I., 482(3), 495(77), 500(16), *506*, *508*, *509*

Amann, H., 367(83), 369(83), 379, 631(89), *641*

Amesz, J., 355, 369, *378*, 383(18), 397, 577, 580(88), 586(156), *604*, *605*, *607*, 611(9), 612(9), *639*

Anderson, A. F. H., 25(112, 113), 26(112), 31(112), 32(112), 33(112), 50(113), *64*, 73(41), *105*, 192, *250*, 273, 274(35), 278(50), *279*, 443(34), *477*

Anderson, D. R., 410, *411*

Anderson, I. C., 17(48), *20*, 384(22), 397

Anderson, J. M., 26(132, 133), 30(132, 133), 32(132, 133), *65*, 356, 376(57), *378*, 403, *410*, 440(25), 442(25), 443(32), 452(54), 453(54), 454(25, 54), *477*, *478*, 500(112), *509*

Andzhaparidze, I. E., 22(29), *62*

Appel, W., 367(83), 369(83), 379, 631(89), *641*

Archibald, J. L., 114(17), 115(21), *118*

Ardao, C., 498, *508*

Arellano, R., 78(84), *106*, 474(113), *479*

Arnold, J. W., 17(38), *20*, 181(108), *184*, 275(39), *279*, 305, *311*, 371(94), 372(95, 96), 344, 373, 377, *379*, 570, *603*, 610, 618(43), 619(49), 620(43, 53), 627(70, 72), 628, 636(43), *639*, *640*, *641*

Arnon, D. I., 285, 297(26), 307, *309*, *310*, *311*, 572(28, 29), 574(25), 575(54), 579(72, 78), 581(91), 585(132, 133, 136), 587(136), *603*, *604*, *605*, *606*

Aron, C., 76(69), *105*, 254, 270, *278*

Aronoff, S., 9(20), 11, 17(44), 15(33), 18, *19*, *20*, 23(56), 24(86, 87), 36(56), *63*, *64*, 95(182), 96(182), *104*, *108*, 236, *251*, 550(154), 566, 575(37), *603*

643

Starnes, W. J., 22(34), 42(34), *62*
Stebbins, E., 506(178), *510*
Steele, C. C., 91(160), *107*
Stein, W. H., 417(9), *425*
Steinmann, E., 291, 301, *310*
Stensby, P. S., 24(90), *64*
Stern, A., 23(57), 24(57), 36(57), 41 (57), 42(57), 44(57), 45(57), 46 (57), 47(57), *63*, 68(8), 69(31), 74 (8), 75, 76(8), 77(31), 78(31), 81 (58), 84(59), 86(31), 90(31), 91 (31), 95, *104, 105*, 120(4), 141(36), *143*, 427, *436*, 439(14), *477*, 497, *508*, 524, *561*
Stewart, A., 180(105), *184*
Stier, E., 124(18), 125(18), *143*
Stocking, C. R., 295(21), *310*, 504(154), *510*
Stockman, D., 551(170), 552(170), *566*
Stoeckenius, W., 328(51), *341*
Stokes, G. G., 4, *19*, 116(25), *118*
Stoll, A., 3, 4(4), *19*, 50(203), *66*, 67 (2, 3), 68, 69, 72, 74(47), 75(11), 76 (47), 77(11), 79, 80(2, 11, 111), 83 (129, 130), 87(147), 90(3), 94(3), 95, 99(129), 102(147), *103, 104, 105, 106, 107*, 128(26), 142(26, 38), *143*, 191(13), *249*, 273(33), *279*, 427, *436*, 497(90), *508*
Storm, C. B., 79(96), *106*
Stoughton, R. W., 536(80), *563*
Strain, H. H., 17(43), *20*, 22(38, 41, 42, 43, 44, 45), 23(38, 41, 42, 43, 44, 45, 66, 71, 72, 73, 74), 24(38, 73, 75, 76, 97, 99, 100), 25(38, 66, 71, 72, 73, 74, 76, 99, 100, 109, 119), 26(38, 109), 27(109), 28(41, 72, 100), 29 (38, 66, 72, 73, 109), 30(38, 71, 109), 31(41, 72), 32(38, 41, 42, 44, 45, 71, 100, 109, 144), 33(38, 72), 34(71, 72), 35(38, 44, 71, 72, 144, 152, 153, 156a), 36(38, 44, 100), 37(74), 38 (41), 41(38, 73, 74, 75, 76, 97, 99, 109, 119), 42(38, 73, 76), 43(38, 73, 76, 99, 100, 119), 44(72, 73, 76), 45 (73, 75, 76, 109), 46(38, 71, 73, 76, 99, 100), 47(38, 73, 76), 48(38, 73), 49(38, 71, 109), 50(71, 72), 53(109), 54(71, 72, 73), 57(71), 58(152, 153), 59(38), *62, 63, 64, 65*, 68(29, 30), 73

(29), 74(48, 49, 51), 79(29, 30, 95), 84(136), 86(30, 51), 94(95), *104, 105, 106, 107*, 111, 116, 117, *117, 118*, 157, *182*, 190(15), 191(15, 16, 17a, 22), 192(17a, 22), 199(22, 28), 200 (22), 204(22), 213 (17a), 215(17a), 223(16, 47, 50), 224(50), 225(16, 47), 227(50), 228(16, 47, 50), 229 (16, 47, 50), 230(50), 231(16, 47, 50), 232(47), 233(47, 50), 234(16, 47), 236(22), 240(47), 241(47), 243 (47), 245(16, 47, 67), 246(67), 247 (16, 50), 248(68), 249(70), *249, 250, 251*, 475(118, 119), *479*, 512, 514(3), 515(3), *518*
Strehler, B. L., 371, *379*, 627, *640*
Strell, M., 68, 69, 80(108, 109, 110), 83 (134), 84(32), 91(110, 168), 93 (168), *104, 106, 107, 108*, 121(6), 122(6), 124(16), 125(16, 20), 126 (23, 24), 127, 140(35), 142(37), *143*
Strickland, J. D. H., 32(145), 48(145), 49(145), *65*
Stryer, L. C., 155(33), 158, *182, 548* (150), *565*
Stupp, R., 152, 155(19), *181*
Subba Rao, V., 560(246), *568*
Sudyina, E. G., 444(42), 455, *477*, 497, 498, *508*
Sud'ina, O. G., 26(127), 32(127), 42 (127), 49(127), *65*
Süs, O., 68(10), 91(10, 162), 94(162), 97(196), 98(10), *104, 108*, 427(1), *436*
Suga, I., 307(51), *311*, 577(66), *604*
Sugita, Y., 494(71), *508*
Susor, W. A., 316(15), *340*
Svec, W. A., 23(73, 74), 24(73), 25(73, 74), 29(73), 35(156a), 37(74), 41 (73, 74), 42(73), 43(73), 44(73), 45 (73), 46(73), 47(73), 48(73), 54 (73), *63, 65*, 79(95), 94(95), *106*, 191(16), 223(16, 50), 224(50), 225 (16), 227(50), 228(16, 50), 229(16, 50), 230(50), 231(16, 50), 233(50), 234(16), 245(16), 247(16, 50), 248 (68), *250, 251*
Svetailo, E. N., 504(153), *510*
Sweers, H. E., 369(90), *379*, 580, *605*
Sweetser, P. B., 372, *379*

Subject Index

A

Absorbancy changes, light induced, 615-618, 636
    475 mμ, 581
    515 mμ, 581
    Chlorella, 351
Absorption spectra, *see also* specific compounds, absorption spectrum
    artifacts, 347
    bacterial chlorophylls *in situ*, 381-396
    Bchl-protein, 416
    in concentrated solution and colloids, 172-175
    derivative spectra, 348-349
    digitonin particle, 404
    effect of solvent, 149
    of substituents, 75
    Euglena cells, 350
    light-scattering samples, 347
    low-temperature spectra, 349
    of monolayers, 259-266
    Scenedesmus, 350
    singlet-triplet, 532
    theory, 147
Accessory pigments, 307, 572, 614
Acetals, 84, 111
Acetylation, 125
Acid number, *see* Hydrochloric acid number
Afterglow, 180, 627
Aggregates
    absorption spectra of, 172-175, 423
    geometry of, 241
    photosensitization by, 556
Aggregation of chlorophylls
    concentration dependence, 243
    from infrared spectrum, 235-239
    from nmr spectrum, 239-245
Aggregation map, 231, 240, 244
Allomerization, 91-94
Allylthiourea, 559
α-Amino β-keto adipic acid, 483
Aminolevulinic acid, biosynthesis of
    ALA dehydrase, 485
    ALA synthetase, 483, 486, 504-505
    ALA transaminase, 486
Aminolysis, 90

Anacystis nidulans
    changes in pigmentation, 332
    lamellae, 314
Analytical methods, 22
Antirrhinum magus, 353

B

Bacterial photochemical system
    cytochrome photooxidation, 586, 616-617
    electron donors, 586-587
    general scheme, 592
Bacteriochlorin, 6, 71, 547
Bacteriochlorin e_6 and esters, 90, *see also* Bacteriochlorophyllin
Bacteriochlorophyll (a), 13, 70
    absorption spectrum, 37, 150, 153, 165, 172, 384-389
    aggregation, 423
    biosynthesis of, 498
    colloids, 172
    dichroism in NH_4^+ oleate, 152
    distribution of, 516-517
    fluorescence *in vivo*, 630, 636
    fluorescence polarization spectrum, 151
    infrared spectrum, 197, 209, 210, 238
    occurrence of, 36
    oxidation, 13, 165, 550
    photobleaching, 616
    photoreduction, 166, 545
    preparation, 59
    rate of biosynthesis, 335
    three absorption bands, 389
    triplet state spectrum, 166
Bacteriochlorophyll *b*, 116, 321, 392
    absorption spectrum, 152
Bacteriochlorophyll *c*, *see* Chlorobium chlorophylls
Bacteriochlorophyllin, absorption spectrum, 164
Bacteriopheophorbide and esters
    infrared spectrum, 209
    nmr spectrum, 224, 226-233
Bacteriopheophytin
    absorption spectrum, 151-153
    fluorescence polarization spectrum, 151
    infrared spectrum, 209